D1747088

Algebraic Riccati Equations

Algebraic Riccati Equations

PETER LANCASTER

Department of Mathematics and Statistics
University of Calgary
Calgary, Alberta

and

LEIBA RODMAN

Department of Mathematics
College of William and Mary
Williamsburg, Virginia

CLARENDON PRESS • OXFORD
1995

Oxford University Press, Walton Street, Oxford OX2 6DP
Oxford New York
Athens Auckland Bangkok Bombay
Calcutta Cape Town Dar es Salaam Delhi
Florence Hong Kong Istanbul Karachi
Kuala Lumpur Madras Madrid Melbourne
Mexico City Nairobi Paris Singapore
Taipei Tokyo Toronto
and associated companies in
Berlin Ibadan

Published in the United States by
Oxford University Press Inc., New York

© P. Lancaster and L. Rodman 1995

All rights reserved. No part of this publication may be
reproduced, stored in a retrieval system, or transmitted, in any
form or by any means, without the prior permission in writing of Oxford
University Press. Within the UK, exceptions are allowed in respect of any
fair dealing for the purpose of research or private study, or criticism or
review, as permitted under the Copyright, Designs and Patents Act, 1988, or
in the case of reprographic reproduction in accordance with the terms of
licences issued by the Copyright Licensing Agency. Enquiries concerning
reproduction outside those terms and in other countries should be sent to
the Rights Department, Oxford University Press, at the address above.

This book is sold subject to the condition that it shall not,
by way of trade or otherwise, be lent, re-sold, hired out, or otherwise
circulated without the publisher's prior consent in any form of binding
or cover other than that in which it is published and without a similar
condition including this condition being imposed
on the subsequent purchaser.

A catalogue record for this book is available from the British Library

Library of Congress Cataloging in Publication Data
(Data available)

ISBN 0–19–853795–6

Typeset by the author using T_EX

Printed in Great Britain by
Bookcraft (Bath) Ltd.
Midsomer Norton, Avon

Dedicated to
ISRAEL GOHBERG

Preface

The first concern of this book is with the solution of quadratic equations in matrices of the form
$$XDX + XA + BX + C = 0, \tag{1}$$
where the coefficients A, B, C, D are real or complex $n \times n$ matrices and $n \times n$ matrix solutions X are to be found. This is our prototype of an "algebraic Riccati equation". Observe that the existence of solutions is immediately a serious isssue as illustrated by simple equations of the form $X^2 + C = 0$ which have no solution.

Riccati equations with the form of equation (1) arise in several problem areas, some of which are described in Part IV of this book. Frequently, they take a symmetric form:
$$XDX + XA + A^*X + C = 0, \tag{2}$$
where C and D are hermitian matrices (A^* is the complex conjugate of the transpose A^T of A). Furthermore, hermitian solutions are generally required for physical reasons. In particular, symmetric Riccati equations arise in the problem of finding solutions of linear differential systems $\dot{x}(t) = Ax(t) + Bu(t)$ which are optimal in the sense of maximizing some quadratic form in $x(t)$ and $u(t)$. It is an interesting fact that analogous problems posed for *difference* equations of the form $x_{k+1} = Ax_k + Bu_k$ ($k = 0, 1, 2, \ldots$) do not lead to quadratic matrix equations, but to equations of "fractional" form:
$$X = A^*XA + Q - A^*XB(B^*XB + R)^{-1}B^*XA. \tag{3}$$

Here A and Q have the size of X, say $n \times n$, but R may have size $m \times m$, say, in which case B is $n \times m$. Equation (3) is also described as an algebraic Riccati equation and, reflecting their origins in the analysis of continuous and discrete systems, equation (2) is described as a "continuous algebraic Riccati equation", or CARE, and equation (3) is known as a "discrete algebraic Riccati equation", or DARE.

Each of the two basic symmetric equations (2) and (3) can be formulated in two ways: (a) with real coefficient matrices when real (and possibly symmetric) matrix solutions are required and (b) with complex coefficient matrices when complex (generally hermitian) matrix solutions are of interest. All four cases are treated here, generally beginning with the "complex case" (b) and then specializing the theory to obtain results for the "real case".

The growth of interest in algebraic (and more general) Riccati equations in the last 35 years has been explosive. Primarily, this has been driven by the important role played by these equations in optimal filter design and control theory, and has resulted in a prodigious number of research publications which have steadily increased our understanding of the equations and their solution sets.

Although the algebraic Riccati equations in our purview are truly nonlinear, they are amenable to solution by methods which rely heavily on linear algebra and the theory of matrices. Also, it transpires that a very important and pervasive role is played by rational matrix functions (i.e. matrices whose entries are real or complex scalar rational functions), their realizations and their factorizations. Our main objective is the description of solutions of algebraic Riccati equations using these methods and, taking advantage of our earlier monographs with I. Gohberg (1982, 1983, and 1986a) to make a presentation which is as self-contained as is reasonably possible. The extensive relevant literature is scattered through many journals and books and is frequently presented in a particular physical context and language. It might also be said that, in the enthusiasm to publish quickly new and useful results, a number of "folk-theorems" have evolved whose proofs are not easy to find and, prompted by some subtle points in the analysis, inaccuracies in the statement of results have not been uncommon. It is our hope that this unified and careful mathematical treatment of the subject with proofs provided (as far as is reasonably possible) will be helpful and timely.

Historically, much of the work of Riccati himself was concerned with ordinary differential equations of the form $\dot{x}(t) = q(x(t)) + f(t)$ where q is a quadratic function of x, and his work dates back to the early 18th century. However, the connections of his work with the issues addressed in this book are tenuous at best. The early work of Kalman (1960a and 1960b) on filtering and control might be said to usher in the new era. At the risk of leaving out many important contributions, we mention here some works which have been influential, at least for the present authors. Papers of Potter (1966) and Kleinman (1968) helped to reveal the role that linear algebra must play, as well as pointing the way to numerical methods of solution. The first edition of the comprehensive monograph by Wonham (1970) was well ahead of its time and is still an oft-quoted reference work. Wide ranging analyses of properties of solutions of continuous algebraic Riccati equations were given by Willems (1971) and Coppel (1974). As mentioned above, the analysis of rational matrix functions is firmly woven into the theory of algebraic Riccati equations and some key developments in this area appeared in the works of Bart et al. (1979 and 1980). Then the importance of the theory of linear pencils of operators (rather than of single operators) was first revealed in a paper of Pappas et al. (1980) and has led to significant advances in theory and numerical methods. More recently, we would mention work of Ando (1988) and draw attention to two recent books: the monograph of Mehrmann (1991) which is the first comprehensive examination of numerical methods of solution of algebraic Riccati equations, and the volume of Bittanti et al. (1991a), which is a useful collection of review and survey articles. Indeed, the present monograph started life as a survey written by the authors for the latter volume.

In order to work with this book it is necessary for the reader to have a good command of linear algebra, and real and complex analysis, all at an undergraduate level. The techniques used are well within the reach of mathematicians and are commonly used throughout engineering, science and applied mathematics. This applies particularly to systems theory and control, signal processing, filter design, and numerical analysis, as well as other fields. Our presentation should therefore be accessible to a broad range of interested users and theoreticians. Selections of the material presented here can be useful for classroom use, at both the undergraduate and graduate levels. In particular, the exercises at the end of many chapters are designed to facilitate classroom use of the material, as well as to fix ideas and provide glimpses into additional material not covered in the main text.

The book is divided into four parts. The first contains necessary material from linear algebra and the analysis of rational matrix functions. Canonical forms of matrices which are either self-adjoint or unitary in an indefinite scalar product play an important role and are the topic of Chapters 2 and 3. This particular collection of results has some novel features and may be more widely useful. Not surprisingly, many properties of solutions of linear matrix equations are required and these are collected in Chapter 5. A necessarily lengthy discussion of rational matrix functions and their realizations appears in Chapter 6. This part also includes results on regular linear pencils of matrices (i.e. of the form $\lambda A - B$, $\lambda \in \mathbb{C}$, with $\det(\lambda A - B) \not\equiv 0$). It is now well recognized that they play an important part in the analysis of algebraic Riccati equations and, in particular, in the study of numerical methods. At the time of writing, this theory is still evolving and the exposition here contains some new features. However, the authors anticipate that for many readers the problems of interest will be in subsequent parts which can be approached directly, with Part I being used simply for easy reference.

Parts II and III are the heart of the monograph and concern the CARE and DARE, respectively. There are considerable parallels between the two cases. Indeed, a Cayley-transform technique allows some results to be transferred from one equation to the other (see Lemma 12.3.2, for example). However, there are also substantial areas where independent treatments are necessary. In both cases solutions can be related to invariant subspaces of an operator formed from the coefficients of the equation, or to deflating subspaces of linear pencils, and these are the primary tools used in Parts II and III. However, characterizations using rational matrix functions arise naturally when the transfer functions of the underlying time-invariant linear systems are introduced, and they form another theme running through these two parts.

Part IV contains accounts of several problem areas in which Riccati equations arise and the general theory developed in Parts II and III is applied in each case. These chapters can be read independently and, with the exception of formal definitions, can be read independently of the rest of the book. They mostly form a part of a more general theory and the exposition may give the reader a foothold in that more general theory, as well as highlighting the special role played therein by an algebraic Riccati equation.

Notes are appended to most chapters which give literature references, often in a historical context, for results developed in the chapter as well as closely related results. In Part IV particularly, pointers are given in the notes to sources where further developments can be found. However, in view of the overwhelming volume of publications on AREs, a complete historical account of development of the theory exceeds the goals of this book.

Some important aspects of the theory and application of Riccati equations are deliberately avoided. There is no attempt at analysis of numerical methods as developed by Mehrmann (1991), for example. Also, we use neither the important geometrical ideas developed by Shayman (1983), nor the approach based on matrix inequalities of Willems (1971) and Trentelman and Willems (1991). Important contributions have also been made by Fuhrmann (1985) using polynomial modules, and are not considered here. There are many important applications in which $D \geq 0$ in equation (2), or $R > 0$ in equation (3), and our attention is focussed on these cases. The cases of indefinite matrices D or R (see Hewer (1993) and Ran and Trentelman (1993)) are not examined. There is substantial literature on differential and difference Riccati equations (see Coppel (1971) and Reid (1972), for example) and, with the exceptions of Section 16.4 and 17.5, they are not considered here in detail. Algebraic Riccati inequalities are also important in theory and application but appear in a relatively casual way here in Chapters 9 and 13 (see, for example, Willems (1971), Faibusovich (1987), and Lindquist et al. (1994)). Other exclusions include disconjugacy of linear Hamiltonian systems (Coppel (1971)), Riccati equations with periodic coefficients (Shayman (1985) and Bittanti et al. (1991b)), differential games (Basar and Bernhard (1991)), descriptor systems (see Bender and Laub (1987a) and (1987b), and Mehrmann (1988a and 1991)), and stochastic systems (Wonham (1968), Lindquist and Picci (1991), Lindquist et al. (1994)). These topics are not omitted because they are unimportant; indeed, they promise some exciting future developments. The major constraints on the choice of material are the size of this volume and the stamina of the authors.

The authors have benefitted from assistance and helpful conversations with many individuals and are grateful to them all, but would like especially to thank D.S. Bernstein, B. DeMoor, L. Elsner, I. Gohberg, M.A. Kaashoek, V. Mehrmann, J.K. Pinter, A.C.M. Ran and P. Zizler. We are also grateful to our host departments at the University of Calgary and the College of William and Mary, respectively, for support during the preparation of this work; also to our respective research granting agencies, the Natural Sciences and Engineering Research Council of Canada, and the National Science Foundation of the U.S.A.

The whole project has been dependent on the skill and dedication of Joanne Longworth, who prepared the TEX master copy, and it is a pleasure to record our sincere thanks to her.

Calgary
May, 1994

P. Lancaster
L. Rodman

Contents

PART I	**MATRIX THEORY**	**1**
1	**PRELIMINARIES FROM THE THEORY OF MATRICES**	**3**
1.1	Similarity and the complex Jordan form	3
1.2	Invariant subspaces	9
1.3	Projectors and invariant subspaces	13
1.4	Real matrices and canonical forms	16
1.5	Square roots of definite matrices	20
1.6	Regular matrix pencils	22
1.7	Functions of matrices	26
1.8	Exercises	29
1.9	Notes	30
2	**INDEFINITE SCALAR PRODUCTS**	**31**
2.1	Definitions and classifications of subspaces	31
2.2	Selfadjoint and unitary matrices	36
2.3	Canonical forms for H-selfadjoint and H-unitary matrices	41
2.4	Invariant nonnegative subspaces	46
2.5	Invariant neutral subspaces	52
2.6	H-symmetric matrices	55
2.7	H-skew-symmetric matrices	60
2.8	H-self-adjoint pencils	62
2.9	H-unitary pencils	64
2.10	Exercises	68
2.11	Notes	70

3 SKEW-SYMMETRIC SCALAR PRODUCTS — 71
- 3.1 Definitions and basic properties — 71
- 3.2 H-skew-symmetric matrices — 74
- 3.3 Invariant neutral subspaces — 78
- 3.4 Connections between sign characteristics — 79
- 3.5 Skew-symmetric and orthogonal pencils — 80
- 3.6 Exercises — 82
- 3.7 Notes — 82

4 MATRIX THEORY AND CONTROL — 83
- 4.1 Controllable pairs — 83
- 4.2 Observable pairs — 86
- 4.3 Two characterizations of controllable pairs — 87
- 4.4 Stabilizable and detectable pairs — 89
- 4.5 The control normal form — 92
- 4.6 Real matrix pairs — 95
- 4.7 Exercises — 95
- 4.8 Notes — 96

5 LINEAR MATRIX EQUATIONS — 97
- 5.1 The Kronecker product — 97
- 5.2 Linear equations in matrices — 99
- 5.3 Lyapunov and symmetric Stein equations — 101
- 5.4 Continuous and analytic dependence — 105
- 5.5 Notes — 106

6 RATIONAL MATRIX FUNCTIONS — 107
- 6.1 Realizations of rational matrix functions — 107
- 6.2 Partial multiplicities and minimal realizations — 113
- 6.3 Locally minimal realizations — 116
- 6.4 Minimal factorization for rational matrix functions — 122
- 6.5 Nonnegative rational matrix functions — 130
- 6.6 Rational matrix functions which are hermitian on the unit circle — 137
- 6.7 The real case — 141
- 6.8 Exercises — 144
- 6.9 Notes — 145

PART II CONTINUOUS ALGEBRAIC RICCATI EQUATIONS — 147

7 GEOMETRIC THEORY: THE COMPLEX CASE — 149
- 7.1 Solutions and invariant subspaces — 149
- 7.2 Symmetric equations and invariant semidefinite subspaces — 151
- 7.3 Existence of hermitian solutions and partial multiplicities — 158
- 7.4 Special and hermitian solutions — 167
- 7.5 Extremal solutions — 168
- 7.6 Partial order — 177
- 7.7 Continuity — 184
- 7.8 Analyticity — 189
- 7.9 Relaxing the controllability condition — 192
- 7.10 The LQR form and matrix pencils — 200
- 7.A Appendix: the metric space of subspaces — 205
- 7.11 Exercises — 210
- 7.12 Notes — 212

8 GEOMETRIC THEORY: THE REAL CASE — 215
- 8.1 Solutions and invariant subspaces — 215
- 8.2 Existence of symmetric solutions — 221
- 8.3 Special and extremal symmetric solutions — 222
- 8.4 Partial order, continuity, and analyticity — 223
- 8.5 Relaxing the controllability assumption — 226
- 8.6 The pencil approach — 227
- 8.7 Exercises — 228
- 8.8 Notes — 229

9 CONSTRUCTIVE EXISTENCE AND COMPARISON THEOREMS — 231
- 9.1 Comparison theorems — 231
- 9.2 The rate of convergence — 236
- 9.3 Stabilizing and almost stabilizing solutions — 238
- 9.4 Comparison theorems: the LQR form — 241
- 9.5 The real case — 244
- 9.6 Exercises — 244
- 9.7 Notes — 245

10 HERMITIAN SOLUTIONS AND FACTORIZATIONS OF RATIONAL MATRIX FUNCTIONS — 247
- 10.1 Nonnegative rational matrix functions — 247
- 10.2 The real case — 253
- 10.3 Notes — 256

11 PERTURBATION THEORY — 257
- 11.1 Existence of hermitian solutions — 257
- 11.2 Extremal solutions: continuous dependence — 260
- 11.3 Extremal solutions: analytic dependence — 263
- 11.4 The real case — 266
- 11.5 Exercises — 266
- 11.6 Notes — 267

PART III DISCRETE ALGEBRAIC RICCATI EQUATIONS — 269

12 GEOMETRIC THEORY FOR THE DISCRETE ALGEBRAIC RICCATI EQUATION — 271
- 12.1 Preliminaries — 271
- 12.2 Hermitian solutions and invariant Lagrangian subspaces — 273
- 12.3 Description of solutions in terms of invariant subspaces — 280
- 12.4 Positive definiteness of Ψ and existence of hermitian solutions — 283
- 12.5 Relaxing the controllability condition — 291
- 12.6 A more general DARE — 295
- 12.7 The real case — 301
- 12.8 Exercises — 303
- 12.9 Notes — 306

13 CONSTRUCTIVE EXISTENCE AND COMPARISON THEOREMS — 307
- 13.1 Existence of maximal hermitian solutions — 307
- 13.2 The rate of convergence — 313
- 13.3 Comparison theorems — 315
- 13.4 Inequalities for partitioned matrices — 319
- 13.5 Stabilizing and positive semidefinite solutions — 324
- 13.6 Exercises — 327
- 13.7 Notes — 327

14 PERTURBATION THEORY FOR DISCRETE ALGEBRAIC RICCATI EQUATIONS 329
14.1 Existence of hermitian solutions 329
14.2 Extremal solutions 330
14.3 Notes 332

15 DISCRETE ALGEBRAIC RICCATI EQUATIONS AND MATRIX PENCILS 333
15.1 The DARE and a symplectic matrix pencil 333
15.2 The DARE and a dilated matrix pencil 336
15.3 Stabilizing solutions 339
15.4 The real case 342
15.5 Exercises 344
15.6 Notes 345

PART IV APPLICATIONS AND CONNECTIONS 347

16 LINEAR-QUADRATIC REGULATOR PROBLEMS 349
16.1 The optimization problem 349
16.2 Properties of the cost functional 351
16.3 Finding an optimal control 357
16.4 The differential and algebraic Riccati equations 359
16.5 The cost functional for the discrete LQR problem 362
16.6 Solution of the discrete LQR problem 366
16.7 Exercises 369
16.8 Notes 369

17 THE DISCRETE KALMAN FILTER 371
17.1 Some concepts and results concerning random vectors 372
17.2 Statement of the problem 375
17.3 The recursive process 377
17.4 Discussion 379
17.5 Time-invariant filters and a Riccati equation 380
17.6 Properties of the observer system 384
17.7 Exercises 385
17.8 Notes 385

18 THE TOTAL LEAST SQUARES TECHNIQUE — 387
18.1 The total least squares problem — 388
18.2 Total least squares with a symmetry constraint — 391
18.3 Notes — 395

19 CANONICAL FACTORIZATION — 397
19.1 Canonical factorization in geometric terms — 397
19.2 The real case — 403
19.3 Spectral factorization — 404
19.4 Notes — 407

20 H^∞ CONTROL PROBLEMS — 409
20.1 The bounded real lemma — 410
20.2 An H^∞ control problem with state feedback — 412
20.3 H^∞ filtering using state estimation — 417
20.4 Notes — 418

21 CONTRACTIVE RATIONAL MATRIX FUNCTIONS — 421
21.1 Realizations of strict contractions — 421
21.2 Inertia of solutions of a "CARE" and poles of strict contractions — 426
21.3 Relaxing the minimality of realizations — 428
21.4 Minimal unitary completions — 430
21.5 The real case — 433
21.6 Notes — 435

22 THE MATRIX SIGN FUNCTION — 437
22.1 Matrix sign function: definition and basic properties — 438
22.2 Connection with the Riccati equation — 440
22.3 The sign function and the DARE — 444
22.4 The real case — 445
22.5 Exercises — 446
22.6 Notes — 446

23	**STRUCTURED STABILITY RADIUS**	**447**
	23.1 Basic properties of the structured stability radius	448
	23.2 The Hamiltonian matrix	451
	23.3 Connection with the Riccati equation	453
	23.4 Exercises	457
	23.5 Notes	457

BIBLIOGRAPHY	459
LIST OF NOTATION AND CONVENTIONS	473
INDEX	477

Part I

MATRIX THEORY

This part of the monograph is designed to cover those parts of the theory of matrices needed for our subsequent investigations of Riccati equations (in Parts II and III) and their applications (in Part IV). Of course, not everything can be done in detail, but more argument and discussion is provided for those topics that are needed in the sequel, and are not so easily found in standard works on matrix theory. Special emphasis is placed on invariant subspaces, indefinite scalar products, controllable and observable matrix pairs and, of course, the theory of linear matrix equations, without which the study of quadratic, and more general, equations in matrices can hardly begin.

Prerequisites for this Part, and hence the whole book, are basic undergraduate linear algebra together with knowledge of rudimentary complex analysis.

1
PRELIMINARIES FROM THE THEORY OF MATRICES

1.1 Similarity and the complex Jordan form

We consider $n \times n$ matrices A with complex entries (i.e. entries in the field of complex numbers $\mathbb{C}$). In a natural way A represents a linear transformation from $\mathbb{C}^n$ into $\mathbb{C}^n$, where $\mathbb{C}^n$ stands for the vector space of n-component columns with complex components. This linear transformation will be denoted by the same letter A. We have therefore the linear transformation $A: \mathbb{C}^n \to \mathbb{C}^n$ defined by $A(x) = Ax$, $x \in \mathbb{C}^n$, where Ax is the matrix vector product of the matrix A and the column x.

A complex number λ_0 is called an *eigenvalue* of a matrix (or linear transformation) A if

$$Ax = \lambda_0 x \qquad (1.1.1)$$

for some non-zero $x \in \mathbb{C}^n$. Non-zero vectors x satisfying (1.1.1) (for given A and its eigenvalue λ_0) are called *eigenvectors* of A associated with the eigenvalue λ_0. Thus, λ_0 is an eigenvalue of A if and only if λ_0 is a root of the *characteristic polynomial* $\det(\lambda I - A)$. Since the characteristic polynomial has degree n, we conclude by the fundamental theorem of algebra that A has n eigenvalues (counted with multiplicities as roots of the characteristic polynomial).

Two $n \times n$ matrices A and B are called *similar* if there exists a nonsingular matrix S (also with complex entries) such that $A = S^{-1}BS$. Geometrically, similar matrices represent the same linear transformation from $\mathbb{C}^n$ into $\mathbb{C}^n$ with respect to different choices of a basis in $\mathbb{C}^n$.

Eigenvalues and eigenvectors behave well under similarity. More precisely we have:

Proposition 1.1.1. *Assume A is similar to B: $A = S^{-1}BS$, where S is nonsingular. Then A and B have the same eigenvalues, and $x \in \mathbb{C}^n$ is an eigenvector of A corresponding to the eigenvalue λ_0 if and only if Sx is an eigenvector of B corresponding to the same λ_0.*

Proof. Since $\det(S^{-1}) = (\det S)^{-1}$, we have

$$\det(\lambda I - A) = \det(S^{-1}(\lambda I - B)S) = \det(\lambda I - B).$$

So A and B have the same characteristic polynomial and therefore the same eigenvalues. Further, the eigenvalue–eigenvector equation for A,

$$Ax = \lambda_0 x, \quad x \neq 0$$

can be rewritten as

$$S^{-1}BSx = \lambda_0 x, \quad \text{or} \quad BSx = \lambda_0(Sx).$$

Since S is nonsingular, $x \neq 0$ implies $Sx \neq 0$, and the second part of this proposition follows. □

An $n \times n$ matrix A is called *diagonable* if it is similar to a diagonal matrix (which necessarily has the eigenvalues of A on the main diagonal). This happens if and only if there exists a basis in $\mathbb{C}^n$ consisting of eigenvectors of A. Thus, the structure of a diagonable matrix is completely described by its eigenvalues and eigenvectors. However, not every matrix is diagonable, and to describe the structure of non-diagonable matrices, we need the concepts of Jordan chains and generalized eigenvectors.

A *Jordan chain* of an $n \times n$ matrix A is an ordered set of vectors $x_1, \ldots, x_r \in \mathbb{C}^n$, such that $x_1 \neq 0$ and for some eigenvalue λ_0 of A the equalities

$$\begin{aligned}(A - \lambda_0 I)x_1 &= 0 \\ (A - \lambda_0 I)x_2 &= x_1, \ \ldots, (A - \lambda_0 I)x_r = x_{r-1}\end{aligned} \quad (1.1.2)$$

hold. Thus, x_1 is an eigenvector of A corresponding to λ_0; the vectors $x_2, \ldots, x_r$ that appear in a Jordan chain are called *generalized eigenvectors* of A associated with the eigenvalue λ_0 and the eigenvector x_1. Jordan chains are also well-behaved with respect to similarity. If $A = S^{-1}BS$ for a nonsingular matrix S, then $x_1, \ldots, x_r \in \mathbb{C}^n$ is a Jordan chain for A corresponding to λ_0 if and only if $Sx_1, \ldots, Sx_r$ is a Jordan chain for B corresponding to λ_0. The proof is analogous to that of Proposition 1.1.1.

Example 1.1.1. Consider the $n \times n$ matrix

$$A = \begin{bmatrix} \lambda_0 & 1 & 0 & \cdots & 0 \\ 0 & \lambda_0 & 1 & \cdots & 0 \\ \vdots & \vdots & & \ddots & \vdots \\ & & & & 1 \\ 0 & 0 & & \cdots & \lambda_0 \end{bmatrix}.$$

This matrix is called a *Jordan block* with eigenvalue λ_0. Clearly, λ_0 is the sole eigenvalue of A, and e_1 is the only eigenvector (up to multiplication by a non-zero complex number). Here and elsewhere in the book we denote by e_i the column having 1 in its i-th component and zeros in all other places (the number of components in e_i is understood from the context). One Jordan chain of A is $e_1, e_2, \ldots, e_n$. This Jordan chain is not unique; a general form for the Jordan chains of A is given by

$$\alpha_1 e_1; \quad \alpha_1 e_2 + \alpha_2 e_1; \quad \alpha_1 e_3 + \alpha_2 e_2 + \alpha_3 e_1; \quad \ldots; \quad \alpha_1 e_r + \alpha_2 e_{r-1} + \ldots + \alpha_r e_1,$$

where $1 \leq r \leq n$; $\alpha_1, \ldots, \alpha_r \in \mathbb{C}$ and $\alpha_1 \neq 0$. □

Example 1.1.2. If A is diagonable, then the Jordan chains of A consist of eigenvectors only (i.e. there are no generalized eigenvectors). □

An important property of a Jordan chain $x_1, \ldots, x_r$ of A is that the vectors $x_1, \ldots, x_r$ are linearly independent. Indeed, let

$$\sum_{i=1}^{r} \alpha_i x_i = 0, \tag{1.1.3}$$

where $\alpha_i \in \mathbb{C}$. Equalities (1.1.2) imply that $(A - \lambda_0 I)^{r-1} x_i = 0$ if $i < r$ and $(A - \lambda_0 I)^{r-1} x_r = x_1$. So, applying $(A - \lambda_0 I)^{r-1}$ to both sides of (1.1.3) we obtain $\alpha_r x_1 = 0$, and since $x_1 \neq 0$, we conclude that $\alpha_r = 0$. Now apply $(A - \lambda_0 I)^{r-2}$ to both sides of (1.1.3), and it is seen in a similar way that $\alpha_{r-1} = 0$. Repeating this procedure, all the coefficients α_i are found to be zero, and therefore $x_1, \ldots, x_r$ are linearly independent.

The subspace span $\{x_1, \ldots, x_r\}$ spanned by a Jordan chain $x_1, \ldots, x_r$ of A will be called a *Jordan subspace* for A.

We now state a version of the Jordan form theorem.

Theorem 1.1.2. *For any $n \times n$ matrix A there exists a basis in $\mathbb{C}^n$ consisting of Jordan chains of A. More precisely: there exist Jordan chains $\{x_{i1}, \ldots, x_{ir_i}\}$, $i = 1, \ldots, k$ of A, corresponding to the (not necessarily distinct) eigenvalues $\lambda_1, \ldots, \lambda_k$, respectively, such that the set of vectors $\{x_{ij}\}$; $1 \leq j \leq r_i$; $1 \leq i \leq k$ forms a basis in $\mathbb{C}^n$.*

We will not prove Theorem 1.1.2 here; this result is classical and can be found, together with its proof, in many linear algebra textbooks, see, e.g., Gantmacher (1959), Lancaster and Tismenetsky (1985), Horn and Johnson (1985), Gohberg et al. (1983), Smith (1984), Finkbeiner (1978).

A statement equivalent to Theorem 1.1.2 can be given in terms of similarity of matrices. To develop this statement, we introduce Jordan matrices. Denote by $J_r(\lambda)$ the $r \times r$ Jordan block with eigenvalue $\lambda \in \mathbb{C}$:

$$J_r(\lambda) = \begin{bmatrix} \lambda & 1 & 0 & \cdots & 0 \\ 0 & \lambda & 1 & \cdots & 0 \\ \vdots & \vdots & & \ddots & \vdots \\ & & & & 1 \\ 0 & 0 & \cdots & 0 & \lambda \end{bmatrix}.$$

Matrices that have the block diagonal form

$$\begin{bmatrix} J_{r_1}(\lambda_1) & 0 & \cdots & 0 \\ 0 & J_{r_2}(\lambda_2) & \cdots & 0 \\ \vdots & \vdots & \ddots & \vdots \\ 0 & 0 & \cdots & J_{r_k}(\lambda_k) \end{bmatrix} \qquad (1.1.4)$$

with Jordan blocks on the main diagonal will be called *Jordan matrices* (the numbers $\lambda_1, \ldots, \lambda_k$ here are not necessarily distinct). The class of Jordan matrices includes the diagonal matrices (they have 1×1 Jordan blocks on the main diagonal). The Jordan matrices are clearly upper triangular. The numbers $\lambda_1, \ldots, \lambda_k$ are the eigenvalues of the Jordan matrix (1.1.4).

Consider now an $n \times n$ matrix A as the linear transformation $\hat{A} : \mathbb{C}^n \to \mathbb{C}^n$ defined by $\hat{A}(x) = Ax$, $x \in \mathbb{C}^n$. Then the matrix A represents $\hat{A}$ in the standard basis $e_1, \ldots, e_n$. Let now

$$X = \{x_{11}, \ldots, x_{1r_1}, x_{21}, \ldots, x_{2r_2}, \ldots, x_{k1}, \ldots, x_{kr_k}\} \qquad (1.1.5)$$

be a basis in $\mathbb{C}^n$ consisting of Jordan chains $\{x_{i1}, \ldots, x_{ir_i}\}$, $i = 1, \ldots, k$ of A, with corresponding eigenvalues $\lambda_1, \ldots, \lambda_k$. Denote by $[x]_X$ the coordinate vector of $x \in \mathbb{C}^n$ with respect to the basis X: if $x = \sum_{i=1}^k \sum_{j=1}^{r_i} \alpha_{ij} x_{ij}$, where $\alpha_{ij} \in \mathbb{C}$, then

$$[x]_X = \begin{bmatrix} \alpha_{11} \\ \vdots \\ \alpha_{1r_1} \\ \vdots \\ \alpha_{k1} \\ \vdots \\ \alpha_{kr_2} \end{bmatrix}.$$

The matrix B representing $\hat{A}$ in the basis X is given by

$$B = [[\hat{A}(x_{11})]_X \cdots [\hat{A}(x_{1r_1})]_X \cdots [\hat{A}(x_{k1})]_X \cdots [\hat{A}(x_{kr_k})]_X].$$

By the definition of Jordan chains we have

$$\hat{A}(x_{i1}) = \lambda_i x_{i1}, \; \hat{A}(x_{i2}) = \lambda_i x_{i2} + x_{i1}, \ldots, \hat{A}(x_{ir_i}) = \lambda_i x_{ir_i} + x_{i,r_i-1}$$

($i = 1, \ldots, k$). Therefore

$$B = \begin{bmatrix} J_{r_1}(\lambda_1) & 0 & \cdots & 0 \\ 0 & J_{r_2}(\lambda_2) & \cdots & 0 \\ \vdots & \vdots & \ddots & \vdots \\ 0 & 0 & \cdots & J_{r_k}(\lambda_k) \end{bmatrix}$$

is a Jordan matrix.

Conversely, it is easy to see that if the matrix representing $\hat{A}$ with respect to some basis X is a Jordan matrix, then X consists of Jordan chains for the matrix A (or, equivalently, Jordan chains for the linear transformation $\hat{A}$). Recalling that matrices representing the same linear transformation with respect to different bases are similar, Theorem 1.1.2 can now be restated as follows:

Theorem 1.1.3. *Every $n \times n$ matrix A is similar to a Jordan matrix.*

It turns out that this Jordan matrix, called the *Jordan form* of A, is essentially uniquely determined by A. More precisely, if

$$B_1 = J_{r_1}(\lambda_1) \oplus \cdots \oplus J_{r_k}(\lambda_k)$$

and

$$B_2 = J_{s_1}(\mu_1) \oplus \cdots \oplus J_{s_\ell}(\mu_\ell)$$

are two Jordan matrices similar to the same $n \times n$ matrix A, then $k = \ell$ (i.e. the numbers of Jordan blocks on the main diagonal in B_1 and B_2 is the same) and the blocks $J_{s_1}(\mu_1), \ldots, J_{s_\ell}(\mu_\ell)$ are obtained by permuting $J_{r_1}(\lambda_1), \ldots, J_{r_k}(\lambda_k)$. Therefore, the properties of the Jordan form of A that are invariant under permutation of the Jordan blocks are actually the properties of A itself. We define several such properties. Given an eigenvalue λ_0 of A, the number of Jordan blocks (including the 1×1 Jordan blocks if any) with eigenvalue λ_0 is called the *geometric multiplicity* of λ_0. The sizes of these Jordan blocks are called the *partial multiplicities* of λ_0, and the sum of partial multiplicities is the *algebraic multiplicity* of λ_0.

Example 1.1.3. Let

$$B = \begin{bmatrix} 1 & 1 & 0 & 0 & 0 & 0 & 0 & 0 & 0 \\ 0 & 1 & 1 & 0 & 0 & 0 & 0 & 0 & 0 \\ 0 & 0 & 1 & 1 & 0 & 0 & 0 & 0 & 0 \\ 0 & 0 & 0 & 1 & 0 & 0 & 0 & 0 & 0 \\ 0 & 0 & 0 & 0 & 1 & 0 & 0 & 0 & 0 \\ 0 & 0 & 0 & 0 & 0 & 1 & 0 & 0 & 0 \\ 0 & 0 & 0 & 0 & 0 & 0 & 0 & 1 & 0 \\ 0 & 0 & 0 & 0 & 0 & 0 & 0 & 0 & 0 \\ 0 & 0 & 0 & 0 & 0 & 0 & 0 & 0 & 0 \end{bmatrix}.$$

This Jordan matrix has eigenvalues 1 and 0. The geometric multiplicity of the eigenvalue 1 is 3, its algebraic multiplicity is 6, and the partial multiplicities are 4, 1 and 1. The eigenvalue 0 has geometric multiplicity 2, algebraic multiplicity 3, and partial multiplicities 2 and 1. □

8 PRELIMINARIES FROM THE THEORY OF MATRICES

To express the multiplicities in terms of the matrix A itself, we introduce the subspaces
$$\text{Ker}\,[(A - \lambda_0 I)^i], \quad i = 1, 2, \ldots . \tag{1.1.6}$$
Here and elsewhere we denote by $\text{Ker}\,X$ the *kernel* of an $m \times n$ matrix X:
$$\text{Ker}\,X = \{y \in \mathbb{C}^n | Xy = 0\};$$
it is clearly a subspace in $\mathbb{C}^n$. The subspaces (1.1.6) form a chain by inclusion:
$$\text{Ker}\,(A - \lambda_0 I) \subseteq \text{Ker}\,(A - \lambda_0 I)^2 \subseteq \ldots \subseteq \text{Ker}\,(A - \lambda_0 I)^p \subseteq \ldots \tag{1.1.7}$$
Indeed, if $(A - \lambda_0 I)^p y = 0$ then obviously also $(A - \lambda_0 I)^{p+1} y = 0$. Since all the subspaces in (1.1.7) are contained in $\mathbb{C}^n$, which is finite dimensional, clearly the infinitely many subspaces $\text{Ker}\,(A - \lambda_0 I)^p$, $p = 1, 2, \ldots$ cannot be all different. Let p_0 be the smallest positive integer such that
$$\text{Ker}\,(A - \lambda_0 I)^{p_0} = \text{Ker}\,(A - \lambda_0 I)^{p_0+1}. \tag{1.1.8}$$
In fact,
$$\text{Ker}\,(A - \lambda_0 I)^{p_0} = \text{Ker}\,(A - \lambda_0 I)^r \tag{1.1.9}$$
for all $r \geq p_0 + 1$. One proves (1.1.9) by induction on r, starting with $r = p_0 + 1$ (which is true by (1.1.8)).

Let us call the number p_0 appearing here the *index* of λ_0. It is not difficult to see that the index is also the size of the largest Jordan block of A with eigenvalue λ_0.

In terms of the subspaces (1.1.7), the multiplicities of λ_0 are given as follows:

Theorem 1.1.4. *Let λ_0 be an eigenvalue of A. Then the geometric multiplicity of λ_0 is equal to*
$$\dim \text{Ker}\,(A - \lambda_0 I),$$
and the algebraic multiplicity of λ_0 is equal to
$$\dim \text{Ker}\,(A - \lambda_0 I)^{p_0}$$
where p_0 is the index of λ_0.

Proof. If A is a Jordan matrix, then Theorem 1.1.4 is verified by an easy inspection. For a general matrix A, write (in view of Theorem 1.1.3) $A = S^{-1}BS$, where B is the Jordan form of A. Denote by $S(\text{Ker}\,(A - \lambda_0 I)^r))$ the set
$$\{Sx | x \in \text{Ker}\,(A - \lambda_0 I)^r\}.$$
It is easy to see that
$$S(\text{Ker}\,(A - \lambda_0 I)^r)) = \text{Ker}\,(B - \lambda_0 I)^r,$$
$r = 1, 2, \ldots$. Since S is invertible, this implies
$$\dim \text{Ker}\,(A - \lambda_0 I)^r = \dim \text{Ker}\,(B - \lambda_0 I)^r, \tag{1.1.10}$$
$r = 1, 2, \ldots$. We have already proved Theorem 1.1.4 for the Jordan form B; since the geometric (resp. algebraic) multiplicity of λ_0 as an eigenvalue of A is,

by definition, the geometric (resp. algebraic) multiplicity of λ_0 as an eigenvalue of B, Theorem 1.1.4 for A now follows from (1.1.10). □

1.2 Invariant subspaces

Let A be an $n \times n$ matrix (with complex entries). A subspace $\mathcal{N} \subseteq \mathbb{C}^n$ is called *invariant* for the matrix A (or in short *A-invariant*) if $Ax \in \mathcal{N}$ for every $x \in \mathcal{N}$. The same definition will be applied to a linear transformation $A : \mathbb{C}^n \longrightarrow \mathbb{C}^n$: A subspace $\mathcal{N} \subseteq \mathbb{C}$ is called *A-invariant* if $x \in \mathcal{N}$ implies $A(x) \in \mathcal{N}$. In this case A can be considered as a linear transformation $\mathcal{N} \longrightarrow \mathcal{N}$; we denote this linear transformation by $A|\mathcal{N}$.

We give some examples of A-invariant subspaces.

Example 1.2.1. The subspaces $\{0\}$ and $\mathbb{C}^n$ are A-invariant for every $n \times n$ matrix A; they are called the trivial invariant subspaces. □

Example 1.2.2. The one-dimensional subspace span $\{x_0\}$ spanned by an eigenvector of A is A-invariant. □

Example 1.2.3. The Jordan subspace span $\{x_1, \ldots, x_r\}$ spanned by a Jordan chain $x_1, \ldots, x_r$ of A is A-invariant. □

In the next example, and elsewhere in the book, we use the notation Im X to denote the *image*, or *range*, of an $m \times n$ matrix X:

$$\text{Im } X = \{Xy \mid y \in \mathbb{C}^n\}.$$

Note that Im X is a subspace in $\mathbb{C}^m$ and may also be described as the "column space" of X. The same notation will be applied to a linear transformation $X : \mathbb{C}^n \longrightarrow \mathbb{C}^m$:

$$\text{Im } X = \{X(y) \mid y \in \mathbb{C}^n\}.$$

Example 1.2.4. Let λ_0 be an eigenvalue of A. Then the subspaces of the form

$$\text{Ker } (A - \lambda_0 I)^r$$

for $r = 0, 1, 2, \ldots$, and the subspaces

$$\text{Im } (A - \lambda_0 I)^r = \{(A - \lambda_0 I)^r x \mid x \in \mathbb{C}^n\}$$

are A-invariant. The verification is easy: If $x \in \text{Ker } (A - \lambda_0 I)^r$, i.e. $(A - \lambda_0 I)^r x = 0$, then also $A(A - \lambda_0 I)^r x = (A - \lambda_0 I)^r Ax = 0$, and hence $Ax \in \text{Ker } (A - \lambda_0 I)^r$. This shows the A-invariance of $\text{Ker } (A - \lambda_0 I)^r$. Further, if $y = (A - \lambda_0 I)^r x$ for some $x \in \mathbb{C}^n$, then $Ay = (A - \lambda_0 I)^r Ax$, and hence $Ay \in \text{Im } (A - \lambda_0 I)^r$. More generally, if B is a matrix commuting with A (i.e. $BA = AB$), then the subspaces Ker B and Im B are A-invariant. □

Example 1.2.5. Let λ_0 be an eigenvalue of A, and let p be the minimal positive integer such that

$$\operatorname{Ker}(A - \lambda_0 I)^p = \operatorname{Ker}(A - \lambda_0 I)^{p+1}.$$

i.e. p is the index of λ_0. Then

$$\operatorname{Ker}(A - \lambda_0 I)^p = \operatorname{Ker}(A - \lambda_0 I)^r \qquad (1.2.1)$$

for all positive integers $r \geq p$, and the subspace (1.2.1) is called the *generalized eigenspace*, or *root subspace*, of A corresponding to λ_0. This subspace will be denoted $\mathcal{R}_{\lambda_0}(A)$; it is A-invariant by Example 1.2.4. □

If $\mathcal{R}_1, \ldots, \mathcal{R}_p$ are subspaces of a linear space $\mathcal{S}$, their *sum* is defined by

$$\mathcal{R}_1 + \cdots + \mathcal{R}_p = \{x_1 + \cdots + x_p : x_j \in \mathcal{R}_j, j = 1, 2, \ldots, p\},$$

and is also a subspace of $\mathcal{S}$.

Example 1.2.6. Let $\lambda_1, \ldots, \lambda_p$ be distinct eigenvalues of A. Then a sum of generalized eigenspaces,

$$\mathcal{R}_{\lambda_1}(A) + \cdots + \mathcal{R}_{\lambda_p}(A)$$

is also A-invariant and is known as a *spectral subspace* of A. (See also Proposition 1.2.4, to follow). □

We now make some useful and well-known observations about invariant subspaces.

Proposition 1.2.1. *The set of A-invariant subspaces (where the matrix or linear transformation A is fixed) is a lattice: If $\mathcal{M}$ and $\mathcal{N}$ are A-invariant subspaces, then so are $\mathcal{M} + \mathcal{N}$ and $\mathcal{M} \cap \mathcal{N}$.*

The easy proof is left to the reader.

Proposition 1.2.2. *Let $x_1, \ldots, x_k \in \mathbb{C}^n$ be a linearly independent set of vectors, and let a linear transformation $A : \mathbb{C}^n \longrightarrow \mathbb{C}^n$, or an $n \times n$ matrix A (understood as a linear transformation $\mathbb{C}^n \longrightarrow \mathbb{C}^n$), be given. The following statements are equivalent:*

(i) *the subspace* $\operatorname{span}\{x_1, \ldots, x_k\}$ *is A-invariant.*

(ii) *with respect to some basis in $\mathbb{C}^n$ the first k vectors of which are $x_1, \ldots, x_k$, the linear transformation A is given by a block triangular matrix*

$$\begin{bmatrix} A_{11} & A_{12} \\ 0 & A_{22} \end{bmatrix} \quad (1.2.2)$$

where the bottom left zero corner is $(n-k) \times k$.

(iii) *the matrix representation of A has the form (1.2.2) for any basis in $\mathbb{C}^n$ having its first k vectors $x_1, \ldots, x_k$.*

The proof follows easily from the definitions of an invariant subspace and of a matrix representation for a linear transformation. Indeed, the A-invariance of span $\{x_1, \ldots, x_k\}$ simply means that Ax_i is a linear combination of $x_1, \ldots, x_k$. So for any basis $X = \{x_1, \ldots, x_k, y_{k+1}, \ldots, y_{n-k}\}$ in $\mathbb{C}^n$ having $x_1 \ldots, x_k$ as its first k vectors we have that the last $n-k$ components in the column vector $[Ax_i]_X$ are zeros.

Proposition 1.2.3. *Let $n \times n$ matrices A and B be similar: $A = S^{-1}BS$. Then:*

(i) *a subspace $\mathcal{N} \subseteq \mathbb{C}^n$ is A-invariant if and only if $S\mathcal{N}$ is B-invariant.*

(ii) *a subspace $\mathcal{N} \subseteq \mathbb{C}^n$ is a root subspace for A if and only if $S\mathcal{N}$ is a root subspace for B.*

(iii) *a subspace $\mathcal{N} \subseteq \mathbb{C}^n$ is a spectral subspace for A if and only if $S\mathcal{N}$ is a spectral subspace for B.*

The proof is left to the reader.

We will show that invariant subspaces can be described in terms of their intersection with generalized eigenspaces. First, we need the concept of a direct sum of subspaces. A subspace $\mathcal{M} \subseteq \mathbb{C}^n$ is called a *direct sum* of the subspaces $\mathcal{N}_1, \ldots, \mathcal{N}_r$ (notation: $\mathcal{M} = \mathcal{N}_1 \dotplus \mathcal{N}_2 \dotplus \cdots \dotplus \mathcal{N}_r$) if

(i) the subspaces $\mathcal{N}_1, \ldots, \mathcal{N}_r$ are linearly independent, i.e. the equality $x_1 + \cdots + x_r = 0$, $x_i \in \mathcal{N}_i$ ($i = 1, \ldots, r$) implies that each x_i is equal to 0; and

(ii) $\mathcal{M}$ is equal to the sum of $\mathcal{N}_1, \ldots, \mathcal{N}_r$, i.e. $\mathcal{M}$ coincides with the set of vectors of the form $x_1 + \cdots + x_r$, where $x_1 \in \mathcal{N}_1, \ldots, x_r \in \mathcal{N}_r$.

Proposition 1.2.4. *Given an $n \times n$ matrix A, the space $\mathbb{C}^n$ is a direct sum of generalized eigenspaces of A:*

$$\mathbb{C}^n = \mathcal{R}_{\lambda_1}(A) \dotplus \cdots \dotplus \mathcal{R}_{\lambda_s}(A), \quad (1.2.3)$$

where $\lambda_1, \ldots, \lambda_s$ are the distinct eigenvalues of A.

We will not prove this proposition here, but refer the reader to Section 2.1 in Gohberg et al. (1986a) where a full proof can be found (see the proof of Theorem 2.1.2 there).

Theorem 1.2.5. *Let A be an $n \times n$ matrix, and let $\mathcal{M} \subseteq \mathbb{C}^n$ be an A-invariant subspace. Then*

$$\mathcal{M} = (\mathcal{R}_{\lambda_1}(A) \cap \mathcal{M}) \dotplus \ldots \dotplus (\mathcal{R}_{\lambda_s}(A) \cap \mathcal{M}),$$

where $\lambda_1, \ldots, \lambda_s$ are the distinct eigenvalues of A.

Proof. Apply Proposition 1.2.4 with A replaced by the restriction $A|\mathcal{M}$. The result is

$$\mathcal{M} = \mathcal{R}_{\mu_1}(A|\mathcal{M}) \dotplus \ldots \dotplus \mathcal{R}_{\mu_r}(A|\mathcal{M}),$$

where $\mu_1, \ldots, \mu_r$ are the distinct eigenvalues of $A|\mathcal{M}$. It remains to observe that each μ_j coincides with one of the λ_i's, say $\mu_j = \lambda_i$, and in this case

$$\mathcal{R}_{\mu_j}(A|\mathcal{M}) = \mathcal{R}_{\lambda_i}(A) \cap \mathcal{M};$$

and for those λ_i's that do not appear as eigenvalues of $A|\mathcal{M}$ we have

$$\mathcal{R}_{\lambda_i}(A) \cap \mathcal{M} = \{0\}. \quad \square$$

We conclude this section with a simple but useful observation concerning invariant subspaces of adjoint matrices. Here and elsewhere in the book the vector space $\mathbb{C}^n$ is considered with the standard euclidean scalar product:

$$\langle x, y \rangle = \sum_{i=1}^{n} x_i \bar{y}_i,$$

where

$$x = \begin{bmatrix} x_1 \\ \vdots \\ x_n \end{bmatrix}, \quad y = \begin{bmatrix} y_1 \\ \vdots \\ y_n \end{bmatrix}.$$

The following concepts are defined in terms of this scalar product:

(1) the *norm*, $\|x\|$, of a vectors $x \in \mathbb{C}^n$:

$$\|x\| = \sqrt{\langle x, x \rangle}.$$

(2) the *angle* α between $x, y \in \mathbb{C}^n$, $x \neq 0$, $y \neq 0$:

$$\cos \alpha = \frac{|\langle x, y \rangle|}{\|x\| \cdot \|y\|}, \quad 0 \leq \alpha \leq \frac{\pi}{2}.$$

(3) the *orthogonal complement* to a nonempty set $\mathcal{K} \subseteq \mathbb{C}^n$:

$$\mathcal{K}^\perp = \{x \in \mathbb{C}^n | \langle x, y \rangle = 0 \text{ for all } y \in \mathcal{K}\}.$$

Observe that the orthogonal complement is always a subspace in $\mathbb{C}^n$ (although $\mathcal{K}$ itself may be any subset of $\mathbb{C}^n$). Note that

$$\mathcal{K} \cap \mathcal{K}^\perp = \{0\};$$

and if $\mathcal{K}$ is a subspace, then also

$$\mathcal{K} \dotplus \mathcal{K}^\perp = \mathbb{C}^n.$$

This explains the usage of the term "complement".

Proposition 1.2.6. *A subspace $\mathcal{M} \subseteq \mathbb{C}^n$ is invariant for an $n \times n$ matrix A if and only if its orthogonal complement is invariant for the adjoint (conjugate transpose) matrix A^*.*

We leave out the easy proof (see, e.g., the proof of Proposition 1.4.4 in Gohberg *et al.* (1986a)).

1.3 Projectors and invariant subspaces

A linear transformation defined by $P : \mathbb{C}^n \to \mathbb{C}^n$ is called a *projector* if $P^2 = P$. The important feature of projectors is that there exists a one-to-one correspondence between the set of all projectors and the set of all pairs of complementary subspaces in $\mathbb{C}^n$. This correspondence is described in Theorem 1.3.1.

Recall first that subspaces $\mathcal{M}, \mathcal{L}$ are called *complementary* (are *direct complements* of each other) if $\mathcal{M} \cap \mathcal{L} = \{0\}$ and $\mathcal{M} \dotplus \mathcal{L} = \mathbb{C}^n$.

The subspaces $\mathcal{M}$ and $\mathcal{L}$ are called *orthogonal* if for each $x \in \mathcal{M}$ and $y \in \mathcal{L}$ we have $\langle x, y \rangle = 0$ and they are *orthogonal complements* if, in addition, they are complementary. In this case, we write $\mathcal{M} = \mathcal{L}^\perp$, $\mathcal{L} = \mathcal{M}^\perp$.

Theorem 1.3.1. *Let P be a projector. Then $(\operatorname{Im} P, \operatorname{Ker} P)$ is a pair of complementary subspaces in $\mathbb{C}^n$. Conversely, for every pair $(\mathcal{L}_1, \mathcal{L}_2)$ of complementary subspaces in $\mathbb{C}^n$, there exists a unique projector P such that $\operatorname{Im} P = \mathcal{L}_1$, $\operatorname{Ker} P = \mathcal{L}_2$.*

Complete proofs of Theorem 1.3.1 and Proposition 1.3.2 below are well-known and found in many sources (see, e.g., Section 1.5 of Gohberg et al. (1986a)). We will not reproduce the proofs here.

We say that P is the projector *on* $\mathcal{L}_1$ *along* $\mathcal{L}_2$ if $\operatorname{Im} P = \mathcal{L}_1$, $\operatorname{Ker} P = \mathcal{L}_2$. A projector P is called *orthogonal* if $\operatorname{Ker} P = (\operatorname{Im} P)^\perp$. Thus the corresponding complementary subspaces are mutually orthogonal. Orthogonal projectors are particularly important and can be characterized as follows.

Proposition 1.3.2. *A projector P is orthogonal if and only if P is self-adjoint, that is, $P^* = P$.*

Note that if P is a projector, so is $I - P$. Indeed, $(I - P)^2 = I - 2P + P^2 = I - 2P + P = I - P$. Moreover, $\operatorname{Ker} P = \operatorname{Im}(I - P)$ and $\operatorname{Im} P = \operatorname{Ker}(I - P)$. It is natural to call the projectors P and $I - P$ *complementary projectors*.

Consider now an invariant subspace $\mathcal{M}$ for a linear transformation $A : \mathbb{C}^n \to \mathbb{C}^n$. For any projector P with $\operatorname{Im} P = \mathcal{M}$ we obtain

$$PAP = AP \tag{1.3.1}$$

Indeed, if $x \in \operatorname{Ker} P$, we obviously have

$$PAPx = APx.$$

If $x \in \operatorname{Im} P = \mathcal{M}$, we see that Ax belongs to $\mathcal{M}$ as well and thus

$$PAPx = PAx = Ax = APx$$

once more. Since $\mathbb{C}^n = \operatorname{Ker} P \dotplus \operatorname{Im} P$, (1.3.1) follows. Conversely, if P is a projector for which (1.3.1) holds, then for every $x \in \operatorname{Im} P$ we have $PAx = Ax$; in other words, $\operatorname{Im} P$ is A invariant. So *a subspace $\mathcal{M}$ is A-invariant if and only if it is the image of a projector P for which (1.3.1) holds.*

Let $\mathcal{M}$ be an A-invariant subspace and let P be a projector on $\mathcal{M}$ [so that (1.3.1) holds]. Denoting the kernel of P by $\mathcal{M}'$, represent A as a 2×2 block matrix

$$\begin{bmatrix} A_{11} & A_{12} \\ A_{21} & A_{22} \end{bmatrix}$$

with respect to the direct sum decomposition $\mathbb{C}^n = \mathcal{M} \dotplus \mathcal{M}'$. Here A_{11} is a linear transformation $PAP|_\mathcal{M} : \mathcal{M} \to \mathcal{M}$, A_{12} is a linear transformation $PA(I - P)|_{\mathcal{M}'} : \mathcal{M}' \to \mathcal{M}$,

$$A_{21} = (I - P)AP|_\mathcal{M} : \mathcal{M} \to \mathcal{M}'$$

$$A_{22} = (I - P)A(I - P)|_{\mathcal{M}'} : \mathcal{M}' \to \mathcal{M}'$$

and all these transformations are written as matrices with respect to some chosen bases in $\mathcal{M}$ and $\mathcal{M}'$. As $\mathcal{M}$ is A-invariant, equation (1.3.1) implies that $(I - P)AP = 0$, that is, $A_{21} = 0$. Hence

$$A = \begin{bmatrix} A_{11} & A_{12} \\ 0 & A_{22} \end{bmatrix}, \qquad (1.3.2)$$

and this is consistent with Proposition 1.2.2.

We conclude this section with a description of spectral subspaces in terms of contour integrals. Let Γ be a simple, closed, rectifiable, and positively oriented contour in the complex plane. In fact, for our purposes polygonal contours will suffice. Given an $n \times n$ matrix $B(\lambda) = [b_{ij}(\lambda)]_{i,j=1}^n$, that depends continuously on the variable $\lambda \in \Gamma$ (this means that each entry $b_{ij}(\lambda)$ in $B(\lambda)$ is a continuous function of λ on Γ) the integral

$$\int_\Gamma B(\lambda) d\lambda = [c_{ij}]_{i,j=1}^n$$

is defined naturally as the $n \times n$ matrix whose entries are the integrals of the entries of $B(\lambda)$:

$$c_{ij} = \int_\Gamma b_{ij}(\lambda) d\lambda; \quad i,j = 1, \ldots, n \, .$$

The same definition of a contour integral applies also for linear transformations $B(\lambda) : \mathbb{C}^n \to \mathbb{C}^n$ that are continuous functions of λ on Γ. We have only to write $B(\lambda)$ as a matrix $[b_{ij}(\lambda)]_{i,j=1}^n$ in a fixed basis, and then interpret $\int_\Gamma B(\lambda) d\lambda$ as a transformation represented by the matrix $[\int_\Gamma b_{ij}(\lambda) d\lambda]_{i,j=1}^n$ in the same basis. One checks easily that this definition is independent of the chosen basis.

Proposition 1.3.3. *Let Λ be a subset of $\sigma(A)$ where A is a linear transformation on $\mathbb{C}^n$, and let Γ be a closed contour having Λ in its interior and $\sigma(A) \backslash \Lambda$ outside Γ. Then the transformation*

$$\frac{1}{2\pi i} \int_\Gamma (\lambda I - A)^{-1} d\lambda$$

is a projector (known as a Riesz projector*) onto the spectral subspace associated with Λ and along the spectral subspace associated with $\sigma(A) \backslash \Lambda$.*

Proof. Using the relation $S(\lambda I - A)^{-1} S^{-1} = (\lambda I - SAS^{-1})^{-1}$ and the Jordan form, we can assume that A is an $n \times n$ matrix given by

$$A = J_{k_1}(\lambda_1) \oplus \cdots \oplus J_{k_p}(\lambda_p)$$

where $J_{k_i}(\lambda_i)$ is the $k_i \times k_i$ Jordan block with λ_i on the main diagonal. One easily verifies that

$$[\lambda I - J_{k_i}(\lambda_i)]^{-1} = \begin{bmatrix} (\lambda - \lambda_i)^{-1} & (\lambda - \lambda_i)^{-2} & \cdots & (\lambda - \lambda_i)^{-k_i} \\ 0 & (\lambda - \lambda_i)^{-1} & & \vdots \\ \vdots & \vdots & & (\lambda - \lambda_i)^{-2} \\ 0 & 0 & \cdots & (\lambda - \lambda_i)^{-1} \end{bmatrix}.$$

As a first consequence of this formula we see immediately that, because $\sigma(A) \cap \Gamma = \emptyset$, $(\lambda I - A)^{-1}$ is indeed continuous on Γ. Further, the Cauchy formula gives

$$\int_\Gamma (\lambda - \lambda_i)^{-m} d\lambda = \begin{cases} 2\pi, & \text{if } m = 1 \text{ and } \lambda_i \text{ is inside } \Gamma \\ 0, & \text{otherwise.} \end{cases}$$

Thus

$$\frac{1}{2\pi i} \int_\Gamma (\lambda I - A)^{-1} d\lambda = S(K_1 \oplus \cdots \oplus K_p) S^{-1} \quad (1.3.3)$$

where $K_i = I$ if $\lambda \in \Lambda$ and $K_i = 0$ if $\lambda \notin \Lambda$. Thus the matrix (1.3.2) is indeed a projector with image and kernel as prescribed by the theorem. □

This proof of Proposition 1.3.3 is standard (see Section 2.4 in Gohberg *et al.* (1986a)).

1.4 Real matrices and canonical forms

Here, we review the main concepts and results of Sections 1.1 and 1.2 for matrices with real entries, or linear transformations from $\mathbb{R}^n$ into $\mathbb{R}^n$. A real $n \times n$ matrix A need not have n real eigenvalues (counted with multiplicities). For example, $\begin{bmatrix} 0 & 1 \\ -1 & 0 \end{bmatrix}$ has no real eigenvalues at all. On the other hand, the non-real eigenvalues of A, if considered as complex numbers, will occur in complex conjugate pairs $\lambda \pm i\mu$, where λ, μ are real and $\mu \neq 0$. A prototype of a real matrix A having non-real complex conjugate eigenvalues is the *real Jordan block*

$$J_{2k}(\lambda \pm i\mu) = \begin{bmatrix} \lambda & \mu & 1 & 0 & \cdots & 0 & 0 \\ -\mu & \lambda & 0 & 1 & \cdots & 0 & 0 \\ 0 & 0 & \lambda & \mu & \cdots & 0 & 0 \\ 0 & 0 & -\mu & \lambda & \cdots & 0 & 0 \\ \vdots & \vdots & \vdots & \vdots & & 1 & 0 \\ & & & & & 0 & 1 \\ 0 & 0 & 0 & 0 & \cdots & \lambda & \mu \\ 0 & 0 & 0 & 0 & \cdots & -\mu & \lambda \end{bmatrix}; \quad \lambda, \mu \in \mathbb{R}, \; \mu > 0$$

of size $2k \times 2k$ (k a positive integer). Clearly, this matrix has eigenvalues $\lambda \pm i\mu$, each with algebraic multiplicity k. A block diagonal matrix

$$B = \begin{bmatrix} K_1 & 0 & \cdots & 0 \\ 0 & K_2 & \cdots & 0 \\ \vdots & \vdots & & \vdots \\ 0 & 0 & \cdots & K_p \end{bmatrix}$$

is called a *real Jordan matrix* if every block K_i is either a Jordan block with a real eigenvalue, or a real Jordan block with a pair of non-real complex conjugate eigenvalues.

Theorem 1.4.1. *Every real $n \times n$ matrix A is similar (with a real similarity matrix) to a real Jordan matrix. Moreover, the real Jordan matrix is uniquely determined by A, up to permutation of the blocks on the main diagonal.*

For the proof of Theorem 1.4.1 we refer the reader to Chapter 12 in Gohberg et al. (1986a). A real Jordan matrix which is similar to a given real $n \times n$ matrix A is called a *real Jordan form* of A.

We will use the concepts of algebraic multiplicity, geometric multiplicity and partial multiplicities of eigenvalues of real matrices in the same sense as for complex matrices. Thus, the real matrix

$$J_6(1 \pm i) \oplus J_{10}(1 \pm i) \oplus J_{14}(1 \pm i) \oplus J_2(2 \pm 3i)$$

has eigenvalues $1 \pm i$ and $2 \pm 3i$; each of the eigenvalues $1 + i$ and $1 - i$ has algebraic multiplicity 15, geometric multiplicity 3 and partial multiplicities 3, 5 and 7. Each of the eigenvalues $2 + 3i$ and $2 - 3i$ has algebraic and geometric multiplicity 1.

Given a real $n \times n$ matrix A, a subspace $\mathcal{M} \subseteq \mathbb{R}^n$ is called A-*invariant* if $Ax \in \mathcal{M}$ for every $x \in \mathcal{M}$. Again, this definition will be applied to linear transformations $A : \mathbb{R}^n \to \mathbb{R}^n$ as well. Many examples of invariant subspaces given in the previous section have counterparts for real matrices:

Example 1.4.1. Let λ_0 be a *real* eigenvalue of a real $n \times n$ matrix A. Let

$$\text{Ker}\,(A - \lambda_0 I)^p = \{x \in \mathbb{R}^n | (A - \lambda_0 I)^p x = 0\},$$

$p = 0, 1, \ldots$, and let p_0 be the index of λ_0. Then

$$\text{Ker}\,(A - \lambda_0 I)^{p_0} = \text{Ker}\,(A - \lambda_0 I)^r$$

for all $r \geq p_0$, and this subspace is called the *generalized eigenspace*, or *root subspace*, of A corresponding to λ_0. We keep the notation $\mathcal{R}_{\lambda_0}(A)$ for this subspace; it is easily seen to be A-invariant.

For non-real eigenvalues of A the root subspaces have to be defined differently to ensure that they are subspaces of $\mathbb{R}^n$. Let $\lambda \pm i\mu$ be a pair of non-real complex conjugate eigenvalues of A. Consider the real matrices $(A^2 - 2\lambda A + (\lambda^2 + \mu^2)I)^p$, $p = 1, \ldots$, and let p_0 be the minimal positive integer such that

$$\mathrm{Ker}\,(A^2 - 2\lambda A + (\lambda^2 + \mu^2)I)^{p_0} = \mathrm{Ker}\,(A^2 - 2\lambda A + (\lambda^2 + \mu^2)I)^{p_0+1}.$$

Again,

$$\mathrm{Ker}\,(A^2 - 2\lambda A + (\lambda^2 + \mu^2)I)^{p_0} = \mathrm{Ker}\,(A^2 - 2\lambda A + (\lambda^2 + \mu^2)I)^r \qquad (1.4.1)$$

for $r \geq p_0$, and the subspace (1.4.1) (which is a subspace in $\mathbb{R}^n$) is A-invariant. This subspace is called the *generalized eigenspace*, or *root subspace* of A corresponding to the pair $\lambda \pm i\mu$ of non-real complex conjugate eigenvalues. We denote this subspace by $\mathcal{R}_{\lambda \pm i\mu}(A)$. □

Example 1.4.2. Let $\lambda_1, \ldots, \lambda_p$ be distinct real eigenvalues of a real $n \times n$ matrix (or linear transformation) A, and let $\mu_1 \pm i\nu_1, \ldots, \mu_q \pm i\nu_q$ be distinct pairs of non-real complex conjugate eigenvalues of A ($\lambda_1, \ldots, \lambda_p$ need not be all of the distinct real eigenvalues, and $\mu_1 \pm i\nu_1, \ldots, \mu_q \pm i\nu_q$ need not be all of the distinct non-real eigenvalues of A). Then the sum of generalized eigenspaces

$$\mathcal{R}_{\lambda_1}(A) + \cdots + \mathcal{R}_{\lambda_p}(A) + \mathcal{R}_{\mu_1 \pm i\nu_1}(A) + \cdots + \mathcal{R}_{\mu_q \pm i\nu_q}(A), \qquad (1.4.2)$$

which is in fact a direct sum, is an A-invariant subspace (in $\mathbb{R}^n$). The A-invariant subspaces that arise in this way are called *spectral invariant subspaces*. As in the complex case (Section 1.3) the subspace (1.4.2) admits an integral representation

$$\mathrm{Im}\left[\frac{1}{2\pi i}\int_\Gamma (\lambda I - A)^{-1} d\lambda\right],$$

where Γ is any closed, simple, rectifiable contour in the complex plane such that $\lambda_1, \ldots, \lambda_p, \mu_1 \pm i\nu_1, \ldots, \mu_q \pm i\nu_q$ are inside Γ, and all other eigenvalues of A are outside Γ. The matrix (or linear transformation)

$$P = (2\pi i)^{-1} \int_\Gamma (\lambda I - A)^{-1} d\lambda \qquad (1.4.3)$$

is a projector on the spectral subspace (1.4.2) along the spectral subspace corresponding to all other eigenvalues of A. The projector P is a real matrix (or linear transformation). Indeed, without loss of generality, choose Γ to be symmetric with respect to the real axis. Then the equality $P = \bar{P}$ follows by taking complex conjugates in (1.4.3) and changing the variable $\lambda \to \bar{\lambda}$. □

The results of Propositions 1.2.1, 1.2.2 and 1.2.3 (with the obvious changes, such as replacing $\mathbb{C}^n$ by $\mathbb{R}^n$, requiring that the similarity matrix S in Proposition 1.2.3 be real, etc.) are valid also for real matrices.

The real analogue of Theorem 1.2.5 looks as follows:

Theorem 1.4.2. *Let A be a real $n \times n$ matrix (or linear transformation), and let $\lambda_1, \ldots, \lambda_p$ be all the real eigenvalues of A (if any), and $\mu_1 \pm i\nu_1, \ldots, \mu_q \pm i\nu_q$ be all the non-real eigenvalues of A (if any), arranged in conjugate pairs. Then any A-invariant subspace $\mathcal{M} \subseteq \mathbb{R}^n$ admits the representation*

$$\mathcal{M} = (\mathcal{R}_{\lambda_1}(A) \cap \mathcal{M}) \dotplus \cdots \dotplus (\mathcal{R}_{\lambda_p}(A) \cap \mathcal{M})$$
$$\dotplus (\mathcal{R}_{\mu_1 \pm i\nu_1}(A) \cap \mathcal{M}) \dotplus \cdots \dotplus (\mathcal{R}_{\mu_q \pm i\nu_q}(A) \cap \mathcal{M}). \quad (1.4.4)$$

Proof. We outline the construction used in the proof of this theorem; more details are found in the work of Shilov (1961) (Section 6.62).

Let $\varphi(\lambda)$ be the characteristic polynomial of A. Then $\varphi(\lambda)$ has real coefficients and therefore can be factorized in the form

$$\varphi(\lambda) = \varphi_1(\lambda) \ldots \varphi_{p+q}(\lambda),$$

where $\varphi_j(\lambda) = (\lambda - \lambda_j)^{\alpha_j}$ for a suitable positive integer α_j ($j = 1, \ldots, p$), and $\varphi_{p+j}(\lambda) = (\lambda^2 - 2\mu_j \lambda + \mu_j^2 + \nu_j^2)^{\beta_j}$ for a suitable positive integer β_j ($j = 1, \ldots, q$) (recall that the roots of $\varphi(\lambda)$ are exactly the eigenvalues of A). Let

$$\psi_j(\lambda) = \frac{\varphi(\lambda)}{\varphi_j(\lambda)} \quad (j = 1, \ldots, p+q).$$

Observe that the polynomials $\psi_j(\lambda)$ have real coefficients. Using the fact that $\varphi(A) = 0$ (the Cayley–Hamilton theorem) we prove that

$$\mathcal{R}_{\lambda_j}(A) = \operatorname{Im}(\psi_j(A)), \quad j = 1, \ldots, p \quad (1.4.5)$$

and

$$\mathcal{R}_{\mu_j \pm i\nu_j}(A) = \operatorname{Im}(\psi_{p+j}(A)), \quad j = 1, \ldots, q. \quad (1.4.6)$$

But the polynomials $\psi_1(\lambda), \ldots, \psi_{p+q}(\lambda)$ have no common zeros (in the complex plane), and therefore there exist polynomials with real coefficients $\tau_1(\lambda), \ldots, \tau_{p+q}(\lambda)$ such that

$$\psi_1(\lambda)\tau_1(\lambda) + \ldots + \psi_{p+q}(\lambda)\tau_{p+q}(\lambda) \equiv 1. \quad (1.4.7)$$

(This is easily proved by induction on $p+q$, using the euclidean algorithm in the case $p + q = 2$). Now for every $x \in \mathcal{M}$, using (1.4.7) we obtain

$$x = \sum_{j=1}^{p+q} \psi_j(A)(\tau_j(A)x).$$

In view of (1.4.5), (1.4.6) and the A-invariance of $\mathcal{M}$ we have

$$\psi_j(A)(\tau_j(A)x) \in \mathcal{R}_{\lambda_j}(A) \cap \mathcal{M} \quad (j = 1, \ldots, p)$$
$$\psi_{p+j}(A)(\tau_{p+j}(A)x) \in \mathcal{R}_{\mu_j + i\nu_j}(A) \cap \mathcal{M} \quad (j = 1, \ldots, q).$$

Therefore x belongs to the right-hand side of (1.4.4), and the inclusion $\subseteq$ in (1.4.4) is established. Since the opposite inclusion $\supseteq$ is trivial, we have proved the equality in (1.4.4).

That the right-hand side of (1.4.4) is a *direct* sum follows from the corresponding direct sum property for complex matrices (Theorem 1.2.5). □

In particular, when $\mathcal{M} = \mathbb{R}^n$, we obtain a decomposition of $\mathbb{R}^n$ as a direct sum of generalized eigenspaces:

$$\mathbb{R}^n = \mathcal{R}_{\lambda_1}(A) \dotplus \cdots \dotplus \mathcal{R}_{\lambda_p}(A) \dotplus \mathcal{R}_{\mu_1 \pm i\nu_1}(A) \dotplus \cdots \dotplus \mathcal{R}_{\mu_q \pm i\nu_q}(a).$$

Finally, we give a version of Proposition 1.2.6 for real matrices. Consider the standard euclidean scalar product in $\mathbb{R}^n$.

$$\langle x, y \rangle = \sum_{i=1}^{n} x_i y_i,$$

where

$$x = \begin{bmatrix} x_1 \\ x_2 \\ \vdots \\ x_n \end{bmatrix}, \quad y = \begin{bmatrix} y_1 \\ y_2 \\ \vdots \\ y_n \end{bmatrix}.$$

Throughout the book $\mathbb{R}^n$ will be considered with this scalar product and the concepts derived from it, such as norm, orthogonal complement, etc.

Proposition 1.4.3. *A subspace* $\mathcal{M} \subseteq \mathbb{R}^n$ *is invariant for a real* $n \times n$ *matrix* A *if and only if* $\mathcal{M}^\perp$ *is invariant for the transposed matrix* A^T.

1.5 Square roots of definite matrices

Recall first of all that a hermitian matrix is diagonable with real eigenvalues. Furthermore, such a matrix is unitarily similar to a diagonal matrix of its eigenvalues. Thus, let A be an $n \times n$ matrix with $A^* = A$ and let its n eigenvalues (with multiplicities) be $\lambda_1, \lambda_2, \ldots, \lambda_n$. Define

$$D = \text{diag}\,[\lambda_1, \lambda_2, \ldots, \lambda_n]. \tag{1.5.1}$$

Then there is an $n \times n$ unitary matrix U (i.e. with $U^*U = I$) such that

$$A = UDU^*. \tag{1.5.2}$$

A hermitian matrix is said to be *positive definite (positive semidefinite)* if $x^*Ax > 0$ ($x^*Ax \geq 0$) for all nonzero $x \in \mathbb{C}^n$. Similar definitions apply for *negative definite*, and *negative semidefinite* matrices. In these four cases we write $A > 0$, $A \geq 0$, $A < 0$ and $A \leq 0$, respectively. In the "positive" case two other characterizations apply and are contained in the following elementary result (for the proof see Section 5.3 of Lancaster and Tismenetsky (1985)).

SQUARE ROOTS OF DEFINITE MATRICES

Theorem 1.5.1. *The following are equivalent:*

(a) $A > 0$ $(A \geq 0)$

(b) $\lambda_j > 0$ $(\lambda_j \geq 0)$ *for every eigenvalue λ_j of A.*

(c) *There is a unique matrix $A_0 > 0$ $(A_0 \geq 0)$ such that $A_0^2 = A$.*

The matrix A_0 of part (c) is naturally written $A^{1/2}$. Note also that when A is a *real* matrix, the matrix $A^{1/2}$ (or A_0) is also real.

Clearly $A \geq 0$ implies that (in equation (1.5.1)) we have $D \geq 0$. Also,

$$D^{1/2} = \operatorname{diag}[\lambda_1^{1/2}, \lambda_2^{1/2}, \ldots, \lambda_n^{1/2}],$$

and, in contrast with (1.5.2),

$$A^{1/2} = U D^{1/2} U^*. \tag{1.5.3}$$

Suppose that $\lambda_1, \ldots, \lambda_r$ are positive and $\lambda_{r+1} = \cdots = \lambda_n = 0$. If $u_1, u_2, \ldots, u_n$ are the columns of U then (1.5.2) and (1.5.3) may be written

$$A = \sum_{j=1}^{r} \lambda_j u_j u_j^*, \quad A^{1/2} = \sum_{j=1}^{r} \lambda_j^{1/2} u_j u_j^*.$$

It follows immediately that

$$\operatorname{Ker}(A^{1/2}) = \operatorname{Ker} A = \operatorname{span}\{u_{r+1}, \ldots, u_n\}, \tag{1.5.4}$$
$$\operatorname{Im}(A^{1/2}) = \operatorname{Im} A = \operatorname{span}\{u_1, \ldots, u_r\}. \tag{1.5.5}$$

In particular, $A^{1/2}$ and A have the same rank. These relations are easily strengthened as follows:

Theorem 1.5.2. *If $A \geq 0$ then $\operatorname{Ker}(A^{1/2}X) = \operatorname{Ker}(AX)$ for any $n \times m$ matrix X, and $\operatorname{Im}(YA^{1/2}) = \operatorname{Im}(YA)$ for any $m \times n$ matrix Y.*

Proof. If $A^{1/2}Xx = 0$ for some $x \in \mathbb{C}^m$, then

$$AXx = A^{1/2}(A^{1/2}Xx) = 0.$$

This proves $\operatorname{Ker}(A^{1/2}X) \subseteq \operatorname{Ker}(AX)$. For the proof of the opposite inclusion, let $AXx = 0$. Then

$$\|A^{1/2}Xx\|^2 = \langle A^{1/2}Xx, A^{1/2}Xx \rangle = \langle AXx, Xx \rangle = 0,$$

and therefore $A^{1/2}Xx = 0$. The proof of $\operatorname{Im}(YA^{1/2}) = \operatorname{Im}(YA)$ is left to the reader. □

1.6 Regular matrix pencils

If A, B are complex (or real) $n \times n$ matrices the family of matrices $\lambda A - B$, $\lambda \in \mathbb{C}$ (or $\mathbb{R}$), is called a *matrix pencil* and a matrix pencil is said to be *regular* if there is a scalar λ_0 such that $\det(\lambda_0 A - B) \neq 0$. Clearly, if $\det A \neq 0$ then a pencil $\lambda A - B$ is regular. However, the purpose of this section is to set up some machinery to handle cases in which both A and B may be singular.

Let $\lambda A - B$ be a regular matrix pencil. Then there are distinct complex numbers $\lambda_1, \lambda_2, \ldots, \lambda_r$, $r \leq n$, for which

$$\det(\lambda_j A - B) = 0, \quad j = 1, 2, \ldots, r. \tag{1.6.1}$$

These numbers are called the *(finite) eigenvalues* of the pencil, and any nonzero vector in $\operatorname{Ker}(\lambda_j A - B)$ is called a *(right) eigenvector* of $\lambda A - B$ corresponding to λ_j. These definitions extend those of Section 1.1 in a natural way and, with a small abuse of language, the classical case is recovered when $A = I$. A matrix pencil is said to have an *eigenvalue at infinity* if the pencil $A - \mu B$ has a zero eigenvalue (i.e. when $\operatorname{Ker} A \neq \{0\}$).

A subspace $\mathcal{S}$ is said to be *deflating* for the pair (A, B) if there is a subspace $\mathcal{T}$ such that $\dim \mathcal{S} = \dim \mathcal{T}$ and $A\mathcal{S} \subseteq \mathcal{T}$, $B\mathcal{S} \subseteq \mathcal{T}$. Clearly, if $A = I$ then $\mathcal{S}$ is B-invariant, and so this concept generalizes that of an invariant subspace. Furthermore, if $\lambda A - B$ is a regular pencil with eigenvectors $x_1, \ldots, x_s$ then the subspace $\mathcal{S}$ spanned by these vectors is deflating for (A, B) and we may take $\mathcal{T} = A\mathcal{S}$. This example is one of a more general class of deflating subspaces described in the first simple lemma:

Lemma 1.6.1. *Let $\lambda A - B$ be a regular matrix pencil. If there is an $n \times r$ matrix X such that $AXK = BX$ (and K is an $r \times r$ matrix) then $\operatorname{Im} X$ is deflating for (A, B).*

Proof. Let $\mathcal{T}$ be any r-dimensional subspace with $\operatorname{Im}(AX) \subseteq \mathcal{T}$. Then $A(\operatorname{Im} X) = \operatorname{Im}(AX) \subseteq \mathcal{T}$. But $AXK = BX$ gives $\operatorname{Im}(BX) \subseteq \operatorname{Im}(AX)$ so that $B(\operatorname{Im} X) \subseteq \mathcal{T}$ as well. Hence the result. □

In Lemma 1.6.1, we can assume without loss of generality that X has rank r. Otherwise, replace X by $\tilde{X}$, where $\tilde{X}$ is an $n \times r_1$ matrix whose columns form a basis in $\operatorname{Im} X$, and replace K by $LKL^{(-1)}$, where L is a (necessarily right invertible) $r_1 \times r$ matrix such that $X = \tilde{X}L$, and where $L^{(-1)}$ is any right inverse of L.

Under the additional hypothesis that X has rank r, reduce K to Jordan canonical form and it is easy to see that, in Lemma 1.6.1, the eigenvalues of K are all finite eigenvalues of $\lambda A - B$. Clearly, similar statements hold if $AX = BXK$ but, in this case, an eigenvalue of $\lambda A - B$ at infinity corresponds to a zero eigenvalue of K.

The importance of deflating subspaces depends on the following fact which also helps to explain the choice of the term "deflating". It is due to Stewart (1972) and can be perceived as a generalization of part of Proposition 1.2.2.

Theorem 1.6.2. *Let $\lambda A - B$ be a regular matrix pencil. If S is an r-dimensional deflating subspace for (A, B) then there are nonsingular matrices X and Y for which*

$$YAX = \begin{bmatrix} A_1 & A_2 \\ 0 & A_3 \end{bmatrix}, \quad YBX = \begin{bmatrix} B_1 & B_2 \\ 0 & B_3 \end{bmatrix}, \quad (1.6.2)$$

A_1, B_1 *are of size $r \times r$, and S is the span of the first r columns of X.*

Conversely, if (1.6.2) holds for nonsingular X and Y and X_1 is the leading $n \times r$ partition of X then $\operatorname{Im} X_1$ is deflating for (A, B).

Proof. Let S be deflating for (A, B) and $AS \subseteq T$, $BS \subseteq T$, $\dim S = \dim T = r$. Define $n \times r$ matrices X_1, Z_1 so that $\operatorname{Im} X_1 = S$, $\operatorname{Im} Z_1 = T$ and complete nonsingular matrices

$$X = [X_1 \ X_2], \quad Y^{-1} = [Z_1 \ Z_2]. \quad (1.6.3)$$

Also, write $Y = \begin{bmatrix} Y_1 \\ Y_2 \end{bmatrix}$ where Y_1 is $r \times n$. Thus, $Y_2 Z_1 = 0$. With this construction, $AS \subseteq T = \operatorname{Im} Z_1$ shows that $Y_2 A X_1 = 0$. Similarly, $Y_2 B X_1 = 0$ and (1.6.2) is obtained.

For the converse, partition X, Y, Y^{-1} as above and (1.6.2) implies

$$A[X_1 \ X_2] = [Z_1 \ Z_2] \begin{bmatrix} A_1 & A_2 \\ 0 & A_3 \end{bmatrix}, \quad B[X_1 \ X_2] = [Z_1 \ Z_2] \begin{bmatrix} B_1 & B_2 \\ 0 & B_3 \end{bmatrix}.$$

Thus $AX_1 = Z_1 A_1$ and $BX_1 = Z_1 B_1$. Defining $S = \operatorname{Im} X_1$ and $T = \operatorname{Im} Z_1$ it is found that $\dim T = \dim S$, $AS \subseteq T$, and $BS \subseteq T$. □

The theorem shows that when $\lambda A - B$ is regular there is a strict equivalence relation:

$$Y(\lambda A - B)X = \begin{bmatrix} \lambda A_1 - B_1 & \lambda A_2 - B_2 \\ 0 & \lambda A_3 - B_3 \end{bmatrix}.$$

Clearly, the "deflated" pencils $\lambda A_1 - B_1$ and $\lambda A_3 - B_3$ must also be regular and the union of their eigenvalues is just the set of eigenvalues of $\lambda A - B$. For a given deflating subspace S, $\lambda A_1 - B_1$ is a reduced pencil corresponding to S and is written $\lambda A_1 - B_1 = (\lambda A - B)|_S$. Clearly, A_1 and B_1 depend on the choice of X and Y, but not the spectrum of $\lambda A_1 - B_1$.

Note also that, in the statement of the theorem, X and Y can be chosen to be unitary matrices. This is achieved by choosing orthonormal bases for S and T as the columns of X_1 and Z_1, respectively, and then completions X_2 and Z_2 in equations (1.6.3) are chosen so that X and Y^{-1} (and hence Y) are unitary.

A converse statement for Lemma 1.6.1 follows readily from the construction used in the last proof.

Corollary 1.6.3. *Let S be an r-dimensional deflating subspace for (A, B) for which $(\text{Ker } A) \cap S = \{0\}$. Then there is an $n \times r$ matrix X_1 of rank r and an $r \times r$ matrix K such that $\text{Im } X_1 = S$ and $AX_1 K = BX_1$.*

Proof. Choose X_1 and define Z_1 as in the above proof. Then $\text{Im } X_1 = S$, $AX_1 = Z_1 A_1$, and $BX_1 = Z_1 B_1$. If $z \in \text{Ker } A_1$ then $X_1 z \in \text{Ker } A$, i.e. $X_1 z \in (\text{Ker } A) \cap S$ so that $X_1 z = 0$ and, hence, $z = 0$. Thus A_1 is invertible, $Z_1 = AX_1 A_1^{-1}$ and hence $BX_1 = AX_1(A_1^{-1} B)$. Thus we may choose $K = A_1^{-1} B_1$. □

The next simple proposition gives an alternate characterization of a deflating subspace which is neat but, perhaps, less intuitive than the definition adopted here.

Proposition 1.6.4. *Let $\lambda A - B$ be a regular matrix pencil. Then S is a deflating subspace for the pair (A, B) if and only if $\dim(AS + BS) \leq \dim S$.*

Proof. If S is deflating the result follows directly from the definition. Conversely, if $\dim(AS + BS) \leq \dim S$, let T be any subspace with the dimension of S for which $T \supseteq AS + BS$. Then we also have $AS \subseteq T$, $BS \subseteq T$ and S is deflating. □

The set of deflating subspaces for a regular pencil can be identified with the lattice of invariant subspaces for a particular $n \times n$ matrix so that no essentially new properties are obtained. To see this, let λ_0 be such that $\det(\lambda_0 A - B) \neq 0$ and make the transformation of parameter $\mu = (\lambda - \lambda_0)^{-1}$. Then

$$\lambda A - B = \mu^{-1}\{\mu(\lambda_0 A - B) + A\} = \mu^{-1}(\lambda_0 A - B)\{\mu I + (\lambda_0 A - B)^{-1} A\}. \quad (1.6.4)$$

Clearly, if λ_j is a finite eigenvalue of $\lambda A - B$ then $\mu_j := (\lambda_j - \lambda_0)^{-1} \neq 0$ is an eigenvalue of the matrix $-(\lambda_0 A - B)^{-1} A$ and $\lambda A - B$ has an eigenvalue at infinity if and only if $(\lambda_0 A - B)^{-1} A$ has a zero eigenvalue. Also, $\lambda_j \neq \lambda_k$ if and only if $\mu_j \neq \mu_k$.

Theorem 1.6.5. *Let $\lambda A - B$ be a regular pencil with $\det(\lambda_0 A - B) \neq 0$. Then a subspace S is invariant under $(\lambda_0 A - B)^{-1} A$ if and only if S is deflating for the pair (A, B).*

Proof. For any subspace S it is apparent that $(\lambda_0 A - B)S + AS \subseteq AS + BS$. Conversely, if $x \in AS + BS$ then $x = Ay - Bz$ where $y, z \in S$ and writing $z_1 = y - \lambda_0 z \in S$,

$$x = A(z_1 + \lambda_0 z) - Bz = (\lambda_0 A - B)z + Az_1 \in (\lambda A - B)S + AS.$$

Thus, $(\lambda_0 A - B)S + AS = AS + BS$, and

$$\dim(AS + BS) = \dim\{(\lambda_0 A - B)S + AS\} = \dim\{S + (\lambda_0 A - B)^{-1} AS\}. \quad (1.6.5)$$

Suppose now that S is deflating for (A, B). Then, by Proposition 1.6.4, $\dim(AS + BS) \leq \dim S$ and, using (1.6.4),

$$\dim\{S + (\lambda_0 A - B)^{-1} AS\} = \dim S. \quad (1.6.6)$$

This implies that $(\lambda_0 A - B)^{-1} AS \subseteq S$, i.e. S is $(\lambda_0 A - B)^{-1} B$-invariant.

REGULAR MATRIX PENCILS   25

Conversely, when $\mathcal{S}$ is $(\lambda_0 A - B)^{-1}A$-invariant (1.6.6) holds; then equation (1.6.5) gives $\dim(A\mathcal{S} + B\mathcal{S}) = \dim \mathcal{S}$, and by Proposition 1.6.4, $\mathcal{S}$ is deflating for (A, B). □

Clearly, the eigenvalues of $\lambda A - B$ and the matrix $-(\lambda_0 A - B)^{-1}A$ are different but there are the same number of distinct eigenvalues and, whatever the choice of λ_0 may be (as long as $\det(\lambda_0 A - B) \neq 0$), the lattice of invariant subspaces of $(\lambda_0 A - B)^{-1}A$ is just the set of deflating subspaces for (A, B). As a result, several important notions introduced earlier in this chapter apply immediately to the eigenvalue problem for a regular pencil $\lambda A - B$; for example, spectral subspaces (Example 1.2.6), generalized eigenvectors, geometric, algebraic and partial multiplicities of eigenvalues. In particular, by combining Theorem 1.6.5 and Proposition 1.2.4, we obtain the following generalization of Proposition 1.2.4 to regular pencils.

Theorem 1.6.6. *Let $\lambda A - B$ be a regular $n \times n$ pencil, and let $\lambda_1, \ldots, \lambda_m$ be all the distinct eigenvalues of $\lambda A - B$, including the eigenvalue at infinity if A is singular. Then $\mathbb{C}^n$ admits a direct sum decomposition.*

$$\mathbb{C}^n = \mathcal{S}_1 \dotplus \ldots \dotplus \mathcal{S}_m, \tag{1.6.7}$$

where $\mathcal{S}_1, \ldots, \mathcal{S}_m$ are deflating subspaces for (A, B), and $(\lambda A - B)|\mathcal{S}_j$ has the sole eigenvalue λ_j $(j = 1, \ldots, m)$.

In fact, if $\lambda_j \in \mathbb{C}$, then

$$\mathcal{S}_j = \operatorname{Im} \frac{1}{2\pi} \int_{\Gamma_j} (\lambda A - B)^{-1} A \, d\lambda, \tag{1.6.8}$$

where Γ_j is a circle of sufficiently small radius centered at λ_j (cf. Exercise 1.8.9). The subspace $\mathcal{S}_j$ from (1.6.8) will be called the *spectral deflating subspace* of (A, B) associated with the eigenvalue λ_j.

Finally, a fundamental theorem giving complete information on the spectrum of a regular pencil will be useful in the sequel. A definition is required; regular $n \times n$ pencils $\lambda A - B$ and $\lambda C - D$ are said to be *strictly equivalent* if there exist nonsingular matrices P and Q such that

$$\lambda A - B = P(\lambda C - D)Q.$$

Observe that strict equivalence determines an equivalence relation on $n \times n$ regular pencils, and that Theorem 1.6.2 concerns a partial reduction using a strict equivalence.

A canonical form under strict equivalence can be described as follows:

Theorem 1.6.7. *If $\lambda A - B$ is a regular pencil then it is strictly equivalent to a regular pencil of the form*

$$\begin{bmatrix} \lambda I - J & 0 \\ 0 & \lambda K - I \end{bmatrix} \tag{1.6.9}$$

where J, K are Jordan matrices and K is nilpotent.

For a complete proof see the Appendix of Gohberg et al. (1986a), or Gantmacher (1959). It is clear that the finite eigenvalues and their multiplicities are just the eigenvalues and the sizes (respectively) of the constituent Jordan blocks of J. Similarly, the partial multiplicities of the eigenvalue at infinity are defined to be the sizes of nilpotent Jordan blocks making up the matrix K.

An important corollary follows immediately from the theorem:

Corollary 1.6.8. *Regular matrix pencils which are strictly equivalent have the same canonical form (1.6.9), and hence the same eigenvalues and partial multiplicities (including the eigenvalue at infinity).*

1.7 Functions of matrices

We recall here some basic facts concerning functions of matrices. More detailed information on functions of matrices can be found in many texts; see especially Gantmacher (1959), Lancaster and Tismenetsky (1985), Gohberg et al. (1986a), and Horn and Johnson (1991).

Let A be an $n \times n$ matrix. If

$$f(\lambda) = \sum_{j=0}^{m} a_j \lambda^j, \quad a_j \in \mathbb{C}$$

is a scalar polynomial, then $f(A)$ is naturally defined by

$$f(A) = \sum_{k=1}^{m} a_k A^k. \tag{1.7.1}$$

Let J be the Jordan form of A; so that $A = S^{-1}JS$ for some invertible matrix S. Then equation (1.7.1) takes the form

$$f(A) = S^{-1} \left(\sum_{k=0}^{m} a_k J^k \right) S. \tag{1.7.2}$$

Denote by $\lambda_1, \ldots, \lambda_p$ all the distinct eigenvalues of A. Then we can write

$$J = K_1 \oplus \ldots \oplus K_p,$$

where

$$K_u = J_{r_{u,1}}(\lambda_u) \oplus \ldots \oplus J_{r_{u,s_u}}(\lambda_u).$$

Here, the terms on the right are Jordan blocks and s_u is the geometric multiplicity of eigenvalue λ_u. The formula (1.7.2) shows that

$$f(A) = S^{-1} \left[\bigoplus_{u=1}^{p} \bigoplus_{q=1}^{s_u} f(J_{r_{uq}}(\lambda_u)) \right] S. \qquad (1.7.3)$$

The polynomial of a Jordan block is easily computed: write

$$f(\lambda) = \sum_{k=1}^{m} (k!)^{-1} f^{(k)}(\lambda_u)(\lambda - \lambda_u)^k;$$

then

$$f(J_r(\lambda)) = \sum_{k=1}^{m} (k!)^{-1} f^{(k)}(\lambda_u)(J_r(\lambda_u)) - \lambda_u I)^k$$

$$= \begin{bmatrix} x_0 & x_1 & \ldots & x_{r-1} \\ 0 & x_0 & \ldots & x_{r-2} \\ \vdots & \vdots & & \vdots \\ 0 & 0 & \ldots & x_0 \end{bmatrix}, \qquad (1.7.4)$$

where $x_k = (k!)^{-1} f^{(k)}(\lambda_u); \ k = 0, \ldots, r-1$.

The formulas (1.7.3), (1.7.4) allow us to extend the definition of $f(A)$ beyond the polynomial functions $f(\lambda)$. Consider the class of (scalar) functions $f(\lambda)$ which are defined and analytic in a (not necessarily connected) open set that contains $\sigma(A)$; the open set may depend on the function $f(\lambda)$. Denote by $\mathcal{A}(\sigma(A))$ the set of all such functions. It is easy to see that $\mathcal{A}(\sigma(A))$ is an algebra, i.e. $f + g$, $f \cdot g$, and $cf \in \mathcal{A}(\sigma(A))$ for every $f, g \in \mathcal{A}(\sigma(A))$ and $c \in \mathbb{C}$. Moreover, $\mathcal{A}(\sigma(A))$ contains the constant function 1 (which is the multiplicative identity in $\mathcal{A}(\sigma(A))$) and $\frac{1}{g} \in \mathcal{A}(\sigma(A))$ provided $g \in \mathcal{A}(\sigma(A))$ and $g(\lambda_0) \neq 0$ for all $\lambda_0 \in \sigma(A)$.

For $f \in \mathcal{A}(\sigma(A))$, define the matrix $f(A)$ by (1.7.3), (1.7.4). Clearly, this extends the definition of $f(A)$ beyond the class of polynomial functions.

We list some elementary properties of functions of matrices:

Proposition 1.7.1. *The correspondence* $f \mapsto f(A)$, $f \in \mathcal{A}(\sigma(A))$ *is an algebra homomorphism:*

$$(\alpha f + \beta g)(A) = \alpha f(A) + \beta g(A); \quad f, g \in \mathcal{A}(\sigma(A)); \quad \alpha, \beta \in \mathbb{C}$$

$$(f \cdot g)(A) = f(A) \cdot g(A); \quad f, g \in \mathcal{A}(\sigma(A))$$

$$1(A) = I.$$

These properties can be checked by a direct verification, using the formulas (1.7.3) and (1.7.4).

Corollary 1.7.2. *Let $h(\lambda) = f(\lambda)/g(\lambda)$ be a rational function where $f(\lambda)$ and $g(\lambda)$ are polynomials and $g(\lambda_0) \neq 0$ for all $\lambda_0 \in \sigma(A)$. Then $h \in \mathcal{A}(\sigma(A))$, $g(A)$ is invertible, and*
$$h(A) = f(A)(g(A))^{-1}.$$

The following fact, known as the *spectral mapping theorem*, follows immediately from the formulas (1.7.3) and (1.7.4):

Proposition 1.7.3. *If $f \in \mathcal{A}(\sigma(A))$, then $\sigma(f(A)) = f(\sigma(A))$; in other words, the eigenvalues of $f(A)$ are precisely the numbers of the form $f(\lambda_0)$, where λ_0 is an eigenvalue of A.*

Next, consider the invariant subspaces of $f(A)$. Proposition 1.2.3 leads to the following observation:

Proposition 1.7.4. *The spectral subspace of $f(A)$ corresponding to a set $\Omega \subseteq \sigma(f(A))$ coincides with $\sum \mathcal{R}_\mu(A)$, where the sum is taken over the set of eigenvalues μ of A such that $f(\mu) \in \Omega$.*

Since $Af(A) = f(A)A$ for every $f \in \mathcal{A}(\sigma(A))$, clearly every A-invariant subspace is also $f(A)$-invariant. But the matrix $f(A)$ can have additional invariant subspaces which are not A-invariant, as the trivial example $f(\lambda) \equiv 0$ shows. The criterion for equality $\text{Inv}(A) = \text{Inv}(f(A))$ is given in the following theorem.

Theorem 1.7.5. *Let $\lambda_1, \ldots, \lambda_p$ be all the distinct eigenvalues of A, and let $f \in \mathcal{A}(\sigma(A))$. Then A and $f(A)$ have exactly the same invariant subspaces if and only if the following conditions hold:*

(i) *$f(\lambda_i) \neq f(\lambda_j)$ for $i \neq j$ $(1 \leq i, j \leq p)$.*

(ii) *$f'(\lambda_j) \neq 0$ for every λ_j whose algebraic multiplicity is greater than its geometric multiplicity.*

Moreover, if (i) and (ii) hold, the partial multiplicities of A at λ_j coincide with the partial multiplicities of $f(A)$ at $f(\lambda_j)$, $j = 1, \ldots, p$.

For the proof of Theorem 1.7.5 we refer the reader to Gohberg et al. (1986a) (see the proof of Theorem 2.11.3 there). The statement concerning multiplicities is in fact a special case of a general result that describes the partial multiplicities of $f(A)$ in terms of those of A, for every $f \in \mathcal{A}(\sigma(A))$ (see Theorem 2.11.1 in Gohberg et al. (1986a), or Theorem 9.4.7 in Lancaster and Tismenetsky (1985)).

1.8 Exercises

1.8.1. Find a Jordan form for the matrix

$$A = \begin{bmatrix} 2 & 0 & 0 & 0 \\ -3 & 2 & 0 & 1 \\ 0 & 0 & 2 & 1 \\ 0 & 0 & 0 & 2 \end{bmatrix},$$

and then find a basis for $\mathbb{R}^4$ consisting of Jordan chains for A.

1.8.2. Find all invariant subspaces of the matrix

$$\begin{bmatrix} 0 & 1 & 0 \\ 0 & 0 & 0 \\ 0 & 0 & 0 \end{bmatrix}.$$

1.8.3. Prove that all invariant subspaces of the Jordan block $J_m(\lambda)$ are given by

$$\{0\}, \quad \text{span}\{e_1, \ldots, e_k\} \quad 1 \le k \le n.$$

1.8.4. Find all real invariant subspaces of the real Jordan block

$$\begin{bmatrix} 0 & 1 & 1 & 0 \\ -1 & 0 & 0 & 1 \\ 0 & 0 & 0 & 1 \\ 0 & 0 & -1 & 0 \end{bmatrix}.$$

1.8.5. Let A be a real matrix with no real eigenvalues. Prove that all A-invariant subspaces are even-dimensional.

1.8.6. Prove that a (complex) $n \times n$ matrix A has only a finite number of invariant subspaces if and only if A is *nonderogatory*, i.e. every eigenvalue of A has geometric multiplicity one. Is this statement true for real matrices and their real invariant subspaces?

1.8.7. Let

$$A = \begin{bmatrix} 1 & 0 & 0 \\ 0 & 1 & 0 \\ 0 & 0 & 0 \end{bmatrix}, \quad B = \begin{bmatrix} 0 & 0 & 0 \\ 0 & 1 & 0 \\ 0 & 0 & 1 \end{bmatrix}.$$

Find all deflating subspaces of the regular matrix pencil $\lambda A + B$.

1.8.8. Let A and B be as in Exercise 1.8.7. Represent each deflating subspace $\mathcal{S}$ of $\lambda A + B$ such that $(\text{Ker } A) \cap \mathcal{S} = \{0\}$ in the form $\mathcal{S} = \text{Im } X_1$, where X_1 and K satisfy the requirements of Corollary 1.6.3.

1.8.9. Let $\lambda A - B$ be a regular pencil. Define

$$P = \frac{1}{2\pi} \int_\Gamma (\lambda A - B)^{-1} A \, d\lambda,$$

where Γ is a closed rectifiable contour in the complex plane such that $\lambda A - B$ is invertible for all $\lambda \in \Gamma$. Prove that P is a projector, and Im P is a deflating subspace for $\lambda A - B$.

1.8.10. Let $\lambda A - B$ be a regular pencil, and let $\hat{A} = XAY$, $\hat{B} = XBY$, where X and Y are invertible matrices. Prove that the subspace $\mathcal{S}$ is deflating for the pencil $\lambda A - B$ if and only if the subspace $Y^{-1}\mathcal{S}$ is deflating for the pencil $\lambda \hat{A} - \hat{B}$.

1.9 Notes

The material of the first five sections of this chapter is standard linear algebra and can be found in innumerable texts. The material on regular pencils is of more recent origin. A thorough treatment in the operator theory setting appeared in the work of Stummel (1971). Stewart's paper of 1972 introduced the idea of "deflating subspaces" to a wide audience. The reader may also wish to consult Section 4.1 of the more recent exposition by Gohberg et al. (1990).

2
INDEFINITE SCALAR PRODUCTS

Our geometric notions of orthogonality and angle extend to $\mathbb{C}^n$ and more general spaces with the aid of a (standard) scalar product. A more general concept has many useful applications, but requires us to abandon some of that geometric intuition. We have in mind the *indefinite* scalar products of the present chapter. In particular, they play vital parts in the analysis of Riccati equations with the symmetries among the coefficients that arise so frequently in practice. Furthermore, the analysis of Riccati equations with complex coefficients, or with real coefficients, demand careful preparations in the context of complex and real spaces, respectively.

Generalized scalar products lead to generalizations of "self-adjoint" and "unitary" matrices, or linear transformations, of course. These notions and properties of the invariant subspaces of such matrices and transformations are the preoccupations of this chapter.

2.1 Definitions and classification of subspaces

Let $\mathbb{C}^n$ be the vector space for all n-dimensional columns whose components are complex numbers, with the standard scalar product $\langle x, y \rangle = \sum_{i=1}^n x_i \bar{y}_i$. We will need basic facts and results concerning other scalar products in $\mathbb{C}^n$, and we will allow these scalar products to be indefinite (i.e. the scalar product of $x \neq 0$ with itself need not be positive), but otherwise satisfying all the usual properties of scalar products. The formal definition runs as follows.

A function $[.,.]$ from $\mathbb{C}^n \times \mathbb{C}^n$ to $\mathbb{C}$ is called an *indefinite scalar product* in $\mathbb{C}^n$ if the following axioms are satisfied:

(i) linearity in the first argument;

$$[\alpha x_1 + \beta x_2, , y] = \alpha [x_1, y] + \beta [x_2, y]$$

for all $x_1, x_2, y \in \mathbb{C}^n$ and all complex numbers α, β;

(ii) antisymmetry;

$$[x, y] = \overline{[y, x]}$$

for all $x, y \in \mathbb{C}^n$;

(iii) non degeneracy; if $[x,y] = 0$ for all $y \in \mathbb{C}^n$, then $x = 0$.

Thus, the function $[.,.]$ satisfies all the properties of the standard scalar product with the possible exception that $[x,x]$ may be non-positive for $x \neq 0$.

It is easily checked that for every $n \times n$ invertible hermitian matrix H the formula

$$[x,y] = \langle Hx, y \rangle, \quad x, y \in \mathbb{C}^n \tag{2.1.1}$$

determines an indefinite scalar product on $\mathbb{C}^n$. Conversely, for every indefinite scalar product $[.,.]$ on $\mathbb{C}^n$ there exists a unique $n \times n$ invertible and hermitian matrix H such that (2.1.1) holds.

Indeed, the matrix $H = [h_{ij}]_{i,j=1}^n$ is defined by $[e_j, e_i] = h_{ij}$, where $e_j = (0,\ldots,0,1,\ldots,0)^T$ is the unit coordinate vector having 1 in the j-th place. Now the antisymmetry and nondegeneracy of $[.,.]$ ensure that H is hermitian and invertible.

Next, we study the properties of subspaces with respect to a given indefinite scalar product $[.,.]$. For any subset $\mathcal{M}$ of $\mathbb{C}^n$ define the *orthogonal companion* of $\mathcal{M}$ in $\mathbb{C}^n$ by

$$\mathcal{M}^{[\perp]} = \{x \in \mathbb{C}^n | [x,y] = 0 \quad \text{for all} \quad y \in \mathcal{M}\}.$$

Note that the symbol $\mathcal{M}^{[\perp]}$ will be reserved for the orthogonal companion with respect to the indefinite scalar product, while the symbol $\mathcal{M}^\perp$ will denote the orthogonal companion in the standard scalar product $\langle .,. \rangle$ in $\mathbb{C}^n$, i.e.

$$\mathcal{M}^\perp = \{x \in \mathbb{C}^n | \langle x,y \rangle = 0 \quad \text{for all} \quad y \in \mathcal{M}\}.$$

Clearly, $\mathcal{M}^{[\perp]}$ is a subspace in $\mathbb{C}^n$, and we will be particularly interested in the case when $\mathcal{M}$ is itself a subspace of $\mathbb{C}^n$. In the latter case, it is not generally true (as experience with the euclidean scalar product might suggest) that $\mathcal{M}^{[\perp]}$ is a direct complement for $\mathcal{M}$. The next example illustrates this point.

Example 2.1.1. Let $[x,y] = \langle Hx, y \rangle$, where

$$H = \begin{bmatrix} 0 & 0 & \cdots & 0 & 1 \\ 0 & 0 & & 1 & 0 \\ \vdots & & & & \vdots \\ 0 & 1 & & 0 & 0 \\ 1 & 0 & \cdots & 0 & 0 \end{bmatrix} \tag{2.1.2}$$

is $n \times n$. Let $\mathcal{M} = \text{span}\{e_1\}$. It is easily seen that $\mathcal{M}^{[\perp]}$ is spanned by $e_1, e_2, \cdots, e_{n-1}$ and is not a direct complement to $\mathcal{M}$ in $\mathbb{C}^n$. □

DEFINITIONS AND CLASSIFICATION OF SUBSPACES

In contrast, it is true that, for any subspace $\mathcal{M}$,

$$\dim \mathcal{M} + \dim \mathcal{M}^{[\perp]} = n. \qquad (2.1.3)$$

To see this observe first that

$$\mathcal{M}^{[\perp]} = H^{-1}(\mathcal{M}^{\perp}). \qquad (2.1.4)$$

For, if $x \in \mathcal{M}^{\perp}$ and $y \in \mathcal{M}$ we have

$$[H^{-1}x, y] = \langle HH^{-1}x, y \rangle = \langle x, y \rangle = 0 \qquad (2.1.5)$$

so that $H^{-1}(\mathcal{M}^{\perp}) \subseteq \mathcal{M}^{[\perp]}$. Conversely, if $x \in \mathcal{M}^{[\perp]}$ and $z = Hx$ then, for any $y \in \mathcal{M}$,

$$0 = [x, y] = [H^{-1}z, y] = \langle z, y \rangle.$$

Thus, $z \in \mathcal{M}^{\perp}$ and $x = H^{-1}z$ so that $\mathcal{M}^{[\perp]} \subseteq H^{-1}(\mathcal{M}^{\perp})$ and (2.1.4) is established. Then (2.1.3) follows immediately.

It follows from (2.1.3) that, for any subspace $\mathcal{M} \subseteq \mathbb{C}^n$,

$$(\mathcal{M}^{[\perp]})^{[\perp]} = \mathcal{M}. \qquad (2.1.6)$$

Indeed, the inclusion $(\mathcal{M}^{[\perp]})^{[\perp]} \supseteq \mathcal{M}$ is evident from the definition of $\mathcal{M}^{[\perp]}$. But (2.1.3) implies that these two subspaces have the same dimension, and so (2.1.6) follows.

A subspace $\mathcal{M}$ is said to be *nondegenerate* (with respect to the indefinite scalar product $[.,.]$) if $x \in \mathcal{M}$ and $[x, y] = 0$ for all $y \in \mathcal{M}$ imply that $x = 0$. Otherwise $\mathcal{M}$ is *degenerate*. For example, the defining property (iii) for $[.,.]$ ensures that $\mathbb{C}^n$ itself is always nondegenerate. In Example 2.1.1 the subspace $\mathcal{M}$ is degenerate because $e_1 \in \mathcal{M}$ and $[e_1, y] = 0$ for all $y \in \mathcal{M}$.

The nondegenerate subspaces can be characterized in another way:

Proposition 2.1.1. $\mathcal{M}^{[\perp]}$ *is a direct complement to* $\mathcal{M}$ *in* $\mathbb{C}^n$ *if and only if* $\mathcal{M}$ *is nondegenerate.*

Proof. By definition, $\mathcal{M}$ is nondegenerate if and only if $\mathcal{M} \cap \mathcal{M}^{[\perp]} = \{0\}$. In view of (2.1.3) this means that $\mathcal{M}^{[\perp]}$ is a direct complement to $\mathcal{M}$. □

In particular, the orthogonal companion of a nondegenerate subspace is again nondegenerate.

A subspace $\mathcal{M}$ of $\mathbb{C}^n$ is called *positive* (with respect to $[.,.]$) if $[x, x] > 0$ for all nonzero x in $\mathcal{M}$, and *nonnegative* if $[x, x] \geq 0$ for all x in $\mathcal{M}$. Clearly, every positive subspace is also nonnegative but the converse is not necessarily true (see Example 2.1.2 below). Observe that a positive subspace is nondegenerate. If the invertible hermitian matrix H is such that $[x, y] = \langle Hx, y \rangle$, $x, y \in \mathbb{C}^n$, we say that a positive (resp. nonnegative) subspace is H-*positive* (resp. H-*nonnegative*).

Example 2.1.2. Let $[x,y] = \langle Hx, y\rangle$, $x, y \in \mathbb{C}^n$, where H is the matrix of size n given by (2.1.2), and assume n is odd. Then the subspace spanned by the first $\frac{1}{2}(n+1)$ unit vectors is nonnegative, but not positive. The subspace spanned by the unit vector with 1 in the $\frac{1}{2}(n+1)$-th position is positive. □

We are to investigate the constraints on the dimensions of positive and nonnegative subspaces. But first a general observation is necessary.

Let $[.,.]_1$ and $[.,.]_2$ be two indefinite scalar products on $\mathbb{C}^n$ with corresponding invertible hermitian matrices H_1, H_2 respectively. Suppose, in addition, that H_1 and H_2 are congruent, i.e. $H_1 = S^* H_2 S$ for some invertible matrix S. In this case, a subspace $\mathcal{M}$ is H_1-positive if and only if $S\mathcal{M}$ is H_2-positive, with a similar statement replacing "positive" by "nonnegative". The proof is direct: take $x \in \mathcal{M}$ and

$$[Sx, Sx]_2 = \langle H_2 Sx, Sx\rangle = \langle S^* H_2 Sx, x\rangle = \langle H_1 x, x\rangle = [x, x]_1.$$

Thus, $[x,x]_1 > 0$ for all nonzero $x \in \mathcal{M}$ if and only if $[y,y]_2 > 0$ for all nonzero y in $S\mathcal{M}$.

Theorem 2.1.2. *The maximal dimension of a positive or of a nonnegative subspace with respect to the indefinite scalar product $[x,y] = \langle Hx, y\rangle$ coincides with the number of positive eigenvalues of H (counted with algebraic multiplicities).*

Note that the maximal possible dimensions of nonnegative and positive subspaces coincide.

Proof. We prove only the nonnegative case (the positive case is analogous). So let $\mathcal{M}$ be a nonnegative subspace, and let $p = \dim \mathcal{M}$. Then

$$\min_{\substack{(x,x)=1 \\ x \in \mathcal{M}}} \langle Hx, x\rangle \geq 0. \qquad (2.1.7)$$

Write all the eigenvalues of H in the nonincreasing order: $\lambda_1 \geq \lambda_2 \geq \cdots \geq \lambda_n$.

At this point we use the well-known max–min characterization of eigenvalues of hermitian matrices (it can be found in many textbooks in linear algebra; see, for example, Noble (1969), or Lancaster and Tismenetsky (1985)). According to this characterization, we have

$$\lambda_p = \max_{\mathcal{L}} \min_{\substack{(x,x)=1 \\ x \in \mathcal{L}}} \langle Hx, x\rangle,$$

where the maximum is taken over all the subspaces $\mathcal{L} \subseteq \mathbb{C}^n$ of dimension p. Then (2.1.7) implies $\lambda_p \geq 0$ and, since H is invertible, $\lambda_p > 0$. So $p \leq k$ where k is the number of positive eigenvalues of H.

DEFINITIONS AND CLASSIFICATION OF SUBSPACES

To find a nonnegative subspace of dimension k, appeal to the observation preceding Theorem 2.1.2. By Sylvester's inertia theorem there exists an invertible matrix S such that S^*HS is a diagonal matrix of $+1$'s and -1's:

$$H_0 = S^*HS = \mathrm{diag}\,[1, \cdots, 1, -1, \cdots, -1], \qquad (2.1.8)$$

where the number of $+1$'s is k. Hence, it is sufficient to find a k-dimensional subspace which is nonnegative with respect to H_0. One such subspace (which is even positive) is spanned by the first k unit vectors in $\mathbb{C}^n$. □

The concepts relating to the negativeness of the quadratic form $[x, x]$ where x belongs to a given subspace are completely analogous. Thus, a subspace $\mathcal{M} \subseteq \mathbb{C}^n$ is called H-negative (where H is such that $[x, y] = \langle Hx, y \rangle$, $x, y \in \mathbb{C}^n$), if $[x, x] < 0$ for all nonzero x in $\mathcal{M}$. Replacing this condition by the requirement that $[x, x] \leq 0$ for all $x \in \mathcal{M}$, we obtain the definition of a *nonpositive* (with respect to $[., .]$), or H-nonpositive subspace. As in Theorem 2.1.2 it can be proved that the maximal possible dimension of an H-negative or of an H-nonpositive subspace is equal to the number of negative eigenvalues of H (counting multiplicities).

Finally we consider the class of subspaces which are peculiar to indefinite scalar product spaces and have no analogues in spaces with the usual euclidean scalar product. A subspace $\mathcal{M} \subseteq \mathbb{C}^n$ is called *neutral* (with respect to $[., .]$), or H-neutral (where H is such that $[x, y] = \langle Hx, y \rangle$, $x, y \in \mathbb{C}^n$) if $[x, x] = 0$ for all $x \in \mathcal{M}$. Sometimes, such subspaces are called *isotropic*. In Example 2.1.2 the subspaces spanned by the first k unit vectors, for $k = 1, \cdots, \frac{n-1}{2}$, are all neutral.

In view of the easily verified polarization identity

$$[x, y] = \frac{1}{4}\{[x + y, x + y] + i[x + iy, x + iy] - [x - y, x - y] - i[x - iy, x - iy]\},$$

a subspace $\mathcal{M}$ is neutral if and only if $[x, y] = 0$ for all $x, y \in \mathcal{M}$. Observe also that a neutral subspace is both nonpositive and nonnegative and (if different from the zero subspace) is necessarily degenerate.

One can easily compute the maximal possible dimension of a neutral subspace.

Theorem 2.1.3. *The maximal possible dimension of an H-neutral subspace is $\min(k, \ell)$, where k (resp. ℓ) is the number of positive (resp. negative) eigenvalues of H, counting multiplicities.*

Proof. In view of the remark preceding Theorem 2.1.2 it may be assumed that $H = H_0$ is given by (2.1.8). The existence of a neutral subspace of dimension $\min(k, \ell)$ is easily seen. A basis for one such subspace can be formed from the unit vectors $e_1, e_2, \cdots$ as follows: $e_1 + e_{k+1}, e_2 + e_{k+2}, \cdots$.

Now let $\mathcal{M}$ be a neutral subspace of dimension p. Since $\mathcal{M}$ is also nonnegative it follows from Theorem 2.1.2 that $p \leq k$. But $\mathcal{M}$ is also nonpositive and so the inequality $p \leq \ell$ also applies. Thus, $p \leq \min(k, \ell)$. □

2.2 Self-adjoint and unitary matrices

Let $[.,.]$ be an indefinite scalar product in $\mathbb{C}^n$. Given an $n \times n$ matrix A, or a linear transformation $A : \mathbb{C}^n \to \mathbb{C}^n$, the *adjoint* matrix (or linear transformation) $A^{[*]}$ with respect to $[.,.]$ is defined by the equality

$$[A^{[*]}x, y] = [x, Ay], \quad \text{for all } x, y \in \mathbb{C}^n. \tag{2.2.1}$$

The equality (2.2.1) does indeed define a matrix $A^{[*]}$ uniquely: the (i,j) entry in $A^{[*]}$ is equal to $[e_j, Ae_i]$.

Let H be the $n \times n$ hermitian invertible matrix inducing the indefinite scalar product $[.,.]$:

$$\langle Hx, y \rangle = [x, y], \quad x, y \in \mathbb{C}^n.$$

Then (2.2.1) can be rewritten in the form

$$\langle HA^{[*]}x, y \rangle = \langle Hx, Ay \rangle,$$

and using the property that

$$\langle z, Ay \rangle = \langle A^*z, y \rangle,$$

where A^* is the conjugate transpose of A, we obtain

$$\langle HA^{[*]}x, y \rangle = \langle A^*Hx, y \rangle.$$

It follows that

$$A^{[*]} = H^{-1}A^*H. \tag{2.2.2}$$

This is a very useful formula for the adjoint matrix; it is basic for all subsequent developments in this chapter. As a first observation, note that $A^{[*]}$ is similar to A^*. Therefore λ_0 is an eigenvalue of A if and only if $\bar{\lambda}_0$ is an eigenvalue of $A^{[*]}$. Also, the partial multiplicities of $\bar{\lambda}_0$ as an eigenvalue of $A^{[*]}$ coincide with the partial multiplicites of λ_0 as an eigenvalue of A. (Using the Jordan form it is easily verified that $\bar{\lambda}_0$ is an eigenvalue of A^* if and only if λ_0 is an eigenvalue of A, and in this case $\bar{\lambda}_0$ has the same partial multiplicities as λ_0.) In particular, the algebraic (resp. geometric) multiplicities of the eigenvalue λ_0 of A, and of the eigenvalue $\bar{\lambda}_0$ of $A^{[*]}$, are the same.

The root subspaces of A and of $A^{[*]}$ enjoy important orthogonality properties.

Theorem 2.2.1. *Let λ, μ be eigenvalues of $A, A^{[*]}$, respectively, such that $\lambda \neq \bar{\mu}$. Then*

$$\mathcal{R}_\lambda(A) \subseteq (\mathcal{R}_\mu(A^{[*]}))^{[\perp]}.$$

SELF-ADJOINT AND UNITARY MATRICES

Proof. Let $x \in \mathcal{R}_\lambda(A)$ and $y \in \mathcal{R}_\mu(A^{[*]})$, so that $(A - \lambda I)^s x = 0$ and $(A - \mu I)^t y = 0$ for some s and t. We are to prove that

$$[x, y] = 0. \qquad (2.2.3)$$

Proceed by induction on $s + t$. For $s = t = 1$ we have $Ax = \lambda x$, $A^{[*]}y = \mu y$. Then

$$\lambda[x, y] = [Ax, y] = [x, A^{[*]}y] = [x, \mu y] = \bar{\mu}[x, y] \qquad (2.2.4)$$

and since $\lambda \neq \bar{\mu}$, we obtain (2.2.3).

Suppose now that (2.2.2) is proved for all $x' \in \mathcal{R}_\lambda(A)$, $y' \in \mathcal{R}_\mu(A^{[*]})$ such that $(A - \lambda I)^{s'} x' = (A - \lambda I)^{t'} y' = 0$ for some s' and t' satisfying $s' + t' < s + t$. Given x and y as above, put $x' = (A - \lambda I)x$, $y' = (A^{[*]} - \mu I)y$. Then, by the induction assumption $[x', y] = [x, y'] = 0$, which means that $\lambda[x, y] = [Ax, y]; \bar{\mu}[x, y] = [x, A^{[*]}y]$. Now use the relations of (2.2.4) once more to complete the proof. $\square$

An $n \times n$ matrix A (or linear transformation $A : \mathbb{C}^n \to \mathbb{C}^n$ is called *self-adjoint* with respect to the indefinite scalar product $[.,.]$ if $A^{[*]} = A$. More briefly, such matrices will also be called *H-self-adjoint*, where H is the $n \times n$ hermitian invertible matrix that induces the indefinite scalar product $[.,.]$.

Example 2.2.1. Let $J_n(\lambda)$ be a Jordan block of size $n \times n$ with a *real* eigenvalue λ. Then $J_n(\lambda)$ is self-adjoint with respect to the scalar product

$$[x, y] = x_n \bar{y}_1 + x_{n-1} \bar{y}_2 + \ldots + x_1 \bar{y}_n \qquad (2.2.5)$$

where

$$x = \begin{bmatrix} x_1 \\ \vdots \\ x_n \end{bmatrix}, \quad y = \begin{bmatrix} y_1 \\ \vdots \\ y_n \end{bmatrix}.$$

Indeed,

$$[x, y] = \langle Hx, y \rangle, \qquad (2.2.6)$$

where H is given by (2.1.2), and the equality $HJ_n(\lambda) = (J_n(\lambda))^* H$ is easily verified. $\square$

Example 2.2.2. Let $K = J_k(\lambda) \oplus J_k(\bar{\lambda})$ be a matrix in the Jordan form of size $2k \times 2k$ having two Jordan blocks, each of size $k \times k$, and with non-real complex conjugate eigenvalues λ and $\bar{\lambda}$. Then K is self-adjoint with respect to the indefinite scalar product (2.2.5) (with $n = 2k$). Indeed, for H given by (2.1.2) we have $HK = K^*H$. $\square$

These examples will serve as building blocks in the canonical form to be presented later (Theorem 2.3.2). In particular, in contrast with hermitian matrices (i.e. matrices that are self-adjoint with respect to the standard scalar product in $\mathbb{C}^n$), the H-self-adjoint matrices need not be diagonable or have real eigenvalues. In fact (see Theorem 3.24 of Gohberg et al. (1983)), any $n \times n$ matrix A which is similar to A^* is self-adjoint with respect to a suitably chosen indefinite scalar product.

Some basic properties of H-self-adjoint matrices are summarized in the following statement.

Theorem 2.2.2. *Let A be an H-self-adjoint $n \times n$ matrix (or linear transformation from $\mathbb{C}^n$ into $\mathbb{C}^n$). Then:*

(i) *The set of eigenvalues of A is symmetric relative to the real axis, i.e. if λ_0 is an eigenvalue of A, then so is $\bar{\lambda}_0$, and the partial multiplicities corresponding to $\bar{\lambda}_0$ are equal to those corresponding to $\bar{\lambda}_0$.*

(ii) *If λ and μ are eigenvalues of A and $\lambda \neq \bar{\mu}$, then*

$$\mathcal{R}_\lambda(A) \subseteq (\mathcal{R}_\mu(A))^{[\perp]}.$$

(iii) *The generalized eigenspace $\mathcal{R}_\lambda(A)$ corresponding to any non-real eigenvalue λ of A is H-neutral.*

Indeed, (i) follows from (2.2.2), while (ii) follows from Theorem 2.2.1. Finally, (iii) is a particular case of (ii) obtained by taking $\lambda = \mu$.

Matrices which are H-unitary form another important class. An $n \times n$ matrix (or linear transformation $A : \mathbb{C}^n \to \mathbb{C}^n$) is called *unitary with respect to an indefinite scalar product* $[x, y] = \langle Hx, y \rangle$, or in short H-*unitary*, if

$$[Ax, Ay] = [x, y] \tag{2.2.7}$$

for all $x, y \in \mathbb{C}^n$. The property (2.2.7) is equivalent to $[Ax, Ax] = [x, x]$ for all $x \in \mathbb{C}^n$; in other words, A is an isometry with respect to $[.,.]$. The equality (2.2.7) can be rewritten in the form

$$\langle A^* H A x, y \rangle = \langle Hx, y \rangle,$$

and so A is H-unitary if and only if the equality

$$A^* H A = H \tag{2.2.8}$$

holds (in particular, A must be nonsingular). Recalling the formula (2.2.2), the characterization (2.2.8) can also be written in the form $A^{[*]} = A^{-1}$.

The following simple examples of H-unitary matrices show that, in contrast with unitary matrices (with respect to the standard scalar product $\langle .,. \rangle$), H-unitary matrices need not be diagonable, and need not have all eigenvalues on the unit circle.

Example 2.2.3. Let $\lambda \in \mathbb{C}$, $|\lambda| = 1$, and let

$$A = \begin{bmatrix} \lambda & 2i\lambda & 2i^2\lambda & \cdots & 2i^{n-1}\lambda \\ 0 & \lambda & 2i\lambda & \cdots & 2i^{n-2}\lambda \\ \vdots & \vdots & \ddots & & \vdots \\ & & & & 2i\lambda \\ 0 & 0 & \cdots & & \lambda \end{bmatrix}, \quad i^2 = -1.$$

Then $A^*HA = H$, where H is given by (2.1.2), and so A is H-unitary. $\square$

Example 2.2.4.
$$[x, y] = \langle Hx, y \rangle, \quad x, y \in \mathbb{C}^{2n},$$

where $H = [\delta_{i+j-1,2n}]_{i,j=1}^{2n}$, and $\delta_{pq} = 1$ if $p = q$, $\delta_{pq} = 0$ if $p \neq q$ is the Kronecker symbol. (Another guise for H of (2.1.2).) For a nonzero $\lambda \in \mathbb{C}$ such that $|\lambda| \neq 1$ put

$$A = \begin{bmatrix} K_1 & 0 \\ 0 & K_2 \end{bmatrix}$$

where

$$K_1 = \begin{bmatrix} \lambda & k_1 & k_2 & \cdots & k_{n-1} \\ 0 & \lambda & k_1 & & \\ \vdots & & \ddots & \ddots & \vdots \\ & & & \lambda & k_1 \\ 0 & & \cdots & 0 & \lambda \end{bmatrix}, \quad K_2 = \begin{bmatrix} \bar{\lambda}^{-1} & \kappa_1 & \kappa_2 & \cdots & \kappa_{n-1} \\ 0 & \bar{\lambda}^{-1} & \kappa_1 & & \\ \vdots & & \ddots & \ddots & \vdots \\ & & & \bar{\lambda}^{-1} & \kappa_1 \\ 0 & & \cdots & 0 & \bar{\lambda}^{-1} \end{bmatrix}$$

and $k_r = \lambda q_1^{r-1}(q_1 - \bar{q}_2)$, $\kappa_r = \bar{\lambda}^{-1} q_2^{r-1}(q_2 - \bar{q}_1)$, $r = 1, 2, \cdots, n-1$, and $q_1 = \frac{1}{2} i(1 + \lambda)$, $q_2 = \frac{1}{2} i(1 + \bar{\lambda}^{-1})$.

A direct computation shows that $A^*HA = H$, so A is H-unitary. Note also that $K_2 = \bar{K}_1^{-1}$. $\square$

There is a strong connection between H-unitary and H-self-adjoint matrices. As in the case of the standard scalar product one way to describe this connection is by the use of Cayley transforms.

Recall that if $|\alpha| = 1$ and $w \neq \bar{w}$ then the map f defined by

$$f(z) = \alpha(z - \bar{w})/(z - w) \tag{2.2.9}$$

maps the real line in the z-plane onto the unit circle in the ζ-plane, where $\zeta = f(z)$. The inverse transformation is

$$z = (w\zeta - \bar{w}\alpha)/(\zeta - \alpha). \tag{2.2.10}$$

Then, if w is not an eigenvalue of A, the matrix $f(A)$ is well-defined and if A is H-self-adjoint one anticipates that $U = f(A)$ is H-unitary. This idea is developed in:

Proposition 2.2.3. *Let A be an H-self-adjoint matrix. Let w be a non-real complex number which is not an eigenvalue of A and let α be any unimodular complex number. Then*

$$U = \alpha(A - \bar{w}I)(A - wI)^{-1} \tag{2.2.11}$$

is H-unitary and α is not an eigenvalue of U.

Conversely, if U is H-unitary, $|\alpha| = 1$ and α is not an eigenvalue of U then for any $w \neq \bar{w}$ the matrix

$$A = (wU - \bar{w}\alpha I)(U - \alpha I)^{-1} \tag{2.2.12}$$

is H-self-adjoint and w is not an eigenvalue of A. Furthermore, formulas (2.2.11) and (2.2.12) are inverse to one another.

Proof. If A is H-self-adjoint and $|\alpha| = 1$ it is easily seen that

$$(A^* - \bar{w}I)H(A - wI) = (\bar{\alpha}A^* - \bar{\alpha}wI)H(\alpha A - \alpha\bar{w}I).$$

Premultiplying by $(A^* - \bar{w}I)^{-1}$ and postmultiplying by $(\alpha A - \alpha \bar{w}I)^{-1}$ it is found that $HU^{-1} = U^*H$ where U is defined by (2.2.11), and this means that U is H-unitary. Furthermore, it follows from (2.2.11) that

$$(U - \alpha I)(A - wI) = \alpha(w - \bar{w})I. \tag{2.2.13}$$

Thus, the hypothesis that w is not real implies that $U - \alpha I$ is invertible and so α is not an eigenvalue of U.

The relation (2.2.13) also gives

$$A = wI + \alpha(w - \bar{w})(U - \alpha I)^{-1} = [w(U - \alpha I) + \alpha(w - \bar{w})I](U - \alpha I)^{-1}$$
$$= (wU - \bar{w}\alpha I)(U - \alpha I)^{-1},$$

so that (2.2.12) and (2.2.11) are, indeed, inverses to each other.

Proof of the converse statement is left to the reader. $\square$

Using Proposition 2.2.3, one easily obtains a complete analogue of Theorem 2.2.2 for H-unitary matrices:

Theorem 2.2.4. *Let A be an H-unitary $n \times n$ matrix (or linear transformation from $\mathbb{C}^n$ into $\mathbb{C}^n$). Then:*

(i) *The set of eigenvalues of A is symmetric relative to the unit circle: i.e. if λ_0 is an eigenvalue of A, then so is $\bar{\lambda}_0^{-1}$, and the partial multiplicities corresponding to $\bar{\lambda}_0^{-1}$ are equal to those corresponding to λ_0.*

(ii) *If λ and μ are eigenvalues of A and $\lambda \neq \bar{\mu}^{-1}$, then*

$$\mathcal{R}_\lambda(A) \subseteq (\mathcal{R}_\mu(A))^{[\perp]}.$$

In particular, the generalized eigenspace of A corresponding to any eigenvalues λ with $|\lambda| \neq 1$ is H-neutral.

2.3 Canonical forms of H-self-adjoint and H-unitary matrices

In this section we present the main results of the chapter — canonical forms for H-self-adjoint and H-unitary matrices. In the study of H-self-adjoint and H-unitary matrices (for example, their invariant subspaces with certain definiteness properties) these canonical forms play roles analogous to that of the Jordan canonical form in the study of general complex matrices.

The starting point is transformations of the indefinite scalar product $[x, y] = \langle Hx, y \rangle$ given by congruence transformations of H. It is clear that, for any invertible matrix S, $H' = SHS^*$ is invertible and hermitian, and therefore H' also defines an indefinite scalar product $[x, y]' = \langle H'x, y \rangle$, $x, y \in \mathbb{C}^n$.

Proposition 2.3.1. *Let H_1, H_2 define indefinite scalar products on $\mathbb{C}^n$ and $H_2 = SH_1S^*$ for some invertible $n \times n$ matrix S. Then A_1 is H_1-self-adjoint (or H_1-unitary) if and only if $A_2 = S^{*-1}A_1S^*$ is H_2-self-adjoint (or H_2-unitary, respectively).*

Proof. We consider the "only if" part of the statement. Proof of the converse statement is analogous. Suppose first that A_1 is H_1-self-adjoint so that, (see (2.2.2)), $H_1A_1 = A_1^*H_1$. Then

$$H_2A_2 = (SH_1S^*)(S^{*-1}A_1S^*) = SH_1A_1S^* = SA_1^*H_1S^* = A_2^*H_2,$$

which implies that A_2 is H_2-self-adjoint.

If A_1 is H_1-unitary then $A_1^{-1} = A_1^{[*]}$, and it follows from (2.2.8) that $H_1A_1^{-1} = A_1^*H$. So we have

$$H_2A_2^{-1} = (SH_1S^*)(S^{*-1}A_1^{-1}S^*) = SH_1A_1^{-1}S^*$$
$$= SH_1^*H_1S^* = (SA_1^*S^{-1})(SH_1S^*) = A_2^*H_2.$$

Thus A_2 is H_2-unitary. $\square$

We now focus on H-self-adjoint matrices. By Proposition 2.3.1, the pair (A, H), where A is H-self-adjoint and $H = H^*$ is nonsingular, is transformed to a pair of the same type via $(A, H) \to (S^{-1}AS, S^*HS)$, where S is nonsingular. So, in seeking the canonical form for the pair (A, H) under this type of transformation, we can always assume that A is reduced to a Jordan form. By Theorem 2.2.2, the set of eigenvalues of A, including the partial multiplicities, is symmetric relative to the real axis. So the Jordan form of A can be taken to consist of Jordan blocks with real eigenvalues (if any) followed by pairs of Jordan blocks of equal size having non-real complex conjugate eigenvalues. This is exactly the canonical form of A given in Theorem 2.3.2 below.

To describe the canonical blocks of H it is convenient to introduce the following definition for matrices of the form (2.1.2): an $n \times n$ hermitian matrix $[\delta_{i+j-1,n}]_{i,j=1}^n$ will be called the *n-sip* matrix (standard involutory permutation).

Theorem 2.3.2. *Let A be an H-self-adjoint matrix of size $n \times n$, having the Jordan form*

$$J = J_{r_1}(\lambda_1) \oplus \cdots \oplus J_{r_\alpha}(\lambda_\alpha) \oplus K_{\alpha+1} \oplus \cdots \oplus K_{\alpha+\beta} \tag{2.3.1}$$

where

$$K_i = \begin{bmatrix} J_{r_i}(\lambda_i) & 0 \\ 0 & J_{r_i}(\bar{\lambda}_i) \end{bmatrix}, \quad i = \alpha+1, \ldots, \alpha+\beta,$$

and where $\lambda_1, \ldots, \lambda_\alpha$ are real; $\lambda_{\alpha+1}, \ldots, \lambda_{\alpha+\beta}$ are non-real with positive imaginary parts (in case A has only real eigenvalues the terms K_i are absent; and in case A has only non-real eigenvalues the terms $J_{r_i}(\lambda_i)$ $(1 \le i \le \alpha)$ are absent). Then there exists a nonsingular matrix S such that $S^{-1}AS = J$ and $S^*HS = P$, where

$$P = \epsilon_1 P_1 \oplus \cdots \oplus \epsilon_\alpha P_\alpha \oplus P_{\alpha+1} \oplus \cdots \oplus P_{\alpha+\beta}. \tag{2.3.2}$$

Here ϵ_i is $+1$ or -1 $(i = 1, \ldots, \alpha)$, and P_i is the r_i-sip matrix for $i = 1, \ldots, \alpha$, and P_i is the $2r_i$-sip matrix for $i = \alpha+1, \ldots, \beta$. Moreover, the form (J, P) of (A, H) is determined uniquely up to permutation of blocks: assume that for some nonsingular matrix T we have

$$T^{-1}AT = L_1 \oplus \cdots \oplus L_\gamma \quad T^*HT = Q_1 \oplus \cdots \oplus Q_\gamma \tag{2.3.3}$$

where, for each pair (L_i, Q_i), EITHER L_i is a Jordan block with a real eigenvalue and $\pm Q_i$ is a sip matrix of the same size as L_i, OR $L_i = M_i \oplus \bar{M}_i$ with M_i a Jordan block having an eigenvalue with positive imaginary part and Q_i is a sip matrix of the same size as L_i.

Then the list of pairs $(L_1, Q_1), \ldots, (L_\gamma, Q_\gamma)$ coincides with the list

$$(J_{r_1}(\lambda_1), \epsilon_1 P_1), \ldots, (J_{r_\alpha}(\lambda_\alpha), \epsilon_\alpha P_\alpha), (K_{\alpha+1}, P_{\alpha+1}), \ldots, (K_{\alpha+\beta}, P_{\alpha+\beta}),$$

possibly after permutation.

The proof of Theorem 2.3.2 is rather long and therefore will not be presented here. A full exposition of the proof can be found in several sources. These include Gohberg et al. (1982a) and (1983), Malcev (1963) and Thompson (1990) (in the last two sources the result is presented in a different form).

Because of the uniqueness assertion in Theorem 2.3.2, the form (J, P) given by (2.3.1) and (2.3.2) is called the *canonical form* of the pair (A, H).

The collection of signs ϵ_α, one sign for every Jordan block with a real eigenvalue in the Jordan form of A (or, equivalently, one sign for every partial multiplicity corresponding to real eigenvalues of A), form the *sign characteristic* of (A, H). The sign characteristic is uniquely determined by A and H, up to permutation of signs attached to equal partial multiplicities corresponding to the same real eigenvalue.

Here are some examples of canonical forms.

CANONICAL FORMS

Example 2.3.1. Let

$$J = \begin{bmatrix} 0 & 1 \\ 0 & 0 \end{bmatrix} \oplus \begin{bmatrix} 0 & 1 \\ 0 & 0 \end{bmatrix} \oplus \begin{bmatrix} 0 & 1 \\ 0 & 0 \end{bmatrix} \oplus \begin{bmatrix} 0 & 1 & 0 \\ 0 & 0 & 1 \\ 0 & 0 & 0 \end{bmatrix} \oplus [2] \oplus \begin{bmatrix} -1 & 1 & 0 \\ 0 & -1 & 1 \\ 0 & 0 & -1 \end{bmatrix}.$$

$$P_1 = \begin{bmatrix} 0 & 1 \\ 1 & 0 \end{bmatrix} \oplus \begin{bmatrix} 0 & 1 \\ 1 & 0 \end{bmatrix} \oplus \begin{bmatrix} 0 & -1 \\ -1 & 0 \end{bmatrix} \oplus \begin{bmatrix} 0 & 0 & -1 \\ 0 & -1 & 0 \\ -1 & 0 & 0 \end{bmatrix} \oplus [-1] \oplus \begin{bmatrix} 0 & 0 & 1 \\ 0 & 1 & 0 \\ 1 & 0 & 0 \end{bmatrix}.$$

Here, the sign characteristic consists of the signs 1, 1 and -1 attached to the partial multiplicities 2 corresponding to the eigenvalues 0, the sign -1 attached to the partial multiplicity 3 corresponding to the eigenvalue 0, the sign -1 attached to the partial multiplicity 1 for the eigenvalue 2, and the sign 1 attached to the partial multiplicity 3 for the eigenvalue -1. The pair (J, P_2), where

$$P_2 = \begin{bmatrix} 0 & -1 \\ -1 & 0 \end{bmatrix} \oplus \begin{bmatrix} 0 & 1 \\ 1 & 0 \end{bmatrix} \oplus \begin{bmatrix} 0 & 1 \\ 1 & 0 \end{bmatrix} \oplus \begin{bmatrix} 0 & 0 & -1 \\ 0 & -1 & 0 \\ -1 & 0 & 0 \end{bmatrix} \oplus [-1] \oplus \begin{bmatrix} 0 & 0 & 1 \\ 0 & 1 & 0 \\ 1 & 0 & 0 \end{bmatrix},$$

is an equivalent canonical form (by Theorem 2.3.2); in other words, there exists an invertible S such that $S^{-1}JS = J$, $S^*P_1S = P_2$. On the other hand, the canonical form (J, P_3), where

$$P_3 = \begin{bmatrix} 0 & -1 \\ -1 & 0 \end{bmatrix} \oplus \begin{bmatrix} 0 & 1 \\ 1 & 0 \end{bmatrix} \oplus \begin{bmatrix} 0 & 1 \\ 1 & 0 \end{bmatrix} \oplus \begin{bmatrix} 0 & 0 & 1 \\ 0 & 1 & 0 \\ 1 & 0 & 0 \end{bmatrix} \oplus [-1] \oplus \begin{bmatrix} 0 & 0 & -1 \\ 0 & -1 & 0 \\ -1 & 0 & 0 \end{bmatrix},$$

is not equivalent to (J, P_1). □

Consider now H-unitary matrices. By analogy with the H-self-adjoint matrices studied above, we would like to find a canonical form for a pair of matrices (U, H), where U is H-unitary, under the transformation $(U, H) \to (S^{-1}US, S^*HS)$, for an invertible matrix S. In contrast with the H-self-adjoint case, it is generally not possible to have the Jordan form for U and a simple hermitian form (such as a block diagonal form with sip matrices as diagonal blocks) in the canonical form for (U, H). We therefore admit a non-Jordan, but still sufficiently transparent, form for U while keeping the sip matrix structure for H.

To state the main result on canonical forms for the pairs (U, H) with H-unitary U, introduce the following Toeplitz matrices:

$$L_r(\mu) = \mu \begin{bmatrix} 1 & 2i & 2i^2 & \cdots & 2i^{r-1} \\ 0 & 1 & 2i & \cdots & 2i^{r-2} \\ \vdots & \vdots & & \ddots & \vdots \\ 0 & 0 & 0 & \cdots & 1 \end{bmatrix} \ ; \ |\mu| = 1,$$

(cf. Example 2.2.3). The size of $L_r(\mu)$ is $r \times r$, and $L_r(\mu)$ is similar to the $r \times r$ Jordan block with eigenvalues μ. Let μ_1, μ_2 be complex numbers such that $\mu_1 \bar{\mu}_2 = 1$, $|\mu_1| > 1$. Define $q_1 = \frac{1}{2}i\,(1+\mu_1)$, $q_2 = \frac{1}{2}i\,(1+\mu_2)$, and

$$\alpha_j = \mu_1 q_1^{j-1}(q_1 - \bar{q}_2), \quad \beta_j = \mu_2 q_2^{j-1}(q_2 - \bar{q}_1),$$

for $j = 1, 2, \ldots$. Finally, we define

$$L_{2r}(\mu_1, \mu_2) = \begin{bmatrix} \mu_1 & \alpha_1 & \alpha_2 & \cdots & \alpha_{r-1} \\ 0 & \mu_1 & \alpha_1 & \cdots & \alpha_{r-2} \\ \vdots & \vdots & \ddots & & \vdots \\ & & & \mu_1 & \alpha_1 \\ 0 & 0 & & \cdots & \mu_1 \end{bmatrix} \oplus \begin{bmatrix} \mu_2 & \beta_1 & \beta_2 & \cdots & \beta_{r-1} \\ 0 & \mu_2 & \beta_1 & \cdots & \beta_{r-2} \\ \vdots & \vdots & \ddots & & \vdots \\ & & & & \\ 0 & 0 & & \cdots & \mu_2 \end{bmatrix}.$$
(2.3.4)

The matrix $L_{2r}(\mu_1, \mu_2)$ is $2r \times 2r$, and since $\alpha_1 \neq 0$, $\beta_1 \neq 0$, it is similar to the direct sum of Jordan blocks $J_r(\mu_1) \oplus J_r(\mu_2)$ (by Theorem 2.3.4 below). It can be verified that $L_{2r}(\mu_1, \mu_2)$ is Q-unitary where Q is the sip matrix of size $2r$.

Theorem 2.3.3. *Let A be an H-unitary $n \times n$ matrix. Then there exists an invertible matrix S such that*

$$\begin{aligned} S^{-1}AS &= L_{r_1}(\mu_1) \oplus \cdots \oplus L_{r_\alpha}(\mu_\alpha) \oplus L_{2r_{\alpha+1}}(\mu_{1,\alpha+1}, \mu_{2,\alpha+1}) \oplus \\ &\quad \cdots \oplus L_{2r_{\alpha+\beta}}(\mu_{1,\alpha+\beta}, \mu_{2,\alpha+\beta}) \end{aligned} \quad (2.3.5)$$

and

$$S^* H S = \epsilon_1 P_1 \oplus \cdots \oplus \epsilon_\alpha P_\alpha \oplus P_{\alpha+1} \oplus \cdots \oplus P_{\alpha+\beta}. \quad (2.3.6)$$

Here $|\mu_i| = 1$ for $i = 1, \ldots, \alpha$; $\mu_{1i}\bar{\mu}_{2i} = 1$ and $|\mu_{1i}| > 1$ for $i = \alpha+1, \ldots, \alpha+\beta$; $\epsilon_i = \pm 1$ for $i = 1, \ldots, \alpha$; and P_i is the r_i-sip matrix for $i = 1, \ldots, \alpha$ and P_i is the $2r_i$-sip matrix for $i = \alpha + 1, \ldots, \beta$. Moreover, the form (2.3.5), (2.3.6) of $S^{-1}AS$ and $S^* H S$ is determined uniquely up to permutation of the pairs of blocks

$$(L_{r_i}(\mu_i), \epsilon_i P_i), \quad i = 1, \ldots, \alpha \quad \text{and} \quad (L_{2r_i}(\mu_{1i}, \mu_{2i}), P_i), \quad i = \alpha+1, \ldots, \beta.$$

The proof of Theorem 2.3.3 is found in Section I.4.2 of Gohberg et al. (1983). As it is rather long, we do not reproduce the proof here.

The signs $\{\epsilon_1, \ldots, \epsilon_\alpha\}$ form the *sign characteristic* of the H-unitary matrix A. Thus, for H-unitary matrices, the sign characteristic consists of a sign ± 1 attached to every partial multiplicity corresonding to eigenvalues on the unit circle (if any). Again, the sign characteristic is uniquely defined up to a permutation of signs corresponding to equal partial multiplicities that correspond to the same eigenvalue.

To work with the sign characteristic of H-unitary matrices in specific examples, the following result is useful:

Theorem 2.3.4. *Let A be H-unitary, and let*

$$g(z) = (wz - \bar{w}\eta)/(z - \eta),$$

where $w \neq \bar{w}$ and η is a unimodular number ($|\eta| = 1$) not in the spectrum of A. Then $g(A)$ is H-self-adjoint. Moreover, the sign $\varepsilon(A)$ in the sign characteristic of (A, H) attached to a partial multiplicity m of a unimodular eigenvalue μ of A, and the sign $\varepsilon(g(A))$ in the sign characteristic of $(g(A), H)$ attached to the same partial multiplicity m of the real eigenvalue $g(\mu)$ of $g(A)$ are related by

$$\varepsilon(A) = \varepsilon(g(A)) \; \text{sign}\{(-\operatorname{Im} w)^{m-1}\},$$

where $\operatorname{Im} w$ is the imaginary part of w.

For the proof of Theorem 2.3.4 see Section I.4.3 in Gohberg et al. (1983), especially the proof of their Theorem I.4.4.

From the form (2.3.5) it is not difficult to see that the Jordan form of A is

$$J_{r_1}(\mu_1) \oplus \cdots \oplus J_{r_\alpha}(\mu_\alpha) \oplus J_{r_{\alpha+1}}(\mu_{1,\alpha+1}) \oplus J_{r_{\alpha+1}}(\mu_{2,\alpha+1}) \oplus \cdots$$
$$\oplus J_{r_{\alpha+\beta}}(\mu_{1,\alpha+\beta}) \oplus J_{r_{\alpha+\beta}}(\mu_{2,\alpha+\beta}),$$

(see the more general Theorem 2.3.5 below). So if A has no eigenvalues on the unit circle, the blocks $(L_{r_i}(\mu_i), \epsilon_i P_i)$ are absent in (2.3.5) and (2.3.6), and if all eigenvalues of A are on the unit circle, the blocks $(L_{2r_i}(\mu_{1i}, \mu_{2i}), P_i)$ are absent.

Because of the form (2.3.5) for H-unitary matrices, we will need to compare the properties of the matrices in the form (2.3.5) and their Jordan forms. For this purpose it will be useful to have the following general statement.

Theorem 2.3.5. *Consider a block diagonal matrix of the form*

$$X = X_1 \oplus X_2 \oplus \cdots \oplus X_p, \qquad (2.3.7)$$

where

$$X_i = X_{i1} \oplus X_{i2} \oplus \cdots \oplus X_{i,k_i},$$

and where X_{ij} is an $\alpha_{ij} \times \alpha_{ij}$ upper triangular Toeplitz matrix

$$X_{ij} = \begin{bmatrix} \lambda_{i1} & \lambda_{i2} & \cdots & \lambda_{i\alpha} \\ 0 & \lambda_{i1} & \cdots & \lambda_{i,\alpha-1} \\ \vdots & \vdots & \ddots & \vdots \\ 0 & 0 & \cdots & \lambda_{i1} \end{bmatrix}, \quad \alpha = \alpha_{ij}$$

with the numbers $\lambda_{i1}, \lambda_{i2}, \ldots$ depending on i but not on j. Assume that the eigenvalues $\lambda_{11}, \ldots, \lambda_{p1}$ of $X_1, \ldots, X_p$, respectively, are distinct, and that $\lambda_{i2} \neq 0$ for $i = 1, \ldots, p$. Then the Jordan form of X is

$$J = J_{\alpha_{11}}(\lambda_{11}) \oplus \cdots \oplus J_{\alpha_{1,k_1}}(\lambda_{11}) \oplus \cdots \oplus J_{\alpha_{p1}}(\lambda_{p1}) \oplus \cdots \oplus J_{\alpha_{p,k_p}}(\lambda_{p1}). \quad (2.3.8)$$

Moreover, the matrices X and J have exactly the same set of invariant subspaces.

We will not present the proof of Theorem 2.3.5 here. It can be found in Chapter 10 of Gohberg et al. (1986a) where, more generally, a description is given for matrices having exactly the same set of invariant subspaces as the Jordan form (2.3.8). We point out that, in general, there are matrices not in the form (2.3.7) (as described in Theorem 2.3.4) with the property that their sets of invariant subspaces coincide with the set of J-invariant subspaces.

2.4 Invariant nonnegative subspaces

Invariant subspaces of an H-self-adjoint matrix A will play a very important role in subsequent parts of this book. More especially, we shall be concerned with A-invariant subspaces which are nonnegative (or nonpositive) with respect to H (see Section 2.1) and of maximal dimension (ref. Theorem 2.1.2). This section is devoted to the analysis of such subspaces.

Some notation and definitions will first be set up. Let A be an $n \times n$ H-self-adjoint matrix, and let $r_1, \ldots, r_\alpha$ be the sizes of Jordan blocks in a Jordan form for A corresponding to the real eigenvalues. Let the corresponding signs in the sign characteristic for (A, H) be $\epsilon_1, \ldots, \epsilon_\alpha$, and let k be the number of positive eigenvalues of H, counted with algebraic multiplicities.

A set $\mathcal{C}$ of non-real eigenvalues of A is called a *c-set* if $\mathcal{C} \cap \bar{\mathcal{C}} = \emptyset$ and $\mathcal{C} \cup \bar{\mathcal{C}}$ is the set of all non-real eigenvalues of A. Here $\bar{\mathcal{C}} = \{\bar{\lambda} | \lambda \in \mathcal{C}\}$. In other words, a c-set is defined by the property that it contains exactly one number from every pair of non-real complex conjugate eigenvalues of A.

Theorem 2.4.1. *For every c-set $\mathcal{C}$ there exists a k-dimensional A-invariant H-nonnegative subspace $\mathcal{N}$ such that $\mathcal{C}$ is the set of non-real eigenvalues of the restriction $A|\mathcal{N}$ (and $\operatorname{In} H = (k, n - k, 0)$).*

Proof. We make use of Theorem 2.3.2 and write

$$A = T^{-1}JT, \quad H = T^*PT, \qquad (2.4.1)$$

for some invertible matrix T, where J and P are given by (2.3.1) and (2.3.2), respectively.

It suffices to prove the theorem for J (in place of A) and P (in place of H). Indeed, if $\mathcal{N}_0$ is a k-dimensional J-invariant P-nonnegative subspace such that $\mathcal{C}$ is the set of non-real eigenvalues of $J|\mathcal{N}_0$, then $\mathcal{N} = T^{-1}\mathcal{N}_0$ satisfies all the requirements of Theorem 2.4.1. (Observe that by the Sylvester inertia theorem the number of positive eigenvalues of H coincides with the number of positive eigenvalues of P, in both cases counted with algebraic multiplicities.)

We let

$$\mathcal{N}_0 = \begin{bmatrix} \mathcal{M}_1 \\ \mathcal{M}_2 \\ \vdots \\ \mathcal{M}_{\alpha+\beta} \end{bmatrix} = \left\{ \begin{bmatrix} x_1 \\ x_2 \\ \vdots \\ x_3 \end{bmatrix} \middle| \; x_1 \in \mathcal{M}_1, x_2 \in \mathcal{M}_2, \ldots, x_{\alpha+\beta} \in \mathcal{M}_{\alpha+\beta} \right\}$$

where $\mathcal{M}_i$ is the $J_{r_i}(\lambda_i)$-invariant subspace (if $i = 1,\ldots,\alpha$) or the K_i-invariant subspace (if $i = \alpha+1,\ldots,\alpha+\beta$) defined as follows:
$\mathcal{M}_i = \mathrm{span}\{e_1,\ldots,e_{r_i/2}\}$ if $1 \le i \le \alpha$ and r_i is even.
$\mathcal{M}_i = \mathrm{span}\{e_1,\ldots,e_{(r_i-1)/2}\}$ if $1 \le i \le \alpha$, r_i is odd, and $\epsilon_i = -1$.
$\mathcal{M}_i = \mathrm{span}\{e_1,\ldots,e_{(r_i+1)/2}\}$ if $1 \le i \le \alpha$, r_i is odd and $\epsilon_i = 1$.
$\mathcal{M}_i$ is the root subspace of K_i corresponding to the eigenvalue of K_i that lies in $\mathcal{C}$, if $\alpha+1 \le i \le \alpha+\beta$.

The subspace $\mathcal{N}_0$ is clearly J-invariant and

$$\dim \mathcal{N}_0 = \sum_{i=1}^{\alpha+\beta} \dim \mathcal{M}_i = \sum_{\substack{i=1 \\ r_i \text{ even}}}^{\alpha} \frac{r_i}{2} + \sum_{\substack{i=1 \\ r_i \text{ odd}}}^{\alpha} \frac{r_i + \epsilon_i}{2} + \sum_{i=\alpha+1}^{\alpha+\beta} r_i.$$

For $i = 1, 2, \ldots, \alpha$ the hermitian matrix $\epsilon_i P_i$ has $\frac{r_i}{2}$ positive eigenvalues if r_i is even and $\frac{r_i+\epsilon_i}{2}$ positive eigenvalues if r_i is odd. For $i = \alpha+1,\ldots,\alpha+\beta$ the hermitian matrix P_i has r_i positive eigenvalues. It follows that $\dim \mathcal{N}_0 = k$, the number of positive eigenvalues of P (the eigenvalues are counted with algebraic multiplicities). Finally, let $f_1,\ldots,f_k$ be the basis in $\mathcal{N}_0$ obtained by putting together the bases in $\mathcal{M}_1,\ldots,\mathcal{M}_{\alpha+\beta}$ consisting of unit coordinate vectors e_j. Then the matrix $[f_i^* H f_j]_{i,j=1}^k$ has the diagonal form

$$D_1 \oplus D_2 \oplus \cdots \oplus D_\alpha \oplus 0, \qquad (2.4.2)$$

where D_i is the zero matrix of size $\frac{r_i}{2}$ if r_i is even, D_i is the zero matrix of size $\frac{r_i+\epsilon_i}{2}$ if r_i is odd and $\epsilon_i = -1$, and $D_i = \mathrm{diag}\,[0,0,\ldots,0,1]$ is of size $\frac{r_i+\epsilon_i}{2}$ if r_i is odd and $\epsilon_i = 1$. It is clear from the diagonal form (2.4.2) that the subspace $\mathcal{N}_0$ is J-nonnegative. □

Since k is the maximal dimension of an H-nonnegative subspace (Theorem 2.1.2), it follows that the subspace $\mathcal{N}$ constructed in Theorem 2.4.1 is maximal H-nonnegative, i.e. any subspace of $\mathbb{C}^n$ that strictly contains $\mathcal{N}$ is not H-nonnegative.

For the case of H-nonpositive invariant subspaces of A there is, of course, another statement dual to that of Theorem 2.4.1. This can be obtained by considering $-H$ in place of H in the last theorem.

In general, the subspace $\mathcal{N}$ is not unique (for a given c-set $\mathcal{C}$).

Example 2.4.1. Let

$$A = \begin{bmatrix} 0 & 1 & 0 & 0 \\ 0 & 0 & 0 & 0 \\ 0 & 0 & 0 & 1 \\ 0 & 0 & 0 & 0 \end{bmatrix}, \quad H = \begin{bmatrix} 0 & 1 & 0 & 0 \\ 1 & 0 & 0 & 0 \\ 0 & 0 & 0 & -1 \\ 0 & 0 & -1 & 0 \end{bmatrix}.$$

The following two-dimensional subspaces of $\mathbb{C}^4$ are A-invariant and H-nonnegative (here the c-set $\mathcal{C}$ is empty):

$$\mathrm{span}\{e_1, e_3\}\,; \quad \mathrm{span}\{e_1 + e_3, e_2 + e_4\}.$$

Note that both subspaces are not only H-nonnegative, but also H-neutral. □

We will be especially interested in the situations when the A-invariant maximal H-nonnegative subspace as in Theorem 2.4.1 is unique (for a given choice of the c-set $\mathcal{C}$). To this end the concept of "sign condition" is needed. A pair of $n \times n$ matrices (A, H), where $H = H^*$ is invertible and A is H-self-adjoint, is said to satisfy the *sign condition* if for every real eigenvalue λ_0 of A (if any) the signs in the sign characteristic of (A, H) attached to the partial multiplicities of A corresponding to λ_0 depend only on the parity of these partial multiplicities. In other words, for every real eigenvalue λ of A the signs in the sign characteristic of (A, H) corresponding to Jordan blocks of even size with eigenvalue λ are all equal, and the signs corresponding to the Jordan blocks of odd size with eigenvalue λ are all equal. In particular, the sign condition is satisfied if A has no real eigenvalues, as well as in the case when A is nonderogatory (i.e., the minimal and characteristic polynomials of A coincide), for then each eigenvalue has just one associated Jordan block.

Theorem 2.4.2. *Let A be H-self-adjoint, and assume that (A, H) satisfies the sign condition. Then for every c-set $\mathcal{C}$ there exists a unique k-dimensional A-invariant H-nonnegative subspace $\mathcal{N}$ such that $\mathcal{C}$ is the set of non-real eigenvalues of the restriction $A|\mathcal{N}$. Here k is the number of positive eigenvalues of H (counted with multiplicities). Also there exists a unique ℓ-dimensional A-invariant H-nonpositive subspace $\mathcal{L}$ such that $\mathcal{C}$ is the set of non-real eigenvalues of the restriction $A|\mathcal{L}$; here ℓ is the number of negative eigenvalues of H (counted with multiplicities).*

Conversely, assume that (A, H) does not satisfy the sign condition. Then for every c-set $\mathcal{C}$ there exists a continuum of k-dimensional A-invariant H-nonnegative subspaces $\mathcal{N}$ and a continuum of ℓ-dimensional A-invariant H-nonpositive subspaces $\mathcal{L}$ such that $\mathcal{C}$ is the set of non-real eigenvalues of the restriction $A|\mathcal{N}$, as well as of the restriction $A|\mathcal{L}$.

The proof of Theorem 2.4.2 is based on the following proposition which admits reduction to the case when A has either only one real eigenvalue (possibly of high multiplicity) or a pair of complex conjugate eigenvalues.

Indeed, an H-self-adjoint $n \times n$ matrix A enjoys the following local property. Let $\mathcal{R}$ be either the spectral subspace of A corresponding to a real eigenvalue or the sum of the spectral subspaces of A corresponding to a pair of non real complex conjugate eigenvalues. Then $P_{\mathcal{R}} H | \mathcal{R}$ is hermitian and invertible, where $P_{\mathcal{R}} : \mathbb{C}^n \to \mathcal{R}$ is the orthogonal projector on $\mathcal{R}$ and $P_{\mathcal{R}} H | \mathcal{R}$ is considered as a matrix with respect to some orthonormal basis in $\mathcal{R}$. Moreover, the restriction $A|\mathcal{R}$ (written as a matrix in the same basis) is $P_{\mathcal{R}} H|\mathcal{R}$-self-adjoint. One can check easily that these properties are independent of the choice of the orthonormal basis in $\mathcal{R}$.

Proposition 2.4.3. *Let A be H-self-adjoint. Then an A-invariant subspace $\mathcal{M}$ is maximal H-nonnegative (maximal H-nonpositive) if and only if for every spectral subspace $\mathcal{R}$ of A corresponding either to a real eigenvalue or to a pair* *of non real conjugate eigenvalues the intersection $\mathcal{M} \cap \mathcal{R}$ is maximal $P_{\mathcal{R}} H | \mathcal{R}$-nonnegative (maximal $P_{\mathcal{R}} H | \mathcal{R}$-nonpositive).*

The proof follows easily from Theorem 2.3.2 upon counting the number of positive eigenvalues of each of the blocks $\epsilon_j P_j$ ($j = 1, \ldots \alpha$) and P_j ($j = \alpha + 1, \ldots, \alpha + \beta$) that appear in (2.3.2).

Proof of Theorem 2.4.2. By Proposition 2.4.3 and Theorem 2.3.2 we may assume that one of the two following cases holds:

(1) $A = J \oplus \bar{J}$, where J is a Jordan matrix with an eigenvalues λ in the open upper half-plane, and

$$H = \begin{bmatrix} 0 & P \\ P & 0 \end{bmatrix},$$

where P is a block diagonal matrix whose blocks on the main diagonal are sip matrices with sizes equal to the sizes of the corresponding Jordan blocks in J.

(2) $A = J = J_{m_1}(\lambda) \oplus \cdots \oplus J_{m_r}(\lambda)$, a Jordan matrix having r Jordan blocks with the same real eigenvalue λ, and $H = \epsilon_1 P_1 \oplus \cdots \oplus \epsilon_r P_r$, where $\epsilon_i = \pm 1$, and for $i = 1, \ldots, r$, P_i is the sip matrix having the same size as J_i.

We consider each of these cases separately, and start with case (2). Assume that (A, H) satisfies the sign condition. We have to prove that there is a unique J-invariant maximal H-nonnegative subspace and a unique J-invariant maximal H-nonpositive subspace. Without loss of generality we may suppose that $\lambda = 0$ (otherwise consider $J - \lambda I$ instead of J). Let $J = J_{m_1}(0) \oplus \cdots \oplus J_{m_r}(0)$. We may also assume that the signs corresponding to the blocks $J_{m_i}(0)$ of odd order are all +1's (otherwise replace H by $-H$). Denote by $e_{ij} \in \mathbb{C}^n$ the vector with 1 in the $(m_1 + \cdots + m_{i-1} + j)$ th coordinate and zeros elsewhere, for $j = 1, \ldots, m_i$ and $i = 1, \ldots, r$, so the vectors $e_{i1}, \ldots, e_{im_i}$ form a Jordan chain for J.

Let $\mathcal{L}$ be a J-invariant maximal H-nonpositive subspace, and let

$$x = \sum_{i=1}^{r} \sum_{j=1}^{m_i} \alpha_{ij} e_{ij} \in \mathcal{L}, \quad \alpha_{ij} \in \mathbb{C}.$$

Denote $p_i = [m_i/2]$. We claim that $\alpha_{ij} = 0$ for $j > p_i$, $i = 1, \ldots, r$. Suppose this not true. For $i = 1, \ldots, r$ define $j_i = \max\{j | \alpha_{ij} \neq 0\}$; then $x = \sum_{i=1}^{r} \sum_{j=1}^{j_i} \alpha_{ij} e_{ij}$. By assumption the number $\gamma = \max\{j_i - p_i\}$ is positive. Note that for all i we have $j_i - \gamma \leq p_i$.

First suppose m_s is odd whenever $\gamma = j_s - p_s$ and $1 \leq s \leq r$. Then it follows that $j_i - \gamma + 1 \leq p_i$ for m_i even. We have

$$J^{\gamma-1}x = \sum_{i=1}^{r}\sum_{j=\gamma}^{j_i} \alpha_{ij}e_{i,j-\gamma+1} \in \mathcal{L}.$$

A computation gives

$$\langle HJ^{\gamma-1}x, J^{\gamma-1}x\rangle = \sum_{i=1}^{r}\sum_{D} \epsilon_i \alpha_{ij}\bar{\alpha}_{ik},$$

where $D = \{(j,k) | j - \gamma + 1 + k - \gamma + 1 = m_i + 1,\ \gamma \leq j \leq j_i,\ \gamma \leq k \leq j_i\}$ and ϵ_i is the sign in the sign characteristic of (J, H) corresponding to the block $J_{m_i}(0)$. For $1 \leq j, k \leq j_i$ we have

$$j - \gamma + 1 + k - \gamma + 1 \leq 2(j_i - \gamma + 1) \leq 2(p_i + 1),$$

and equality holds if and only if $k = j = j_i$ and $j_i - \gamma = p_i$. In that case m_i is odd, and hence $\epsilon_i = +1$ and $2(p_i + 1) = m_i + 1$. So

$$j - \gamma + 1 + k - \gamma + 1 = m_i + 1$$

if and only if $j_i - \gamma = p_i$, and then $\epsilon_i = +1$. Hence

$$\langle HJ^{\gamma-1}x, J^{\gamma-1}x\rangle = \sum_{\{i | j_i - \gamma = p_i\}} |\alpha_{ij_i}|^2 > 0$$

by assumption. On the other hand $\mathcal{L}$ is H-nonpositive, so $\langle HJ^{\gamma-1}x, J^{\gamma-1}x\rangle \leq 0$, and a contradiction is obtained.

Next suppose there exists s such that $\gamma = j_s - p_s$ and m_s is even. In this case an analogous computation shows that

$$\langle HJ^{\gamma}x, J^{\gamma}x\rangle = 0, \quad <HJ^{\gamma}x, J^{\gamma-1}x> = \pm\sum_{E}|\alpha_{ij_i}|^2 \neq 0. \qquad (2.4.3)$$

where $E = \{i | m_i$ is even and $j_i - \gamma = p_i\}$, and the sign $\pm$ coincides with the sign of (J, H) corresponding to the blocks of even size. Now (2.4.3) is contradictory, because the Cauchy–Schwartz inequality, which holds on the H-nonpositive subspace $\mathcal{L}$, gives

$$|\langle HJ^{\gamma}x, J^{\gamma-1}x\rangle|^2 \leq \langle HJ^{\gamma}x, J^{\gamma}x\rangle\langle HJ^{\gamma-1}x, J^{\gamma-1}x\rangle.$$

We have proved that $\alpha_{ij} = 0$ if $j > p_i$. Since $\mathcal{L}$ is maximal H-nonpositive, we have $\dim \mathcal{L} = \sum_{i=1}^{r} p_i$, as one easily sees using the assumption that $\epsilon_i = +1$ for blocks of odd size. This leaves only the possibility

$$\mathcal{L} = \left\{\sum_{i=1}^{r}\sum_{j=1}^{p_i}\alpha_{ij}e_{ij} \,\middle|\, \alpha_{ij} \in \mathbb{C}\right\}.$$

So $\mathcal{L}$ is unique.

Now let $\mathcal{N}$ be a J-invariant maximal H-nonnegative subspace. Then $(H\mathcal{N})^\perp$ is a J-invariant maximal H-nonpositive subspace. Indeed, J-invariance of $(H\mathcal{N})^\perp$ is easily checked. Further, assume that $\langle Hx, x\rangle > 0$ for some $x \in (H\mathcal{N})^\perp$. Then the subspace $\text{span}\{x\} + \mathcal{N}$ is H-nonnegative, because for $m \in \mathcal{N}$, $\alpha \in \mathbb{C}$ we have

$$\langle H(\alpha x + m), (\alpha x + m)\rangle = |\alpha|^2 \langle Hx, x\rangle + \langle Hm, m\rangle \geq 0.$$

Since $\mathcal{N}$ is maximal H-nonnegative, it follows that $x \in \mathcal{N}$. Hence $x \in \mathcal{N} \cap (H\mathcal{N})^\perp$ which gives $\langle Hx, x\rangle = 0$. This is a contradiction, so $\langle Hx, x\rangle \leq 0$ for all $x \in (H\mathcal{N})^\perp$. Finally, the maximality of $(H\mathcal{N})^\perp$ follows from a dimension argument. Now $(H\mathcal{N})^\perp$ is unique, as we have shown above. So $\mathcal{N}$ is unique as well.

Assume now that (A, H) does not satisfy the sign condition (we are still considering the case (2)). We may assume that one of the following two cases holds:

(3) $m_1 = 2p_1$ and $m_2 = 2p_2$ are even, $\epsilon_1 = 1$, $\epsilon_2 = -1$;

(4) $m_1 = 2p_1 + 1$ and $m_2 = 2p_2 + 1$ are odd, $\epsilon_1 = 1$, $\epsilon_2 = -1$.

Let $\mathcal{M}$ be a fixed A'-invariant maximal H'-nonpositive subspace, where

$$A' = J_{m_3}(\lambda) \oplus \cdots \oplus J_{m_r}(\lambda), \quad H' = \epsilon_3 P_3 \oplus \cdots \oplus \epsilon_r P_r.$$

Put

$$A'' = J_{m_1}(\lambda) \oplus J_{m_2}(\lambda), \quad H'' = \epsilon_1 P_1 \oplus \epsilon_2 P_2.$$

A dimension argument shows that for every A''-invariant maximal P''-nonpositive subspace $\mathcal{L}$ the sum $\mathcal{L} + \mathcal{M}$ is maximal H-nonpositive (and clearly A-invariant). So it is sufficient to find a continuum of A''-invariant maximal H''-nonpositive subspaces. Assuming $p_1 \geq p_2$, in both cases (3) and (4) such a continuum is given by

$$\mathcal{L}(\theta) = \text{span}\{e_1, \ldots, e_{p_1-p_2}, e_{p_1-p_2+1} + \theta e_{m_1+1},$$
$$e_{p_1-p_2+2} + \theta e_{m_1+2}, \ldots, e_{p_1-p_2+m_2} + \theta e_{m_1+m_2}\}, \quad |\theta| = 1.$$

In fact the subspaces $\mathcal{L}(\theta)$ are H''-neutral.

It remains to consider the case (1). In this case (A, H) clearly satisfy the sign condition. Without loss of generality we can assume $\mathcal{S} = \{\lambda\}$. Then clearly, $\text{span}\{e_1, \ldots, e_{n/2}\}$ is the only n-dimensional A-invariant subspace $\mathcal{L}$ such that $A|\mathcal{L}$ has the sole eigenvalue λ (here $n \times n$ is the size of A). As the subspace $\text{span}\{e_1, \ldots, e_{n/2}\}$ happens to be H-neutral, the proof of Theorem 2.4.2 in case (1) is completed. $\square$

2.5 Invariant neutral subspaces

In this section we will study cases when the A-invariant maximal H-nonnegative subspace described in Theorem 2.4.1 is actually H-neutral and, moreover, when the A-invariant maximal H-neutral subspace is unique (for a given choice of the c-set $\mathcal{C}$). As Example 2.4.1 shows, extra hypotheses are needed to ensure the uniqueness. In particular, it will be required that the H-self-adjoint matrix A has only even partial multiplicities corresponding to its real eigenvalues (if any). As the form (2.3.2) shows, in this case the sizes of all sip matrices P_i ($1 \leq i \leq \alpha+\beta$) are even, and therefore P (and consequently H) has as many positive eigenvalues (counted with multiplicities) as negative eigenvalues. Thus, the size n of A (and H) is even, and by Theorems 2.1.2 and 2.1.3 the maximal dimension of an H-nonnegative subspace and the maximal dimension of an H-neutral subspace are equal to $\frac{n}{2}$.

Theorem 2.5.1. *Let A be an H-self-adjoint $n \times n$ matrix such that the partial multiplicities of A corresponding to its real eigenvalues (if any) are all even. Then for every c-set $\mathcal{C}$ there exists a unique A-invariant H-neutral subspace $\mathcal{N}$ with the properties that $\dim \mathcal{N} = \frac{n}{2}$, the set of non real eigenvalues of $A|\mathcal{N}$ coincides with $\mathcal{C}$ and the partial multiplicities of $A|\mathcal{N}$ corresponding to real eigenvalues are exactly the halves of the partial multiplicities of A corresponding to real eigenvalues.*

A detailed proof of this theorem is given by Gohberg *et al.* (1983), Section I.3.12. See also Lancaster *et al.* (1994).

Note that in Theorem 2.5.1 the requirement that the partial multiplicities of $A|\mathcal{N}$ corresponding to real eigenvalues are exactly the halves of those for A cannot be omitted in general, as Example 2.4.1 shows. However, it turns out that if we require an additional condition that the signs in the sign characteristic of (A, H) which correspond to Jordan blocks with the same eigenvalue are equal, then for every c-set $\mathcal{C}$ there exists a unique H-neutral A-invariant $\frac{n}{2}$-dimensional subspace $\mathcal{N}$ with $\mathcal{C}$ the set of non real eigenvalues of $A|\mathcal{N}$. (Obviously, this additional condition is violated in Example 2.4.1.) Moreover, in this case we are able to describe all H-neutral A-invariant $\frac{n}{2}$-dimensional subspaces, as follows.

Theorem 2.5.2. *Let A be an H-self-adjoint matrix such that the partial multiplicities of A corresponding to the real eigenvalues are all even, and the signs in the sign characteristic of (A, H) corresponding to the same real eigenvalue are all equal. Fix a c-set $\mathcal{C}$, and let $\mathcal{M}_+$ be the spectral subspace of A corresponding to its eigenvalues in $\mathcal{C}$. Then for every A-invariant subspace $\mathcal{N}_+ \subseteq \mathcal{M}_+$ there exists a unique A-invariant H-neutral $\frac{n}{2}$-dimensional subspace $\mathcal{N}$ such that*

$$\mathcal{N} \cap \mathcal{M}_+ = \mathcal{N}_+. \tag{2.5.1}$$

Moreover, the partial multiplicities of the restriction $A|\mathcal{N}$ corresponding to the real eigenvalue λ_0 of $A|\mathcal{N}$ are exactly the halves of the partial multiplicities of A corresponding to λ_0 (considered as an eigenvalue of A).

INVARIANT NEUTRAL SUBSPACES

In other words, Theorem 2.5.2 gives a one-to-one correspondence between the set of A-invariant subspaces $\mathcal{N}_+$ of $\mathcal{M}_+$ and the set of A-invariant H-neutral $\frac{n}{2}$-dimensional subspaces $\mathcal{N}$, which is given by equality (2.5.1).

Proof. By Theorem 1.2.5 every A-invariant subspace $\mathcal{M}$ is the direct sum of the intersections $\mathcal{M} \cap \mathcal{R}_i$, where $\mathcal{R}_i$ is either the root subspace of A corresponding to one real eigenvalue or a direct sum of two root subspaces of A corresponding to a pair of non real complex conjugate eigenvalues. Also, by Theorem 2.2.2 if $\mathcal{R}_i \neq \mathcal{R}_j$, then $\mathcal{R}_i$ and $\mathcal{R}_j$ are H-orthogonal. Thus, the proof is reduced to two cases:

(i) $\sigma(A) = \{\lambda_0\}$, λ_0 is real (in this case we can assume $\lambda_0 = 0$ without loss of generality);

(ii) $\sigma(A) = \{\lambda_0, \bar{\lambda}_0\}$, $\lambda_0 \notin \mathbb{R}$.

Consider case (i). In this case $\mathcal{M}_+ = 0$, so Theorem 2.5.2 asserts that there is a unique A-invariant H-neutral $\frac{n}{2}$-dimensional subspace $\mathcal{N}$. The uniqueness of such a subspace follows from Theorem 2.4.2, as the sign condition of that theorem is obviously satisfied. To prove the existence, it can be assumed without loss of generality that (A, H) is in the canonical form described in Theorem 2.3.2:

$$A = J = J_1 \oplus \cdots \oplus J_r; \quad H = P = \epsilon P_1 \oplus \cdots \oplus \epsilon P_n$$

where J_i is the nilpotent Jordan block of size m_i ($i = 1, \ldots, r$); P_i is the $m_i \times m_i$ sip matrix; and $\epsilon = \pm 1$.

Denote by x_{ij} the $(m_1 + \cdots + m_{i-1})$-th unit coordinate vector in $\mathbb{C}^n$, $j = 1, \ldots, m_i$ (recall that $n = m_1 + \cdots + m_r$); so the vectors $x_{i1}, \ldots, x_{im_i}$ form a Jordan chain of $\lambda I - J$. Then $\mathcal{N} = \text{span}\{x_{ij} | 1 \leq j \leq \frac{1}{2}m_i;\ i = 1, \ldots, r\}$ is A-invariant, H-neutral, and has dimension $n/2$.

Now consider case (ii). Again, assume (A, H) is in the canonical form described in Theorem 2.3.2. Rearranging blocks $K_{\alpha+1}, \ldots, K_{\alpha+\beta}$ and $P_{\alpha+1}, \ldots, P_{\alpha+\beta}$ (which amounts to a simultaneous similarity-congruence $(A, H) \to (S^{-1}AS, S^*HA)$ for a suitable invertible S) we can write A and H in the following form:

$$A = J = \begin{bmatrix} J_+ & 0 \\ 0 & \bar{J}_+ \end{bmatrix}$$

where $J_+ = J_1 \oplus \cdots \oplus J_r$ and J_i is the Jordan block of size p_i with eigenvalue λ_0;

$$H = \tilde{P} = \begin{bmatrix} 0 & P \\ P & 0 \end{bmatrix}$$

where $P = P_1 \oplus \cdots \oplus P_r$, and P_i is the sip matrix of size p_i. Observe that $\bar{J}_+ = P^{-1} J_+^* P$; so the pair $(J, \tilde{P})$ can, in turn, be reduced by simultaneous similarity-congruence to the pair (K, Q), where $K = \begin{bmatrix} J_+ & 0 \\ 0 & J_+^* \end{bmatrix}$, $Q = \begin{bmatrix} 0 & I \\ I & 0 \end{bmatrix}$. It is sufficient to verify Theorem 2.5.2 for $A = K$, $H = Q$.

Given a J_+-invariant subspace $\mathcal{N}_+$, put $\mathcal{N} = \left\{ \begin{bmatrix} x \\ y \end{bmatrix} \in \mathbb{C}^n \mid x \in \mathcal{N}_+,\ y \in \mathcal{N}_+^\perp \right\}$. Clearly, $\mathcal{N}$ is K-invariant, $\dim \mathcal{N} = \frac{n}{2}$ and $\mathcal{N} \cap \mathcal{R}_{\lambda_0}(K) = \begin{bmatrix} \mathcal{N}_+ \\ 0 \end{bmatrix}$. Further, $\mathcal{N}$ is Q-neutral. Indeed, for $x_1, x_2 \in \mathcal{N}_+$, $y_1, y_2 \in \mathcal{N}_+^\perp$ we have

$$\left\langle \begin{bmatrix} 0 & I \\ I & 0 \end{bmatrix} \begin{bmatrix} x_1 \\ y_1 \end{bmatrix}, \begin{bmatrix} x_2 \\ y_2 \end{bmatrix} \right\rangle = \langle y_1, x_2 \rangle + \langle x_1, y_2 \rangle = 0.$$

Now let $\mathcal{N}'$ be a K-invariant, $\frac{n}{2}$-dimensional, Q-neutral subspace such that

$$\mathcal{N}' \cap \mathcal{R}_{\lambda_0}(K) = \begin{bmatrix} \mathcal{N}_+ \\ 0 \end{bmatrix}. \tag{2.5.2}$$

As $\mathcal{N}'$ is K-invariant, we have $\mathcal{N}' = (\mathcal{N}' \cap \mathcal{R}_{\lambda_0}(K)) \dotplus (\mathcal{N}' \cap \mathcal{R}_{\bar{\lambda}_0}(K))$. In fact

$$\mathcal{N}' \cap \mathcal{R}_{\bar{\lambda}_0}(K) \subseteq \left\{ \begin{bmatrix} 0 \\ x \end{bmatrix} \,\Big|\, x \in \mathcal{N}_+^\perp \right\}. \tag{2.5.3}$$

Indeed, if $\begin{bmatrix} 0 \\ x \end{bmatrix} \in \mathcal{N}'$ for some $x \in \mathbb{C}^{n/2} \setminus \mathcal{N}_+^\perp$, then there exists $y \in \mathcal{N}_+$ such that $\langle x, y \rangle \neq 0$, and therefore

$$\left\langle \begin{bmatrix} 0 & I \\ I & 0 \end{bmatrix} \begin{bmatrix} 0 \\ x \end{bmatrix}, \begin{bmatrix} y \\ 0 \end{bmatrix} \right\rangle = \langle x, y \rangle \neq 0;$$

a contradiction with Q-neutrality of $\mathcal{N}'$. Now (2.5.2) and (2.5.3) together with $\dim \mathcal{N}' = \frac{n}{2}$, imply that $\mathcal{N}' = \mathcal{N}$.

Theorem 2.5.2 is proved. □

Let A, H be as in Theorem 2.5.2. Observe that for every A-invariant H-neutral $\frac{n}{2}$-dimensional subpsace $\mathcal{N}$ the partial multiplicities of the restriction $A|_\mathcal{N}$ which correspond to a real eigenvalue are just half the partial multiplicities of A corresponding to the same eigenvalue (see Theorem 2.5.1). In particular, the restriction of $A|_\mathcal{N}$ to the sum of the root subspaces of $A|_\mathcal{N}$ corresponding to the real eigenvalues is independent of $\mathcal{N}$ (in the sense that any two such restrictions, for different subspaces $\mathcal{N}$, are the same).

Later in this book the most important case of Theorem 2.5.2 will be that in which the c-set $\mathcal{C}$ consists of all eigenvalues of A (if any) in the open upper half-plane.

A special case of Theorem 2.5.2 is sufficiently important to justify a separate statement:

Corollary 2.5.3. *Let A be an H-self-adjoint matrix as in Theorem 2.5.2 and suppose, in addition, that $\sigma(A) \subseteq \mathbb{R}$. Then there exists a unique A-invariant, H-neutral, $\frac{n}{2}$-dimensional subspace.*

Still another result on the existence of invariant neutral subspaces applies in a situation requiring no hypotheses on the structure of real eigenvalues:

Theorem 2.5.4. *Let A be an H-self-adjoint $n \times n$ matrix, where n is even and assume that there exists an $\frac{n}{2}$-dimensional A-invariant H-neutral subspace $\mathcal{M}_0$. Then for every c-set $\mathcal{C}$ there exists an $\frac{n}{2}$-dimensional A-invariant H-neutral subspace $\mathcal{M}$ such that the set of non real eigenvalues of $A|\mathcal{M}$ coincides with $\mathcal{C}$.*

Proof. By the hypotheses and by Proposition 2.4.3 for every real eigenvalue λ_0 the root subspace $\mathcal{R}_{\lambda_0}(A)$ is even dimensional and

$$\dim(\mathcal{M}_0 \cap \mathcal{R}_{\lambda_0}(A)) = \frac{1}{2} \dim \mathcal{R}_{\lambda_0}(A).$$

Now the subspace

$$\mathcal{M} = \sum_{\lambda_0 \in \mathcal{C}} (A|\mathcal{R}_{\lambda_0}(A) + \sum_{\lambda_0 \in \mathcal{R}} (\mathcal{M}_0 \cap \mathcal{R}_{\lambda_0}(A))$$

clearly satisfies the requirements of Theorem 2.5.4. □

By combining Theorems 2.3.3 and 2.3.4, all the results presented in this section are valid also for H-unitary matrices. The only difference from the H-self-adjoint case is that now a *c-set* is defined as a set $\mathcal{C}$ of nonunimodular eigenvalues (i.e. having absolute value different from 1) of an H-unitary matrix A such that

$$\mathcal{C} \cap \{\bar{\lambda}^{-1} | \lambda \in \mathcal{C}\} = \emptyset \quad \text{and} \quad \mathcal{C} \cup \{\bar{\lambda}^{-1} | \lambda \in \mathcal{C}\}$$

consist of all nonunimodular eigenvalues of A. We leave the statements of the corresponding results for H-unitary matrices for interested readers. In the sequel, when necessary, these results will be used with reference to the corresponding results for H-self-adjoint matrices.

2.6 H-symmetric matrices

We now turn attention to the real space $\mathbb{R}^n$ and to an indefinite scalar product $[.,.]$ on $\mathbb{R}^n$. It is defined as a function from $\mathbb{R}^n \times \mathbb{R}^n$ to $\mathbb{R}$ satisfying the following axioms:

(i) *Linearity in the first argument:*

$$[\alpha x_1 + \beta x_2, y] = \alpha[x_1, y] + \beta[x_2, y]$$

for all $x_1, x_2, y \in \mathbb{R}^n$ and all real numbers α and β.

(ii) *Symmetry:* $[x, y] = [y, x]$ for all $x, y \in \mathbb{R}^n$.

(iii) *Nondegeneracy:* if $[x, y] = 0$ for all $y \in \mathbb{R}^n$, then $x = 0$.

It is easily seen that every indefinite scalar product $[.,.]$ on $\mathbb{R}^n$ is induced by an $n \times n$ invertible real symmetric matrix H:

$$[x, y] = \langle Hx, y \rangle, \quad x, y \in \mathbb{R}^n,$$

where $\langle x, y \rangle = \sum_{i=1}^{n} x_i y_i$ is the standard (euclidean) scalar product in $\mathbb{R}^n$.

All the definitions, results and proofs of Section 2.1 concerning classifications of subspaces are valid also in the real case, with one small exception. As in the complex case a subspace $\mathcal{M} \subseteq \mathbb{R}^n$ is said to be *neutral* with respect to the indefinite scalar product $[.,.]$ (or H-neutral when $[.,.]$ is induced by H) if $[x, x] = 0$ for all $x \in \mathcal{M}$. To see that this is equivalent to $[x, y] = 0$ for all $x, y \in \mathcal{M}$, the polarization identity used in Section 2.1 is replaced by the simple relation

$$[x, y] = \frac{1}{2} \{[x + y, x + y] - [x, x] - [y, y]\}$$

which now holds because $[.,.]$ is linear in *both* arguments.

Let $[.,.]$ be an indefinite scalar product in $\mathbb{R}^n$ induced by an invertible real symmetric matrix H. Thus, $[x, y] = \langle Hx, y \rangle$ for all $\mathcal{X}, y \in \mathbb{R}^n$. A real $n \times n$ matrix $A^{[*]}$ is called the *adjoint* of A (with respect to $[.,.]$), or the H-*adjoint* of A if

$$[A^{[*]}x, y] = [x, Ay]$$

for all $x, y \in \mathbb{R}^n$. The analogues of H-self-adjoint and H-unitary matrices are natural: a real $n \times n$ matrix A is called H-symmetric if $A^{[*]} = A$, and H-orthogonal if $A^{[*]} = A^{-1}$. All the results of Section 2.2 remain valid for the real case as well, with the exception of the results concerning the Cayley transforms which, in general, do not transform real matrices to real matrices.

Now let us study real H-symmetric matrices A and, in particular, canonical forms for such pairs (A, H). This is the main concern of this section.

The building blocks of these canonical forms are based on the real Jordan form of A (described in Section 1.3) and on the sip matrices (as in the complex case). One such building block is given by the following example.

Example 2.6.1. Let $J_{2r}(\lambda \pm i\mu)$ be a real Jordan block of size $2r$ (see Section 1.4) and let H be the $2r$-sip matrix. Then $J_{2r}(\lambda \pm i\mu)$ is H-symmetric. □

We now state the canonical form for real H-symmetric matrices. It is analogous to the canonical form for complex H-self-adjoint matrices (Theorem 2.3.2), with the Jordan form replaced by the real Jordan form.

Theorem 2.6.1. *Let H be a nonsingular real $n \times n$ symmetric matrix, and let A be a real H-symmetric matrix. Then there exists an invertible real matrix S such that $J := S^{-1}AS$ and $P := S^*HS$ have the following forms:*

$$J = J_{r_1}(\lambda_1) \oplus \cdots \oplus J_{r_\alpha}(\lambda_\alpha) \oplus J_{2r_{\alpha+1}}(\lambda_{\alpha+1} \pm i\mu_{\alpha+1}) \oplus \cdots \oplus J_{2r_{\alpha+\beta}}(\lambda_{\alpha+\beta} \pm i\mu_{\alpha+\beta}) \quad (2.6.1)$$

and

$$P = \epsilon_1 P_1 \oplus \cdots \oplus \epsilon_\alpha P_\alpha \oplus P_{\alpha+1} \oplus \cdots \oplus P_{\alpha+\beta}, \quad (2.6.2)$$

where $\lambda_1, \ldots, \lambda_{\alpha+\beta}$ and $\mu_{\alpha+1}, \ldots, \mu_{\alpha+\beta}$ are positive numbers; $\epsilon_j = \pm 1$ ($j = 1, \ldots, \alpha$); P_j is the r_j-sip matrix if $j = 1, \ldots, \alpha$, and P_j is the $2r_j$-sip matrix if $j = \alpha+1, \ldots, \alpha+\beta$. Moreover, the canonical form (J, P) of (A, H) is uniquely determined by (A, H) up to permutation of blocks $(J_{r_j}(\lambda_j), \epsilon_j P_j)$ for $j = 1, \ldots, \alpha$, and $(J_{2r_j}(\lambda_j \pm i\mu_j), P_j)$ for $j = \alpha+1, \ldots, \alpha+\beta$.

The proof of Theorem 2.6.1 can be found in Section I.5.2 of Gohberg et al. (1983).

As in the complex case, the numbers $\{\epsilon_i\}_{i=1}^{\alpha}$ are said to form the *sign characteristic* of the pair (A, H). Thus, the sign characteristic is the set of numbers $+1$ or -1 attached to each partial multiplicity of A corresponding to a real eigenvalue. Note also that J is simply the real Jordan form of A. So, if A has no real eigenvalues the terms $(J_{2r_j}(\lambda_j \pm i\mu_j), P_j)$ are absent in (2.6.1), (2.6.2), and if A has no non real eigenvalues, the terms $(J_{r_j}(\lambda_j), \epsilon_j P_j)$ are absent.

Now let us consider A-invariant H-nonnegative subspaces in $\mathbb{R}^n$, where A is a real H-symmetric matrix. A simple example shows that Theorem 2.4.1 is not generally valid for real matrices.

Example 2.6.2. Let

$$H = \begin{bmatrix} 0 & 1 \\ 1 & 0 \end{bmatrix}, \quad A = \begin{bmatrix} 0 & 1 \\ -1 & 0 \end{bmatrix}.$$

Clearly, A is H-symmetric and H has one positive eigenvalue, but there is no one-dimensional A-invariant subspaces in $\mathbb{R}^2$. (In contrast, observe that there is a one-dimensional A-invariant H-neutral subspace in $\mathbb{C}^2$. This is ensured by Theorem 2.5.2. Indeed, span $\begin{bmatrix} 1 \\ i \end{bmatrix}$ is such a subspace.) □

We will obtain a real analogue of Theorem 2.4.1. Note from the very beginning that the notion of a c-set (as introduced in Theorem 2.4.1) is not useful in the framework of real matrices. Indeed, the set of non real eigenvalues of the restriction of a real matrix to a real invariant subspace may contain pairs of complex conjugate numbers, and therefore cannot be a c-set.

Now let us determine the maximal possible dimension of an A-invariant H-nonnegative (or H-neutral) subspace. Assume first of all that the eigenvalues of A consist of a pair of non real complex conjugate eigenvalues. In this case, by Theorem 2.6.1, the size of A is even, say $m = 2k$, and H has k positive eigenvalues (counted with algebraic multiplicities) and k negative eigenvalues.

Lemma 2.6.2. *Let A be a real $m \times m$ H-symmetric matrix, and assume that A has $\lambda \pm i\mu$, (λ, μ real and $\mu \neq 0$) as its only eigenvalues. Let $m = 2k$. Then*

  (i) *the largest dimension of an A-invariant H-nonnegative subspace in $\mathbb{R}^m$ is k if k is even and $k - 1$ if k is odd.*

  (ii) *the largest dimension of an A-invariant H-neutral subspace in $\mathbb{R}^m$ is k if k is even and $k - 1$ if k is odd.*

Proof. We dispose of the easy part of Lemma 2.6.2 first. Let $\mathcal{M}$ be an A-invariant H-nonnegative subspace in $\mathbb{R}^m$. Then $\dim \mathcal{M} \leq k$ by Theorem 2.1.2. Now A has no odd-dimensional invariant subspace (see Exercise 1.8.5). So $\dim \mathcal{M} \leq k$ if k is even and $\dim \mathcal{M} \leq k - 1$ if k is odd.

It remains to construct an A-invariant H-neutral subspace $\mathcal{M}$ in $\mathbb{R}^n$ of dimension k (if k is even), or $k - 1$ (if k is odd). By Theorem 2.6.1 we can assume, without loss of generality, that

$$A = J_{2r_1}(\lambda \pm i\mu) \oplus \cdots \oplus J_{2r_\gamma}(\lambda \pm i\mu) \qquad (2.6.3)$$

$$H = P_1 \oplus \cdots \oplus P_\gamma, \qquad (2.6.4)$$

where P_j is the $2r_j$-sip matrix. We can also assume that $r_1, \ldots, r_\alpha$ are even while $r_{\alpha+1}, \ldots, r_\gamma$ are odd. (If all r_j's are even, or if all of them are odd, the changes in subsequent reasoning are self-evident.) The blocks $J_{2r_j}(\lambda \pm i\mu)$ with odd r_j are collected in pairs:

$$K_1 = J_{2r_{\alpha+1}}(\lambda \pm i\mu) \oplus J_{2r_{\alpha+2}}(\lambda \pm i\mu);$$
$$K_2 = J_{2r_{\alpha+3}}(\lambda \pm i\mu) \oplus J_{2r_{\alpha+4}}(\lambda \pm i\mu);$$
$$\vdots$$
$$K_\beta = J_{2r_{\alpha+2\beta-1}}(\lambda \pm i\mu) \oplus J_{2r_{\alpha+2\beta}}(\lambda \pm i\mu),$$

where $2\beta = \gamma - \alpha$ if $\gamma - \alpha$ is even, and $2\beta = \gamma - \alpha - 1$ if $\gamma - \alpha$ is odd. So we can rewrite A in the form

$$A = J_{2r_1}(\lambda \pm i\mu) \oplus \cdots \oplus J_{2r_\alpha}(\lambda \pm i\mu) \oplus K_1 \oplus K_2 \oplus \cdots \oplus K_\beta \oplus J_{2r_\gamma}(\lambda \pm i\mu)$$

(if $\gamma - \alpha$ is even then the term $J_{2r_\gamma}(\lambda \pm i\mu)$ is absent in the above formula). We now construct an A-invariant H-neutral subspace $\mathcal{M}$ of requisite dimension in the following form (assuming first that $\gamma - \alpha$ is odd):

$$\mathcal{M} = \left\{ \begin{bmatrix} x_1 \\ x_2 \\ \vdots \\ x_{\alpha+\beta+1} \end{bmatrix} \middle| \begin{array}{l} x_1 \in \mathcal{M}_1, \ldots, x_\alpha \in \mathcal{M}_\alpha, x_{\alpha+1} \in \mathcal{N}_1, \ldots, x_{\alpha+\beta} \in \mathcal{N}_\beta, \\ x_{\alpha+\beta+1} \in \mathcal{L} \end{array} \right\}.$$

Here $\mathcal{M}_j$ is the subspace in $\mathbb{R}^{2r_j}$ defined by

$$\mathcal{M}_j = \text{span}\{e_1, \ldots, e_{r_j}\} \quad (j = 1, \ldots, \alpha). \tag{2.6.5}$$

For $j = 1, \ldots, \beta$, $\mathcal{N}_j$ is the subspace in $\mathbb{R}^{2(r_{\alpha+2j-1})+2(r_{\alpha+2j})}$ defined by (for typographical reasons we write here $e(j)$ rather than e_j):

$$\mathcal{N}_j = \text{span}\{e(1), \ldots, e(r_{\alpha+2j-1} - 1), e(2r_{\alpha+2j-1} + 1), \ldots,$$
$$e(2r_{\alpha+2j-1} + r_{\alpha+2j} - 1), e(r_{\alpha+2j-1}) + e(2r_{\alpha+2j-1} + r_{\alpha+2j} + 1), \tag{2.6.6}$$
$$e(r_{\alpha+2j-1} + 1) - e(2r_{\alpha+2j-1} + r_{\alpha+2j})\};$$

and finally, $\mathcal{L}$ is the subspace in $\mathbb{R}^{2r_\gamma}$ spanned by $e_1, \ldots, e_{r_j-1}$. A straightforward verification shows that $\mathcal{M}$ is A-invariant H-neutral of dimension $k-1$.

If $\gamma - \alpha$ is even, then

$$A = J_{2r_1}(\lambda \pm i\mu) \oplus \cdots \oplus J_{2r_\alpha}(\lambda \pm i\mu) \oplus K_1 \oplus K_2 \oplus \cdots \oplus K_\beta,$$

and we take $\mathcal{M}$ in the form

$$\mathcal{M} = \left\{ \begin{bmatrix} x_1 \\ x_2 \\ \vdots \\ x_{\alpha+\beta} \end{bmatrix} \middle| x_1 \in \mathcal{M}_1, \ldots, x_\alpha \in \mathcal{M}_\alpha; \, x_{\alpha+1} \in \mathcal{N}_1, \ldots, x_{\alpha+\beta} \in \mathcal{N}_{\alpha+\beta} \right\},$$

where $\mathcal{M}_j$ and $\mathcal{N}_j$ are given by (2.6.5) and (2.6.6), respectively. □

Theorem 2.6.3. *Let A be a real $n \times n$ H-symmetric matrix. Then:*

(i) *The largest dimension of an A-invariant H-nonnegative subspace in $\mathbb{R}^n$ is equal to $k - p$, where k is the number of positive eigenvalues of H (counted acccording to algebraic multiplicities) and p is the number of distinct pairs of non real complex conjugate eigenvalues $\{\lambda_j \pm i\mu_j\}_{j=1}^p$ of A such that the algebraic multiplicity of $\lambda_j + i\mu_j$ is odd $(j = 1, \ldots, p)$.*

(ii) *Assume that the partial multiplicities of A corresponding to the real eigenvalues (if any) are all even. Then there exists an A-invariant H-neutral subspace $\mathcal{M} \subseteq \mathbb{R}^n$ of dimension $k - p$, where k and p are defined as in (i), and $\mathcal{M}$ has the additional property that the partial multiplicities of $A|\mathcal{M}$ corresponding to any real eigenvalue λ_0 are exactly the halves of the partial multiplicities of A corresponding to λ_0.*

Proof. By Theorems 2.6.1 and 1.4.2 the analysis is reduced to two cases: 1) A has a single real eigenvalue; 2) A has only one pair of non real complex conjugate numbers as eigenvalues. In the case 2) the result follows from Lemma 2.6.2. So assume the case 1) holds. The statement (i) is proved in the same way as Theorem 2.4.1 (the complex analogue). Similarly, the statement (ii) is proved as in the complex case (i.e. the proof of Theorem I.3.21 in Gohberg et al. (1983)). □

2.7 H-skew-symmetric matrices

As in the preceding section, we consider the indefinite scalar product $[x,y] = \langle Hx, y\rangle$ defined in $\mathbb{R}^n$ by a real invertible symmetric $n \times n$ matrix H. An $n \times n$ real matrix A is called *H-skew-symmetric* if

$$[Ax, y] = -[x, Ay]$$

for all $x, y \in \mathbb{R}^n$. This equality can be expressed in the form $HA = -A^T H$.

To state the canonical form for H-skew-symmetric matrices under the transformation $A \to S^{-1}AS$, $H \to S^T HS$, we introduce the following notation:

$$F_j = \begin{bmatrix} 0 & & 1 \\ & \reflectbox{$\ddots$} & -1 \\ (-1)^{j-1} & & 0 \end{bmatrix}, \qquad (2.7.1)$$

so that F_j is a $j \times j$ matrix which is symmetric if j is odd and skew-symmetric if j is even;

$$G_{2j} = \begin{bmatrix} 0 & & F_2^{j-1} \\ & -F_2^{j-1} & \\ (-1)^{j-1}F_2^{j-1} & & 0 \end{bmatrix}, \qquad (2.7.2)$$

so that G_{2j} is a $2j \times 2j$ matrix which is symmetric for all j. Recall also the notation for real Jordan blocks introduced in Section 1.4.

Theorem 2.7.1. *Given an H-skew-symmetric matrix A there exists an invertible real matrix S such that*

$$S^{-1}AS = \bigoplus_{i=1}^{m} A_i, \qquad S^T HS = \bigoplus_{i=1}^{m} H_i,$$

where each pair of matrices (A_i, H_i) has one of the following four types, either

$$\left. \begin{aligned} A_i &= \bigoplus_{j=1}^{p} J_{2n_j+1}(0) \oplus \bigoplus_{j=1}^{q} (J_{n_{p+j}}(0) \oplus -J_{n_{p+j}}(0)^T), \\ H_i &= \bigoplus_{j=1}^{p} \epsilon_j F_{2n_j+1} \oplus \bigoplus_{j=1}^{q} \begin{bmatrix} 0 & I_{n_{p+j}} \\ I_{n_{p+j}} & 0 \end{bmatrix} \end{aligned} \right\} \qquad (2.7.3)$$

where $n_{p+1}, \ldots, n_{p+q}$ are even integers and $\epsilon_1, \ldots, \epsilon_p$ are equal to ± 1, or

$$A_i = \bigoplus_{j=1}^{p}[J_{n_j}(a) \oplus -J_{n_j}(a)^T], \quad H_i = \bigoplus_{j=1}^{p}\begin{bmatrix} 0 & I_{n_j} \\ I_{n_j} & 0 \end{bmatrix} \qquad (2.7.4)$$

where $a > 0$, or

$$A_i = \bigoplus_{j=1}^{p} J_{2n_j}(\pm ib), \quad H_i = \bigoplus_{j=1}^{p} \epsilon_j G_{2n_j} \qquad (2.7.5)$$

where $b > 0$ and $\epsilon_1, \ldots, \epsilon_p$ are equal to ± 1, or

$$A_i = \bigoplus_{j=1}^{p}(J_{2n_j}(a \pm ib) \oplus -(J_{2n_j}(a \pm ib))^T), \quad H_i = \bigoplus_{j=1}^{p}\begin{bmatrix} 0 & I_{2n_j} \\ I_{2n_j} & 0 \end{bmatrix} \qquad (2.7.6)$$

where $a, b > 0$.

(Of course, the numbers a, b as well as the signs ϵ_j and the numbers p, q may be different in different pairs of matrices (A_i, H_i).)

Moreover this form is uniquely determined by the matrices A and H, up to simultaneous permutations of pairs of blocks in A_i, H_i.

Again, the collection of signs ϵ_j from (2.7.3) (for every Jordan block of odd size having eigenvalues 0 in the Jordan form of A) and from (2.7.5) (for every real Jordan block with a pair $\pm ib$ of pure imaginary nonzero eigenvalues) will be called the *sign characteristic* of (A, H).

In the sequel there will be special interest in subspaces of $\mathbb{R}^n$ which are A-invariant and H-neutral and also maximal subspaces of this type (i.e. maximal with respect to inclusion). It can be shown (see Lancaster and Rodman (1994)) that all such subspaces have the same dimension and this dimension can be specified in terms of the parameters appearing in the canonical form of Theorem 2.7.1. The argument is technical and will not be reproduced here.

To state the appropriate result it is convenient to introduce some notation. With (A, H) as in Theorem 2.7.1 let $ib_1, \ldots, ib_r$ be all the nonzero pure imaginary eigenvalues of A in the upper half-plane, and let ib_k ($k = 1, 2, \ldots, r$) have partial multiplicities $\{n_j\}$ and the associated members of the sign characteristic $\{\epsilon_j\}$ (as in equation (2.7.5)). Then, for $k = 1, 2, \ldots, r$, define

$$\delta_k = \left|\sum \epsilon_j\right| \qquad (2.7.7)$$

where the summation is over those j for which n_j is odd.

Theorem 2.7.2. *Let $H = H^T$ be invertible and $HA = -A^T H$. Then all maximal A-invariant H-neutral subspaces of $\mathbb{R}^n$ have the same dimension.*

If the distinct nonzero pure imaginary eigenvalues of A are $\pm ib_k$, $k = 1, 2, \ldots, r$, then this dimension is

$$\frac{1}{2}n - \sum_{k=1}^{r}\delta_k - \frac{1}{2}\left|\sum_{j=1}^{p}\epsilon_j \kappa_j\right| \qquad (2.7.8)$$

where $\epsilon_1, \ldots, \epsilon_p$ are defined as in (2.7.3) and $\kappa_j = +1$ or -1 according as $2n_j + 1$ (in (2.7.3)) is an odd integer of the form $4k + 1$ or $4k - 1$, respectively.

Note that, if A is invertible then $0 \notin \sigma(A)$ and the last term of (2.7.8) does not appear. Also, if A has no eigenvalues at all on the imaginary axis, then there is a maximal A-invariant and H-neutral subspace of dimension $\frac{1}{2}n$. When $0 \in \sigma(A)$ it is easily seen that n and p have the same parity. It follows from this that the dimension of (2.7.8) is, indeed, an integer.

"Signed" subspaces will also be of interest. In particular, we have:

Theorem 2.7.3. *Let $H = H^T$ be invertible and assume there is an invertible A such that $HA = -A^T H$. Then the dimensions of maximal H-nonnegative and H-nonpositive subspaces are*

$$\frac{1}{2}n + \sum \epsilon_j, \quad \frac{1}{2}n - \sum \epsilon_j, \qquad (2.7.9)$$

respectively, where the summations are over those ϵ_j in the sign characteristic of (A, H) associated with odd partial multiplicities. Furthermore, there are A-invariant subspaces with the dimensions of (2.7.9) which are H-nonnegative and H-nonpositive, respectively.

Proof. We sketch the proof. Note first of all that, because the product HA is invertible and skew-symmetric, n is even and so the numbers (2.7.9) are well-defined. Note that, in the canonical form of Theorem 2.7.1, the terms (2.7.3) do not appear. Hence, evaluation of the inertia of H depends on the real symmetric matrices $\epsilon_j G_{2n_j}$ (of size $2n_j$) appearing in (2.7.5).

It is easily verified that if n_j is even then G_{2n_j} has equal numbers of positive and negative eigenvalues. When n_j is odd the number of positive eigenvalues of G_{2n_j} exceeds the number of negative eigenvalues by two. Recalling Theorem 2.1.2, these facts lead readily to the conclusion of (2.7.9).

Finally, the canonical form for the pair (A, H) of (2.7.4–6) can be used to establish the existence of the corresponding subspaces. □

2.8 H-self-adjoint pencils

Let H be an invertible $n \times n$ hermitian matrix and, as in Section 2.1 define a corresponding indefinite scalar product $[.,.]$ on $\mathbb{C}^n$ by $[x, y] = \langle Hx, y \rangle$ for all $x, y \in \mathbb{C}^n$. A matrix pencil $\lambda A - B$ will be called *H-self-adjoint* if

$$A^* H B = B^* H A. \qquad (2.8.1)$$

Notice that this provides a natural generalization of H-self-adjoint matrices, as defined in Section 2.2 because, when $A = I$, (2.8.1) implies that B is an H-self-adjoint matrix. Our first result, together with Corollary 1.6.8, leads immediately to the conclusion that, like H-self-adjoint matrices, the spectrum of a regular H-self-adjoint pencil is symmetric with respect to the real line (cf. Theorem 2.2.2).

H-SELF-ADJOINT PENCILS

Theorem 2.8.1. *If $\lambda A - B$ is a regular H-self-adjoint pencil then it is strictly equivalent to the pencil $\lambda A^* - B^*$.*

Proof. Using Theorem 1.6.7, we know there are Jordan matrices J and K, with K nilpotent, such that

$$(\lambda A - B)Q^{-1} = P \begin{bmatrix} \lambda I - J & 0 \\ 0 & \lambda K - I \end{bmatrix} \qquad (2.8.2)$$

for some nonsingular matrices P and Q. Form the partition $P = [P_1 \ P_0]$ in such a way that the products $P_1 J$ and $P_0 K$ exist. Then it follows from equation (2.8.2) that

$$AQ^{-1} = [P_1 \ P_0 K], \qquad BQ^{-1} = [P_1 J \ P_0].$$

Multiplying on the left with $Q^{-*}B^*H$ and $Q^{-*}A^*H$, respectively, yields

$$Q^{-*}(B^*HA)Q^{-1} = Q^{-*}[B^*HP_1 \ B^*HP_0 K],$$

and

$$Q^{-*}(A^*HB)Q^{-1} = Q^{-*}[A^*HP_1 J \ A^*HP_0].$$

But now the defining property (2.8.1) implies that

$$A^*HP_1 J = B^*HP_1, \qquad A^*HP_0 = B^*HP_0 K. \qquad (2.8.3)$$

Now we have

$$[A^*HP_1 \ B^*HP_0] \begin{bmatrix} \lambda I - J & 0 \\ 0 & \lambda K - I \end{bmatrix}$$
$$= \lambda [A^*HP_1 \ B^*HP_0 K] - [A^*HP_1 J \ B^*HP_0]$$

and, using equations (2.8.3),

$$[A^*HP_1 \ B^*HP_0] \begin{bmatrix} \lambda I - J & 0 \\ 0 & \lambda K - I \end{bmatrix} = (\lambda A^* - B^*)H[P_1 \ P_0]$$
$$= (\lambda A^* - B^*)HP. \qquad (2.8.4)$$

Because $\lambda A - B$ is regular, it follows that the two matrix pencils in this equation must also be regular and, consequently, $[A^*HP_1 \ B^*HP_0]$ is nonsingular. Thus, equation (2.8.4) determines a strict equivalence and the result follows. $\square$

Since $\lambda A - B$ and $\lambda A^* - B^*$ have the same canonical matrices J and K of equation (2.8.2) the following properties of their spectra are obtained:

Corollary 2.8.2. *Let $\lambda A - B$ be an H-self-adjoint pencil. Then:*

(a) *If λ_0 is an eigenvalue of $\lambda A - B$ then so is $\bar{\lambda}_0$, and $\lambda_0, \bar{\lambda}_0$ have the same partial multiplicities.*

(b) *If $\det A = 0$ then $\lambda A - B$ and $\lambda A^* - B^*$ have an eigenvalue at infinity with the same partial multiplicities.*

2.9 H-unitary pencils

As in the preceding section, let H be an invertible hermitian matrix which generates the indefinite scalar product $[.,.]$ on $\mathbb{C}^n$. A matrix pencil $\lambda L - M$ is said to be *H-unitary* if

$$L^* H L = M^* H M. \tag{2.9.1}$$

If $L = I$ then we observe that M is an H-unitary matrix, so we are continuing with generalizations of the ideas developed in Section 2.2.

Cayley transforms can still be used to connect the properties of H-self-adjoint and H-unitary pencils. Thus, using the same inverse functions of equations (2.2.9) and (2.2.10), an extension of Proposition 2.2.3 can be formulated:

Proposition 2.9.1. (i) *If $\lambda A - B$ is an H-self-adjoint pencil and $w, \alpha \in \mathbb{C}$ with $w \neq \bar{w}$, $|\alpha| = 1$,*

$$L = wA - B, \qquad M = \alpha(\bar{w}A - B), \tag{2.9.2}$$

then $\lambda L - M$ is an H-unitary pencil.
(ii) *If $\lambda L - M$ is an H-unitary pencil, $w \neq \bar{w}$, $|\alpha| = 1$, and*

$$A = \alpha L - M \qquad B = \bar{w}\alpha L - wM \tag{2.9.3}$$

then $\lambda A - B$ is an H-self-adjoint pencil.

The proof is a simple verification which is left to the reader. Observe that the Cayley transform seems to be well suited to this context because, in contrast to the classical eigenvalue problem, no inversions are required in its use. Another important observation is:

Proposition 2.9.2. *The pencils $\lambda A - B$ and $\lambda L - M$ of the first (or the second) statement of Proposition* 2.9.1 *have the same lattice of deflating subspaces.*

Again, this is a simple verification; it follows readily from the definition of a deflating subspace (see Section 1.6).

We turn now to an analogue of Theorem 2.8.1.

Theorem 2.9.3. *Let $\lambda L - M$ be a regular H-unitary pencil. Then $\lambda L - M$ and $\lambda M^* - L^*$ are strictly equivalent.*

Proof. Using Theorem 1.6.7 we may write

$$(\lambda L - M)P = T \begin{bmatrix} \lambda I - J & 0 \\ 0 & \lambda K - I \end{bmatrix} \tag{2.9.4}$$

for some invertible matrices P and T, and J, K are Jordan matrices with K nilpotent. Partition T conformally with (2.9.4) in the form $T = [T_1 \ T_0]$. Then

$$MP = [T_1 J \ T_0], \qquad LP = [T_1 \ T_0 K]$$

and consequently

$$M^*HM = [M^*HT_1 J \ M^*HT_0]P^{-1}, \qquad L^*HL = [L^*HT_1 \ L^*HT_0 K]P^{-1}.$$

The defining property (2.9.1) now gives

$$M^*HT_1 J = L^*HT_1, \qquad M^*HT_0 = L^*HT_0 K,$$

and therefore

$$[M^*HT_1 \ L^*HT_0]\begin{bmatrix} \lambda I - J & 0 \\ 0 & \lambda K - I \end{bmatrix} = (\lambda M^* - L^*)H[T_1 \ T_0],$$
$$= (\lambda M^* - L^*)T. \qquad (2.9.5)$$

Taking determinants and using the regularity of $\lambda L - M$, it follows that

$$\det[M^*HT_1 \ L^*HT_0] \ne 0,$$

and the result follows from equation (2.9.5). □

Important properties of the spectrum of H-unitary pencils follow immediately from the theorem.

Corollary 2.9.4. *Let $\lambda L - M$ be a regular H-unitary pencil. Then:*

(a) *If λ_0 is a nonzero eigenvalue of $\lambda L - M$, then so is $(\bar{\lambda}_0)^{-1}$, and $\lambda_0, (\bar{\lambda}_0)^{-1}$ have the same partial multiplicities.*

(b) *If $\lambda L - M$ has the zero eigenvalue then $\lambda L - M$ also has an eigenvalue at infinity and the zero and infinite eigenvalues have the same partial multiplicities.*

It will be useful to examine a second equivalence relation on unitary pencils; but one which connects pencils which may be unitary in different (but congruent) indefinite scalar products. Thus, for $j = 1, 2$ let $\lambda L_j - M_j$ be H_j-unitary pencils. Such a pair of pencils is said to be H_1, H_2-*unitarily equivalent* if there exist invertible matrices P and Q such that

$$\lambda L_1 - M_1 = P(\lambda L_2 - M_2)Q \qquad (2.9.6)$$

and

$$H_1 = (P^{-1})^* H_2 P^{-1}. \qquad (2.9.7)$$

Equation (2.9.6) is, of course, a strict equivalence so that unitarily equivalent pencils share the same eigenvalues and partial multiplicities. Equation (2.9.7) may be interpreted as saying that P^{-1} is (H_1, H_2)-unitary. Of course, the matrix P^{-1} is H-unitary if $H_1 = H_2 = H$. Notice that for unitary pencils to be equivalent in this sense it is necessary that H_1 and H_2 be congruent. In particular, H_1 and H_2 must have the same number of positive (or negative) eigenvalues, counted with multiplicities. For an H-unitary pencil $\lambda L - M$ this notion provides a context for the simultaneous reduction of L, M and H.

Theorem 2.9.5. *Let $\lambda L - M$ be a regular H-unitary pencil with a zero eigenvalue of algebraic multiplicity m (i.e. $\det(\lambda L - M)$ has a zero of order m at the origin). Then $\lambda L - M$ is (H, G)-unitarily equivalent to a G-unitary pencil of the form*

$$\lambda \begin{bmatrix} I & 0 & 0 \\ 0 & I_m & 0 \\ 0 & 0 & N \end{bmatrix} - \begin{bmatrix} S & 0 & 0 \\ 0 & N^* & 0 \\ 0 & 0 & I_m \end{bmatrix} \qquad (2.9.8)$$

where N is an $m \times m$ nilpotent matrix and the invertible hermitian matrix G has the form

$$G = \begin{bmatrix} G_0 & 0 & 0 \\ 0 & 0 & iI_m \\ 0 & -iI_m & 0 \end{bmatrix}. \qquad (2.9.9)$$

Moreover, the nilpotent matrix N is unique up to similarity and the G_0-unitary pencil $\lambda I - S$ is unique up to unitary equivalence.

We will not give a proof of this theorem here, but refer the reader to a paper of Wimmer (1991) where a proof can be found (see the proof of Theorem 2.1 there). It should be noted that a pencil $\lambda L - M$ is called symplectic in the paper of Wimmer if the pencil $\lambda L^* - M^*$ is H-unitary according to our definition.

The G_0-unitary pencil $\lambda I - S$ of Theorem 2.8.5 can, in turn, be reduced to the canonical form described in Theorem 2.3.3. Indeed, we have $G_0 = S^* G_0 S$ (by definition of the G_0-unitary property of $\lambda I - S$). Therefore S is G_0-unitary and, by Theorem 2.3.3, there exists an invertible matrix Q such that

$$Q^* G_0 Q = H_0; \qquad Q^{-1} S Q = K_0,$$

where K_0 and H_0 are given by (2.3.5) and (2.3.6), respectively. With $P = Q^{-1}$ we now obtain

$$\lambda I - K_0 = P(\lambda I - S)Q \quad \text{and} \quad H_0 = (P^{-1})^* G_0 P^{-1}.$$

Thus, $\lambda I - K_0$ and $\lambda I - S$ are (H_0, G_0)-unitarily equivalent and, in Theorem 2.9.5 we may use K_0 and H_0 in place of S and G_0, respectively. The uniqueness statement in Theorem 2.3.3 guarantees that the pairs of blocks in K_0 and H_0 are unique up to permutation. Thus, the *sign characteristic* of a regular H-unitary pencil $\lambda L - M$ is defined via Theorem 2.3.3 and the reduction of the G_0-unitary pencil $\lambda I - S$ to the canonical form given by Theorem 2.3.3. The concepts based

H-UNITARY PENCILS

on sign characteristic and developed in Section 2.4 for H-self-adjoint matrices (such as the sign condition used in Theorem 2.4.2) can now be extended in the obvious way to regular H-unitary pencils.

The next theorem shows that the deflating and H-positive (or H-negative) subspaces transform well under unitary equivalence.

Theorem 2.9.6. *Let $\lambda L_1 - M_1$ and $\lambda L_2 - M_2$ be H_1- and H_2-unitary pencils, respectively, and assume that $\lambda L_1 - M_1$, and $\lambda L_2 - M_2$ are (H_1, H_2)-unitarily equivalent, so that (2.9.6) and (2.9.7) hold. Then:*

(i) *A subspace $\mathcal{S} \subseteq \mathbf{C}^n$ is deflating for (L_2, M_2) if and only if $Q^{-1}\mathcal{S}$ is deflating for (L_1, M_1).*

(ii) *A subspace $\mathcal{S} \subseteq \mathbf{C}^n$ is a spectral deflating subspace for (L_2, M_2) associated with the eigenvalue $\lambda_0 \in \mathbf{C} \cup \{\infty\}$ if and only if $Q^{-1}\mathcal{S}$ is a spectral deflating subspace for (L_1, M_1) associated with the same λ_0.*

(iii) *A subspace $\mathcal{S}$ is H_2-neutral if and only if $P\mathcal{S}$ is H_1-neutral. This statement is also valid when "neutral" is replaced by any one of the adjectives: "positive", "negative", "nonpositive", "nonnegative", "nondegenerate".*

(iv) *Subspaces $\mathcal{S}, \mathcal{T} \subseteq \mathbf{C}^n$ are H_2-orthogonal (i.e. $\langle H_2 x, y \rangle = 0$ for all $x \in \mathcal{S}$, $y \in \mathcal{T}$) if and only if $P\mathcal{S}, P\mathcal{T}$ are H_1-orthogonal.*

Proof. Parts (iii) and (iv) follow easily from equation (2.9.7). For the proof of part (ii) we use the formula (1.6.8) (assuming $\lambda_0 \in \mathbf{C}$). Let Γ be a circle of sufficiently small radius centered at λ_0. Then

$$\frac{1}{2\pi i} \int_\Gamma (\lambda L_1 - M_1)^{-1} L_1 \, d\lambda = Q^{-1} \left[\frac{1}{2\pi i} \int_\Gamma (\lambda L_2 - M_2)^{-1} L_2 \, d\lambda \right] Q,$$

and (ii) follows for $\lambda_0 \in \mathbf{C}$. If $\lambda_0 = \infty$, we apply the already proved part of (ii) to the pencils $L_1 - \lambda M_1$ and $L_2 - \lambda M_2$.

Finally, we prove (i). Let $\mathcal{S}$ be deflating for (L_2, M_2). Then there is a subspace $\mathcal{T}$ having the same dimension as $\mathcal{S}$ and such that

$$L_2 \mathcal{S} \subseteq \mathcal{T}, \qquad M_2 \mathcal{S} \subseteq \mathcal{T}. \tag{2.9.10}$$

These inclusions imply (in view of the relations $L_1 = PL_2Q$, $M_1 = PM_2Q$)

$$L_1(Q^{-1}\mathcal{S}) \subseteq P\mathcal{T}, \qquad M_1(Q^{-1}\mathcal{S}) \subseteq P\mathcal{T}. \tag{2.9.11}$$

So $Q^{-1}\mathcal{S}$ is deflating for (L_1, M_1). Conversely, (2.9.11) implies (2.9.10), and (i) is proved. □

Using Theorem 2.9.5 and the canonical form of the G_0-unitary pencil $\lambda I - S$ (see the remarks after Theorem 2.9.5), as well as parts (ii) and (iii) of Theorem 2.9.6, the following corollary is easily obtained.

Corollary 2.9.7. Let σ_1 and σ_2 be sets of eigenvalues of a regular H-unitary pencil $\lambda L - M$ such that

$$\sigma_1 \cap \{\bar{\lambda}^{-1} | \lambda \in \sigma_2\} = \emptyset$$

(we formally put $\bar{0}^{-1} = \infty$ and $\bar{\infty}^{-1} = 0$). Then the sum of the spectral deflating subspaces of $\lambda L - M$ corresponding to the eigenvalues in σ_1 is H-orthogonal to the sum of the spectral deflating subspaces corresponding to σ_2. In particular, if σ is a set of eigenvalues of $\lambda L - M$ such that

$$\sigma \cap \{\bar{\lambda}^{-1} | \lambda \in \sigma\} = \emptyset,$$

then the sum of spectral deflating subspaces corresponding to σ is H-neutral.

2.10 Exercises

2.10.1. Verify that $L_{2r}(\mu_1, \mu_2)$ given by (2.3.4) is unitary with respect to the $2r$-sip matrix. Hint: Write $L_{2r}(\mu_1, \mu_2) = K_1 \oplus K_2$ as in (2.3.4). Check first that $K_2 = \bar{K}_1^{-1}$.

2.10.2. Show that the set of H-self-adjoint matrices forms a real vector space of dimension $\frac{1}{2}[p(p+1) + q(q+1)] + 2pq$, where H has p (resp. q) positive (resp. negative) eigenvalues.

2.10.3. Assume H is hermitian and $H^2 = I$. Prove that a matrix A is H-unitary if and only if A^* is H-unitary.

2.10.4. Show that the set of H-unitary matrices is a multiplicative group which is unbounded, i.e. contains matrices of arbitrary large norm (unless H is positive definite or negative definite).

2.10.5. Find all invariant maximal nonnegative subspaces for the following pairs (A, H), where A is H-self-adjoint:

(a) $A = J_2(0) \oplus J_3(0) \oplus J_2(1)$;

$$H = \begin{bmatrix} 0 & 1 \\ 1 & 0 \end{bmatrix} \oplus \begin{bmatrix} 0 & 0 & -1 \\ 0 & -1 & 0 \\ -1 & 0 & 0 \end{bmatrix} \oplus \begin{bmatrix} 0 & -1 \\ -1 & 0 \end{bmatrix}.$$

(b) $A = \begin{bmatrix} J_2(i) & 0 \\ 0 & J_2(-i) \end{bmatrix}; H = \begin{bmatrix} 0 & 0 & 0 & 1 \\ 0 & 0 & 1 & 0 \\ 0 & 1 & 0 & 0 \\ 1 & 0 & 0 & 0 \end{bmatrix}.$

2.10.6. Provide the details of the proof of Theorem 2.6.3.

2.10.7. Give an example showing that the property of Exercise 2.10.2 does not extend to regular H-unitary pencils, i.e. there exist regular H-unitary pencils $\lambda L - M$, where $H^2 = I$, such that the pencil $\lambda L^* - M^*$ is not H-unitary. (Hint: Consider (15.1.3)).

2.10.8. Consider a self-adjoint matrix polynomial of the form $L(\lambda) = \sum_{j=0}^{l} \lambda^j A_j$ with $A_j^* = A_j$, $j = 0, 1, 2, \ldots, l$ and A_l invertible. If $\hat{A}_j = A_l^{-1} A_j$, $j = 0, 1, 2, \ldots,$ and

$$H = \begin{bmatrix} A_1 & A_2 & \cdots & A_{l-1} & A_l \\ A_2 & & & A_l & 0 \\ \vdots & & & & \\ A_{l-1} & A_l & & & \\ A_l & 0 & & & 0 \end{bmatrix}, \quad C = \begin{bmatrix} 0 & I & 0 \cdots & 0 \\ 0 & 0 & I & \vdots \\ \vdots & & & \ddots & 0 \\ 0 & & & & I \\ -\hat{A}_0 & -\hat{A}_1 & \cdots & & -\hat{A}_{l-1} \end{bmatrix}$$

show that the companion matrix C is H-self-adjoint and find the signature of H in terms of the signature of A_l. (The self-adjoint pencil $\lambda H - HC$ is a "linearization" of $L(\lambda)$. See Chapter II.2 of Gohberg et al. (1983).)

2.10.9. Consider a regular self-adjoint matrix polynomial of the form $L(\lambda) = \lambda^2 A_2 + \lambda A_1 + A_0$. Thus, $\det L(\lambda) \not\equiv 0$, $A_2^* = A_2$, $A_1^* = A_1$ and $A_0^* = A_0$. (Note that A_2 may be singular.) Define

$$A = \begin{bmatrix} I & 0 \\ 0 & A_2 \end{bmatrix}, \quad B = \begin{bmatrix} 0 & I \\ -A_0 & -A_1 \end{bmatrix}.$$

It is known that $\lambda A - B$ is a "linearization" of $L(\lambda)$. (See Chapter 7 of Gohberg et al. (1982a).) Show that $\lambda A - B$ is a regular H-self-adjoint pencil where

$$H = \begin{bmatrix} A_1 & I \\ I & 0 \end{bmatrix}.$$

Is there a generalization of this result to regular self-adjoint matrix polynomials of higher degree (with singular leading coefficient)?

2.10.10. Consider a matrix polynomial of the form $L(\lambda) = \sum_{j=0}^{2k} \lambda^j A_j$ where $A_j^* = A_{2k-j}$, $j = 0, \ldots, 2k$ and A_{2k} is invertible. Show that if $\hat{L}(\lambda) = \lambda^{-k} L(\lambda)$ then $|\lambda| = 1$ implies $\hat{L}(\lambda)^* = \hat{L}(\lambda)$. Let C be the companion matrix of L (as defined in Exercise 2.10.8 above) and show that C is G-unitary where

$$G = i \begin{bmatrix} 0 & G_1 \\ -G_1^* & 0 \end{bmatrix}$$

and G_1 is the block lower-triangular Toeplitz matrix

$$G_1 = \begin{bmatrix} A_{2k} & 0 & \cdots & 0 \\ \vdots & \ddots & \ddots & \vdots \\ A_{k+2} & & & 0 \\ A_{k+1} & A_{k+2} & \cdots & A_{2k} \end{bmatrix}.$$

2.10.11. Let q be a rational function of the form

$$q(z) = a_0 + \sum_{j=1}^{k}(a_j z^j + \bar{a}_k z^{-j}).$$

Show that, if U is H-unitary, then $q(U)$ is H-self-adjoint.

2.11 Notes

The text of Malcev (1963) (translation of the 1956 Russian original) contains an early systematic discussion of canonical forms for H-self-adjoint matrices. A more recent development, on which much of this chapter is based, can be found in the monograph of Gohberg et al. (1983). Theorem 2.4.2 is more recent and was obtained by Ran and Rodman (1984c). The real canonical form of Theorem 2.6.1 was obtained by Thompson (1990). Lemma 2.6.2 and Theorem 2.6.3 are new. Results related to Theorems 2.6.3 and 2.7.2 concerning the dimension of maximal invariant neutral subspaces have been obtained by Lancaster et al. (1994) and Lancaster and Rodman (1994). The material on pencils in the last two sections has evolved in recent years. The notion of "H-self-adjoint pencils" introduced in Section 2.8 seems to be new. Theorems 2.9.3 and 2.9.5 are due to Wimmer (1991). The remainder of this section appears here for the first time.

3
SKEW-SYMMETRIC SCALAR PRODUCTS

Chapter 2 was concerned with complex spaces having an indefinite scalar product. The detailed study of Riccati equations with real coefficients requires that we also consider scalar products on $\mathbb{R}^n$ generated by real, invertible, skew-symmetric matrices; a further step away from the familiar euclidean situation.

Nevertheless, the connections with the indefinite scalar products introduced in Chapter 2 are strong. For example, if K and A are real $n \times n$ matrices with K nonsingular and skew-symmetric, A will be said to be skew-symmetric with respect to K if KA is symmetric (see Section 3.2 below). On the other hand, iK is hermitian and because $(iK)(iA)$ is also real and symmetric, it follows that iA is iK-self-adjoint in the sense of Section 2.2. Thus, the theory of Chapter 2 applies as well, but the investigations of this chapter in terms of real matrices yield more detailed information in situations relevant to our subsequent study of real Riccati equations.

3.1 Definitions and basic properties

In this chapter we study scalar products $[.,.]$ on $\mathbb{R}^n$ that are skew-symmetric i.e. for which $[x, y] = -[y, x]$ for all x and y. The exact definition is as follows. A function $[.,.]$ from $\mathbb{R}^n \times \mathbb{R}^n$ to $\mathbb{R}$ is called a *skew-symmetric scalar product* if the following axioms are satisfied:

(i) Linearity:
$$[\alpha x_1 + \beta x_2, y] = \alpha[x_1, y] + \beta[x_2, y]$$
for all $x_1, x_2, y \in \mathbb{R}^n$ and all real numbers α, β;

(ii) skew-symmetry:
$$[x, y] = -[y, x]$$
for all $x, y \in \mathbb{R}^n$;

(iii) nondegeneracy: if $[x, y] = 0$ for all $y \in \mathbb{R}^n$, then $x = 0$.

In particular, (i) and (ii) imply that $[.,.]$ is linear with respect to the second component as well:

$$[x, \alpha y_1 + \beta y_2] = \alpha[x, y_1] + \beta[x, y_2]$$

for all $x, y_1, y_2 \in \mathbb{R}^n$ and all $\alpha, \beta \in \mathbb{R}$.

We could also define a skew-symmetric scalar product in $\mathbb{C}^n$, in which case axiom (ii) takes the form $[x, y] = -\overline{[y, x]}$ for all $x, y \in \mathbb{C}^n$. However, there is nothing essentially new here when compared with the indefinite scalar products studied in Chapter 2. Indeed, it is easy to see that a skew-symmetric scalar product in $\mathbb{C}^n$ is of the form

$$[x, y] = i\{x, y\}, \quad x, y \in \mathbb{C}^n,$$

where $\{x, y\}$ is an indefinite scalar product in $\mathbb{C}^n$. In contrast, the skew-symmetric scalar products in $\mathbb{R}^n$ have properties distinct from those of indefinite scalar products in $\mathbb{R}^n$, as we shall see in this chapter.

Observe first of all that in a skew-symmetric scalar product $[.,.]$ we have

$$[x, x] = 0 \qquad (3.1.1)$$

for every $x \in \mathbb{R}^n$.

Let us translate the notion of skew-symmetric scalar products into the language of matrices. It is easy to see that if $[.,.]$ is a skew-symmetric scalar product in $\mathbb{R}^n$, then there exists a unique invertible real skew-symmetric matrix H such that

$$[x, y] = \langle Hx, y \rangle, \qquad (3.1.2)$$

where as usual $\langle .,. \rangle$ is the standard scalar product in $\mathbb{R}^n$. (Recall that an $n \times n$ matrix H is called *skew-symmetric* if $H = -H^T$.) Conversely, every invertible real skew-symmetric matrix H defines a skew-symmetric scalar product by the formula (3.1.2). Thus, real skew-symmetric matrices and their properties will play a key role in this chapter. Some of their basic properties are given in the next proposition.

Proposition 3.1.1. *Let H be an $n \times n$ real skew-symmetric matrix. Then:*

(a) *H is diagonable, i.e. the partial multiplicities corresponding to any eigenvalue are all equal to 1.*

(b) *The eigenvalues of H are on the imaginary axis and symmetric relative to the real axis; i.e. if $\lambda_0 \neq 0$ is an eigenvalue of H, then the real part of λ_0 is zero, and the number $\bar{\lambda}_0 = -\lambda_0$ is also an eigenvalue of H having the same algebraic multiplicity as λ_0.*

(c) *The rank of H is even.*

DEFINITIONS AND BASIC PROPERTIES

(d) *There exists an invertible real matrix S such that $S^T H S$ has the form*

$$S^T H S = \begin{bmatrix} 0 & 1 \\ -1 & 0 \end{bmatrix} \oplus \cdots \oplus \begin{bmatrix} 0 & 1 \\ -1 & 0 \end{bmatrix} \oplus 0, \qquad (3.1.3)$$

where the block $\begin{bmatrix} 0 & 1 \\ -1 & 0 \end{bmatrix}$ *appears* $\frac{1}{2}($rank $H)$ *times.*

In connection with part (d) observe that if H is skew-symmetric, then $S^T H S$ is skew-symmetric for any invertible real matrix S. Also, if H is invertible, then n is even.

Proof. (a) Observe that the matrix iH is hermitian and therefore diagonable. Consequently, H is diagonable as well.

(b) As iH is hermitian, all eigenvalues of iH are real. But it is easy to see that λ_0 is an eigenvalues of iH if and only if $-i\lambda_0$ is an eigenvalue of H. Thus, all eigenvalues of H have zero real parts. The symmetry of eigenvalues relative to the real axis is a general property of every real matrix, in particular H has this property.

(c) By (b), the sum of algebraic multiplicities of all nonzero eigenvalues of H is even; but this sum coincides with the rank of H (as one can verify using the Jordan form of H).

(d) If $H = 0$, then any real invertible S satisfies (3.1.2). So assume $H \neq 0$. Then there exist $x_0, y_0 \in \mathbb{R}^n$ such that $\langle Hx_0, y_0 \rangle \neq 0$. Multiplying x_0 by a nonzero number (if necessary) we can assume $\langle Hx_0, y_0 \rangle = 1$. Observe that because of (3.1.1) the vectors x_0 and y_0 are linearly independent. Now choose a real invertible S such that

$$Se_1 = x_0, \quad Se_2 = y_0. \qquad (3.1.4)$$

Then $\langle S^T H S e_1, e_2 \rangle = 1$, and since

$$\langle S^T H S e_1, e_1 \rangle = -\langle S^T H S e_2, e_2 \rangle = 0, \quad \langle S^T H S e_2, e_1 \rangle = -\langle S^T H S e_1, e_2 \rangle,$$

the top left 2×2 corner of $S^T H S$ has the form $\begin{bmatrix} 0 & 1 \\ -1 & 0 \end{bmatrix}$.

Consider now the subspace

$$\mathcal{M}_1 = \{x \in \mathbb{R}^n | \langle Hx, y \rangle = 0 \quad \text{for all } y \in \text{span}\{x_0, y_0\}\}.$$

Clearly, $\mathcal{M}_1 = \text{span}\{Hx_0, Hy_0\}^\perp$ and therefore $\dim \mathcal{M}_1 = n - 2$. It is also claimed that

$$\mathcal{M}_1 \cap \text{span}\{x_0, y_0\} = \{0\}.$$

Indeed, if $x' = \alpha x_0 + \beta y_0$ belongs to the intersection on the left-hand side, then

$$0 = \langle H(\alpha x_0 + \beta y_0), x_0 \rangle = -\beta, \quad 0 = \langle H(\alpha x_0 + \beta y_0), y_0 \rangle = \alpha,$$

and so $x' = 0$. Thus, $\mathcal{M}_1$ is a direct complement to $\text{span}\{x_0, y_0\}$.

If $\langle Hx, y\rangle = 0$ for all $x, y \in \mathcal{M}_1$, then any real invertible S satisfying (3.1.3) will do for the proof of (d). Otherwise, let $x_1, y_1 \in \mathcal{M}_1$ be such that $\langle Hx_1, y_1\rangle = 1$. In addition to (3.1.4), S must also satisfy $Se_3 = x_1$, $Se_4 = y_1$. Let

$$\mathcal{M}_2 = \{x \in \mathbb{R}^n | \langle Hx, y\rangle = 0 \quad \text{for all } y \in \text{span}\{x_0, y_0, x_1, y_1\}\}.$$

Verify that $\mathcal{M}_2$ is a direct complement to $\text{span}\{x_0, y_0, x_1, y_1\}$, and (if $\langle Hx, y\rangle \neq 0$ for some $x, y \in \mathcal{M}_2$) choose $x_2, y_2 \in \mathcal{M}_2$ so that $\langle Hx_2, y_2\rangle = 1$, and require that $Se_5 = x_2$, $Se_6 = y_2$. Continuing this process, we eventually find a real invertible S such that (3.1.3) holds. Clearly, the rank of the right-hand side in (3.1.3) is equal to $2q$, where q is the number of appearances of the block $\begin{bmatrix} 0 & 1 \\ -1 & 0 \end{bmatrix}$. Thus, rank $H = 2q$. □

We now consider classification of subspaces. Let $[x, y] = \langle Hx, y\rangle$ be a skew-symmetric scalar product in $\mathbb{R}^n$. A subspace $\mathcal{M} \subseteq \mathbb{R}^n$ is called *neutral* with respect to $[.,.]$ or *H-neutral* if $[x, y] = 0$ for all $x, y \in \mathcal{M}$. Observe that $\mathcal{M}$ is H-neutral if and only if $S^{-1}\mathcal{M}$ is $S^T HS$-neutral (where S is any real invertible matrix).

Theorem 3.1.2. *The maximal dimension of an H-neutral subspace is $\frac{n}{2}$.*

Proof. In view of Proposition 3.1.1(d) and the remark preceding this theorem, we can assume that

$$H = \begin{bmatrix} 0 & 1 \\ -1 & 0 \end{bmatrix} \oplus \cdots \oplus \begin{bmatrix} 0 & 1 \\ -1 & 0 \end{bmatrix}$$

($n/2$ times). Recall that H must be invertible, so the zero direct summand in (3.1.3) does not appear. Now the subspace $\text{span}\{e_1, e_3, \ldots, e_{n-1}\}$ is $\frac{n}{2}$-dimensional and H-neutral.

Conversely, let $\mathcal{M}$ be a k-dimensional H-neutral subspace. Let $x_1, \ldots, x_k$ be a basis in $\mathcal{M}$, and let S be any real invertible matrix such that $Se_i = x_i$ ($i = 1, \ldots, k$). Then the top left $k \times k$ corner in the $n \times n$ matrix $S^T HS$ is zero. Since the matrix $S^T HS$ is invertible, we must have $k \leq \frac{n}{2}$. □

3.2 H-skew-symmetric matrices

Let $[x, y]$ be a skew-symmetric scalar product in $\mathbb{R}^n$. Given an $n \times n$ matrix (or linear transformation in $\mathbb{R}^n$) A, the *skew-adjoint* matrix (or linear transformation) $A^{[T]}$ with respect to $[.,.]$ is defined by the equality

$$[A^{[T]}x, y] = -[x, Ay] \tag{3.2.1}$$

for all $x, y \in \mathbb{R}^n$. It is easy to see that $A^{[T]}$ exists and is uniquely defined by (3.2.1).

We express $A^{[T]}$ in terms of the real skew-symmetric matrix H that induces the skew-symmetric scalar product $[.,.]$ defined by equation (3.1.2). The equality (3.2.1) can be rewritten in the form

$$\langle HA^{[T]}x, y\rangle = \langle -Hx, Ay\rangle, \quad x, y \in \mathbb{R}^n,$$

which in turn can be rewritten as the matrix equality $HA^{[T]} = -A^T H$, or

$$A^{[T]} = -H^{-1}A^T H. \tag{3.2.2}$$

In particular, $A^{[T]}$ is similar to $(-A^T)$, and we obtain the first properties of skew-adjoint matrices:

Proposition 3.2.1. *If λ_0 is an eigenvalue of A, then $-\lambda_0$ is an eigenvalue of $A^{[T]}$. Moreover, the partial multiplicities of $-\lambda_0$ as an eigenvalue of $A^{[T]}$ coincide with the partial multiplicities of λ_0 as an eigenvalue of A.*

An orthogonality relation analogous to Theorem 2.2.1 holds as well:

Theorem 3.2.2. *Let $\mathcal{M}$ be one of the subspaces $\mathcal{R}_\lambda(A)$ or $\mathcal{R}_{\mu\pm i\nu}(A)$ (where $\lambda \in \mathbb{R}$ and $\mu, \nu \in \mathbb{R}$, $\nu > 0$), and let $\mathcal{N}$ be one of the subspaces $\mathcal{R}_\alpha(A^{[T]})$ or $\mathcal{R}_{\beta\pm i\gamma}(A^{[T]})$ (where $\alpha \in \mathbb{R}$, and $\beta, \gamma \in \mathbb{R}$, $\gamma > 0$). Assume further $\lambda \neq -\alpha$ (if $\mathcal{M} = \mathcal{R}_\lambda(A)$ and $\mathcal{N} = \mathcal{R}_\alpha(A^{[T]})$), or $\mu + i\nu \neq -\beta + i\gamma$ (if $\mathcal{M} = \mathcal{R}_{\mu\pm i\nu}(A)$ and $\mathcal{N} = \mathcal{R}_{\beta\pm i\gamma}(A^{[T]})$). Then $\mathcal{M}$ and $\mathcal{N}$ are orthogonal to each other with respect to $[.,.]$:*

$$\mathcal{M} \subseteq \mathcal{N}^{[\perp]},$$

where

$$\mathcal{N}^{[\perp]} = \{x \in \mathbb{R}^n \mid [x, y] = 0 \quad \text{for all } y \in \mathcal{N}\}.$$

Proof. We give the proof for the case $\mathcal{M} = \mathcal{R}_{\mu\pm i\nu}(A)$, $\mathcal{N} = \mathcal{R}_{\beta\pm i\gamma}(A^{[T]})$, where it is assumed $\mu + i\nu \neq -\beta + i\gamma$. (In all other cases the proof is analogous.)

The proof will proceed by reduction to the complex case. Consider the real matrix A as an $n \times n$ complex matrix (or a linear transformation $\mathbb{C}^n \to \mathbb{C}^n$) in the natural way:

$$A(x + iy) = Ax + i(Ay), \quad x, y \in \mathbb{R}^n.$$

Let H be the real skew-symmetric matrix defining the skew-symmetric scalar product:

$$[x, y] = \langle Hx, y\rangle, \tag{3.2.3}$$

and let $\{x, y\} = \langle iHx, y\rangle$ be the indefinite scalar product induced by the hermitian matrix iH on $\mathbb{C}^n$. We consider the scalar product $[.,.]$ also as extended to $\mathbb{C}^n$ by the formula (3.2.3). For every $x, y \in \mathbb{R}^n$ we have

$$\{Ax, y\} = i[Ax, y] = -i[x, A^{[T]}y] = i[x, -A^{[T]}y] = \{x, -A^{[T]}y\},$$

and therefore $A^{\{*\}} = -A^{[T]}$, where by $A^{\{*\}}$ we denote the adjoint of A with respect to the indefinite scalar product $\{.,.\}$. By Theorem 2.2.1, we have

$$\{x,y\} = 0 \qquad (3.2.4)$$

for every $x \in \mathcal{R}_{\mu+i\nu}(A) + \mathcal{R}_{\mu-i\nu}(A) \subseteq \mathbb{C}^n$ and every $y \in \mathcal{R}_{-\beta+i\gamma}(A^{\{*\}}) + \mathcal{R}_{-\beta-i\gamma}(A^{\{*\}}) \subseteq \mathbb{C}^n$. But the real subspace $\mathcal{R}_{\mu\pm i\nu}(A)$ is contained in $\mathcal{R}_{\mu+i\nu}(A) + \mathcal{R}_{\mu-i\nu}(A)$. On the other hand,

$$\mathcal{R}_{-\beta+i\gamma}(A^{\{*\}}) = \mathcal{R}_{-\beta+i\gamma}(-A^{[T]}) = \mathcal{R}_{\beta-i\gamma}(A^{[T]})$$

and similarly for $\mathcal{R}_{-\beta-i\gamma}(A^{\{*\}})$. Thus, the real subspace $\mathcal{R}_{\beta\pm i\gamma}(A^{[T]})$ is contained in

$$\mathcal{R}_{-\beta+i\gamma}(A^{\{*\}}) + \mathcal{R}_{-\beta-i\gamma}(A^{\{*\}}).$$

So from (3.2.4) we obtain $[x,y] = 0$ for every $x \in \mathcal{R}_{\mu\pm i\nu}(A)$ and $y \in \mathcal{R}_{\beta\pm i\gamma}(A^{[T]})$. □

The concept of an H-skew-symmetric matrix is now natural. A real $n \times n$ matrix A is called *skew-symmetric* with respect to $[.,.]$, or H-*skew-symmetric*, if the equality

$$[Ax, y] = -[x, Ay]$$

holds for all $x, y \in \mathbb{R}^n$. In view of (3.2.2) A is H-skew-symmetric if and only if the equality $A = -H^{-1}A^T H$, or, equivalently,

$$HA = -A^T H = (HA)^T,$$

holds.

Note that the same phrase "H-skew-symmetric" is used in Section 2.7 and has a different meaning. To avoid confusion it must be understood *a priori* whether H is symmetric (as in Section 2.7), or H is skew-symmetric, as in this case.

As an immediate consequence of Proposition 3.2.1 we see that the eigenvalues of an H-skew-symmetric matrix A are located symmetrically with respect to the origin: if λ_0 is an eigenvalue of A, then $-\lambda_0$ is also an eigenvalue of A having the same partial multiplicities as λ_0. Being a real matrix, the eigenvalues of A are symmetric with respect to the real axis. Combining these two symmetries, we see that if λ_0 is an eigenvalue of A, then so are $-\lambda_0$, $\bar{\lambda}_0$ and $-\bar{\lambda}_0$, and each of them has the same partial multiplicities as λ_0.

Theorem 3.2.2 leads to the orthogonality properties of root subspaces of H-skew-symmetric matrices:

Theorem 3.2.3. *Let A be H-skew-symmetric and let $\mathcal{M}$ be one of the root subspaces $\mathcal{R}_\lambda(A)$ or $\mathcal{R}_{\mu\pm i\nu}(A)$. Let $\mathcal{N}$ be one of the subspaces $\mathcal{R}_\alpha(A)$ or $\mathcal{R}_{\beta\pm i\gamma}(A)$ (here $\lambda, \mu, \nu, \alpha, \beta$ and γ are all real and ν, γ are positive). Assume further $\lambda \neq -\alpha$ (if $\mathcal{M} = \mathcal{R}_\lambda(A)$ and $\mathcal{N} = \mathcal{R}_\alpha(A)$) or $\mu + i\nu \neq -\beta + i\gamma$ (if $\mathcal{M} = \mathcal{R}_{\mu\pm i\nu}(A)$ and $\mathcal{N} = \mathcal{R}_{\beta\pm i\gamma}(A)$). Then $\mathcal{M}$ and $\mathcal{N}$ are orthogonal with respect to $[.,.]$.*

H-SKEW-SYMMETRIC MATRICES

Basic examples of H-skew-symmetric matrices with respect to a skew-symmetric scalar product will now be given. We use the matrices F_j defined by equation (2.7.1).

Example 3.2.1. Let $A = J_{2n}(0)$, $H = \epsilon F_{2n}$, where $\epsilon = \pm 1$. It is easy to see that A is H-skew-symmetric. □

Example 3.2.2. Let a be a real number; put

$$A = \begin{bmatrix} J_n(a) & 0 \\ 0 & -J_n(a)^T \end{bmatrix}, \quad H = \begin{bmatrix} 0 & I_n \\ -I_n & 0 \end{bmatrix}.$$

Again, A is H-skew-symmetric. □

Example 3.2.2 can easily be generalized. Namely, for any real $n \times n$ matrix B let

$$A = \begin{bmatrix} B & 0 \\ 0 & -B^T \end{bmatrix}, \quad H = \begin{bmatrix} 0 & I_n \\ -I_n & 0 \end{bmatrix}.$$

Then A is H-skew-symmetric. Of special interest to us for the canonical form to be presented later will be the case when B is a real Jordan block corresponding to a pair of non-real complex conjugate eigenvalues (see Section 1.4).

Example 3.2.3. For this example, we consider the powers of $F_2 = \begin{bmatrix} 0 & 1 \\ -1 & 0 \end{bmatrix}$. Clearly, F_2^k is skew-symmetric if k is odd and F_2^k is symmetric if k is even. Therefore, the $2n \times 2n$ matrix G_{2n} defined by equation (2.7.2) is skew-symmetric for all positive integers n. Let $A = J_{2n}(\pm ib)$, where b is a nonzero real number. Then A is ϵG_{2n}-skew-symmetric (a straightforward calculation will verify this), where $\epsilon = 1$ or $\epsilon = -1$. □

It is easy to see that the property of H-skew-symmetry for A is preserved under the simultaneous transformation of A and H: $A \to S^{-1}AS$, $H \to S^T HS$, where S is a real invertible matrix. In other words, if $H = -H^T$ is invertible and A is H-skew-symmetric, then $S^T HS$ is also skew-symmetric and invertible and $S^{-1}AS$ is $S^T HS$-skew-symmetric. Thus, questions arise concerning existence and uniqueness of a canonical form for pairs (A, H) under such a transformation (when A is H-skew-symmetric). This is provided by the following theorem. Basically, Examples 3.2.1–3.2.3 serve as the building blocks for this canonical form.

Theorem 3.2.4. *Let H be a real $n \times n$ skew-symmetric invertible matrix, and let A be a real $n \times n$ matrix which is H-skew-symmetric. Then there exists an invertible real matrix S such that*

$$S^{-1}AS = A_1 \oplus \cdots \oplus A_r, \quad S^T HS = H_1 \oplus \cdots \oplus H_r, \qquad (3.2.5)$$

and where each pair (A_j, H_j) of diagonal blocks has one of the following five types:

I. $A_j = J_{2n_j}(0)$, $\quad H_j = \epsilon_j F_{2n_j}$, where F_{2n_j} is given by (2.7.1) and $\epsilon_j = \pm 1$;

II. $A_j = J_{k_j}(0) \oplus -(J_{k_j}(0))^T$, $\quad H_j = \begin{bmatrix} 0 & I_{k_j} \\ -I_{k_j} & 0 \end{bmatrix}$, where k_j is an odd positive integer;

III. $A_j = J_{m_j}(a_j) \oplus -(J_{m_j}(a_j))^T$, $\quad H_j = \begin{bmatrix} 0 & I_{m_j} \\ -I_{m_j} & 0 \end{bmatrix}$, where a_j is a positive real number;

IV. $A_j = J_{2\ell_j}(\pm ib_j)$, $\quad H_j = \delta_j G_{2\ell_j}$, where $G_{2\ell_j}$ is given by (2.7.2), $\delta_j = \pm 1$ and $b_j > 0$;

V. $A_j = J_{2p_j}(a_j \pm ib_j) \oplus -(J_{2p_j}(a_j \pm ib_j))^T$, $\quad H_j = \begin{bmatrix} 0 & I_{2p_j} \\ -I_{2p_j} & 0 \end{bmatrix}$, where a_j, b_j are real numbers and $b_j > 0$.

The decomposition (3.2.5) is uniquely defined by H and A, up to a permutation of the pairs $(A_1, H_1), \ldots, (A_r, H_r)$.

The *sign characteristic* consists of the signs of the numbers ϵ_j and δ_j appearing in canonical pairs of types I and IV, respectively; thus, a plus-sign or a minus-sign is attached to every even partial multiplicity corresponding to the eigenvalue 0, and to every pair of equal partial multiplicities corresponding to nonzero pure imaginary eigenvalues $\pm ib$ ($b \in \mathbb{R}$, $b \neq 0$).

The proof of Theorem 3.2.4 is involved, and therefore is not presented here. A full proof (including more general results) is provided by Thompson (1990).

3.3 Invariant neutral subspaces

Let $[x, y] = \langle Hx, y \rangle$ be a skew-symmetric scalar products in $\mathbb{R}^n$, and let A be an $n \times n$ H-skew-symmetric matrix. In this section we study A-invariant H-neutral subspaces of maximal possible dimension. Of particular interest are A-invariant H-neutral subspaces of dimension $\frac{n}{2}$ (if they exist). As seen by Theorem 3.1.2, H-neutral subspaces of dimension $\frac{n}{2}$ are maximal, i.e. any strictly bigger subspace in $\mathbb{R}^n$ is not H-neutral.

As in Section 2.7 (when A denoted an H-skew-symmetric matrix but H itself was symmetric), it turns out that all maximal A-invariant H-neutral subspaces have the same dimension and we can write down a formula for this dimension using the canonical form for (A, H) of Theorem 3.2.4. The result has much the same appearance as Theorem 2.7.2 and depends on the nature of the nonzero pure imaginary eigenvalues and, in particular, on parameters δ_k defined as in equation (2.7.6). Again we refer the reader to the paper of Lancaster and Rodman (1994) for the proof.

Theorem 3.3.1. *Let $H = -H^T$ be invertible and $HA = (HA)^T$. Then all maximal A-invariant H-neutral subspaces have the same dimension.*

If the distinct nonzero pure imaginary eigenvalues of A are $\pm ib_k$, $k = 1, 2, \ldots, r$, then this dimension is

$$\frac{1}{2}n - \sum_{k=1}^{r} \eta_k$$

where $\eta_k = |\sum_j \delta_j|$ and (for fixed k) $\delta_1, \ldots, \delta_r$ are the signs corresponding to the blocks $J_{2\ell_j}(\pm ib_k)$ with odd ℓ_j in the canonical form of (A, H).

As a special case of this theorem we see immediately that there are A-invariant H-neutral subspaces of dimension $\frac{1}{2}n$ if there are no nonzero pure imaginary eigenvalues or if such eigenvalues exist but have only even partial multiplicities.

3.4 Connections between sign characteristics

Let H be an $n \times n$ real skew-symmetric matrix which is also invertible, and let A be a real H-skew-symmetric matrix in the sense of Sections 3.2 and 3.3, i.e. $H^T = -H$ and $(HA)^T = HA$. Then there is a sign characteristic associated with the pair (A, H) which is determined by the canonical form of Theorem 3.2.4. However, we also have $(HA)A = -A^T(HA)$ so that, if HA is also invertible, A is HA-skew-symmetric in the sense of Section 2.7 and there is another sign characteristic associated with the pair (A, HA) and determined by the canonical form of Theorem 2.7.1.

But this is not all! The matrix iH is hermitian and invertible and therefore induces an indefinite scalar product on $\mathbb{C}^n$ in which the matrix iA is selfadjoint, i.e. $(iHiA)^* = (iH)(iA)$. So the pair (iA, iH) has a third sign characteristic determined by Theorem 2.3.3. For brevity, we refer to the three sign characteristics as those of (A, H), (A, HA) and (iA, iH), respectively.

It is clear that these sign characteristics must be intimately connected, and it is our purpose in this section to state the appropriate results. For brevity, the proofs are omitted.

Theorem 3.4.1. *Let H be a real $n \times n$ skew-symmetric invertible matrix, and let A be a real H-skew-symmetric matrix (i.e. $(HA)^T = HA$). If δ is the sign in the sign characteristic of (A, H) corresponding to the Jordan block $J_{2\ell}(\pm ib)$ ($b > 0$) in the real Jordan form of A, then $(-1)^{\ell-1}\delta$ and $-\delta$ are the signs in the sign characteristic of (iA, iH) corresponding to the Jordan blocks $J_\ell(b)$ and $J_\ell(-b)$, respectively, in the Jordan form of iA.*

Furthermore, if ϵ is the sign in the sign characteristic of (A, H) corresponding to the Jordan block $J_{2m}(0)$, then $(-1)^{m-1}\epsilon$ is the sign in the sign characteristic of (iA, iH), corresponding to the same Jordan block.

This theorem can be proved by careful comparison of the canonical forms of Theorems 3.2.4 and 2.3.3 and involves some lengthly calculations. When A is invertible the comparison of canonical forms for (A, H) (of Theorem 3.2.4) and (A, HA) (of Theorem 2.7.1) is less formidable. This is because the transforming matrix S of equation (3.2.5) is closely related to the corresponding matrix, say S_0, of Theorem 2.7.1. Thus, if $S^T H S$, $S^{-1} A S$ are in canonical form for the pair (A, H), then

$$S^T(HA)S = (S^T HS)(S^{-1} AS)$$

and it remains to reduce the product of canonical forms on the right. The following simple result is obtained.

Theorem 3.4.2. *Let H be an $n \times n$ real invertible skew-symmetric matrix and let A be a real invertible H-skew-symmetric matrix (i.e. $(HA)^T = HA$). Then the sign characteristics of the pairs (A, H) and (A, HA) coincide.*

3.5 Skew-symmetric and orthogonal pencils

In this section we briefly describe the notions of H-skew-symmetric and H-orthogonal pencils, in the framework of a skew-symmetric scalar product on $\mathbb{R}^n$.

Let H be an $n \times n$ real skew-symmetric invertible matrix. An $n \times n$ real matrix pencil $\lambda A - B$ is called H-skew-symmetric if $A^T H B = -B^T H A$. If $A = I$, then the pencil $\lambda A - B$ is H-skew-symmetric precisely when B is H-skew-symmetric (as defined in Section 3.2). Thus, the concept of an H-skew-symmetric pencil is a generalization of H-skew-symmetric matrices.

Notice also that, if we define $H_0 = iH$ and $B_0 = iB$ then $H_0^* = H_0$ and $A^T H B = -B^T H A$ implies $A^* H_0 B_0 = B_0^* H_0 A$. Thus, a real skew-symmetric pencil will share some spectral properties with H_0-self-adjoint pencils, as discussed in Section 2.8.

Proposition 3.5.1. *The spectrum of a regular H-skew-symmetric pencil $\lambda A - B$ has Hamiltonian symmetry, i.e. if $\lambda_0 \in \sigma(\lambda A - B)$, then also $\pm \bar{\lambda}_0 \in \sigma(\lambda A - B)$, and the partial multiplicities of $\pm \bar{\lambda}_0$ coincide with those of λ_0.*

Proof. The symmetry of $\sigma(\lambda A - B)$ relative to the real axis follows because A and B are real matrices. Next, observe that $\lambda A - iB$ is iH-self-adjoint (in the sense of Section 2.8); so by Theorem 2.8.1 $\sigma(\lambda A - iB)$ is also symmetric relative to the real axis. It remains to observe that $\lambda_0 \in \sigma(\lambda A - iB)$ if and only if $-i\lambda_0 \in \sigma(\lambda A - B)$. □

In complete analogy with the H-unitary pencils studied in Section 2.9, we define an H-orthogonal pencil $\lambda L - M$ by the equality

$$L^T H L = M^T H M.$$

Here, L and M are $n \times n$ real matrices.

The eigenvalues of a regular H-orthogonal pencil $\lambda L - M$ are symmetric relative to the unit circle, as well as being symmetric relative to the real axis; the latter fact is simply a consequence of the matrices L and M being real. Thus, if $\lambda_0 \in \sigma(\lambda L - M)$, then also λ_0^{-1}, $\bar{\lambda}_0^{-1}$, $\bar{\lambda}_0 \in \sigma(\lambda L - M)$, and each of these eigenvalues of $\lambda L - M$ has the same partial multiplicities as λ_0. Here, the case $\lambda_0 = \infty$ is not excluded, with the obvious understanding that $\infty^{-1} = 0$. The proof is easily reduced to Corollary 2.9.4; indeed, $\lambda L - M$ is an iH-unitary pencil in the sense of Section 2.9.

The $n \times n$ real matrix pencils $\lambda L_1 - M_1$ and $\lambda L_2 - M_2$, where $\lambda L_j - M_j$ is H_j-orthogonal ($j = 1, 2$), are called (H_1, H_2)-*orthogonally equivalent* if there exist invertible real matrices P and Q such that

$$\lambda L_1 - M_1 = P(\lambda L_2 - M_2)Q; \quad H_1 = (P^{-1})^T H_2 P^{-1}.$$

The real analogue of Theorem 2.9.5 for (H_1, H_2)-orthogonal equivalence runs as follows:

Theorem 3.5.2. *Let $\lambda L - M$ be a regular H-orthogonal pencil, and let m be the algebraic multiplicity of the zero eigenvalue. Then $\lambda L - M$ is (H, G)-orthogonally equivalent to a G-orthogonal pencil of the form*

$$\lambda \begin{bmatrix} I & 0 & 0 \\ 0 & I_m & 0 \\ 0 & 0 & N \end{bmatrix} - \begin{bmatrix} S & 0 & 0 \\ 0 & N^T & 0 \\ 0 & 0 & I_m \end{bmatrix}, \quad (3.5.1)$$

where N is an $m \times m$ nilpotent matrix, and where the real invertible skew-symmetric matrix G has the form

$$G = \begin{bmatrix} G_0 & 0 & 0 \\ 0 & 0 & I_m \\ 0 & -I_m & 0 \end{bmatrix}. \quad (3.5.2)$$

Moreover, in the forms (3.5.1) and (3.5.2) the nilpotent matrix N is unique up to similarity, and the G_0-orthogonal pencil $\lambda I - S$ is unique up to orthogonal equivalence.

The proof of Theorem 3.5.2 is obtained in the same way as that of Theorem 2.9.5 (see the proof of Theorem 2.1 in Wimmer (1991)).

When $L = I$, the orthogonal equivalence of H-orthogonal pencils $\lambda L - M$ specializes to orthogonal equivalence of H-orthogonal matrices. Let H be an $n \times n$ real invertible skew-symmetric matrix. An $n \times n$ real matrix U is called H-*orthogonal* if $U^T H U = H$. Two pairs (U_1, H_1) and (U_2, H_2), where U_j is H_j-orthogonal for $j = 1, 2$, are called *orthogonally equivalent* if $U_1 = T^{-1}U_2 T$ and $H_1 = T^T H_2 T$ for some real invertible matrix T. We are not aware of a canonical form under the orthogonal equivalence of pairs of real matrices (U, H) as above.

However, note that U is iH-unitary, therefore the canonical form of Theorem 2.3.3 can (and will) be used to describe the Jordan form and sign characteristic of H-orthogonal matrices U under the orthogonal equivalence of the pair (U, H). Here, a sign ± 1 is attached to every Jordan block corresponding to a unimodular eigenvalue in the Jordan form of U.

3.6 Exercises

In this set of exercises H is a real $n \times n$ skew-symmetric invertible matrix.

3.6.1. Let A and H be as in Example 3.2.2. Assume $a \neq 0$. Find all (real) n-dimensional A-invariant H-neutral subspaces.

3.6.2. Let A be a real H-skew-symmetric matrix without eigenvalues on the imaginary axis. Prove that there exists a real invertible matrix S such that

$$S^{-1}AS = \begin{bmatrix} B & 0 \\ 0 & -B^T \end{bmatrix}, \quad S^T H S = \begin{bmatrix} 0 & I_{n/2} \\ -I_{n/2} & 0 \end{bmatrix}$$

for some matrix B.

3.6.3. Give an example showing that the hypothesis on the eigenvalues of A is essential in Exercise 3.6.2.

3.6.4. Prove Theorem 3.4.1 for the case when A is nilpotent (i.e. $A^n \equiv 0$ for some integer n).

3.7 Notes

The canonical forms discussed here have been obtained elsewhere (generally with different terminology). See, for example, Djokovic *et al.* (1983), Thompson (1990), and Wimmer (1991).

4
MATRIX THEORY AND CONTROL

The ideas and theory presented in this chapter will play important parts in our study of Riccati equations. They revolve around the concepts of controllable (and observable) matrix pairs. The importance of these concepts was recognized in the early years of development of control theory and it was in that context that the terms "controllable" and "observable" arose and have some physical meaning. Our approach uses linear algebra in the development of these concepts because it is this understanding that is required in the sequel. Their significance in the context of systems theory and control is mentioned as the chapter progresses, but is not given complete mathematical development.

4.1 Controllable pairs

Consider a pair of matrices $A \in \mathbb{C}^{n \times n}$, $B \in \mathbb{C}^{n \times m}$; so the product AB is always well-defined. Define a sequence of subspaces $\mathcal{C}_0, \mathcal{C}_1, \mathcal{C}_2, \ldots$ of $\mathbb{C}^n$ in terms of A and B as follows:

$$\mathcal{C}_0 = \text{Im } B, \quad \mathcal{C}_1 = \text{Im } [B \ \ AB], \quad \mathcal{C}_2 = \text{Im } [B \ \ AB \ \ A^2 B], \ldots .$$

Inductively, we have $\mathcal{C}_0 = \text{Im } B$ and then

$$\mathcal{C}_{p+1} = \mathcal{C}_p + \text{Im } (A^{p+1} B), \quad p = 0, 1, 2, \ldots .$$

Clearly, we have

$$\mathcal{C}_0 \subseteq \mathcal{C}_1 \subseteq \mathcal{C}_2 \subseteq \ldots . \tag{4.1.1}$$

We first show that these inclusions are proper up to a certain index k (dependent on A and B) and thereafter equality obtains. In any case, as we work in n-dimensional space, there are not more than n proper inclusions.

Proposition 4.1.1. *If $\mathcal{C}_{k+1} = \mathcal{C}_k$ for some integer k, then $\mathcal{C}_j = \mathcal{C}_k$ for all $j \geq k$.*

Proof. It is sufficient to prove $C_{k+2} = C_{k+1}$ and the result then follows by induction. Let $x \in C_{k+2}$, then for some $x_0, x_1, \ldots, x_{k+2} \in \mathbb{C}^m$,

$$x = [B \ AB \ \ldots \ A^{k+2}B] \begin{bmatrix} x_0 \\ x_1 \\ \vdots \\ x_{k+2} \end{bmatrix} = Bx_0 + A[B \ AB \ \ldots \ A^{k+1}B] \begin{bmatrix} x_1 \\ x_2 \\ \vdots \\ x_{k+2} \end{bmatrix}.$$

Since $C_{k+1} = C_k$ there exist $y_1, \ldots, y_{k+1} \in \mathbb{C}^m$ such that

$$[B \ AB \ \ldots \ A^{k+1}B] \begin{bmatrix} x_1 \\ x_2 \\ \vdots \\ x_{k+2} \end{bmatrix} = [B \ AB \ \ldots \ A^k B] \begin{bmatrix} y_1 \\ y_2 \\ \vdots \\ y_{k+1} \end{bmatrix}.$$

Consequently,

$$x = [B \ AB \ \ldots \ A^{k+1}B] \begin{bmatrix} x_0 \\ y_1 \\ y_2 \\ \vdots \\ y_{k+1} \end{bmatrix} \in C_{k+1}.$$

Thus, $C_{k+2} \subseteq C_{k+1}$ and so $C_{k+2} = C_{k+1}$. □

The smallest integer k for which Proposition 4.1.1 holds is a characteristic of the pair (A, B). Using this k we write $C_k = C_{A,B}$ and call this the *controllable subspace* of the pair (A, B). If $C_{A,B} = \mathbb{C}^n$ then the pair (A, B) is said to be *controllable*.

Since $k \leq n-1$ we may write

$$C_{A,B} = \text{Im} \, [B \ AB \ A^2B \ \ldots \ A^{n-1}B] = \sum_{r=0}^{n-1} \text{Im} \, (A^r B). \qquad (4.1.2)$$

From this, the classical defining property of a controllable pair can be deduced: the pair (A, B) is controllable if and only if

$$\text{rank} \, [B \ AB \ A^2B \ \ldots \ A^{n-1}B] = n, \qquad (4.1.3)$$

i.e. when the rows of this $n \times mn$ matrix are linearly independent.

The next proposition gives a useful geometric characterization of the controllable subspace.

Proposition 4.1.2. *The controllable subspace $\mathcal{C}_{A,B}$ is the smallest A-invariant subspace containing* Im B.

Proof. Let the sequence $\{\mathcal{C}_r\}_{r=0}^{\infty}$ and integer k be as in Proposition 4.1.1. If $x \in \mathcal{C}_r$ it is apparent that $Ax \in \mathcal{C}_{r+1}$. Thus, if $x \in \mathcal{C}_k$ then $Ax \in \mathcal{C}_{k+1} = \mathcal{C}_k$, and so $\mathcal{C}_k = \mathcal{C}_{A,B}$ is A-invariant. Also, Im $B \subseteq \mathcal{C}_{A,B}$ trivially.

Finally, let $\mathcal{S}$ be any subspace of $\mathbb{C}^n$ for which $A\mathcal{S} \subseteq \mathcal{S}$ and Im $B \subseteq \mathcal{S}$. It is to be shown that $\mathcal{C}_{A,B} \subseteq \mathcal{S}$. For $r = 1, 2, \ldots$ we have

$$\text{Im } (A^{r-1}B) = A^{r-1} \text{Im } B \subseteq A^{r-1}\mathcal{S} \subseteq \mathcal{S}.$$

From equation (4.1.2) it follows that $\mathcal{C}_{A,B} \subseteq \mathcal{S}$. □

It is clear that for matrices B_1 and B_2 with Im B_1 = Im B_2 we have $\mathcal{C}_{A,B_1} = \mathcal{C}_{A,B_2}$. In particular, if $C > 0$ we have Im (BCB^*) = Im B or, what is equivalent, Ker (BCB^*) = Ker B^*.

To see this, observe first that the inclusion Ker $B^* \subseteq$ Ker (BCB^*) is apparent. Then, if $x \in$ Ker (BCB^*) we have $x^*BCB^*x = 0$. Introduce the positive definite square root $C^{1/2}$ of C (see Theorem 1.5.1), and we obtain $\|C^{1/2}B^*x\| = 0$. Thus $C^{1/2}B^* = 0$ and, as $C^{1/2}$ is nonsingular, $x \in$ Ker B^*. It follows that Ker B^* = Ker (BCB^*).

Using this result we get a useful corollary to Proposition 4.1.2:

Corollary 4.1.3. *If $C > 0$ then (A, B) and (A, BCB^*) have the same controllable subspace.*

In particular, (A, B) is controllable if and only if (A, BCB^*) is controllable, where $C > 0$.

Now let us briefly indicate the origin of the notion of controllability in systems theory. Let $A \in \mathbb{C}^{n \times n}$, $B \in \mathbb{C}^{n \times m}$ and consider the differential initial value problem

$$\dot{x}(t) = Ax(t) + Bu(t), \quad x(0) = x_0. \tag{4.1.4}$$

Here the vector function $u(t)$ is seen as a *control* or *input* function. Such a system is said to be controllable if, given any initial vector x_0, terminal vector x_1 and time $T > 0$ there is a continuous control function $u(t)$ for which the state vector $x(t)$ takes the value x_1 at time T.

Theorem 4.1.4. *The system (4.1.4) is controllable if and only if the matrix pair (A, B) is controllable.*

The reader is referred to one of the many available texts and monographs for the proof of this cornerstone of systems theory (see Wonham (1979) or Lancaster and Tismenetsky (1985), for example).

When (A, B) is not a controllable pair, the subspace $\mathcal{C}_{A,B}$ has an important physical meaning as the *reachable subspace*, i.e. the set of all terminal vectors x_1 that can be attained from the origin $x_0 = 0$ in some finite time.

4.2 Observable pairs

Let $C \in \mathbb{C}^{m \times n}$, $A \in \mathbb{C}^{n \times n}$, and define a sequence of subspaces $\mathcal{K}_0, \mathcal{K}_1, \mathcal{K}_2, \ldots$ of $\mathbb{C}^n$ in terms of the pair (C, A) by

$$\mathcal{K}_0 = \operatorname{Ker} C, \quad \mathcal{K}_1 = \operatorname{Ker} \begin{bmatrix} C \\ CA \end{bmatrix}, \quad \mathcal{K}_2 = \operatorname{Ker} \begin{bmatrix} C \\ CA \\ CA^2 \end{bmatrix}, \ldots .$$

In other words

$$\mathcal{K}_p = \bigcap_{r=0}^{p} \operatorname{Ker}(CA^r), \quad p = 0, 1, 2, \ldots,$$

and we have

$$\mathcal{K}_0 \supseteq \mathcal{K}_1 \supseteq \mathcal{K}_2 \supseteq \ldots . \tag{4.2.1}$$

Proposition 4.2.1. *If $\mathcal{K}_{k+1} = \mathcal{K}_k$ for some integer k, then $\mathcal{K}_j = \mathcal{K}_k$ for all $j \geq k$.*

Proof. We have

$$\mathcal{K}_p = \operatorname{Ker} \begin{bmatrix} C \\ CA \\ \vdots \\ CA^p \end{bmatrix} = (\operatorname{Im} [C^* A^* C^* \; \ldots \; (A^*)^p C^*])^\perp = \mathcal{C}_p^\perp,$$

where $\mathcal{C}_p$ is the subspace defined as in 4.1.1, but for the pair (A^*, C^*). As $\mathcal{K}_p$ is the orthogonal complement of $\mathcal{C}_p$ the result follows on applying Proposition 4.1.1 to (A^*, C^*). □

The subspace

$$\mathcal{U}_{C,A} = \bigcap_{r=0}^{\infty} \operatorname{Ker}(CA^r) = \bigcap_{r=0}^{n-1} \operatorname{Ker}(CA^r) = (\mathcal{C}_{A^*, C^*})^\perp \tag{4.2.2}$$

is known as the *unobservable* subspace of the pair (C, A) and when $\mathcal{U}_{C,A} = \{0\}$, the pair (C, A) is said to be *observable*.

A certain duality is seen to be emerging between observable and controllable pairs which depends on the relation $\mathcal{K}_p = \mathcal{C}_p^\perp$, $p = 0, 1, 2, \ldots$. In particular, it follows immediately that:

Proposition 4.2.2. *The pair (C, A) is observable if and only if (A^*, C^*) is controllable.*

The dual statement for Proposition 4.1.2 is then found to be:

Proposition 4.2.3. *The unobservable subspace $\mathcal{U}_{C,A}$ is the maximal A-invariant subspace contained in $\operatorname{Ker} C$.*

To find physical interpretations of these ideas consider a system of the form

$$\dot{x}(t) = Ax(t), \quad x(0) = x_0; \quad y(t) = Cx(t), \tag{4.2.3}$$

where $C \in \mathbb{C}^{m \times n}$, $A \in \mathbb{C}^{n \times n}$. Here, $y(t)$ is described as the *output* vector function. This *system* is said to be observable when $y(t) \equiv 0$ if and only if $x_0 = 0$. Then it can be proved that:

Theorem 4.2.4. *The system (4.2.3) is observable if and only if the matrix pair (C, A) is observable.*

Thus, for an observable system, $\mathcal{U}_{C,A} = \{0\}$ and every nonzero initial vector x_0 determines a nonzero output $y(t) = Ce^{At}x_0$. When $\mathcal{U}_{C,A} \neq \{0\}$ then every nonzero initial vector $x_0 \in \mathcal{U}_{C,A}$ is unobservable in the sense that it produces an identically zero output $y(t)$. Again, we refer to Wonham (1979) or Lancaster and Tismenetsky (1985) for proofs of Theorem 4.2.4.

4.3 Two characterizations of controllable pairs

In this section two theorems are presented which give alternate characterizations of controllable pairs. Both will be useful for us subsequently and are well-known results of systems theory. The first concerns pairs (A, B) (where A is $n \times n$ and B is $n \times m$) and the associated matrix function:

$$W(t) := \int_0^t e^{As} BB^* e^{A^*s} \, ds, \tag{4.3.1}$$

known as the *controllability Gramian* of (A, B).

Theorem 4.3.1. *For any $t > 0$, $\operatorname{Im} W(t) = \mathcal{C}_{A,B}$.*

Proof. Clearly $W(t)$ is a hermitian matrix-valued function so that $\operatorname{Im} W(t) = (\operatorname{Ker} W(t))^\perp$. Using equation (4.2.2) it is therefore equivalent to prove that $\operatorname{Ker} W(t) = \mathcal{U}_{B^*,A^*}$.

Let $x \in \operatorname{Ker} W(t)$. Then

$$x^* W(t) x = \int_0^t \|B^* e^{A^*s} x\|^2 \, ds = 0.$$

Hence, for all $s \in [0, t]$, $B^* e^{A^*s} x = 0$. Differentiate this relation repeatedly with respect to s and take the limit as $s \to 0_+$ to obtain

$$B^* (A^*)^r x = 0, \quad r = 0, 1, 2, \ldots, n-1. \tag{4.3.2}$$

Using (4.2.2) again we obtain $\operatorname{Ker} W(t) \subseteq \mathcal{U}_{B^*,A^*}$.

Conversely, if $x \in \mathcal{U}_{B^*, A^*}$ then (4.3.2) holds and (using the Cayley–Hamilton theorem) $B^*(A^*)^r x = 0$ for all nonnegative integers r. Hence $B^* e^{A^* s} x = 0$, $x^* W(t) x = 0$ and, as $W(t) \geq 0$, $x \in \operatorname{Ker} W(t)$. □

Corollary 4.3.2. *The pair (A, B) is controllable if and only if, for any $t > 0$, the matrix $W(t)$ of (4.3.1) is positive definite.*

The following result originates with Hautus (1969).

Theorem 4.3.3. *The following statements are equivalent for a pair of matrices $A \in \mathbb{C}^{n \times n}$ and $B \in \mathbb{C}^{n \times m}$:*

  (i) *The pair (A, B) is controllable.*

  (ii) $\operatorname{Ker} B^* \cap \operatorname{Ker} (\lambda I - A)^* = \{0\}$ *for all $\lambda \in \mathbb{C}$.*

  (iii) $\operatorname{rank} [\lambda I - A, B] = n$ *for all $\lambda \in \mathbb{C}$.*

(Note that conditions (ii) and (iii) are significant only when λ is an eigenvalue of A.)

Proof. Since
$$\operatorname{Ker} B^* \cap \operatorname{Ker} (\lambda I - A)^* = \operatorname{Ker} \begin{bmatrix} B^* \\ (\lambda I - A)^* \end{bmatrix} = (\operatorname{Im} B + \operatorname{Im} (\lambda I - A))^\perp$$
it is easily seen that (ii) and (iii) are equivalent.

Now we show that (i) ⇒ (ii). If $y \neq 0$ and
$$y \in \operatorname{Ker} B^* \cap \operatorname{Ker} (\lambda I - A)^*,$$
then $y^* B = 0$ and $y^* A^r = \lambda^r y^*$ for $r = 1, 2, 3, \ldots$. Hence
$$y^* [B \quad AB \quad \ldots \quad A^{n-1} B] = 0$$
and, from (4.1.3), we see that (A, B) cannot be controllable. Thus (i) ⇒ (ii).

Now suppose (4.1.3) is false and
$$\operatorname{rank} [B \quad AB \quad \ldots \quad A^{n-1} B] < n.$$
Then there is a $z \neq 0$ such that
$$z^* A^r B = 0, \quad r = 0, 1, 2, \ldots, n - 1. \tag{4.3.3}$$
Let ψ be a polynomial of minimal degree, say d, such that $z^* \psi(A) = 0$. Then ψ has at least one zero λ, and $\psi(\mu) = \varphi(\mu)(\mu - \lambda)$ for all $\mu \in \mathbb{C}$ and some polynomial φ. Let $y^* = z^* \varphi(A)$ so that $y \neq 0$ by definition of ψ. Then
$$y^*(A - \lambda I) = z^* \varphi(A)(A - \lambda I) = z^* \psi(A) = 0.$$
But also, using (4.3.3),
$$y^* B = z^* \varphi(A) B = 0,$$
and $y \in \operatorname{Ker} B^* \cap \operatorname{Ker} (\lambda I - A)^*$. Hence (ii) ⇒ (i). □

STABILIZABLE AND DETECTABLE PAIRS

Using the duality of Proposition 4.2.2 corresponding criteria for observability are obtained:

Corollary 4.3.4. *The following statements are equivalent for a pair of matrices $C \in \mathbb{C}^{m \times n}$ and $A \in \mathbb{C}^{n \times n}$:*
  (i) *The pair (C, A) is observable.*
  (ii) $\operatorname{Ker} C \cap \operatorname{Ker}(\lambda I - A) = \{0\}$ *for all $\lambda \in \mathbb{C}$.*
  (iii) $\operatorname{rank} \begin{bmatrix} C \\ \lambda I - A \end{bmatrix} = n$ *for all $\lambda \in \mathbb{C}$.*

4.4 Stabilizable and detectable pairs

The concept of "feedback" plays an important role in systems theory. Consider, in particular, a system of the form

$$\dot{x}(t) = Ax(t) + Bu(t)$$

(as in (4.1.4)) where A is $n \times n$ and B is $n \times m$. If the input $u(t)$ is made to depend on the state $x(t)$ then "state feedback" is obtained. If time-invariance and linearity are to be preserved, a natural choice for $u(t)$ is, say

$$u(t) = Kx(t) + v(t)$$

for some $m \times n$ matrix K. Then the system takes the form

$$\dot{x}(t) = (A + BK)x(t) + Bv(t).$$

The effect has been to transform the system matrix A to $A + BK$ and now algebraic questions naturally arise concerning properties of $A + BK$ that can be achieved by choice of K when the pair (A, B) is fixed.

When feedback is employed in this way it is important to note that the resulting pair $(A + BK, B)$ has the same controllable subspace as (A, B).

Lemma 4.4.1. *Let A, B, K be matrices of sizes $n \times n$, $n \times m$, and $m \times n$ respectively and write $\hat{A} = A + BK$. Then:*
  (i) *For $r = 0, 1, 2, \ldots$, $\operatorname{Im} [B \quad \hat{A}B \quad \ldots \quad \hat{A}^r B] = \operatorname{Im} [B \quad AB \quad \ldots \quad A^r B]$.*
  (ii) $\mathcal{C}_{\hat{A}, B} = \mathcal{C}_{A, B}$.
  (iii) $(\hat{A}, B)$ *is controllable if and only if (A, B) is controllable.*

Proof. If we can prove part (i), then parts (ii) and (iii) follow immediately from the definitions of controllable subspaces and controllable pairs. Write

$$\mathcal{C}_r = \mathrm{Im}\,[B \quad AB \quad \ldots \quad A^r B] = \sum_{j=0}^{r} \mathrm{Im}\,(A^j B),$$

$$\hat{\mathcal{C}}_r = \mathrm{Im}\,[B \quad \hat{A}B \quad \ldots \quad \hat{A}^r B] = \sum_{j=0}^{r} \mathrm{Im}\,(\hat{A}^j B).$$

We prove by induction that $\hat{\mathcal{C}}_r = \mathcal{C}_r$. Trivially, we have $\hat{\mathcal{C}}_0 = \mathcal{C}_0$. Assume $\hat{\mathcal{C}}_{s-1} = \mathcal{C}_{s-1}$ and let $x \in \hat{\mathcal{C}}_s$. Then there are vectors $x_0, \ldots, x_s$ such that

$$x = \sum_{j=0}^{s-1} \hat{A}^j B x_j + \hat{A}^s B x_s. \qquad (4.4.1)$$

By hypothesis, the first term on the right is in $\mathcal{C}_{s-1}$ and for the second term

$$\begin{aligned}\hat{A}^s B x_s &= (A+BK)(A+BK)^{s-1} B x_s \\ &= A(A+BK)^{s-1} B x_s + BK(A+BK)^{s-1} B x_s. \\ &\in \mathrm{Im}\,\sum_{j=1}^{s} A^j B + \mathrm{Im}\, B = \mathcal{C}_s.\end{aligned}$$

Thus, it follows from (4.4.1) that $\hat{\mathcal{C}}_s \subseteq \mathcal{C}_s$.

Now apply this inclusion again with A replaced by $A + BK$ and K replaced by $-K$. We obtain $\mathcal{C}_s \subseteq \hat{\mathcal{C}}_s$. Hence the result. □

The notion of "stabilizability" takes two forms originating with the theory of continuous (differential) systems and discrete (difference) systems. In these two cases an $n \times n$ matrix is said to be *stable* according as its eigenvalues are all in the open left half-plane, or the open unit disc, respectively. The same word, "stable", may be used in either case when its meaning is clear from the context. For example, the first and second definitions apply throughout Part II and Part III of this work, respectively.

Where it is necessary to make a distinction the respective terms *c-stable* and *d-stable* will be used. With these understandings we now define a pair (A, B) (where A is $n \times n$ and B is $n \times m$) to be *stabilizable* if there is a (feedback) matrix K such that $A+BK$ is stable. Where necessary, we use the terms "*c*-stabilizable" and "*d*-stabilizable" with the obvious meanings.

Theorem 4.4.2. *Let the pair (A, B) be controllable. Then:*

(a) *The pair (A, B) is both c-stabilizable and d-stabilizable.*

(b) *For any $t > 0$ the matrix*

$$V = \int_0^t e^{-As} BB^* e^{-A^*s}\, ds$$

is positive definite and, if $K = -B^*V^{-1}$, the matrix $A + BK$ is c-stable.

Only the proof of part (b) is given here, and hence the c-stabilizability of (A, B). In fact, given any n complex numbers, there is a matrix K for which the eigenvalues of $A + BK$ take these pre-assigned values (so K solves the so-called "spectral assignment" problem; see Gohberg et al. (1986a), for example). A proof of this more general result is outlined in the exercises of Section 4.7.

Proof of (b). Clearly, $(-A, B)$ is a controllable pair and, from Theorem 4.3.1, we see that $V > 0$ and so K is well-defined. Write $\hat{A} = A + BK$ and we have

$$\hat{A}V + V\hat{A}^* = AV + VA^* - 2BB^*.$$

But

$$AV + VA^* = -\int_0^t \frac{d}{ds}(e^{-As} BB^* e^{-A^*s})\, ds$$
$$= -e^{-At} BB^* e^{-A^*t} + BB^*,$$

and so

$$\hat{A}V + V\hat{A}^* = -e^{-At} BB^* e^{-A^*t} - BB^*. \qquad (4.4.2)$$

By Lemma 4.4.1, (A, B) controllable implies that the pair $(\hat{A}, B)$ is also controllable.

Now we crave the reader's indulgence and refer forward to Theorem 5.3.2. Applying this to equation (4.4.2) it follows that $\hat{A}$ is c-stable. $\square$

"Detectability" of matrix pairs (C, A) with sizes $m \times n$ and $n \times n$, respectively, is defined as a concept dual to the idea of stabilizability and, similarly, turns out to be a weaker concept than observability. To be precise, a pair (C, A) (as above) is said to be *detectable* if and only if (A^*, C^*) is stabilizable.

Now we have

(C, A) observable $\implies (A^*, C^*)$ controllable (by Proposition 4.2.2)
$\implies (A^*, C^*)$ stabilizable (by Theorem 4.4.2)
$\implies (C^*, A^*)$ detectable (by definition).

Thus, observable pairs are necessarily detectable, but not conversely.

4.5 The control normal form

Let $A \in \mathbb{C}^{n \times n}$ and $B \in \mathbb{C}^{n \times m}$ ($B \neq 0$). Even though the pair (A, B) may not be controllable, it is possible to generate an associated, or residual controllable pair, possibly of smaller size. To see this, let $x_1, x_2, \ldots, x_r$ be a basis for the controllable subspace $\mathcal{C}_{A,B}$ and then extend this set to a basis $x_1, \ldots, x_r, x_{r+1}, \ldots, x_n$ for $\mathbb{C}^n$. As $\mathcal{C}_{A,B}$ is A-invariant and $\operatorname{Im} B \subseteq \mathcal{C}_{A,B}$ (Proposition 4.1.2), the representations of A and B in this basis (and the standard basis for $\mathbb{C}^m$) have the form

$$\begin{bmatrix} A_1 & A_3 \\ 0 & A_2 \end{bmatrix}, \quad \begin{bmatrix} B_1 \\ 0 \end{bmatrix} \qquad (4.5.1)$$

where A_1 is $r \times r$ and B_1 is $r \times m$. We claim that the pair (A_1, B_1) is controllable.

Proposition 4.5.1. *For any matrix pair (A, B) with $A \in \mathbb{C}^{n \times n}$ and $B \in \mathbb{C}^{n \times m}$ there is a nonsingular matrix X such that*

$$X^{-1}AX = \begin{bmatrix} A_1 & A_3 \\ 0 & A_2 \end{bmatrix}, \quad X^{-1}B = \begin{bmatrix} B_1 \\ 0 \end{bmatrix} \qquad (4.5.2)$$

where A_1 is $r \times r$, B_1 is $r \times m$ and r is the dimension of $\mathcal{C}_{A,B}$. Furthermore, the pair (A_1, B_1) is controllable.

Proof. Only the last statement remains to be proved. It is easily seen that

$$[B \; AB \; \ldots \; A^{n-1}B] = X \begin{bmatrix} B_1 & A_1 B_1 & \ldots & A_1^{n-1} B_1 \\ 0 & 0 & & 0 \end{bmatrix},$$

and (from (4.1.2)) r is the rank of the matrix on the left. Consequently

$$\operatorname{rank} [B_1 \; A_1 B_1 \; \ldots \; A_1^{n-1} B_1] = \operatorname{rank} [B_1 \; A_1 B_1 \; \ldots \; A_1^{r-1} B_1] = r,$$

and so (using (4.1.3)) (A_1, B_1) is controllable. □

The pair of matrices

$$\begin{bmatrix} A_1 & A_3 \\ 0 & A_2 \end{bmatrix}, \quad \begin{bmatrix} B_1 \\ 0 \end{bmatrix}$$

of (4.5.2) is known as a *control (or Kalman) normal form* of (A, B). This admits a useful characterization of stabilizable pairs.

Proposition 4.5.2. *Assume that the pair (A, B) is not controllable. Then the pair (A, B) is stabilizable if and only if the matrix A_2 of a control normal form (4.5.2) is stable.*

Proof. For any matrix $K \in \mathbb{C}^{m \times n}$ we have, using (4.5.2),

$$A + BK = X \begin{bmatrix} A_1 + B_1 K & A_3 \\ 0 & A_2 \end{bmatrix} X^{-1}.$$

Since (A_1, B_1) is controllable it is also stabilizable and K can be chosen so that $A_1 + B_1 K$ is stable. The result follows from the fact that $\sigma(A + BK) = \sigma(A_1 + B_1 K) \cup \sigma(A_2)$. $\square$

Clearly, this proposition holds whether "stability" refers to "c-stability" or "d-stability". A similar remark applies to the following analogue of Lemma 4.4.1 which will also be useful.

Lemma 4.5.3. *Let A be $n \times n$, and B be $n \times m$. Then (A, B) is stabilizable if and only if $(A + BK, BL)$ is stabilizable for any $m \times n$ matrix K and any $m \times p$ matrix L for which $\mathrm{Im}\,(BL) = \mathrm{Im}\,B$.*

Proof. Using the construction of Proposition 4.5.1 it may be assumed without loss of generality that

$$A = \begin{bmatrix} A_1 & A_3 \\ 0 & A_2 \end{bmatrix}, \quad B = \begin{bmatrix} B_1 \\ 0 \end{bmatrix}$$

where (A_1, B_1) is controllable. Using the same decomposition write $K = [K_1 \ K_2]$. Thus,

$$A + BK = \begin{bmatrix} A_1 + B_1 K_1 & A_3 + B_1 K_2 \\ 0 & A_2 \end{bmatrix}, \quad BL = \begin{bmatrix} B_1 L \\ 0 \end{bmatrix}. \quad (4.5.3)$$

By Lemma 4.4.1, $(A_1 + B_1 K_1, B_1)$ is controllable. However, as $\mathrm{Im}\,(B_1 L) = \mathrm{Im}\,B_1$, it follows from Proposition 4.1.2 that $(A_1 + B_1 K_1, B_1 L)$ is controllable as well. Furthermore, (A, B) and $(A + BK, BL)$ have the same controllable subspace and (4.5.3) is a control normal form. Now it follows from Proposition 4.5.2 that both (A, B) and $(A + BK, BL)$ are stabilizable if and only if A_2 is stable. $\square$

The following variation on the theme will also be useful in the sequel.

Lemma 4.5.4. *If $D \geq 0$ and (A, D) is c-stabilizable, then there is an $X \geq 0$ such that $A - DX$ is c-stable.*

Proof. Let $\mathcal{C} = \mathcal{C}_{A,D}$ and consider orthonormal bases for $\mathcal{C}$ and $\mathcal{C}^\perp$. Using the decomposition $\mathbb{C}^n = \mathcal{C} \oplus \mathcal{C}^\perp$ we obtain representations which (with a small abuse of notation) we write as equalities:

$$A = \begin{bmatrix} A_1 & A_{12} \\ 0 & A_2 \end{bmatrix}, \quad D = \begin{bmatrix} D_1 & 0 \\ 0 & 0 \end{bmatrix}$$

where $D_1 \geq 0$, (A_1, D_1) is controllable, and A_2 is c-stable (see Propositions 4.5.1 and 4.5.2).

As (A_1, D_1) is a controllable pair, it follows from Theorem 4.4.2 that $A_1 - D_1 Y^{-1}$ is c-stable, where

$$Y = \int_0^1 e^{-A_1 t} D_1 e^{-A_1^* t} \, dt > 0.$$

Choosing $X = \begin{bmatrix} Y^{-1} & 0 \\ 0 & 0 \end{bmatrix} \geq 0$ it is found that $A - DX$ is c-stable. □

Corollary 4.5.5. *If (A, B) is a c-stabilizable pair then there is an $X \geq 0$ such that $A - (BRB^*)X$ is c-stable for any $m \times m$ matrix $R > 0$.*

Proof. Choose $K = 0$ and $L = RB^*$ in Lemma 4.5.3. Since $R > 0$ implies $\text{Im}\,(BL) = \text{Im}\,(BRB^*) = \text{Im}\,B$, it follows that (A, BRB^*) is c-stabilizable. But $BRB^* \geq 0$ and so, by Lemma 4.5.4, there is an $X \geq 0$ such that $A - (BRB^*)X$ is c-stable. □

The control normal form also leads to a useful generalization of the Hautus criterion (Theorem 4.3.3) to pairs that are merely stabilizable, rather than controllable.

Theorem 4.5.6. (a) *A pair (A, B), where A is $n \times n$ and B is $n \times m$, is c-stabilizable if and only if*

$$\text{rank}\,[\lambda I - A \ \ B] = n$$

for every $\lambda \in \mathbb{C}$ with $\text{Re}\,\lambda \geq 0$.
(b) *A pair (A, B) is d-stabilizable if and only if*

$$\text{rank}\,[\lambda I - A \ \ B] = n$$

for every $\lambda \in \mathbb{C}$ with $|\lambda| \geq 1$.

Proof. We prove part (a) only (the proof of (b) is similar). Without loss of generality we may assume that A and B are in the control normal form:

$$A = \begin{bmatrix} A_1 & A_3 \\ 0 & A_2 \end{bmatrix}, \quad B = \begin{bmatrix} B_1 \\ 0 \end{bmatrix}.$$

Since (A_1, B_1) is c-controllable, by Theorem 4.3.3 we have

$$\text{rank}\,[\lambda I - A_1 \ \ B_1] = p,$$

where A_1 is of size $p \times p$. Therefore,

$$\text{rank}\,[\lambda I - A \ \ B] = \text{rank}\,\begin{bmatrix} \lambda I - A_1 & -A_3 & B_1 \\ 0 & \lambda I - A_2 & 0 \end{bmatrix} < n$$

if and only if $\text{rank}\,(\lambda I - A_2) < n - p$, i.e. λ is an eigenvalue of A_2. It remains to apply Proposition 4.5.2. □

REAL MATRIX PAIRS

It will be useful to note that a pair (A, B) is controllable if and only if $(-A, B)$ is controllable. This follows immediately from the definition of $\mathcal{C}_{A,B}$, for example. Then Proposition 4.5.2 shows that (A, B) c-stabilizable does not imply that $(-A, B)$ is c-stabilizable. In contrast, (A, B) d-stabilizable *does* imply that $(-A, B)$ is d-stabilizable, because A_2 has all its eigenvalues in the open unit disc if and only if $-A_2$ does so.

4.6 Real matrix pairs

For convenience, the discussion of this chapter has been in the context of complex matrices and spaces $\mathbb{C}^n$. However, it is easily seen that all statements are readily re-phrased for real matrices and spaces $\mathbb{R}^n$. For example, if A is $n \times n$, B is $n \times m$, and both are real then the controllable subspace $\mathcal{C}_{A,B}$ is well-defined as a subspace of $\mathbb{R}^n$ and the geometric characterizations of Propositions 4.1.2 and 4.2.3 hold verbatim. Similarly, construction of the control normal form can be completed entirely in real arithmetic. The only possible exception to these remarks arises in Theorems 4.3.3 and 4.5.6. In these theorems it is necessary to retain the phrase "for all $\lambda \in \mathbb{C}$" because, of course, though A is real it may have non-real eigenvalues.

4.7 Exercises

4.7.1. Let (A, B) be a controllable pair, with B a single column ($n \times 1$). Prove that in a suitable basis in $\mathbb{C}^n$ the pair (A, B) takes the form

$$A_0 = \begin{bmatrix} 0 & 1 & 0 & \cdots & 0 \\ 0 & 0 & 1 & \cdots & 0 \\ \vdots & \vdots & \vdots & & \vdots \\ 0 & 0 & 0 & \cdots & 1 \\ a_0 & a_1 & a_2 & \cdots & a_{n-1} \end{bmatrix}, \quad B_0 = \begin{bmatrix} 0 \\ \vdots \\ 0 \\ 1 \end{bmatrix} \quad (4.7.1)$$

(in other words, there exists an invertible $n \times n$ matrix S such that $A_0 = S^{-1}AS$, $B_0 = S^{-1}B$). Hint: One such basis is $B, AB, \ldots, A^{n-1}B$.

4.7.2. Let (A, B) be as in Exercise 4.7.1. Prove that for every n-tuple of (not necessarily distinct) complex numbers $\lambda_1, \ldots, \lambda_n$ there is a $1 \times n$ row F such that $\lambda_1, \ldots, \lambda_n$ are the eigenvalues of $A + BF$, each eigenvalue repeated according to its algebraic multiplicity. Hint: Assuming $A = A_0$, $B = B_0$, where (A_0, B_0) is given by (4.7.1), define $F = [f_1, \ldots, f_n]$ by

$$\lambda^n - \sum_{j=0}^{n-1}(a_j + f_j)\lambda^j = (\lambda - \lambda_1)\ldots(\lambda - \lambda_n).$$

4.7.3. Let (A, B) be controllable, where A is $n \times n$ and B is $n \times m$. Prove that for every n-tuple of complex numbers $\lambda_1, \ldots, \lambda_n$ there is an $m \times n$ matrix F such that $\lambda_1, \ldots, \lambda_n$ are the eigenvalues of $A + BF$ (counted with their algebraic multiplicities). Hint: Using the following lemma, reduce the proof to the case of a one-column matrix B.

Lemma 4.7.1. *If (A, B) is controllable, then for every $x \in \text{Im } B$, $x \neq 0$, there exists an F such that the pair $(A + BF, x)$ is controllable.*

(A proof of Lemma 4.7.1 can be found, for example, in Wonham (1979), Section 2.1).

4.7.4. Let (A, B) be a controllable pair of real matrices. Prove that for every n-tuple of complex numbers $\alpha = \{\lambda_1, \ldots, \lambda_n\}$ which is *symmetric* (in the sense that if λ appears in α, then so does $\bar{\lambda}$, and the number of appearances of λ and of $\bar{\lambda}$ in α is the same) there is a real matrix F such that $\lambda_1, \ldots, \lambda_n$ are the eigenvalues of $A + BF$ (counted with their algebraic multiplicities). Hint: Lemma 4.7.1 is valid also in the framework of real matrices.

4.7.5. Prove the converse of Exercise 4.7.3: If A and B are $n \times n$ and $n \times m$ matrices, respectively, and the pair (A, B) is not controllable, then there exists a $\lambda_0 \in \mathbb{C}$ such that $\lambda_0 \in \sigma(A + BF)$ for every $n \times m$ matrix F. Hint: Use the control normal form.

4.7.6. State and prove the real analogue of Exercise 4.7.5.

4.7.7. Let $A \in \mathbb{C}^{n \times n}$ and $R = \frac{1}{2}(A + A^*)$. Show that, if $R \leq 0$ and (A, R) is a controllable pair, then A is c-stable.

4.8 Notes

All the material in this chapter is fundamental in the analysis of linear systems and control and can be found in many texts and monographs. We mention the following samples: Brockett (1970), Gohberg et al. (1986a), Kailath (1980), Kalman et al. (1969), Kwakernaak and Sivan (1972), Russell (1979), Sontag (1990), and Wonham (1970 and 1979).

5
LINEAR MATRIX EQUATIONS

In this chapter we present a self-contained account of some well-known facts concerning solutions of linear matrix equations. Our attention is restricted mainly to the Lyapunov and Stein equations:

$$SA - BS = \Gamma, \quad \text{and} \quad S - BSA = \Gamma,$$

respectively (see equations (5.2.3) and (5.2.4) below). In particular, their symmetric forms (when $B = A^*$, $\Gamma^* = \Gamma$) are important in the sequel. Conditions are developed which ensure the (suitably defined) "stability" of solution matrices S of these equations.

5.1 The Kronecker product

Let us first introduce a matrix product which is a useful tool in the study of linear equations in matrices. If A, B are complex matrices of sizes $m \times m$ and $n \times n$, respectively, their *Kronecker product* is defined to be the partitioned $mn \times mn$ matrix

$$A \otimes B = \begin{bmatrix} a_{11}B & a_{12}B & \cdots & a_{1m}B \\ a_{21}B & a_{22}B & & a_{2m}B \\ \vdots & & & \\ a_{m1}B & a_{m2}B & \cdots & a_{mm}B \end{bmatrix}.$$

(This product is sometimes called a *right* Kronecker product, or a *direct*, or *tensor* product.) First note some easily verified algebraic properties (see Chapter 12 of Lancaster and Tismenetsky (1985) for more details).

If A_1, A_2 are $m \times m$ and B_1, B_2 are $n \times n$ then we have

$$(A_1 A_2) \otimes (B_1 B_2) = (A_1 \otimes B_1)(A_2 \otimes B_2). \tag{5.1.1}$$

If A and B are invertible then $A \otimes B$ is invertible and

$$(A \otimes B)^{-1} = A^{-1} \otimes B^{-1}. \tag{5.1.2}$$

Consider a complex polynomial in two variables, say

$$p(x,y) = \sum_{i,j=0}^{\ell} c_{ij} x^i y^j$$

for some complex coefficients c_{ij}. The value of this polynomial at (A, B) is defined to be

$$p(A;B) = \sum_{i,j=0}^{\ell} c_{ij} A^i \otimes B^j. \qquad (5.1.3)$$

A fundamental theorem on Kronecker products expresses the eigenvalues of $p(A;B)$ in terms of those of A and B.

Theorem 5.1.1. *If A and B are square matrices and $p(A;B)$ is defined by (5.1.3), then*

$$\sigma(p(A;B)) = \{p(\lambda_r, \mu_s) : \lambda_r \in \sigma(A), \mu_s \in \sigma(B)\}.$$

Proof. Using Theorem 1.1.3, there are nonsingular matrices P and Q such that $J_1 := PAP^{-1}$ and $J_2 := QBQ^{-1}$ are Jordan matrices. Using properties (5.1.1) and (5.1.2) it is found that

$$\begin{aligned}
J_1^i \otimes J_2^j &= (PA^i P^{-1}) \otimes (QB^j Q^{-1}) \\
&= (P \otimes Q)(A^i \otimes B^j)(P^{-1} \otimes Q^{-1}) \\
&= (P \otimes Q)(A^i \otimes B^j)(P \otimes Q)^{-1}.
\end{aligned}$$

Hence

$$p(J_1; J_2) = (P \otimes Q) p(A;B) (P \otimes Q)^{-1},$$

and this similarity shows that $\sigma(p(A;B))$ coincides with $\sigma(p(J_1;J_2))$. But it is easily seen that $p(J_1;J_2)$ is upper triangular with diagonal entries equal to $p(\lambda_r, \mu_s)$. The result follows. □

Two important special cases of the theorem justify a separate statement:

Corollary 5.1.2. *If A is $m \times m$ with eigenvalues $\lambda_1, \ldots, \lambda_m$ and B is $n \times n$ with eigenvalues $\mu_1, \ldots, \mu_n$, then*

(a) *the eigenvalues of $A \otimes B$ are the mn numbers $\lambda_r \mu_s$, $1 \leq r \leq m$, $1 \leq s \leq n$.*

(b) *the eigenvalues of $(A \otimes I_n) + (I_m \otimes B)$ are the mn numbers $\lambda_r + \mu_s$, $1 \leq r \leq m$, $1 \leq s \leq n$.*

Proof. Parts (a) and (b) are obtained on setting $p(x,y) = xy$ and $p(x,y) = x+y$ in the theorem, respectively. □

5.2 Linear equations in matrices

Before considering matrix equations we introduce an operation on matrices which has the effect of "stacking" the columns of an $m \times n$ matrix to form a column vector with mn entries. Thus, if A is $m \times n$ with columns $A_{*1}, A_{*2}, \ldots, A_{*n}$ then the *vec-function* acting on A has the form

$$\operatorname{vec} A := \begin{bmatrix} A_{*1} \\ A_{*2} \\ \vdots \\ A_{*n} \end{bmatrix} \in \mathbb{C}^{mn}.$$

It is apparent that the vec-function is linear in the sense that

$$\operatorname{vec}(\alpha A_1 + \beta A_2) = \alpha \operatorname{vec} A_1 + \beta \operatorname{vec} A_2$$

for any $m \times n$ matrices A_1 and A_2 and any numbers α and β.

The vital property of the vec-function for our purposes is contained in the next proposition.

Proposition 5.2.1. *If A is $m \times m$, B is $n \times n$ and X is $m \times n$, then*

$$\operatorname{vec}(AXB) = (B^T \otimes A) \operatorname{vec} X$$

Proof. The jth column of AXB, say $(AXB)_{*j}$ is

$$(AXB)_{*j} = AXB_{*j} = \sum_{k=1}^{n} b_{kj}(AX)_{*k} = \sum_{k=1}^{n} (b_{kj}) AX_{*k}$$
$$= [b_{1j}A \quad b_{2j}A \quad \ldots \quad b_{nj}A] \operatorname{vec} X.$$

It remains only to identify the right-hand side with the jth block row of $(B^T \otimes A) \operatorname{vec} X$. $\square$

Now consider a linear equation in matrices of the form

$$A_1 X B_1 + A_2 X B_2 + \cdots + A_p X B_p = C \tag{5.2.1}$$

where $A_1, \ldots, A_p, B_1, \ldots, B_p, C$ are given (and their sizes are compatible), and X is to be found. Using Proposition 5.2.1 and the linearity of the vec-function it is clear that (5.2.1) can be reformulated as

$$\{(B_1^T \otimes A_1) + \cdots + (B_p^T \otimes A_p)\} \operatorname{vec} X = \operatorname{vec} C. \tag{5.2.2}$$

This reformulation, together with Corollary 5.1.2 is the key to the following basic properties of the so-called Sylvester and Stein equations.

Theorem 5.2.2. *For given matrices* $A \in \mathbb{C}^{m \times m}$, $B \in \mathbb{C}^{n \times n}$ *and* $\Gamma \in \mathbb{C}^{n \times m}$ *the Sylvester equation*

$$SA - BS = \Gamma \qquad (5.2.3)$$

has a unique solution if and only if A *and* B *have no eigenvalues in common.*

Proof. Write $SA - BS = I_n SA - BSI_m$ and rewrite equation (5.2.3) using the device of (5.2.2) in the form

$$(A^T \otimes I_n - I_m \otimes B) \operatorname{vec} S = \operatorname{vec} \Gamma.$$

It follows from Corollary 5.1.2(b) that the coefficient matrix has eigenvalues $\lambda_r - \mu_s$ where $\lambda_r \in \sigma(A)$ and $\mu_s \in \sigma(B)$. Thus, there is a unique solution of (5.2.3) if and only if $\lambda_r \neq \mu_s$ for any r and s. □

Theorem 5.2.3. *For given matrices* $A \in \mathbb{C}^{m \times m}$, $B \in \mathbb{C}^{n \times n}$ *and* $\Gamma \in \mathbb{C}^{n \times m}$ *the Stein equation*

$$S - BSA = \Gamma \qquad (5.2.4)$$

has a unique solution if and only if $\lambda_r \mu_s \neq 1$ *for any* $\lambda_r \in \sigma(A)$, $\mu_s \in \sigma(B)$.

Proof. Using (5.2.2) equation (5.2.4) is first reformulated as

$$(I_{mn} - A^T \otimes B) \operatorname{vec} S = \operatorname{vec} \Gamma$$

and, using Corollary 5.1.2(a), it is found that the coefficient matrix is nonsingular if and only if $\lambda_r \mu_s \neq 1$ for all $\lambda_r \in \sigma(A)$ and $\mu_s \in \sigma(B)$. □

Under the conditions of Theorem 5.2.2, or of 5.2.3, it is possible to use contour integrals to give explicit formulae for the solutions of the Sylvester equation, or Stein equation, respectively.

Theorem 5.2.4. (a) *If the Sylvester equation (5.2.3) has a unique solution* S *then*

$$S = \frac{1}{2\pi i} \int_{\mathcal{C}_1} (\lambda I - B)^{-1} \Gamma (\lambda I - A)^{-1} d\lambda \qquad (5.2.5)$$

where $\mathcal{C}_1$ *is a simple closed contour with* $\sigma(A)$ *inside* $\mathcal{C}_1$ *and* $\sigma(B)$ *outside* $\mathcal{C}_1$.
(b) *If the Stein equation (5.2.4) has a unique solution* S *then*

$$S = \frac{1}{2\pi i} \int_{\mathcal{C}_2} (\lambda I - B)^{-1} \Gamma (I - \lambda A)^{-1} d\lambda \qquad (5.2.6)$$

where $\mathcal{C}_2$ *is a simple closed contour with* $\sigma(B)$ *inside* $\mathcal{C}_2$ *and, for each nonzero* $w \in \sigma(A)$, w^{-1} *is outside* $\mathcal{C}_2$.

Proof. (a) The proof is by verification. Substitute the expression (5.2.5) for S in the left side of the Sylvester equation (5.2.3) to get

$$SA - BS = \frac{1}{2\pi i} \int_{\mathcal{C}_1} [(\lambda I - B)^{-1}\Gamma(\lambda I - A)^{-1}A - B(\lambda I - B)^{-1}\Gamma(\lambda I - A)^{-1}]\, d\lambda.$$

Next add and subtract $\lambda(\lambda I - B)^{-1}\Gamma(\lambda I - A)^{-1}$ in the integrand to get

$$SA - BS = \frac{1}{2\pi i} \int_{\mathcal{C}_1} [(\lambda I - B)^{-1}\Gamma(\lambda I - A)^{-1}(A - \lambda I)$$
$$+ (\lambda I - B)(\lambda I - B)^{-1}\Gamma(\lambda I - A)^{-1}]\, d\lambda$$
$$= -\frac{1}{2\pi i} \int_{\mathcal{C}_1} (\lambda I - B)^{-1}\Gamma\, dz + \frac{1}{2\pi i} \int_{\mathcal{C}_1} \Gamma(\lambda I - A)^{-1}\, d\lambda.$$

Since $\sigma(B)$ is outside $\mathcal{C}_1$, the first term is 0 by Cauchy's theorem. Since $\sigma(A)$ is outside $\mathcal{C}_1$, it follows from Proposition 1.3.3 that the second term reduces to Γ. Thus S is the solution of (5.2.3) as asserted.

(b) This is verified in a similar way. □

5.3 Lyapunov and symmetric Stein equations

Consider the Lyapunov equation

$$SA^* + AS = W, \qquad (5.3.1)$$

where A and W are given $n \times n$ complex matrices and W is hermitian. We focus on hermitian solutions S. (It is easy to see that S solves (5.3.1) if and only if its hermitian part $\frac{1}{2}(S + S^*)$ solves (5.3.1) and its skew hermitian part $\frac{1}{2i}(S - S^*)$ solves the homogeneous equation $TA^* + AT = 0$.)

It follows from Theorem 5.2.2 that (5.3.1) has a unique solution (necesarily hermitian) provided A has no pairs of eigenvalues symmetric relative to the imaginary axis (i.e. $z_0 \in \sigma(A)$ implies $-\bar{z}_0 \notin \sigma(A)$). In particular, the solution is unique if all eigenvalues of A are in the open right half-plane, and for such a matrix A the solution is given by

$$S = \int_{-\infty}^{0} e^{At} W e^{A^*t}\, dt \qquad (5.3.2)$$

(the integral converges because all eigenvalues of A and A^* have positive real parts). Indeed, substituting the right-hand side of (5.3.2) in (5.3.1) we have

$$SA^* + AS = \int_{-\infty}^{0} (e^{At}We^{A^*t}A^* + Ae^{At}We^{A^*t})\,dt$$
$$= \int_{-\infty}^{0} (e^{At}We^{A^*t})'\,dt = (e^{At}We^{A^*t})\,|\,_{t=-\infty}^{t=0} = W.$$

Similarly, the solution S of (5.3.1) is unique if all eigenvalues of A are in the open left half-plane and then S is given by the formula

$$S = -\int_{0}^{\infty} e^{At}We^{A^*t}\,dt. \tag{5.3.3}$$

Theorem 5.3.1. *Assume that all eigenvalues of A lie in the open right half-plane and let S be the unique solution of (5.3.1). Then:*

(a) *If $W > 0$ then $S > 0$ and, if $W \geq 0$, then $S \geq 0$.*

(b) *If, moreover, $W \geq BB^*$ where (A, B) is a controllable pair, then we have $S > 0$.*

Proof. Part (a) follows immediately from equation (5.3.2). Now assume the hypotheses of part (b) and let $Sx = 0$ for some $x \in \mathbb{C}^n$. Then

$$0 = x^*(SA^* + AS)x = x^*Wx \geq x^*BB^*x \geq 0.$$

It follows that $Wx = 0$ and also $x^*B = 0$. Furthermore, using equation (5.3.1), $SA^*x = 0$. Hence

$$0 = x^*A(SA^* + AS)A^*x = x^*AWA^*x \geq x^*ABB^*A^*x \geq 0.$$

Thus $x^*B = 0$ and also $x^*AB = 0$ (together with $WA^*x = 0$ and $S(A^*)^2x = 0$).

Repeating this procedure it is found that $x^*A^jB = 0$ for $j = 0, 1, 2, \ldots$, and so

$$x^*[B \quad AB \quad \ldots \quad A^{n-1}B] = 0.$$

Using equation (4.1.3) this contradicts the controllability of (A, B) unless $x = 0$. Hence $S > 0$. □

As a converse of Theorem 5.3.1 we have:

Theorem 5.3.2. *Suppose $W \geq 0$ and (5.3.1) has a solution $S \geq 0$.*

(a) *If $W > 0$ then A has all its eigenvalues in the open right half-plane and we have $S > 0$.*

(b) *If (A, B) is a controllable pair and $W \geq BB^*$ then A has all its eigenvalues in the open right half-plane and we have $S > 0$.*

Proof. Let $\lambda = \mu + i\omega$ ($\mu, \omega \in \mathbb{R}$) be an eigenvalue of A. Then there is a nonzero x such that $x^*A = (\mu + i\omega)x^*$. Also,

$$x^*Wx = x^*(SA^* + AS)x = (\bar{\lambda} + \lambda)x^*Sx = 2\mu(x^*Sx). \quad (5.3.4)$$

Clearly, $W > 0$ now implies $\mu > 0$ for all eigenvalues of A, as required. Then $S > 0$ follows from Theorem 5.3.1. This completes the proof of part (a).

Now suppose that (A, B) is controllable and $W \geq BB^*$. Define $\lambda = \mu + i\omega$ and x as in (5.3.4), and $\mu \leq 0$ implies

$$0 = x^*Wx \geq x^*BB^*x \geq 0$$

and $B^*x = 0$. Thus $x \in (\operatorname{Ker} B^*) \cap (\operatorname{Ker}(A - i\omega I)^*)$, which contradicts the controllability of (A, B) as described in Theorem 4.3.3. Hence $\mu > 0$. Now $S > 0$ follows from Theorem 5.3.1. □

Part (a) of this theorem can be extended by a continuity argument to give:

Corollary 5.3.3. *If $W \geq 0$ and (5.3.1) has a solution $S > 0$ then all the eigenvalues of A are in the closed right half-plane.*

Proof. Let $A_\epsilon = A + \epsilon S^{-1}$, $\epsilon > 0$. Then (5.3.1) gives

$$SA_\epsilon^* + A_\epsilon S = W + 2\epsilon I.$$

Since $W + 2\epsilon I > 0$ it follows from Theorem 5.3.2(a) that $\sigma(A_\epsilon)$ is in the open right half-plane. Since the eigenvalues of A_ϵ depend continuously on ϵ the result is obtained on taking the limit as $\epsilon \to 0$. □

The controllability condition of part (b) of Theorem 5.3.2 can be relaxed to some extent. Proposition 4.5.1 is used in the next proof and so the stronger result depends on the logical sequence of statements: $5.3.2 \Rightarrow 4.4.2 \Rightarrow 4.5.1 \Rightarrow 5.3.4$. In particular Theorem 4.4.2 shows that (A, B) controllable implies that $(-A, B)$ is stabilizable.

Theorem 5.3.4. *Let $(-A, B)$ be a c-stabilizable pair with $A \in \mathbb{C}^{n \times n}$, $B \in \mathbb{C}^{n \times m}$ and suppose $W \geq BB^*$. If the equation $SA^* + AS = W$ has a solution $S \geq 0$ then A has all its eigenvalues in the open right half-plane.*

Proof. Let A have an eigenvalue $\lambda = \mu + i\omega$ where $\mu, \omega \in \mathbb{R}$ and $\mu \leq 0$. We seek a contradiction.

Take advantage of Proposition 4.5.1 and its proof and, without loss of generality, write A and B in control normal form:

$$A = \begin{bmatrix} A_1 & A_3 \\ 0 & A_2 \end{bmatrix}, \quad B = \begin{bmatrix} B_1 \\ 0 \end{bmatrix}$$

where $\mathbb{C}^n = \mathcal{C}_{A,B} \dotplus \mathcal{S}$, say, (A_1, B_1) is controllable, and $-A_2$ is c-stable.

There is a nonzero x such that $x^*A = \lambda x^*$. Write $x = x_1 + x_2$ where $x_1 \in \mathcal{C}_{A,B}$ and $x_2 \in \mathcal{S}$. Then we obtain

$$x_1^* A_1 = \lambda x_1^*, \quad x_1^* A_3 + x_2^* A_2 = \lambda x_2^*.$$

If $x_1 = 0$ then $x_2 \neq 0$ and $\mu + i\omega$ (with $\mu \leq 0$) is an eigenvalue of A_2. But this contradicts the c-stability of $-A_2$. Hence $x_1 \neq 0$ and λ is an eigenvalue of A_1.

Now equation (5.3.1) implies

$$(\bar{\lambda} + \lambda) x^* S x = x^* W x$$

and $\mu \leq 0$, $S \geq 0$ imply $x^* W x = 0$. Thus

$$0 \leq x^* B B^* x \leq x^* W x = 0,$$

and so $x \in \operatorname{Ker} B^*$. But this gives $x_1 \in \operatorname{Ker} B_1^*$ and then

$$x_1 \in (\operatorname{Ker} B_1^*) \cap \operatorname{Ker}(A_1 - \bar{\lambda}I)^*.$$

From Theorem 4.3.3 we obtain a contradiction with the controllability of (A_1, B_1). Hence $\mu > 0$. $\square$

Assuming A has the Jordan form, criteria for the existence of hermitian solutions S of (5.3.1) are given in the Appendix of Ball et al. (1990). Also, formulas describing the hermitian solutions S (when they exist) in terms of the Jordan structure of A can be found in the same appendix.

Consider now the symmetric Stein equation

$$S - A^* S A = V, \tag{5.3.5}$$

where A and V are given $n \times n$ matrices and V is hermitian. By Theorem 5.2.3 the equation (5.3.5) has a unique solution S (necessarily hermitian) provided $z\bar{w} \neq 1$ for every pair of eigenvalues z, w of A. In particular, this is the case when all eigenvalues of A lie in the open unit disc $\{\lambda \in \mathbb{C} \mid |\lambda| < 1\}$ and then the unique solution is given by

$$S = \sum_{j=0}^{\infty} A^{*j} V A^j \tag{5.3.6}$$

(the series converges because $|z_0| < 1$ for every $z_0 \in \sigma(A)$).

We need an analogue of Theorem 5.3.1.

CONTINUOUS AND ANALYTIC DEPENDENCE

Theorem 5.3.5. *Assume that all eigenvalues of matrix A are in the open unit disc and let S be the unique solution of* (5.3.5).

(a) *If $V \geq 0$, then $S \geq 0$.*

(b) *If $V \geq C^*C$ where (C, A) is an observable pair, then $S > 0$.*

Proof. Part (a) follows immediately from the representation (5.3.6). Given the hypotheses of part (b) we have $S \geq 0$ from part (a). If $Sx = 0$ for some $x \in \mathbb{C}^n$, then using (5.3.6) again, it follows that

$$0 = \sum_{j=0}^{\infty} x^* A^{*j} V A^j x \geq \sum_{j=0}^{\infty} (x^* A^{*j} C^*)(C A^j x) \geq 0.$$

Hence $CA^j x = 0$ for $j = 0, 1, 2, \ldots$. But (C, A) observable implies that $\cap_{j=0}^{\infty} \operatorname{Ker}(CA^j) = \{0\}$ (see equation (4.2.2)). Hence $x = 0$ and S must be positive definite. □

5.4 Continuous and analytic dependence

We present here some results on parameter dependence of solutions of linear equations that will be needed in later chapters.

Let Ω be an open set in the k-dimensional real space $\mathbb{R}^k$. The set Ω will be interpreted as the admissible values for a set of k parameters. A matrix (or vector) valued function $A(t)$, $t = (t_1, \ldots, t_k) \in \Omega$, is said to belong to the class $\mathcal{C}^p$, where p is a nonnegative integer if every entry of $A(t)$ has continuous partial derivatives with respect to t_j up to, and including, the order p. As usual, when $p = 0$, the class $\mathcal{C}^0$ is interpreted as the class of continuous functions. The class $\mathcal{C}^\infty$ consists of matrix functions having continuous partial derivatives of all orders. We introduce also the class $\mathcal{A}$ of matrix functions $A(t)$ all of whose entries are analytic in Ω, i.e. for every $t_0 = (t_{01}, \ldots, t_{0k}) \in \Omega$ the function $A(t)$ is represented in some neighborhood of t_0 by a convergent power series

$$A(t) = \sum (t_1 - t_{01})^{\alpha_1} \cdots (t_k - t_{0k})^{\alpha_k} A_{\alpha_1 \ldots \alpha_k},$$

where the sum is taken over all k-tuples of nonnegative integers $(\alpha_1, \ldots, \alpha_k)$.

The basic result of this section is

Theorem 5.4.1. *Given an $n \times n$ matrix function $A(t)$ and $n \times 1$ column $b(t)$ both of class $\mathcal{C}^p$ for some p, or both of class $\mathcal{A}$; if the equation*

$$A(t)x = b(t) \tag{5.4.1}$$

has a unique solution $x = x(t)$ for every $t \in \Omega$, then $x(t)$ is also in the class $\mathcal{C}^p$ or $\mathcal{A}$ (as appropriate).

The proof of this theorem is trivial: just write $x(t) = A(t)^{-1}b(t)$, and observe that the function $A(t)^{-1} = (\det A)^{-1}(\operatorname{Adj} A(t))$, where $\operatorname{Adj} A(t)$ is the algebraic adjoint of $A(t)$, is of the class $\mathcal{C}^p$ or $\mathcal{A}$, as appropriate.

Despite the trivial proof, Theorem 5.4.1 is an important result. In fact, it is valid for any class $\mathcal{X}$ of complex valued functions defined on a set X with the property that the class $\mathcal{X}$ is closed under the operations of addition, multiplication, and taking the inverse of a nowhere zero function.

Theorem 5.4.1 has obvious corollaries concerning the continuous and analytic dependence of the solutions to the Sylvester equation $SA - BS = \Gamma$, Stein equation $S - BSA = \Gamma$, Lyapunov equation $SA^* + AS = W$ and symmetric Stein equation $S - A^*SA = V$ that have been studied in Sections 5.1–5.3. We state such a corollary concerning the Sylvester equation only, and leave the statement of other relevant corollaries to the reader.

Corollary 5.4.2. *Consider the Sylvester equation*

$$SA - BS = \Gamma, \tag{5.4.2}$$

where the given matrices A, B and Γ are of sizes $n \times n$, $m \times m$ and $m \times n$, respectively, and S is an unknown $m \times n$ matrix. Assume that $A = A(t)$, $B = B(t)$ and $\Gamma = \Gamma(t)$ are functions of $t \in \Omega$ of the class $\mathcal{C}^p$ (for some p) or $\mathcal{A}$. If $\sigma(A(t)) \cap \sigma(B(t)) = \emptyset$ for all $t \in \Omega$, then the unique solution $S = S(t)$ of (5.4.2) is also of the class $\mathcal{C}^p$ (for some p) or $\mathcal{A}$, as appropriate.

Again, the result of Corollary 5.4.2 can be extended to the class $\mathcal{X}$ of functions described above.

5.5 Notes

The material of this chapter is fundamental for linear algebra and the theory of matrices. See, for example, Gantmacher (1959), Horn and Johnson (1991), or Lancaster and Tismenetsky (1985). For extensions to the operator case see Daleckii and Krein (1974). The basic idea behind the results of Section 5.3 are due to Chen (1973) and Wimmer (1974).

6
RATIONAL MATRIX FUNCTIONS

In this chapter we study $r \times n$ matrices $W(\lambda)$ whose elements are rational functions of a complex variable λ. Thus we may write

$$W(\lambda) = \left[\frac{p_{ij}(\lambda)}{q_{ij}(\lambda)}\right]_{i,j=1}^{r,n}$$

where $p_{ij}(\lambda)$ and $q_{ij}(\lambda)$ are scalar polynomials and $q_{ij}(\lambda)$ are not identically zero. Such functions $W(\lambda)$ are called *rational matrix functions*.

For most of the chapter it will be assumed that the coefficients of $W(\lambda)$ are complex matrices (i.e. the coefficients of $p_{ij}(\lambda)$ and $q_{ij}(\lambda)$ are complex numbers). We consider rational matrix functions with real coefficients in the last section.

6.1 Realizations of rational matrix functions

Let

$$W(\lambda) = \left[\frac{p_{ij}(\lambda)}{q_{ij}(\lambda)}\right]_{i,j=1}^{r,n}$$

be an $r \times n$ rational matrix function. We assume that $W(\lambda)$ is finite at infinity; that is, in each entry $p_{ij}(\lambda)/q_{ij}(\lambda)$ of $W(\lambda)$ the degree of the polynomial $p_{ij}(\lambda)$ is less than or equal to the degree of $q_{ij}(\lambda)$.

A *realization* of the rational matrix function $W(\lambda)$ is a representation of the form

$$W(\lambda) = D + C(\lambda I_m - A)^{-1}B, \quad \lambda \notin \sigma(A) \tag{6.1.1}$$

where for some m, A, B, C, D are (complex) matrices of sizes $m \times m$, $m \times n$, $r \times m$, $r \times n$, respectively. Observe that

$$\lim_{\lambda \to \infty} (\lambda I - A)^{-1} = 0. \tag{6.1.2}$$

Indeed, to verify (6.1.2) we can assume, without loss of generality, that A is in the Jordan form:

$$A = J_{r_1}(\lambda_1) \oplus \cdots \oplus J_{r_m}(\lambda_m).$$

Then

$$(\lambda I - A)^{-1} = \begin{bmatrix} (\lambda-\lambda_1)^{-1} & (\lambda-\lambda_1)^{-2} & \cdots & (\lambda-\lambda_1)^{-r_1} \\ 0 & (\lambda-\lambda_1)^{-1} & \cdots & (\lambda-\lambda_1)^{-r_1+1} \\ \vdots & \vdots & \ddots & \vdots \\ 0 & 0 & \cdots & (\lambda-\lambda_1)^{-1} \end{bmatrix}$$

$$\oplus \cdots \oplus \begin{bmatrix} (\lambda-\lambda_m)^{-1} & (\lambda-\lambda_m)^{-2} & \cdots & (\lambda-\lambda_m)^{-r_m} \\ 0 & (\lambda-\lambda_m)^{-1} & \cdots & (\lambda-\lambda_m)^{-r_m+1} \\ \vdots & \vdots & \ddots & \vdots \\ 0 & 0 & \cdots & (\lambda-\lambda_m)^{-1} \end{bmatrix}$$

and (6.1.2) follows. So if there exists a realization (6.1.1), then necessarily $D = W(\infty)$. We may thus identify such a realization with the triple (A, B, C). The following lemma is useful in the proof of existence of a realization.

Lemma 6.1.1. *Let*
$$H(\lambda) = \sum_{j=0}^{l-1} \lambda^j H_j$$

and
$$L(\lambda) = \lambda^l I + \sum_{j=0}^{l-1} \lambda^j A_j$$

be polynomials with matrix coefficients, where the H_j are $r \times n$ matrices and A_j are $n \times n$ matrices. Put

$$B = \begin{bmatrix} 0 \\ \vdots \\ 0 \\ I \end{bmatrix}, \quad A = \begin{bmatrix} 0 & I & \cdots & 0 \\ \vdots & & \ddots & \vdots \\ 0 & 0 & \cdots & I \\ -A_0 & -A_1 & \cdots & -A_{l-1} \end{bmatrix}, \quad C = [H_0 \; \cdots \; H_{l-1}]$$

Then
$$H(\lambda)L(\lambda)^{-1} = C(\lambda I - A)^{-1}B.$$

Proof. We start with the formula for $L(\lambda)^{-1}$. Define $nl \times nl$ matrices with polynomial entries $E(\lambda)$ and $F(\lambda)$ as follows:

$$F(\lambda) = \begin{bmatrix} I & 0 & \cdots & 0 & 0 \\ -\lambda I & I & \cdots & 0 & 0 \\ \vdots & \vdots & & \vdots & \vdots \\ 0 & 0 & \cdots & I & 0 \\ 0 & 0 & \cdots & -\lambda I & I \end{bmatrix}$$

(where $I = I_n$),

REALIZATIONS OF RATIONAL MATRIX FUNCTIONS

$$E(\lambda) = \begin{bmatrix} B_{l-1}(\lambda) & B_{l-2}(\lambda) & \cdots & B_0(\lambda) \\ -I & 0 & \cdots & 0 \\ 0 & -I & \cdots & \\ \vdots & \vdots & & \vdots \\ 0 & 0 & \cdots & -I & 0 \end{bmatrix}$$

where $B_0(\lambda) = I$ and $B_{r+1}(\lambda) = \lambda B_r(\lambda) + A_{l-r-1}$ for $r = 0, 1, \ldots, l-2$. Direct multiplication shows that

$$E(\lambda)(\lambda I - A) = \begin{bmatrix} L(\lambda) & 0 \\ 0 & I \end{bmatrix} F(\lambda),$$

and hence

$$\begin{bmatrix} L(\lambda)^{-1} & 0 \\ 0 & I \end{bmatrix} = F(\lambda)(\lambda I - A)^{-1}(E(\lambda))^{-1}. \tag{6.1.3}$$

The structure of $E(\lambda)$ shows that the first n columns of the matrix $E(\lambda)^{-1}$ have the form

$$B = \begin{bmatrix} 0 \\ \vdots \\ 0 \\ I \end{bmatrix}.$$

Now, multiplying equation (6.1.3) on the left by $P = [I \ 0 \ \cdots \ 0]$ and on the right by P^T and, using the relation $PF(\lambda) = P$, we obtain the formula

$$L(\lambda)^{-1} = P(\lambda I - A)^{-1} B. \tag{6.1.4}$$

We may define $C_1(\lambda), \ldots, C_l(\lambda)$ for all $\lambda \notin \sigma(A)$ by

$$\begin{bmatrix} C_1(\lambda) \\ \vdots \\ C_l(\lambda) \end{bmatrix} = (\lambda I - A)^{-1} B.$$

From equation (6.1.4) we see that $C_1(\lambda) = L(\lambda)^{-1}$. As

$$(\lambda I - A) \begin{bmatrix} C_1(\lambda) \\ C_2(\lambda) \\ \vdots \\ C_l(\lambda) \end{bmatrix} = B,$$

the special form of A yields

$$C_i(\lambda) = \lambda^{i-1} C_1(\lambda), \quad 1 \le i \le l.$$

It follows that $C(\lambda I - A)^{-1} B = \sum_{j=0}^{l-1} H_j C_{j+1}(\lambda) = H(\lambda) L(\lambda)^{-1}$, and the proof is complete. $\square$

We are now ready to prove the existence of realizations.

Theorem 6.1.2. *Every $r \times n$ rational matrix function that is finite at infinity has a realization.*

Proof. Let $W(\lambda)$ be an $r \times n$ rational matrix function with finite value at infinity. There exists a monic scalar polynomial $l(\lambda)$ such that $l(\lambda)W(\lambda)$ is a (matrix) polynomial. For example, $l(\lambda)$ could be the least common multiple of the denominators of entries in $W(\lambda)$. Put $H(\lambda) = l(\lambda)(W(\lambda) - W(\infty))$. Then $H(\lambda)$ is an $r \times n$ matrix polynomial. Clearly, $L(\lambda) = l(\lambda)I_n$ is monic, i.e. its leading coefficient is the identity matrix, and $W(\lambda) = W(\infty) + H(\lambda)L(\lambda)^{-1}$. Further
$$\lim_{\lambda \to \infty} H(\lambda)L(\lambda)^{-1} = \lim_{\lambda \to \infty} [W(\lambda) - W(\infty)] = 0.$$
So the degree of $H(\lambda)$ is strictly less than the degree of $L(\lambda)$ and Lemma 6.1.1 can be applied to find A, B, C for which
$$W(\lambda) = W(\infty) + H(\lambda)L(\lambda)^{-1} = W(\infty) + C(\lambda I - A)^{-1}B.$$
This is a realization of $W(\lambda)$. □

A realization for $W(\lambda)$ is far from being unique. This can be seen from our construction of a realization because there are many choices for $l(\lambda)$. In general, if (A, B, C) is a realization of $W(\lambda)$, then so is $(\hat{A}, \hat{B}, \hat{C})$, where

$$\hat{A} = \begin{bmatrix} A_{11} & A_{12} & A_{13} \\ 0 & A & A_{23} \\ 0 & 0 & A_{33} \end{bmatrix}, \quad \hat{B} = \begin{bmatrix} B_1 \\ B \\ 0 \end{bmatrix}, \quad \hat{C} = [0 \ C \ C_1] \qquad (6.1.5)$$

for any matrices A_{ij}, B_1, and C_1 with suitable sizes (in other words, the matrices $\hat{A}, \hat{B}, \hat{C}$ are of size $s \times s$, $s \times n$, $r \times s$, respectively, and partitioned with respect to the orthogonal sum $\mathbb{C}^s = \mathbb{C}^p \oplus \mathbb{C}^m \oplus \mathbb{C}^q$, where m is the size of A; for instance, A_{13} is a $p \times q$ matrix).

In addition, a realization (A, B, C) can always be replaced by a realization (SAS^{-1}, SB, CS^{-1}), which is said to be *similar* to (A, B, C).

We will be especially interested in realizations (A, B, C) of $W(\lambda)$ with additional properties. In particular those for which the pair (C, A) is observable and the pair (A, B) is controllable (see Chapter 4). The next result shows that, in some sense, any realization "contains" a realization with these two properties. To make this precise it is convenient to introduce another definition.

Let (A, B, C) be a realization of $W(\lambda)$ and let A be of size $m \times m$. Consider a direct sum decomposition

$$\mathbb{C}^m = \mathcal{L} \dotplus \mathcal{M} \dotplus \mathcal{N} \qquad (6.1.6)$$

where the subspaces $\mathcal{L}$ and $\mathcal{L} \dotplus \mathcal{M}$ are A- invariant with the property that $C|\mathcal{L} = 0$ and $\text{Im } B \subseteq \mathcal{L} \dotplus \mathcal{M}$. A realization $(P_\mathcal{M} A | \mathcal{M}, P_\mathcal{M} B, C|\mathcal{M})$, where

$P_\mathcal{M} : \mathbb{C}^m \to \mathcal{M}$ is a projector on $\mathcal{M}$ with $\operatorname{Ker} P_\mathcal{M} \supseteq \mathcal{L}$, is called a *reduction* of (A, B, C). Note that $(P_\mathcal{M} A|\mathcal{M}, P_\mathcal{M} B, C|\mathcal{M})$ is again a realization for the same $W(\lambda)$. [See the proof that (6.1.5) is a realization of $W(\lambda)$ if (A, B, C) is.] We shall also say that (A, B, C) is a *dilation* of $(P_\mathcal{M} A|\mathcal{M}, P_\mathcal{M} B, C|\mathcal{M})$.

Theorem 6.1.3. *Any realization (A, B, C) of $W(\lambda)$ is a dilation of a realization (A_0, B_0, C_0) of $W(\lambda)$ with the pair (C_0, A_0) observable and the pair (A_0, B_0) controllable.*

For the proof of Theorem 6.1.3 we refer the reader to Theorem 7.1.3 in Gohberg et al. (1986a). It turns out that a realization (A, B, C) with the controllability and observability condition is essentially unique. To state this result precisely, we need some observations concerning one-sided invertibility of matrices. If the pair (C, A) is observable then (see Section 4.2)

$$\bigcap_{j=0}^{p-1} \operatorname{Ker}(CA^j) = \{0\} \tag{6.1.7}$$

for some positive integer p. Indeed, (6.1.7) holds for any p greater than or equal to the degree of the minimal polynomial of A. Similarly, for a controllable pair (A, B) the matrix

$$[B, AB, \ldots, A^{p-1}B] \tag{6.1.8}$$

has linearly independent rows for every integer p greater than or equal to the degree of the minimal polynomial of A. Condition (6.1.7) can be written in the form

$$\operatorname{Ker} \begin{bmatrix} C \\ CA \\ \vdots \\ CA^{p-1} \end{bmatrix} = \{0\},$$

and therefore there exists a *left inverse* of $\begin{bmatrix} C \\ \vdots \\ CA^{p-1} \end{bmatrix}$, i.e. a matrix G such that

$$G \begin{bmatrix} C \\ \vdots \\ CA^{p-1} \end{bmatrix} = I_m$$

(where $m \times m$ is the size of A). The linear independence of the rows of (6.1.8) means that

$$\operatorname{Im}[B, AB, \ldots, A^{p-1}B] = \mathbb{C}^m,$$

and therefore there exists a *right inverse* of $[B, AB, \ldots, A^{p-1}B]$, i.e. a matrix F such that

$$[B, AB, \ldots, A^{p-1}B]F = I_m.$$

We will denote by X^{-L} a left inverse (if one exists) of the matrix X (note that the left inverse is not unique unless X is square). A right inverse (if one exists) of the matrix Y will be denoted Y^{-R}.

Theorem 6.1.4. *Let (A_1, B_1, C_1) and (A_2, B_2, C_2) be realizations for a rational matrix function $W(\lambda)$ for which (C_1, A_1) and (C_2, A_2) are observable pairs and (A_1, B_1), (A_2, B_2) are controllable pairs. Then the sizes of A_1 and A_2 coincide, and there exists a nonsingular matrix S such that*

$$A_1 = S^{-1}A_2 S, \quad B_1 = S^{-1}B_2, \quad C_1 = C_2 S \qquad (6.1.9)$$

Moreover, the matrix S is unique and is given by

$$\begin{aligned} S &= \begin{bmatrix} C_2 \\ C_2 A_2 \\ \vdots \\ C_2 A_2^{p-1} \end{bmatrix}^{-L} \begin{bmatrix} C_1 \\ C_1 A_1 \\ \vdots \\ C_1 A_1^{p-1} \end{bmatrix} \\ &= [B_2, A_2 B_2, \ldots, A_2^{p-1} B_2][B_1, A_1 B_1, \ldots, A_1^{p-1} B_1]^{-R} \qquad (6.1.10) \end{aligned}$$

Here p is any integer such that

$$\mathrm{Ker} \begin{bmatrix} C_i \\ C_i A_i \\ \vdots \\ C_i A_i^{p-1} \end{bmatrix} = \{0\}, \quad (i = 1, 2),$$

and the matrices

$$[B_i, A_i B_i, \ldots, A_i^{p-1} B_i], \quad (i = 1, 2)$$

have linearly independent rows.

For the proof we again refer to Gohberg et al. (1986a). (See Theorem 7.1.4 there.)

Theorems 6.1.3 and 6.1.4 allow us to deduce the following important fact.

Theorem 6.1.5. *In a realization (A, B, C) of $W(\lambda)$, (C, A) and (A, B) are observable pairs and controllable pairs, respectively, if and only if the size of A is minimal among all possible realizations of $W(\lambda)$.*

Proof. Assume that the size m of A is minimal. By Theorem 6.1.3, there is a reduction (A', B', C') of (A, B, C) that is a realization for $W(\lambda)$ and such that (A', B') is controllable and (C'', A') is observable. But because of the minimality of m the realizations (A', B', C') and (A, B, C) must be similar, and this implies that (A, B, C) also has these properties.

Conversely, assume that in a realization (A, B, C) of $W(\lambda)$, the pairs (A, B) and (C, A) are controllable and observable, respectively. Arguing by contradiction, suppose that there is a realization (A'', B'', C'') of (A', B', C') with controllable (A'', B'') and observable (C'', A''). But then the size of A'' is not bigger than that of A, which contradicts Theorem 6.1.4 unless (A'', B'', C'') and (A, B, C) are similar. □

Realizations of the kind described in this theorem are, naturally, called *minimal realizations* of $W(\lambda)$. That is, they are those realizations for which the size of A is as small as possible.

6.2 Partial multiplicities and minimal realizations

Here we present a brief discussion of partial multiplicities of rational matrix functions and their formulation in terms of minimal realizations. In this discussion we drop the assumption (made at the beginning of the previous section) that the rational matrix functions involved are finite at infinity.

The starting point is the *local Smith form* for rational matrix functions.

Theorem 6.2.1. *Let $W(\lambda)$ be an $m \times n$ rational matrix function, and let $\lambda_0 \in \mathbb{C}$. Then $W(\lambda)$ admits a representation*

$$W(\lambda) = E(\lambda) \operatorname{diag}[(\lambda - \lambda_0)^{\kappa_1}, \ldots, (\lambda - \lambda_0)^{\kappa_r}, 0, \ldots, 0] F(\lambda), \qquad (6.2.1)$$

where $E(\lambda)$ is an $m \times m$ rational matrix function which is defined (i.e. has no pole) at λ_0 and $E(\lambda_0)$ is nonsingular; $F(\lambda)$ is an $n \times n$ rational matrix function which is defined at λ_0 with nonsingular $F(\lambda_0)$, and

$$\kappa_1 \leq \ldots \leq \kappa_r \qquad (6.2.2)$$

are integers (positive, negative or zero). Moreover, the number r of the integers (6.2.2) and the integers κ_j themselves are uniquely determined by $W(\lambda)$ and λ_0. In fact

$$r = \max \operatorname{rank} W(\lambda) \qquad (6.2.3)$$

where the maximum is taken over all $\lambda \in \mathbb{C}$ that are not poles of $W(\lambda)$.

The representation (6.2.1) is called the *local Smith form* of $A(\lambda)$ at λ_0 and the integers $\kappa_1 \leq \ldots \leq \kappa_r$ are called the *partial multiplicities* of $W(\lambda)$ at λ_0.

Proof. We apply elementary operations to $W(\lambda)$ each of which amounts to multiplication (on the left or on the right) by a rational matrix function without poles at λ_0 and having nonsingular value at λ_0. After a finite number of such operations we will obtain the form (6.2.1).

It may be assumed that $W(\lambda) \not\equiv 0$ (otherwise Theorem 6.2.1 is trivial). If $f(\lambda)$ is a scalar rational function and $\lambda_0 \in \mathbb{C}$, we define the *multiplicity* of $f(\lambda)$ at λ_0 as the unique integer α for which $g(\lambda) := f(\lambda)(\lambda - \lambda_0)^{-\alpha}$ is defined at λ_0 and $g(\lambda_0) \neq 0$. Let $w_{ij}(\lambda)$ be a nonzero entry in $W(\lambda)$ such that the multiplicity of $w_{ij}(\lambda)$ at λ_0 is minimal among all nonzero entries in $W(\lambda)$. Permuting rows and columns, if necessary, it may be assumed that $w_{ij}(\lambda) = w_{11}(\lambda)$ is in the $(1,1)$ position. Add the first column of $W(\lambda)$ mutliplied by $-w_{12}(\lambda)(w_{11}(\lambda))^{-1}$ to the second column. Because of the choice of $w_{11}(\lambda)$ this operation corresponds to the postmultiplication by a rational matrix function with constant determinant which is defined at λ_0. The matrix that results from this operation has 0 in the $(1,2)$ position. By similar operations we make zeros of all entries in the first row and first column of $W(\lambda)$, except for the $(1,1)$ entry. Repeat this procedure for the $(m-1) \times (n-1)$ matrix formed by all but the first row and first column of $W(\lambda)$, and so on. Eventually we represent $W(\lambda)$ in the form

$$W(\lambda) = E_0(\lambda) \operatorname{diag}[w_1(\lambda), \ldots, w_r(\lambda), 0, \ldots, 0] F_0(\lambda),$$

where $E_0(\lambda)$ and $F_0(\lambda)$ are rational matrix functions which are defined and nonsingular at λ_0, and $w_1(\lambda), \ldots, w_r(\lambda)$ are scalar rational functions. Let $\kappa_1, \ldots, \kappa_r$ be the multiplicities of $w_1(\lambda), \ldots, w_r(\lambda)$, respectively, at λ_0. Permuting rows and columns of $W(\lambda)$, if necessary, we can assume that $\kappa_1 \leq \ldots \leq \kappa_r$. Now put

$$E(\lambda) = E_0(\lambda) \operatorname{diag}(w_1(\lambda)(\lambda - \lambda_0)^{-\kappa_1}, \ldots, w_r(\lambda)(\lambda - \lambda_0)^{-\kappa_r}, 1, \ldots, 1)$$

$$F(\lambda) = F_0(\lambda)$$

to satisfy (6.2.1).

From the formula (6.2.1) it follows easily that

$$\operatorname{rank} W(\lambda) = r \qquad (6.2.4)$$

for all λ with $0 < |\lambda - \lambda_0| < \epsilon$, where $\epsilon > 0$ is chosen sufficiently small. Let $r_0 = \max \operatorname{rank} W(\lambda)$, where the maximum is taken over all $\lambda \in \mathbb{C}$ that are not poles of $W(\lambda)$, and let $f(\lambda)$ be the determinant of an $r_0 \times r_0$ submatrix of $W(\lambda)$ chosen so that $f(\lambda) \not\equiv 0$. Since $f(\lambda)$ is a rational function, it has only finitely many poles and zeros and, therefore, $f(\lambda) \neq 0$ for $0 < |\lambda - \lambda_0| < \epsilon_0$ and some $\epsilon_0 > 0$. Comparing with (6.2.4) we see that (6.2.3) follows.

Finally, we prove the uniqueness of κ_j. Indeed, from the Binet–Cauchy formulas (see, for example, Theorem A.2.1 in Gohberg et al. (1986a) or Theorem 2.5.1 of Lancaster and Tismenetsky (1985)) it follows that for $j = 1, \ldots, r$ the integer $\kappa_1 + \cdots + \kappa_j$ coincides with the minimal multiplicity at λ_0 among all not identically zero determinants of the $j \times j$ submatrices of $W(\lambda)$. Thus, the sums $\kappa_1 + \cdots + \kappa_j$ ($1 \leq j \leq r$), and therefore also $\kappa_1, \ldots, \kappa_r$ themselves, are determined in terms of $W(\lambda)$ and so they are unique. $\square$

From now on we will assume that the rational matrix function $W(\lambda)$ is *regular*, i.e. $W(\lambda)$ is of square size and $\det W(\lambda)$ is not identically zero.

Note that $\lambda_0 \in \mathbb{C}$ is a *pole* of $W(\lambda)$ [i.e., a pole of at least one entry in $W(\lambda)$] if and only if $W(\lambda)$ has a negative partial multiplicity at λ_0. Indeed, the minimal partial multiplicity of $W(\lambda)$ at λ_0 coincides with the minimal multiplicity at λ_0 of the not identically zero entries of $W(\lambda)$. Also, $\lambda_0 \in \mathbb{C}$ is a *zero* of $W(\lambda)$ [by definition, this means that λ_0 is a pole of $W(\lambda)^{-1}$] if and only if $W(\lambda)$ has a positive partial multiplicity. In particular, for every $\lambda_0 \in \mathbb{C}$, except for a finite number of points, all partial multiplicities are zeros.

There is a close relationship between the partial multiplicities of $W(\lambda)$ and the minimal realization of $W(\lambda)$. Namely, let $W(\lambda)$ be a regular rational $n \times n$ matrix. Let

$$W(\lambda) = \sum_{j=-\infty}^{q} \lambda^j W_j \qquad (6.2.5)$$

be the Laurent series of $W(\lambda)$ at infinity (here q is some nonnegative integer and the coefficients W_i are $n \times n$ matrices): write $U(\lambda) = \sum_{j=0}^{q} \lambda^j W_j$ for the polynomial part of $W(\lambda)$. Thus $W(\lambda) - U(\lambda)$ takes the value 0 at infinity, and we may write

$$W(\lambda) = C(\lambda I - A)^{-1} B + U(\lambda) \qquad (6.2.6)$$

where $C(\lambda I - A)^{-1} B$ is a minimal realization of the rational matrix function $W(\lambda) - U(\lambda)$. We say that (6.2.6) is a *minimal realization* of $W(\lambda)$.

The connection between the partial multiplicities of $W(\lambda)$ and the minimal realizations for $W(\lambda)$ and $W(\lambda)^{-1}$ are given by the following theorem.

Theorem 6.2.2. *Let (6.2.6) and*

$$W(\lambda)^{-1} = C_1(\lambda I - A_1)^{-1} B_1 + V(\lambda)$$

be minimal realizations of $W(\lambda)$ and $W(\lambda)^{-1}$ (here $V(\lambda)$ is a matrix polynomial). Then, for a fixed pole λ_0 of $W(\lambda)$, the absolute values of negative partial multiplicities of $W(\lambda)$ at λ_0 coincide with the partial multiplicities of λ_0 as an eigenvalue of A. For a fixed zero μ_0 of $W(\lambda)$, the positive partial multiplicities of $W(\lambda)$ at μ_0 coincide with the partial multiplicities of μ_0 as an eigenvalue of A_1.

We will not prove Theorem 6.2.2 here. For a complete proof the reader is referred to Section 7.2 of Gohberg et al. (1986a).

6.3 Locally minimal realizations

The concept of a minimal realization can be developed locally as well, i.e. with respect to the poles and zeros of a rational matrix function which are located in a certain fixed part of the complex plane. In contrast, the minimality of realizations (introduced in Section 6.1) is of a global nature, i.e. reflects properties of all the poles and zeros of a rational matrix function. The local minimality of realizations is to be developed in this section. It is needed in this book in Chapter 21 only, although it is of independent interest.

Let $W(\lambda)$ be an $n \times n$ rational matrix function with a realization

$$W(\lambda) = D + C(\lambda I - A)^{-1}B. \tag{6.3.1}$$

For a given set $\sigma \subset \mathbb{C}$ let $P(\sigma)$ be the Riesz projector for A associated with $\sigma \cap \sigma(A)$ (see Proposition 1.3.3). If $\sigma \cap \sigma(A) = \emptyset$, we formally put $P(\sigma) = 0$. The realization (6.3.1) is called *minimal on σ* if the pair

$$(C|\operatorname{Im} P(\sigma), A|\operatorname{Im} P(\sigma))$$

is observable and the pair $(A|\operatorname{Im} P(\sigma), P(\sigma)B)$ is controllable. Here, $P(\sigma)$ is understood as a linear transformation onto $\operatorname{Im} P(\sigma)$.

If $\sigma \supseteq \sigma(A)$, then $P(\sigma) = I$ and, in view of Theorem 6.1.5, a realization (6.3.1) is minimal on σ if and only if it is minimal. This explains our choice of the terminology "minimal on σ".

Proposition 6.3.1. *Let $\sigma_1, \ldots, \sigma_s$ be mutually disjoint sets in $\mathbb{C}$. Then a realization (6.3.1) is minimal on $\bigcup_{j=1}^{s} \sigma_j$ if and only if it is minimal on each σ_j ($j = 1, \ldots, s$).*

Proof. Passing, if necessary, to other bases, it is easily seen that Proposition 6.3.1 follows from the following two statements:

(I) Let pairs of matrices (C_j, A_j), $j = 1, \ldots, s$, be given, where C_j is of size $n \times m_j$, and A_j is of size $m_j \times m_j$. Assume that

$$\sigma(A_i) \cap \sigma(A_j) = \emptyset \quad \text{for } i \neq j. \tag{6.3.2}$$

Then the pair

$$(C = [C_1 \; C_2 \; \ldots \; C_s], A = A_1 \oplus \cdots \oplus A_s)$$

is observable if and only if each of the pairs (C_j, A_j) ($j = 1, \ldots, s$) is observable.

(II) Let pairs of matrices (A_j, B_j), $j = 1, \ldots, s$, be given, where B_j is of size $m_j \times n$, and A_j is of size $m_j \times m_j$. Assume that (6.3.2) holds. Then the pair

$$(A = A_1 \oplus \cdots \oplus A_s, B = \begin{bmatrix} B_1 \\ B_2 \\ \vdots \\ B_s \end{bmatrix}) \qquad (6.3.3)$$

is controllable if and only if each of the pairs (A_j, B_j) $(j = 1, \ldots, s)$ is controllable.

We prove statement (II) only. The proof of (I) is left as an exercise. If (6.3.3) is controllable, then for p sufficiently large

$$\mathbb{C}^{m_1} \oplus \cdots \oplus \mathbb{C}^{m_s} = \operatorname{Im} [B, AB, \ldots, A^{p-1}B]$$
$$= \operatorname{Im} [B_1, A_1 B_1, \ldots, A_1^{p-1} B_1] \oplus \cdots \oplus \operatorname{Im} [B_s, A_s B_s, \ldots, A_s^{p-1} B_s].$$

In particular, for $j = 1, 2, \ldots, s$,

$$\mathbb{C}^{m_j} = \operatorname{Im} [B_j, A_j B_j, \ldots, A_j^{p-1} B_j],$$

and hence (A_j, B_j) is controllable for $j = 1, \ldots, s$.

Assume now that each (A_j, B_j) is controllable. Then

$$\operatorname{span} \{p(A_j) B_j x : p(\lambda) \text{ is a scalar polynomial}, x \in \mathbb{C}^n\} = \mathbb{C}^{m_j}.$$

In view of (6.3.2), for $j = 1, \ldots, s$ there exists a polynomial $p_j(\lambda)$ such that $p_j(A_j) = I$ and $p_j(A_k) = 0$ for $k \neq j$. It follows that

$$\operatorname{span} \{p_j(A) p(A) B x : p(\lambda) \text{ is a polynomial}, x \in \mathbb{C}^n\}$$
$$= 0 \oplus \cdots \oplus 0 \oplus \mathbb{C}^{m_j} \oplus 0 \oplus \cdots \oplus 0. \qquad (6.3.4)$$

Since the subspace

$$\operatorname{span} \{q(A) B x : q(\lambda) \text{ is a polynomial}, x \in \mathbb{C}^n\} \qquad (6.3.5)$$

contains (6.3.4), we find that the subspace (6.3.5) also contains

$$\mathbb{C}^{m_1} \oplus \cdots \oplus \mathbb{C}^{m_j} \oplus \cdots \oplus \mathbb{C}^{m_s} = \mathbb{C}^m \quad (m = m_1 + \cdots + m_s),$$

which amounts to the controllability of (A, B). □

In particular, it follows from Proposition 6.3.1 that the representation of equation (6.3.1) is minimal if and only if it is minimal on each eigenvalue of A.

Some elementary properties of locally minimal realizations are summarized in the next theorem:

Theorem 6.3.2. *Let $\sigma \subseteq \mathbb{C}$. The following statements are equivalent for a realization (6.3.1), where A is $m \times m$, C is $n \times m$ and B is $m \times n$:*

(i) *The realization (6.3.1) is minimal on σ.*

(ii) $\operatorname{rank}[\lambda I - A, B] = m$ *for every* $\lambda \in \sigma$ *and* $\operatorname{rank} \begin{bmatrix} \lambda I - A \\ C \end{bmatrix} = m$ *for every* $\lambda \in \sigma$.

(iii) *if*

$$A = \begin{bmatrix} A_{11} & A_{12} & A_{13} \\ 0 & A_0 & A_{23} \\ 0 & 0 & A_{33} \end{bmatrix}, \quad B = \begin{bmatrix} B_1 \\ B_0 \\ 0 \end{bmatrix}, \quad C = [0 \ \ C_0 \ \ C_1] \quad (6.3.6)$$

and the realization (A_0, B_0, C_0) is minimal, then

$$\sigma \cap \sigma(A_{11}) = \emptyset, \quad \sigma \cap \sigma(A_{33}) = \emptyset. \tag{6.3.7}$$

Observe that the realization (A, B, C) can always be brought to the form (6.3.6) by a change of basis in which the linear transformations represented by A, B and C are written in matrix form (see Theorem 6.1.3).

Proof. Assume that (ii) holds. Using the form (6.3.6) we have

$$\operatorname{rank} \begin{bmatrix} \lambda I - A_{11} & -A_{12} & -A_{13} & B_1 \\ 0 & \lambda I - A_0 & -A_{23} & B_0 \\ 0 & 0 & \lambda I - A_{33} & 0 \end{bmatrix} = m \quad \text{for } \lambda \in \sigma$$

$$\operatorname{rank} \begin{bmatrix} \lambda I - A_{11} & -A_{12} & -A_{13} \\ 0 & \lambda I - A_0 & -A_{23} \\ 0 & 0 & \lambda I - A_{33} \\ 0 & C_0 & C_1 \end{bmatrix} = m \quad \text{for } \lambda \in \sigma$$

These conditions imply that $\lambda I - A_{33}$ and $\lambda I - A_{11}$ are invertible for $\lambda \in \sigma$, i.e., statement (iii) holds.

Assume that statement (i) holds. Applying a similarity, if necessary $((A, B, C) \longrightarrow (S^{-1}AS, S^{-1}B, CS)$ for some invertible S), it can be assumed without loss of generality that

$$A = \begin{bmatrix} A_\sigma & 0 \\ 0 & A' \end{bmatrix},$$

where $\sigma(A_\sigma) \subseteq \sigma$ and $\sigma(A') \cap \sigma = \emptyset$. Partition B and C accordingly:

$$B = \begin{bmatrix} B_\sigma \\ B' \end{bmatrix}, \quad C = [C_\sigma \ \ C'].$$

By (i), the realization $(A_\sigma, B_\sigma, C_\sigma)$ is minimal, and therefore (by Theorem 6.1.5) the pair (A_σ, B_σ) is controllable and the pair (C_σ, A_σ) is observable. Thus (using Theorem 4.3.3 and Proposition 4.2.2), for every $\lambda \in \sigma$ we have

$$\operatorname{rank}[\lambda I - A_\sigma, B_\sigma] = m_\sigma$$

$$\operatorname{rank}\begin{bmatrix} \lambda I - A_\sigma \\ C_\sigma \end{bmatrix} = m_\sigma, \tag{6.3.8}$$

where $m_\sigma \times m_\sigma$ is the size of A_σ. Since $\lambda I - A'$ is invertible for every $\lambda \in \sigma$, it follows from (6.3.8) that

$$\operatorname{rank}\begin{bmatrix} \lambda I - A_\sigma & 0 \\ 0 & \lambda I - A' \\ C_\sigma & C' \end{bmatrix} = \operatorname{rank}\begin{bmatrix} \lambda I - A_\sigma & 0 & B_\sigma \\ 0 & \lambda I - A' & B' \end{bmatrix} = m$$

for every $\lambda \in \sigma$, and (ii) follows.

It remains to prove that (iii) implies (i). Applying, if necessary, a similarity to the realization (A_0, B_0, C_0) it can be assumed that

$$A_0 = \begin{bmatrix} A_{0\sigma} & 0 \\ 0 & A'_0 \end{bmatrix}, \quad B_0 = \begin{bmatrix} B_{0\sigma} \\ B'_0 \end{bmatrix}, \quad C_0 = [C_{0\sigma} \ C'_0], \tag{6.3.9}$$

where all eigenvalues of $A_{0\sigma}$ are in σ and all eigenvalues of A'_0 are outside of σ (the case when A'_0 is empty is not excluded; in this case the subsequent proof should be adjusted in an obvious way). Clearly the realization $(A_{0\sigma}, B_{0\sigma}, C_{0\sigma})$ is minimal.

Write

$$A = \begin{bmatrix} A_{11} & A'_{12} & A''_{12} & A'_{13} \\ 0 & A_{0\sigma} & 0 & A'_{23} \\ 0 & 0 & A'_0 & A''_{23} \\ 0 & 0 & 0 & A_{33} \end{bmatrix}, \quad B = \begin{bmatrix} B_1 \\ B_{0\sigma} \\ B'_0 \\ 0 \end{bmatrix},$$

$$C = [0 \ C_{0\sigma} \ C'_0 \ C_1]. \tag{6.3.10}$$

The partitions here are obtained by superimposing the partitions of equations (6.3.9) on the original partitions (6.3.6). Let X and Y be the unique matrices satisfying

$$YA_{0\sigma} - A_{11}Y = -A'_{12},$$

$$X\begin{bmatrix} A'_0 & A''_{23} \\ 0 & A_{33} \end{bmatrix} - A_{0\sigma}X = -[0 \ A'_{23}]. \tag{6.3.11}$$

Their existence and uniqueness are ensured by Theorem 5.2.2 since

$$\sigma(A_{0\sigma}) \cap \sigma(A_{11}) = \emptyset, \quad \sigma(A_{0\sigma}) \cap \sigma\begin{bmatrix} A'_0 & A''_{23} \\ 0 & A_{33} \end{bmatrix} = \emptyset.$$

The form of the right-hand side of (6.3.11) together with the uniqueness of X shows that X has the form
$$X = [0 \ X_0],$$
where X_0 satisfies the equation
$$X_0 A_{33} - A_{0\sigma} X_0 = -A'_{23}.$$

A calculation shows that
$$\begin{bmatrix} I & Y & 0 & 0 \\ 0 & I & 0 & X_0 \\ 0 & 0 & I & 0 \\ 0 & 0 & 0 & I \end{bmatrix} A \begin{bmatrix} I & Y & 0 & 0 \\ 0 & I & 0 & X_0 \\ 0 & 0 & I & 0 \\ 0 & 0 & 0 & I \end{bmatrix}^{-1} = \begin{bmatrix} A_{11} & 0 & * & * \\ 0 & A_{0\sigma} & 0 & 0 \\ 0 & 0 & A'_0 & A''_{23} \\ 0 & 0 & 0 & A_{33} \end{bmatrix}, \quad (6.3.12)$$

where $*$ denotes entries whose value is irrelevant to our discussion. Using the equality (6.3.12) we pass to a realization which is similar to (A, B, C):

$$\begin{bmatrix} A_{11} & 0 & * & * \\ 0 & A_{0\sigma} & 0 & 0 \\ 0 & 0 & A'_0 & A''_{23} \\ 0 & 0 & 0 & A_{33} \end{bmatrix}, \begin{bmatrix} B_1 + Y B_{0\sigma} \\ B_{0\sigma} \\ B'_0 \\ 0 \end{bmatrix}, [0, C_{0\sigma}, C'_0, -C_{0\sigma} X_0 + C_1]. \quad (6.3.13)$$

The Riesz projector $P(\sigma)$ of the matrix (6.3.13) associated with σ is easily found by inspection:
$$P(\sigma) = \begin{bmatrix} 0 & 0 & 0 & 0 \\ 0 & I & 0 & 0 \\ 0 & 0 & 0 & 0 \\ 0 & 0 & 0 & 0 \end{bmatrix}.$$

Thus, the minimality on σ of the realization (6.3.1), which is equivalent to the minimality on σ of the realization (6.3.13), amounts to the minimality of $(A_{0\sigma}, B_{0\sigma}, C_{0\sigma})$. But we have observed already that the realization $(A_{0\sigma}, B_{0\sigma}, C_{0\sigma})$ is minimal. □

We now make connections with the properties of rational matrix functions.

Combining the equivalence statements (i) and (iii) of Theorem 6.3.2 with Theorem 6.2.2, we obtain

Theorem 6.3.3. *Let*
$$W(\lambda) = C(\lambda I - A)^{-1} B + U(\lambda),$$
where $U(\lambda)$ is a matrix polynomial, and assume that the realization (A, B, C) is minimal on σ. Then, for a fixed pole $\lambda_0 \in \sigma$ of $W(\lambda)$, the absolute values of negative partial multiplicities of $W(\lambda)$ at λ_0 coincide with the partial multiplicities of λ_0 as an eigenvalue of A.

LOCALLY MINIMAL REALIZATIONS

An analogous statement holds for zeros of $W(\lambda)$ in σ and a realization

$$W(\lambda)^{-1} = C'(\lambda I - A')^{-1} B' + V(\lambda),$$

where $V(\lambda)$ is a matrix polynomial and (A', B', C') is minimal on σ.

If the matrix D in the representation of equation (6.3.1) is invertible, then we obtain a realization for $W(\lambda)^{-1}$:

$$W(\lambda)^{-1} = D^{-1} - D^{-1} C(\lambda I - A^{\times})^{-1} B D^{-1}, \qquad (6.3.14)$$

where $A^{\times} = A - BD^{-1}C$. Indeed, a straightforward multiplication gives

$$[D + C(\lambda I - A)^{-1} B][D^{-1} - D^{-1} C(\lambda I - A^{\times})^{-1} BD^{-1}$$
$$= I + C\{(\lambda I - A)^{-1} - (\lambda I - A^{\times})^{-1} - (\lambda I - A)^{-1} BD^{-1} C(\lambda I - A^{\times})^{-1}\} BD^{-1}$$
$$= I + C(\lambda I - A)^{-1}\{\lambda I - A^{\times} - (\lambda I - A) - BD^{-1}C\}(\lambda I - A^{\times})^{-1} BD^{-1} = I.$$

Proposition 6.3.4. *The realization (6.3.1) is minimal on σ if and only if the realization (6.3.14) is minimal on σ.*

Proof. Without loss of generality we may assume that (A, B, C) are given in the forms of equations (6.3.6). Then

$$A - BD^{-1}C = \begin{bmatrix} A_{11} & * & * \\ 0 & A_0 - B_0 D^{-1} C_0 & * \\ 0 & 0 & A_{33} \end{bmatrix}, \qquad (6.3.15)$$

$$BD^{-1} = \begin{bmatrix} B_1 D^{-1} \\ B_0 D^{-1} \\ 0 \end{bmatrix}, \quad -D^{-1}C = [0, -D^{-1} C_0, -D^{-1} C_1] \qquad (6.3.16)$$

and, as usual, a $*$ denotes a block entry whose value is immaterial at this moment. Since the realization (A_0, B_0, C_0) is minimal, so is $(A_0 - B_0 D^{-1} C_0, B_0 D^{-1}, -D^{-1} C_0)$. By the equivalence of statements (i) and (iii) in Theorem 6.3.2 (applied to both (6.3.6) and (6.3.15), (6.3.16)), the minimality on σ of both (6.3.1) and (6.3.14) amounts to the same conditions

$$\sigma \cap \sigma(A_{11}) = \emptyset, \quad \sigma \cap \sigma(A_{33}) = \emptyset. \quad \square$$

6.4 Minimal factorization for rational matrix functions

In this section we describe the minimal factorizations of a rational matrix function in terms of certain invariant subspaces. To make the presentation more transparent, we restrict ourselves to the case when the rational matrix functions involved are $n \times n$ and have value I at infinity. (The same analysis applies to the case when the matrix functions have invertible value at infinity.)

We start with the definition. The *McMillan degree* of a rational $n \times n$ matrix function $W(\lambda)$ [with $W(\infty) = I$], denoted $\delta(W)$, is the size of the matrix A in a minimal realization

$$W(\lambda) = I + C(\lambda I - A)^{-1}B. \qquad (6.4.1)$$

A special case of equation (6.3.14) yields

$$W(\lambda)^{-1} = I - C(\lambda I - A^\times)^{-1}B \qquad (6.4.2)$$

where $A^\times = A - BC$.

Proposition 6.4.1. *The realization* (6.4.1) *of* $W(\lambda)$ *is minimal if and only if the realization* (6.4.2) *of* $W(\lambda)^{-1}$ *is minimal.*

Proof. This is a special case of Proposition 6.3.4. However, we give a direct proof as well. It will be proved that minimality of (6.4.1) implies the minimality of (6.4.2) (the converse direction follows by interchanging the roles of $W(\lambda)$ and $W(\lambda)^{-1}$). By Theorem 6.1.5 the following statement has to be proved: if (A, B) is controllable and (C, A) is observable, then $(A - BC, B)$ is controllable and $(C, A - BC)$ is observable. Note that (C, A) is observable if and only if (A^*, C^*) is controllable. Thus, the above statement follows from Lemma 4.4.1. □

It follows from Proposition 6.4.1 that $W(\lambda)$ and $W(\lambda)^{-1}$ have the same McMillan degree, i.e. $\delta(W) = \delta(W^{-1})$.

We now consider factorizations of $W(\lambda)$ of the form

$$W(\lambda) = W_1(\lambda)W_2(\lambda)\cdots W_p(\lambda) \qquad (6.4.3)$$

where, for $j = 1,\ldots,p$, $W_j(\lambda)$ are $n \times n$ rational matrix functions with minimal realizations

$$W_j(\lambda) = I + C_j(\lambda I - A_j)^{-1}B_j \qquad (6.4.4)$$

It is not difficult to obtain a realization for $W(\lambda)$ in terms of the minimal realizations (6.4.4). Consider first the case of two rational matrix functions $W_1(\lambda)$ and $W_2(\lambda)$. Then the product $W_1(\lambda)W_2(\lambda)$ has a realization

$$W_1(\lambda)W_2(\lambda) = I + [C_1 \ C_2]\left(\lambda I - \begin{bmatrix} A_1 & B_1C_2 \\ 0 & A_2 \end{bmatrix}\right)^{-1}\begin{bmatrix} B_1 \\ B_2 \end{bmatrix}. \qquad (6.4.5)$$

Indeed, the following formula is easily verified by multiplication:

$$\left(\lambda I - \begin{bmatrix} A_1 & B_1C_2 \\ 0 & A_2 \end{bmatrix}\right)^{-1} = \begin{bmatrix} (\lambda I - A_1)^{-1} & (\lambda I - A_1)^{-1}B_1C_2(\lambda I - A_2)^{-1} \\ 0 & (\lambda I - A_2)^{-1} \end{bmatrix}$$

so the right-hand side of (6.4.5) is equal to

$$I + C_1(\lambda I - A_1)^{-1}B_1 + [C_1(\lambda I - A_1)^{-1}B_1C_2(\lambda I - A_2)^{-1}$$
$$+ C_2(\lambda I - A_2)^{-1}]B_2$$
$$= I + C_1(\lambda I - A_1)^{-1}B_1 + C_1(\lambda I - A_1)^{-1}B_1C_2(\lambda I - A_2)^{-1}B_2$$
$$+ C_2(\lambda I - A_2)^{-1}B_2$$
$$= W_1(\lambda) + (W_1(\lambda) - I)(W_2(\lambda) - I) + (W_2(\lambda) - I)$$
$$= W_1(\lambda)W_2(\lambda).$$

Formula (6.4.5) applied several times yields a realization for $W(\lambda)$:

$$W(\lambda) = I + [C_1 \; C_2 \; \cdots \; C_p]\left(\lambda I - \begin{bmatrix} A_1 & B_1C_2 & \cdots & B_1C_p \\ 0 & A_2 & \cdots & B_2C_p \\ 0 & 0 & & \vdots \\ \vdots & \vdots & & \vdots \\ 0 & 0 & \cdots & A_p \end{bmatrix}\right)^{-1} \begin{bmatrix} B_1 \\ B_2 \\ \vdots \\ B_p \end{bmatrix}$$
(6.4.6)

This realization is not necessarily minimal, even though each of the realizations (6.4.4) is minimal. So we have (in view of Theorem 6.1.3)

$$\delta(W) \leq \delta(W_1) + \cdots + \delta(W_p).$$

We say that the factorization (6.4.3) is *minimal* if actually $\delta(W) = \delta(W_1) + \cdots + \delta(W_p)$, that is, realization (6.4.6) is minimal as well. In informal terms, minimality of the factorization (6.4.3) means that zero-pole cancellation does not occur between the factors $W_j(\lambda)$. Because the McMillan degrees of a rational matrix function (with value I at infinity) and of its inverse are the same, the factorization (6.4.3) is minimal if and only if the corresponding factorization for the inverse matrix function

$$W(\lambda)^{-1} = W_p(\lambda)^{-1}W_{p-1}(\lambda)^{-1}\ldots W_1(\lambda)^{-1}$$

is minimal.

Now consider minimal factorizations (6.4.3) with just two factors ($p = 2$). A description of all such factorizations in terms of certain subspace decompositions can be given. Let A be taken from a minimal realization $W(\lambda) = I + C(\lambda I - A)^{-1}B$. Write $A^\times = A - BC$, and let A and $A^\times$ be of size m.

We say that a direct sum decomposition

$$\mathbb{C}^m = \mathcal{L} \dotplus \mathcal{N} \tag{6.4.7}$$

is a *supporting decomposition* for $W(\lambda)$ if the subspace $\mathcal{L}$ is A-invariant, and the subspace $\mathcal{N}$ is $A^\times$-invariant.

Note that a supporting decomposition for $W(\lambda)$ depends on the choice of minimal realization. We assume, however, that the minimal realization of $W(\lambda)$ is fixed, and thereby suppress the dependence of supporting decompositions on this choice. (In view of Theorems 6.1.4 and 6.1.5, there is no loss of generality in making this assumption.)

Given the supporting decomposition (6.4.7) a unique projector π can be defined so that $\mathcal{L} = \operatorname{Ker} \pi$ and $\mathcal{N} = \operatorname{Im} \pi$. Such a projector will be called a *supporting projector*. Thus, supporting projectors π are characterized by the property that $\operatorname{Ker} \pi$ is A-invariant and $\operatorname{Im} \pi$ is $A^\times$-invariant. The language of supporting projectors is often more convenient than that of supporting decompositions.

The role of supporting decompositions in the minimal factorization problem is revealed in the next theorem, which is the main result of this section.

Theorem 6.4.2. *Let (6.4.7) be a supporting decomposition for $W(\lambda)$. Then $W(\lambda)$ admits a minimal factorization*

$$\begin{aligned} W(\lambda) &= [I + C\pi_\mathcal{L}(\lambda I - A)^{-1}\pi_\mathcal{L} B][I + C\pi_\mathcal{N}(\lambda I - A)^{-1}\pi_\mathcal{N} B] \\ &= [I + C(\lambda I - A)^{-1}\pi_\mathcal{L} B][I + C\pi_\mathcal{N}(\lambda I - A)^{-1} B] \end{aligned} \tag{6.4.8}$$

where $\pi_\mathcal{L}$ is the projector on $\mathcal{L}$ along $\mathcal{N}$, and $\pi_\mathcal{N} = I - \pi_\mathcal{L}$.

Conversely, for every minimal factorization $W(\lambda) = W_1(\lambda)W_2(\lambda)$, where the factors are rational matrix functions with value I at infinity there exists a unique supporting decomposition $\mathbb{C}^m = \mathcal{L} \dotplus \mathcal{N}$ such that

$$\begin{aligned} W_1(\lambda) &= I + C\pi_\mathcal{L}(\lambda I - A)^{-1}\pi_\mathcal{L} B \\ W_2(\lambda) &= I + C\pi_\mathcal{N}(\lambda I - A)^{-1}\pi_\mathcal{N} B. \end{aligned} \tag{6.4.9}$$

Note that the second equality in (6.4.8) follows from the relations $\pi_\mathcal{L} A \pi_\mathcal{L} = A \pi_\mathcal{L}$ and $\pi_\mathcal{N} A \pi_\mathcal{N} = \pi_\mathcal{N} A$, which express the A invariance of $\mathcal{L}$ (see Section 1.3).

Proof. With respect to the direct sum decomposition (6.4.7), write

$$A = \begin{bmatrix} A_{11} & A_{12} \\ 0 & A_{22} \end{bmatrix}, \quad A^\times = \begin{bmatrix} A_{11}^\times & 0 \\ A_{21}^\times & A_{22}^\times \end{bmatrix},$$

$$C = [C_1 \ C_2], \quad B = \begin{bmatrix} B_1 \\ B_2 \end{bmatrix}.$$

For example, A_{12} is understood as the linear transformation from $\mathcal{N}$ into $\mathcal{L}$ induced by A (in other words, $A_{12} = \pi_\mathcal{L} A \pi_\mathcal{N}$), C_2 is understood as the linear

MINIMAL FACTORIZATION

transformation from $\mathcal{M}$ into $\mathbb{C}^n$ induced by C (in other words, $C_2 = C\pi_\mathcal{M}$), and other block entries in this representation of $A, A^\times, C$ and B are interpreted in a similar way. Note that A has block upper triangular form because of the A-invariance of $\mathcal{L}$ and $A^\times$ has block lower triangular form because of the $A^\times$-invariance of $\mathcal{N}$.

Note, in particular, that the triangular form of $A^\times$ implies $A_{12} = B_1 C_2$. Applying formula (6.4.5), we now see that the product on the right-hand side of (6.4.8) is indeed $W(\lambda)$. Further, denoting $W_\mathcal{K}(\lambda) = I + C\pi_\mathcal{K}(\lambda I - A)^{-1}\pi_\mathcal{K} B$, for $\mathcal{K} = \mathcal{L}$, or $\mathcal{K} = \mathcal{N}$, we obviously have $\delta(W_\mathcal{K}) \leq \dim \mathcal{K}$. Hence

$$\delta(W) \leq \delta(W_\mathcal{L}) + \delta(W_\mathcal{N}) \leq \dim \mathcal{L} + \dim \mathcal{N} = m.$$

Since, by definition, $m = \delta(W)$, it follows that

$$\delta(W_\mathcal{L}) + \delta(W_\mathcal{N}) = m = \delta(W)$$

and the factorization (6.4.8) is minimal.

Next assume that $W = W_1 W_2$ is a minimal factorization of W, and for $i = 1, 2$ let

$$W_i(\lambda) = I + C_i(\lambda I - A_i)^{-1} B_i$$

be a minimal realization of $W_i(\lambda)$. By the multiplication formula (6.4.5)

$$W(\lambda) = I + \tilde{C}(\lambda I - \tilde{A})^{-1} \tilde{B} \qquad (6.4.10)$$

where

$$\tilde{C} = [C_1 \ C_2], \quad \tilde{A} = \begin{bmatrix} A_1 & B_1 C_2 \\ 0 & A_2 \end{bmatrix}, \quad \tilde{B} = \begin{bmatrix} B_1 \\ B_2 \end{bmatrix}.$$

Note that

$$\tilde{A} - \tilde{B}\tilde{C} = \begin{bmatrix} A_1 - B_1 C_1 & 0 \\ -B_2 C_1 & A_2 - B_2 C_2 \end{bmatrix}.$$

As the factorization $W = W_1 W_2$ is minimal, the realization (6.4.10) is minimal. Hence, by Theorem 6.1.4, for some invertible matrix S we have

$$\tilde{C} = CS, \quad \tilde{A} = S^{-1} AS, \quad \tilde{B} = S^{-1} B.$$

To satisfy (6.4.9), put $\mathcal{L} = S\tilde{\mathcal{L}}$ and $\mathcal{N} = S\tilde{\mathcal{N}}$, where

$$\tilde{\mathcal{L}} = \mathrm{span}\{e_1, \ldots, e_{p_1}\}, \quad \tilde{\mathcal{N}} = \mathrm{span}\{e_{p_1+1}, \ldots, e_{p_1+p_2}\}$$

and A_i has size p_i for $i = 1$ and 2.

It remains to prove the uniqueness of $\mathcal{L}$ and $\mathcal{N}$. Assume that $\mathbb{C}^m = \mathcal{L}' \dotplus \mathcal{N}'$ is also a supporting decomposition such that

$$W_1(\lambda) = I + C\pi_{\mathcal{L}'}(\lambda I - A)^{-1}\pi_{\mathcal{L}'}B$$
$$W_2(\lambda) = I + C\pi_{\mathcal{N}'}(\lambda I - A)^{-1}\pi_{\mathcal{N}'}B. \qquad (6.4.11)$$

As the realizations (6.4.9) and (6.4.11) are minimal (see the first part of the proof), there exist invertible linear transformations $T_{\mathcal{L}} : \mathcal{L}' \to \mathcal{L}$, $T_{\mathcal{N}} : \mathcal{N}' \to \mathcal{N}$ such that

$$C\pi_{\mathcal{K}'} = C\pi_{\mathcal{K}}T_{\mathcal{K}}, \quad \pi_{\mathcal{K}'}A\pi_{\mathcal{K}'} = (T_{\mathcal{K}})^{-1}\pi_{\mathcal{K}}A\pi_{\mathcal{K}}T_{\mathcal{K}}$$
$$\pi_{\mathcal{K}'}B = (T_{\mathcal{K}})^{-1}\pi_{\mathcal{K}}B, \quad \mathcal{K} = \mathcal{L} \text{ or } \mathcal{N}.$$

Therefore, the invertible linear transformation $T : \mathbb{C}^m \to \mathbb{C}^m$ defined by $T|\mathcal{K}' = T_{\mathcal{K}}$ for $\mathcal{K} = \mathcal{L}$ or $\mathcal{N}$ is a similarity between the minimal realization $W(\lambda) = I + C(\lambda I - A)^{-1}B$ and itself:

$$C = CT; \quad A = T^{-1}AT; \quad B = T^{-1}B.$$

Because of the uniqueness of such a similarity (Theorem 6.1.4), we must have $T = I$. So $\mathcal{L}' = \mathcal{L}$, $\mathcal{N}' = \mathcal{N}$. □

Using formula (6.4.2), we can rewrite the minimal factorization (6.4.8) in terms of the minimal factorization of the inverse matrix function:

$$W(\lambda)^{-1} = [I - C\pi_{\mathcal{N}}(\lambda I - A^\times)^{-1}\pi_{\mathcal{N}}B][I - C\pi_{\mathcal{L}}(\lambda I - A^\times)^{-1}\pi_{\mathcal{L}}B]$$
$$= [I - C(\lambda I - A^\times)^{-1}\pi_{\mathcal{N}}B][I - C\pi_{\mathcal{L}}(\lambda I - A^\times)^{-1}B] \qquad (6.4.12)$$

where the second equality follows from $\pi_{\mathcal{N}}A^\times\pi_{\mathcal{N}} = A^\times\pi_{\mathcal{N}}$ and $\pi_{\mathcal{L}}A^\times\pi_{\mathcal{L}} = \pi_{\mathcal{L}}A^\times$, expressing the $A^\times$ invariance of $\mathcal{N}$.

Using Theorem 6.2.2, together with the formulas (6.4.9) and (6.4.12), we can describe the partial multiplicities of the factors in these minimal realizations in terms of the partial multiplicities of the eigenvalues of A and $A^\times$:

Theorem 6.4.3. *Let $W(\lambda) = W_1(\lambda)W_2(\lambda)$ be the minimal factorization given by (6.4.8) which corresponds to a supporting decomposition $\mathbb{C}^m = \mathcal{L} \dotplus \mathcal{N}$. Then the poles (resp. zeros) of $W_1(\lambda)$ coincide with the eigenvalues of $A|\mathcal{L}$ (resp. of $\pi_{\mathcal{L}}A^\times|\mathcal{L}$), and the absolute values of negative (resp. positive) partial multiplicities of $W_1(\lambda)$ at λ_0 coincide with the partial multiplicities of λ_0 as an eigenvalue of $A|\mathcal{L}$ (resp. of $\pi_{\mathcal{L}}A^\times|\mathcal{L}$).*

The poles (resp. zeros) of $W_2(\lambda)$ coincide with the eigenvalues of $\pi_{\mathcal{N}}A|\mathcal{N}$ (resp. of $A^\times|\mathcal{N}$), and the absolute values of negative (resp. positive) partial multiplicities of $W_2(\lambda)$ at λ_0 coincide with the partial multiplicities of λ_0 as an eigenvalue of $\pi_{\mathcal{N}}A|\mathcal{N}$ (resp. of $A^\times|\mathcal{N}$).

MINIMAL FACTORIZATION

We state explicitly a fact concerning minimal factorizations that will be used in subsequent chapters.

Theorem 6.4.4. *Given the minimal factorization (6.4.8), which is based on the supporting decomposition (6.4.7), both realizations in (6.4.9) are minimal. Similarly, given minimal factorization (6.4.12), both of the realizations*

$$W_2(\lambda)^{-1} = I - C(\lambda I - A^\times)^{-1} \pi_{\mathcal{N}} B, \quad W_1(\lambda)^{-1} = I - C\pi_{\mathcal{L}}(\lambda I - A^\times) B$$

are minimal.

Proof. The minimality of (6.4.8) implies

$$\delta(W) = \delta(W_1) + \delta(W_2) \tag{6.4.13}$$

where $W_1(\lambda)$ and $W_2(\lambda)$ are the factors in (6.4.8). Now (6.4.9) gives

$$\delta(W_1) \leq \dim \mathcal{L}, \quad \delta(W_2) \leq \dim \mathcal{N}, \tag{6.4.14}$$

and therefore

$$\delta(W_1) + \delta(W_2) \leq \dim \mathcal{L} + \dim \mathcal{N} = m = \delta(W), \tag{6.4.15}$$

where the last equality is ensured because $W(\lambda) = I + C(\lambda I - A)^{-1}B$ is a minimal realization of $W(\lambda)$. Combining (6.4.13) and (6.4.15) we see that the equalities must hold in (6.4.14). This means that the realizations (6.4.9) are minimal. The second part of the theorem is proved in a similar way. □

We conclude this section with an instructive and detailed example illustrating the contents of Theorem 6.4.2.

Example 6.4.1. (adapted from Gohberg *et al.* (1986a)). Let

$$W(\lambda) = \begin{bmatrix} 1 + \lambda^{-1}(\lambda - 1)^{-1} & \lambda^{-1} \\ 0 & 1 + \lambda^{-1} \end{bmatrix}.$$

Then $W(\lambda)$ has a realization

$$W(\lambda) = I + C(\lambda I - A)^{-1} B \tag{6.4.16}$$

where

$$A = \begin{bmatrix} 1 & 0 & 0 \\ 1 & 0 & 0 \\ 0 & 0 & 0 \end{bmatrix}, \quad B = \begin{bmatrix} 1 & 0 \\ 0 & 1 \\ 0 & 1 \end{bmatrix}, \quad C = \begin{bmatrix} 0 & 1 & 0 \\ 0 & 0 & 1 \end{bmatrix}.$$

This realization is minimal. Indeed, the matrix

$$\begin{bmatrix} C \\ CA \end{bmatrix} = \begin{bmatrix} 0 & 1 & 0 \\ 0 & 0 & 1 \\ 1 & 0 & 0 \\ 1 & 0 & 0 \end{bmatrix}$$

has rank 3 and hence zero kernel. The matrix

$$[B, AB] = \begin{bmatrix} 1 & 0 & 1 & 0 \\ 0 & 1 & 1 & 0 \\ 0 & 1 & 0 & 0 \end{bmatrix}$$

has rank 3, and hence its image is $\mathbb{C}^3$. Further

$$A^\times = A - BC = \begin{bmatrix} 1 & -1 & 0 \\ 1 & 0 & -1 \\ 0 & 0 & -1 \end{bmatrix}.$$

Let us find all invariant subspaces for A and $A^\times$. We will use the notation $\langle x, y, z \rangle$ for the column vector $\begin{bmatrix} x \\ y \\ z \end{bmatrix}$. It is easy to see that $\langle 1, 1, 0 \rangle$ is an eigenvector of A corresponding to the eigenvalue 1, whereas the vectors $\langle 0, 0, 1 \rangle$, $\langle 0, 1, 0 \rangle$ are the eigenvectors of A corresponding to the eigenvalue 0. Hence all one-dimensional A-invariant subspaces are of the form span$\{\langle 1, 1, 0 \rangle\}$; span$\{\langle 0, 1, 0 \rangle\}$; span$\{\langle 0, \alpha, 1 \rangle\}$, $\alpha \in \mathbb{C}$. All two-dimensional A-invariant subspaces are of the form

$$\text{span}\{\langle 1, 0, 0 \rangle, \langle 0, 1, 0 \rangle\}; \quad \text{span}\{\langle 1, 1, 0 \rangle, \langle 0, \alpha, 1 \rangle\}, \quad \alpha \in \mathbb{C}$$

$$\text{span}\{\langle 0, 1, 0 \rangle, \langle 0, 0, 1 \rangle\}.$$

Passing to $A^\times$, we find that $A^\times$ has three eigenvalues -1, $\gamma = \frac{1}{2}(1 + i\sqrt{3})$, and $\bar\gamma$ with corresponding eigenvectors $\langle 1, 2, 3 \rangle$, $\langle 1, \bar\gamma, 0 \rangle$, and $\langle 1, \gamma, 0 \rangle$, respectively. There are three one-dimensional $A^\times$-invariant subspaces span$\{\langle 1, 2, 3 \rangle\}$, span$\{\langle 1, \bar\gamma, 0 \rangle\}$, span$\{\langle 1, \gamma, 0 \rangle\}$, and three two-dimensional $A^\times$-invariant subspaces span$\{\langle 1, 2, 3 \rangle, \langle 1, \bar\gamma, 0 \rangle\}$, span$\{\langle 1, 2, 3 \rangle, \langle 1, \gamma, 0 \rangle\}$, and span$\{\langle 1, 0, 0 \rangle, \langle 0, 1, 0 \rangle\}$.

We now write down all supporting decompositions

$$\mathbb{C}^3 = \mathcal{L} \dotplus \mathcal{N}, \tag{6.4.17}$$

where $\mathcal{L}$ is one-dimensional, $\mathcal{N}$ is two-dimensional and by definition of a supporting decomposition, $\mathcal{L}$ is A-invariant and $\mathcal{N}$ is $A^\times$-invariant. By inspection of the list of A-invariant and $A^\times$-invariant subspaces we see that the supporting decompositions (6.4.17) with $\dim \mathcal{L} = 1$, $\dim \mathcal{N} = 2$ are given by

(1) $\mathcal{L} = \text{span}\{\langle 1,1,0 \rangle\}$; $\mathcal{N} = \text{span}\{\langle 1,2,3 \rangle, \langle 1,\bar{\gamma},0 \rangle\}$.

(2) $\mathcal{L} = \text{span}\{\langle 1,1,0 \rangle\}$; $\mathcal{N} = \text{span}\{\langle 1,2,3 \rangle, \langle 1,\gamma,0 \rangle\}$.

(3) $\mathcal{L} = \text{span}\{\langle 0,1,0 \rangle\}$; $\mathcal{N} = \text{span}\{\langle 1,2,3 \rangle, \langle 1,\bar{\gamma},0 \rangle\}$.

(4) $\mathcal{L} = \text{span}\{\langle 0,1,0 \rangle\}$; $\mathcal{N} = \text{span}\{\langle 1,2,3 \rangle, \langle 1,\gamma,0 \rangle\}$.

(5) $\mathcal{L} = \text{span}\{\langle 0,\alpha,1 \rangle\}$; $\mathcal{N} = \text{span}\{\langle 1,2,3 \rangle, \langle 1,\bar{\gamma},0 \rangle\}$, where $\alpha \in \mathbb{C}$ is such that $\bar{\gamma} + 3\alpha - 2 \neq 0$.

(6) $\mathcal{L} = \text{span}\{\langle 0,\alpha,1 \rangle\}$; $\mathcal{N} = \text{span}\{\langle 1,2,3 \rangle, \langle 1,\gamma,0 \rangle\}$, where $\alpha \in \mathbb{C}$ is such that $\gamma + 3\alpha - 2 \neq 0$.

(7) $\mathcal{L} = \text{span}\{\langle 0,\alpha,1 \rangle\}$; $\mathcal{N} = \text{span}\{\langle 1,0,0 \rangle, \langle 0,1,0 \rangle\}$.

We shall compute in detail the minimal factorization of $W(\lambda)$ corresponding to the supporting decomposition (6.4.17), where $\mathcal{L}$ and $\mathcal{N}$ are given by (2). It is convenient to write the matrices A, B, C (understood as transformations in the standard orthonormal bases in $\mathbb{C}^2$ and $\mathbb{C}^3$) with respect to the basis $\langle 1,1,0 \rangle$, $\langle 1,\gamma,0 \rangle$, $\langle 1,2,3 \rangle$ in $\mathbb{C}^3$ and the standard basis $\langle 1,0 \rangle$, $\langle 0,1 \rangle$ in $\mathbb{C}^2$:

$$A = \begin{bmatrix} 1 & 1 & 1 \\ 0 & 0 & 0 \\ 0 & 0 & 0 \end{bmatrix} \; ; \quad C = \begin{bmatrix} 1 & \gamma & 2 \\ 0 & 0 & 3 \end{bmatrix} ;$$

$$B = \begin{bmatrix} -\gamma(1-\gamma)^{-1} & -\frac{1}{3}(\gamma+1)(\gamma-1)^{-1} \\ (1-\gamma)^{-1} & \frac{2}{3}(\gamma-1)^{-1} \\ 0 & \frac{1}{3} \end{bmatrix}.$$

The corresponding minimal factorization is now easily found:

$$W(\lambda) = W_1(\lambda)W_2(\lambda),$$

where

$$W_1(\lambda) = I + \begin{bmatrix} 1 \\ 0 \end{bmatrix} (\gamma-1)^{-1} \begin{bmatrix} -\gamma(1-\gamma)^{-1}, & -\frac{1}{3}(\gamma+1)(\gamma-1)^{-1} \end{bmatrix}$$

$$= \begin{bmatrix} 1 - \frac{\gamma}{(1-\gamma)(\lambda-1)} & \frac{-(\gamma+1)}{3(\gamma-1)(\lambda-1)} \\ 0 & 1 \end{bmatrix}$$

and

$$W_2(\lambda) = \begin{bmatrix} 1 + \frac{\gamma}{(1-\gamma)\lambda} & \frac{2}{3\lambda} + \frac{2\gamma}{3(\gamma-1)\lambda} \\ 0 & \frac{1+\lambda}{\lambda} \end{bmatrix}.$$

Replacing γ by $\bar{\gamma}$ we obtain the minimal factorization corresponding to the choice (1) of $\mathcal{L}$ and $\mathcal{N}$. □

6.5 Nonnegative rational matrix functions

An $n \times n$ rational matrix function $W(\lambda)$ is called *nonnegative* on the real axis if $W(\lambda) \geq 0$, i.e. $W(\lambda)$ is positive semidefinite hermitian for every real λ which is not a pole of W. We will assume throughout this section that $W(\lambda)$ is analytic at infinity and $W(\infty) = I$. In particular, $\det W(\lambda) \not\equiv 0$ and $W(\lambda_0)$ is in fact positive definite for all real λ_0 that are not zeros or poles of $W(\lambda)$.

We start with the properties of partial multiplicities of nonnegative rational matrix functions.

Lemma 6.5.1. *If $W(\lambda)$ is an $n \times n$ rational matrix function which is nonnegative on the real axis, then the partial multiplicities of $W(\lambda)$ at every real point λ_0 are all even.*

Proof. Without loss of generality, we assume that $W(\lambda)$ is analytic at λ_0 (otherwise multiply $W(\lambda)$ by $(\lambda - \lambda_0)^k$, for a suitable even integer k). We use a local version of Rellich's theorem. This asserts that, for any $n \times n$ rational matrix function $V(\lambda)$ satisfying $V(\lambda) = V(\lambda)^*$ for real λ (not poles of $V(\lambda)$) and analytic at $\lambda_0 \in \mathbb{R}$, there exists a neighborhood $U \subseteq \mathbb{R}$ of λ_0 and an analytic $n \times n$ matrix function $Q(\lambda)$ defined for $\lambda \in U$ such that, for $\lambda \in U$, $Q(\lambda)(Q(\lambda))^* = I$ (i.e. $Q(\lambda)$ is unitary for $\lambda \in U$) and

$$V(\lambda) = (Q(\lambda))^* \operatorname{diag}[\mu_1(\lambda), \ldots, \mu_n(\lambda)] Q(\lambda), \quad \lambda \in U, \qquad (6.5.1)$$

where the scalar functions $\mu_1(\lambda), \ldots, \mu_n(\lambda)$ are analytic on U. A transparent proof of Rellich's theorem is found in Gohberg *et al.* (1982a), Chapter A.6, and therefore will not be reproduced here.

Apply Rellich's theorem for $V(\lambda) = W(\lambda)$. For each λ the n numbers, $\mu_1(\lambda), \ldots, \mu_n(\lambda)$ are just the eigenvalues of $W(\lambda)$, and since $W(\lambda)$ is nonnegative on the real axis, we must have

$$\mu_j(\lambda) \geq 0, \quad \lambda \in U, \quad \text{for } j = 1, \ldots, n. \qquad (6.5.2)$$

Therefore, the multiplicitiy α_j of λ_0 as a zero of $\mu_j(\lambda)$ is an even nonnegative integer. In other words, we can write

$$\mu_j(\lambda) = (\lambda - \lambda_0)^{\alpha_j} \nu_j(\lambda),$$

where $\nu_j(\lambda)$ are analytic at λ_0 and $\nu_j(\lambda_0) \neq 0$. Permuting the rows and columns of $W(\lambda)$ (if necessary), we can assume that $\alpha_1 \leq \ldots \leq \alpha_n$. Observe also that, in view of (6.5.2), $\nu_j(\lambda) > 0$ in some neighborhood $U_1 \subseteq U$ of λ_0. We now rewrite (6.5.1) (with $V(\lambda) = W(\lambda)$) in the form

$$W(\lambda) = (R(\lambda))^* \operatorname{diag}[(\lambda - \lambda_0)^{\alpha_1}, \ldots, (\lambda - \lambda_0)^{\alpha_n}] R(\lambda), \quad \lambda \in U_0,$$

where $R(\lambda) = \operatorname{diag}[\nu_1(\lambda)^{1/2}, \ldots, \nu_n(\lambda)^{1/2}] Q(\lambda)$. Clearly, $R(\lambda)$ is analytic and invertible for $\lambda \in U_0$. Now, just as in the proof of uniqueness of partial multiplicities (see the proof of Theorem 6.2.1) we conclude that, for $j = 1, \ldots, n$, the sum $\alpha_1 + \cdots + \alpha_j$ coincides with the minimal multiplicity at λ_0, among all not identically zero determinants of the $j \times j$ submatrices of $W(\lambda)$. Therefore,

$$\alpha_1 + \cdots + \alpha_j = \kappa_1 + \cdots + \kappa_j, \quad j = 1, \ldots, n$$

where $\kappa_1 \leq \ldots \leq \kappa_n$ are the partial multiplicities of $W(\lambda)$ at λ_0. Thus, $\kappa_j = \alpha_j$ ($j = 1, \ldots, n$), and since all α_j's are even, the same is true for the κ_j's. □

It turns out that the symmetric minimal factorizations of a nonnegative rational matrix function $W(\lambda)$ (i.e. of the form $W(\lambda) = (L(\bar{\lambda}))^* L(\lambda)$) exist and can be conveniently described in terms of supporting decompositions. We need a preliminary analysis of minimal realizations for such functions $W(\lambda)$. In this analysis, the essential property of $W(\lambda)$ is that it is *hermitian* (on the real line): $W(\lambda) = (W(\lambda))^*$ when λ is real and not a pole of $W(\lambda)$. Clearly, every nonnegative rational matrix function is hermitian.

Let $W(\lambda)$ be a hermitian rational matrix function, and let

$$W(\lambda) = I + C(\lambda I - A)^{-1} B \tag{6.5.3}$$

be a minimal realization. By taking adjoints in (6.5.3) we obtain

$$W(\lambda) = I + B^*(\lambda I - A^*)^{-1} C^*. \tag{6.5.4}$$

Although (6.5.4) was obtained only for real λ (not poles of $W(\lambda)$), in fact it is true for all $\lambda \in \mathbb{C}$ (not poles of $W(\lambda)$). This is because both sides of (6.5.4) represent rational matrix functions which are equal on an infinite set (i.e. on the real line with the poles of $W(\lambda)$ deleted), and therefore are equal everywhere. By Theorem 6.1.5 we see that the realization (6.5.4) is minimal as well, and by Theorem 6.1.4 there is a unique invertible matrix H such that

$$HA = A^* H, \quad C = B^* H, \quad B = H^{-1} C^*. \tag{6.5.5}$$

Now these equations remain true if H is replaced by H^*. In view of the uniqueness of H we obtain $H = H^*$. Moreover, the first equality in (6.5.5) means that A is H-self-adjoint.

Let $A^\times = A - BC$; then, using (6.5.5):

$$HA^\times = H(A - BC) = A^* H - HBC = A^* H - C^* B^* H = A^{\times *} H,$$

and so $A^\times$ is H-self-adjoint as well. We can rewrite the minimal realization (6.5.3) in the form

$$W(\lambda) = I + B^* H(\lambda I - A)^{-1} B. \tag{6.5.6}$$

Now let

$$W(\lambda) = K(\lambda) L(\lambda) \tag{6.5.7}$$

be a minimal factorization of $W(\lambda)$, where $K(\infty) = L(\infty) = I$. Then (by taking adjoints and using the assumption that $W(\lambda)$ is hermitian)

$$W(\lambda) = L_*(\lambda)K_*(\lambda), \qquad (6.5.8)$$

where $L_*(\lambda) = (L(\bar{\lambda}))^*$ and $K_*(\lambda) = (K(\bar{\lambda}))^*$ are also rational matrix functions with $L_*(\infty) = K_*(\infty) = I$. It is easy to see (by using Theorem 6.1.5 for example) that

$$\delta(L_*) = \delta(L), \quad \delta(K_*) = \delta(K),$$

and therefore the factorization (6.5.8) is minimal as well. Now we can make the formal statement:

Lemma 6.5.2. *Let $W(\lambda)$ be a hermitian rational matrix function with minimal realization (6.5.3) and with $H = H^*$ defined uniquely by (6.5.5). Let π be the supporting projector corresponding to the minimal factorization (6.5.7). Then $H^{-1}(I - \pi^*)H$ is the supporting projector corresponding to the minimal factorization (6.5.8) (with respect to the same minimal realization (6.5.3)).*

Proof. Denote $\tau = H^{-1}(I - \pi^*)H$. Let us check first that τ is indeed a supporting projector. Since π is a projector, so is $I - \pi^*$ and, consequently, so is τ. Further, observe that

$$H(\operatorname{Ker}\tau) = \operatorname{Ker}(I - \pi^*) = \operatorname{Im}\pi^* = [\operatorname{Ker}\pi]^\perp,$$

where "$\perp$" denotes the orthogonal complement with respect to the usual scalar product in $\mathbb{C}^m$. Pick $x \in \operatorname{Ker}\tau$; then $Hx \in \operatorname{Im}\pi^*$, say $Hx = \pi^*y$. Now

$$\tau Ax = Ax - H^{-1}\pi^* HAx = H^{-1}(I - \pi^*)A^*Hx$$
$$= H^{-1}(I - \pi^*)A^*\pi^*y = H^{-1}[\pi A(I - \pi)]^*y,$$

and since $\operatorname{Ker}\pi$ is A-invariant, we have $A(I - \pi) = (I - \pi)A(I - \pi)$, and therefore $\pi A(I - \pi) = 0$. So $\tau Ax = 0$, and $Ax \in \operatorname{Ker}\tau$. Hence $\operatorname{Ker}\tau$ is A-invariant.

We check now that

$$A^\times(\operatorname{Im}\tau) \subseteq \operatorname{Im}\tau. \qquad (6.5.9)$$

Choose $x \in \operatorname{Im}\tau$; as $H(\operatorname{Im}\tau) = \operatorname{Im}(I - \pi^*)$, we have $Hx = (I - \pi^*)y$. Now $\tau A^\times x = H^{-1}(A^\times(I-\pi))^*Hx = H^{-1}(A^\times(I-\pi))^*(I-\pi^*)y = H^{-1}((I-\pi)A^\times(I-\pi))^*y$ and since $\operatorname{Im}\pi$ is $A^\times$-invariant, i.e. $A^\times\pi = \pi A^\times\pi$, we get

$$\tau A^\times x = H^{-1}((I - \pi)A^\times)^*y = H^{-1}(A^\times)^*Hx = A^\times x,$$

in view of the H-self-adjointness of $A^\times$. So $A^\times x \in \operatorname{Im}\tau$, (6.5.9) is checked, and it follows that τ is a supporting projector.

We now calculate the factors in the minimal factorization $W(\lambda) = W_1(\lambda)W_2(\lambda)$ that corresponds to τ, and are given by the formula (6.4.8). Thus

$$\begin{aligned}W_2(\lambda) &= I + C\tau(\lambda I - A)^{-1}\tau B \\ &= I + CH^{-1}(I - \pi^*)H(\lambda I - A)^{-1}H^{-1}(I - \pi^*)HB \\ &= I + B^*(I - \pi^*)(\lambda I - HAH^{-1})^{-1}(I - \pi^*)C^* \\ &= I + B^*(I - \pi^*)(\lambda I - A^*)^{-1}(I - \pi^*)C^*.\end{aligned}$$

Taking adjoints gives

$$W_{2*}(\lambda) = I + C(I - \pi)(\lambda I - A)^{-1}(I - \pi)B.$$

In view of the same formula (6.4.8) applied to the minimal factorization (6.5.7), it is seen that $W_{2*}(\lambda) = K(\lambda)$. □

We are now ready to state and prove the main result of this section.

Theorem 6.5.3. *Let $W(\lambda)$ be an $n \times n$ nonnegative rational matrix function having value I at infinity. Then $W(\lambda)$ admits a minimal factorization of the form*

$$W(\lambda) = (L(\bar{\lambda}))^* L(\lambda), \qquad (6.5.10)$$

where $L(\lambda)$ is a rational matrix function having value I at infinity. Furthermore, let

$$W(\lambda) = I + C(\lambda I - A)^{-1}B \qquad (6.5.11)$$

be a minimal realization of $W(\lambda)$ and let H be the unique hermitian matrix determined by the equations

$$HA = A^*H, \quad C = B^*H, \quad B = H^{-1}C^*. \qquad (6.5.12)$$

Then the set of all minimal factorizations of the form (6.5.10) is in one-to-one correspondence with the set Q of all ordered pairs of subspaces $\{\mathcal{L}, \mathcal{N}\}$, where $\mathcal{L}$ is A-invariant, $\mathcal{N}$ is $(A - BB^*H)$-invariant, and both $\mathcal{L}$ and $\mathcal{N}$ are $\frac{m}{2}$-dimensional and H-neutral (here m is the McMillan degree of $W(\lambda)$, which coincides with the size of the matrices A and H).

The minimal factorization of the form (6.5.10) which corresponds to an ordered pair of subspaces $\{\mathcal{L}, \mathcal{N}\}$ as above is given by the formula

$$W(\lambda) = [I + B^*H(\lambda I - A)^{-1}(I - \pi)B][I + B^*H\pi(\lambda I - A)^{-1}B], \qquad (6.5.13)$$

where π is the projector on $\mathcal{N}$ along $\mathcal{L}$. In particular, for any ordered pair of subspaces $\{\mathcal{L}, \mathcal{N}\} \in Q$ the subspaces $\mathcal{L}$ and $\mathcal{N}$ are direct complements to each other.

Proof. First of all we observe that, in view of Lemma 6.5.1 and Theorem 6.2.2, the partial multiplicities of A corresponding to its real eigenvalues (if any) are all even. But, as we have seen before, A is H-self-adjoint. Thus, the canonical form (2.3.1), (2.3.2) shows that the size m of A (and H) must be even. So the usage of $\frac{m}{2}$ in the statement of Theorem 6.5.3 is justified.

Assume now that (6.5.10) is the minimal factorization corresponding to a supporting projector π. As $\delta(L(\lambda)) = \delta((L(\bar{\lambda}))^*) = \frac{m}{2}$, clearly Im π and Ker π are $\frac{m}{2}$-dimensional. By Lemma 6.5.2 the same minimal factorization corresponds also to the supporting projector $\tau = H^{-1}(I - \pi^*)H$. By uniqueness of the supporting projector corresponding to the same minimal factorization, we have

$$\pi = H^{-1}(I - \pi^*)H. \qquad (6.5.14)$$

Denote by $[.,.]$ the indefinite scalar product in $\mathbb{C}^m$ induced by H. For $x \in $ Ker π we have

$$0 = \pi x = H^{-1}(I - \pi^*)Hx = (I - \pi^*)Hx,$$

or equivalently $Hx = \pi^* Hx$. Now for $x, y \in $ Ker π

$$[x, y] = (Hx, y) = (\pi^* Hx, y) = (Hx, \pi y) = 0.$$

This proves that Ker π is H-neutral. Observe that $I - \pi = H^{-1}\pi^* H$, i.e. $I - \pi$ satisfies the equation (6.5.14) (with π replaced by $I - \pi$). So, repeating the above argument for $I - \pi$ (rather than π) we find that Ker $(I - \pi) = $ Im π is H-neutral as well.

Thus, in view of Theorem 6.4.2, it remains only to prove that if $\mathcal{L}$ and $\mathcal{N}$ are $\frac{m}{2}$-dimensional H-neutral subspaces such that $\mathcal{L}$ is A-invariant and $\mathcal{N}$ is $(A - BB^*H)$-invariant, then necessarily $\mathcal{L}$ and $\mathcal{N}$ are direct complements to each other. Clearly, it will suffice to show that $\mathcal{L} \cap \mathcal{N} = \{0\}$. Let $x \in \mathcal{L} \cap \mathcal{N}$; then by the H-neutrality of $\mathcal{L}$ and $\mathcal{N}$ we have

$$[Ax, x] = 0, \quad [(A - BB^*H)x, x] = 0.$$

So

$$0 = [BB^*Hx, x] = (HBB^*Hx, x) = (C^*Cx, x) = (Cx, Cx),$$

and therefore $Cx = 0$. Now

$$(A - BB^*H)x = Ax \in \mathcal{L} \cap \mathcal{N}$$

for any $x \in \mathcal{L} \cap \mathcal{N}$, and therefore the subspace $\mathcal{L} \cap \mathcal{N}$ is A-invariant. We conclude that $\mathcal{L} \cap \mathcal{N}$ is an A-invariant subspace contained in Ker C. Therefore,

$$\mathcal{L} \cap \mathcal{N} \subseteq \sum_{i=0}^{\infty} \text{Ker}\,(CA^i),$$

and by the observability of the pair (C, A) we conclude that $\mathcal{L} \cap \mathcal{N} = \{0\}$. □

NONNEGATIVE RATIONAL MATRIX FUNCTIONS   135

We can say more about the poles and zeros of the factor $L(\lambda)$ in (6.5.10). Indeed, by Theorem 6.4.3 the poles and zeros of $L(\lambda)$ coincide with the eigenvalues of $\pi_{\mathcal{N}} A | \mathcal{N}$ and $A^{\times} | \mathcal{N}$, respectively, where $\pi_{\mathcal{N}}$ is the supporting projector corresponding to (6.5.10), and we denote $A^{\times} = A - BB^*H$. Choose c-sets $\mathcal{C}_A$ and $\mathcal{C}_{A^{\times}}$ for A and $A^{\times}$, respectively (recall that a set of nonreal eigenvalues of A is called a c-set if it is a maximal set of non-real eigenvalues of A that does not contain any λ_0 and its complex conjugate $\bar{\lambda}_0$ simultaneously). By Theorem 2.5.1, there exist an A-invariant subspace $\mathcal{L}$ and an $A^{\times}$-invariant subspace $\mathcal{N}$ such that both $\mathcal{L}$ and $\mathcal{N}$ are H-neutral and $\frac{m}{2}$-dimensional, and the nonreal eigenvalues of $\pi_{\mathcal{N}} A | \mathcal{N}$ and $A^{\times} | \mathcal{N}$ coincide with $\mathcal{C}_A$ and $\mathcal{C}_{A^{\times}}$, respectively; here $\pi_{\mathcal{N}}$ is the projector on $\mathcal{N}$ along $\mathcal{L}$. So in view of Theorem 6.5.3 we obtain the following result (except for the uniqueness statement).

Theorem 6.5.4. *Let $W(\lambda)$ be as in Theorem 6.5.3, with the minimal realization (6.5.11), and let $A^{\times} = A - BB^*H$. Then for any choice of c-sets $\mathcal{C}_A$ and $\mathcal{C}_{A^{\times}}$ there exists a unique minimal factorization (6.5.10) such that the nonreal poles of $L(\lambda)$ coincide with $\mathcal{C}_A$ and the nonreal zeros of $L(\lambda)$ coincide with $\mathcal{C}_{A^{\times}}$.*

The uniqueness part of Theorem 6.5.4 follows from deeper properties, which will not be proved here, concerning A and $A^{\times}$ as H-self-adjoint matrices. Namely, it turns out that the sign characteristic of (A, H) and of $(A^{\times}, H)$ consists of $+1$'s only (see, for example, Theorem II.3.11 in Gohberg et al. (1983)). Now the uniqueness part of Theorem 6.5.4 follows by appealing to Theorem 2.5.2.

We conclude this section with a result concerning local minimality of realizations of nonnegative rational matrix functions. Having in mind subsequent applications (in Chapter 21), we consider here rational $n \times n$ matrix functions that are allowed to have a zero (but not a pole) at infinity. Such functions admit a realization

$$W(\lambda) = D + C(\lambda I - A)^{-1} B. \qquad (6.5.15)$$

Together with this realization consider the realization of the adjoint rational matrix function

$$(W(\bar{\lambda}))^* = D^* + B^*(\lambda I - A^*)^{-1} C^*.$$

The product formula (6.4.5) (adapted to the present case when D is not necessarily the identity matrix) gives

$$W(\lambda)(W(\bar{\lambda}))^* = DD^* + [C, DB^*]\left(\lambda I - \begin{bmatrix} A & BB^* \\ 0 & A^* \end{bmatrix}\right)^{-1}\begin{bmatrix} BD^* \\ C^* \end{bmatrix}. \qquad (6.5.16)$$

Theorem 6.5.5. *If the realization (6.5.15) is minimal on the real line $\mathbb{R}$, then (6.5.16) is minimal on $\mathbb{R}$ as well.*

Proof. Fix $\lambda \in \mathbb{R}$, and let S be an invertible matrix such that

$$S(\lambda I - A)S^{-1} = \begin{bmatrix} 0 & 0 \\ A_1 & A_2 \end{bmatrix}, \qquad (6.5.17)$$

where the rows of $[A_1 \ A_2]$ are linearly independent. The existence of form (6.5.17) follows easily using the Jordan form of A. Partition SB and CS^{-1} conformally with (6.5.17):

$$SB = \begin{bmatrix} B_1 \\ B_2 \end{bmatrix}, \quad CS^{-1} = [C_1 \ C_2].$$

Since (A, B, C) is minimal on $\mathbb{R}$, using the equivalence of (i) and (ii) in Theorem 6.3.2, we obtain

$$\operatorname{rank} \begin{bmatrix} 0 & 0 & B_1 \\ A_1 & A_2 & B_2 \end{bmatrix} = \operatorname{rank}[S(\lambda I - A)S^{-1}, SB]$$

$$= \operatorname{rank} S[\lambda I - A \ B] \begin{bmatrix} S^{-1} & 0 \\ 0 & I \end{bmatrix} = \operatorname{rank}[\lambda I - A \ B] = m,$$

where A has size $m \times m$. It follows that the rows of B_1 are linearly independent. Also,

$$\operatorname{rank} \begin{bmatrix} A_1^* & C_1^* \\ A_2^* & C_2^* \end{bmatrix} = \operatorname{rank} S^{*-1}[\lambda I - A^*, C^*] \begin{bmatrix} S^* & 0 \\ 0 & I \end{bmatrix}$$

$$= \operatorname{rank}[\lambda I - A^*, C^*] = m. \qquad (6.5.18)$$

Once more, the minimality on $\mathbb{R}$ of (A, B, C) and the equivalence of statements (i) and (ii) in Theorem 6.3.2 have been used. Now

$$\operatorname{rank} \begin{bmatrix} \lambda I - A & -BB^* & BD^* \\ 0 & \lambda I - A^* & C^* \end{bmatrix}$$

$$= \operatorname{rank} \begin{bmatrix} S & 0 \\ 0 & S^{*-1} \end{bmatrix} \begin{bmatrix} \lambda I - A & -BB^* & BD^* \\ 0 & \lambda I - A^* & C^* \end{bmatrix} \begin{bmatrix} S^{-1} & 0 & 0 \\ 0 & S^* & 0 \\ 0 & 0 & I \end{bmatrix}$$

$$= \operatorname{rank} \begin{bmatrix} 0 & 0 & -B_1 B_1^* & * & * \\ A_1 & A_2 & * & * & * \\ 0 & 0 & 0 & A_1^* & C_1^* \\ 0 & 0 & 0 & A_2^* & C_2^* \end{bmatrix},$$

which is equal to $2m$ in view of the linear independence of rows of each of the three matrices $-B_1 B_1^*$, $[A_1 \ A_2]$ and $\begin{bmatrix} A_1^* & C_1^* \\ A_2^* & C_2^* \end{bmatrix}$ (see (6.5.18)). Also

$$\text{rank} \begin{bmatrix} \lambda I - A & -BB^* \\ 0 & \lambda I - A^* \\ C & DB^* \end{bmatrix}$$

$$= \text{rank} \begin{bmatrix} 0 & I & 0 \\ I & 0 & 0 \\ 0 & 0 & I \end{bmatrix} \begin{bmatrix} \lambda I - A & -BB^* & BD^* \\ 0 & \lambda I - A^* & C^* \end{bmatrix}^* \begin{bmatrix} 0 & I \\ I & 0 \end{bmatrix} = 2m.$$

By the equivalence of (i) and (ii) in Theorem 6.3.2 (applied to the realization (6.5.16)) we find that the realization (6.5.16) is minimal on $\mathbb{R}$. $\square$

6.6 Rational matrix functions which are hermitian on the unit circle

In this section we study rational matrix functions $W(\lambda)$ which take hermitian values on the unit circle:

$$W(\lambda) = (W(\lambda))^* \quad (6.6.1)$$

for all λ with $|\lambda| = 1$ which are not poles of $W(\lambda)$. Our goal is to describe minimal realizations and factorizations of such functions. It turns out that their minimal realizations have certain symmetries which are reflected in their descriptions in terms of supporting decompositions (as in Theorem 6.4.2).

To start with, observe that the property (6.6.1) is equivalent to

$$W(\lambda) = (W(\bar{\lambda}^{-1}))^* \quad (6.6.2)$$

for every $\lambda \in \mathbb{C}\backslash\{0\}$ which is not a pole of $W(\lambda)$. Indeed, assume (6.6.1) holds. Then both sides of (6.6.2) represent rational matrix functions which coincide on the unit circle (with the possible exception of a finite number of points — the poles of $W(\lambda)$). But if two rational matrix functions coincide on an infinite set, they must coincide everywhere. Thus (6.6.2) holds. Conversely, (6.6.2) obviously implies (6.6.1).

We assume also that $W(\lambda)$ is analytic and invertible at infinity. Suppose that

$$W(\lambda) = D + C(\lambda I - A)^{-1}B \quad (6.6.3)$$

is a minimal realization of the rational $n \times n$ matrix function $W(\lambda)$ which is hermitian on the unit circle. Since $W(\infty) = W(0)^* = D$ it follows that $W(0)$ is invertible, and hence A is also invertible. Further

$$W(0) = D^* = D - CA^{-1}B \quad (6.6.4)$$

From (6.6.2) it follows that

$$W(\lambda) = D^* + B^*(\lambda^{-1}I - A^*)^{-1}C^* = D^* - B^*A^{*-1}(\lambda I - A^{*-1})^{-1}\lambda C^*$$

$$= D^* - B^*A^{*-1}C^* - B^*A^{*-1}(\lambda I - A^{*-1})^{-1}A^{*-1}C^*.$$

Taking into account (6.6.4) one sees that

$$W(\lambda) = D - B^*A^{*-1}(\lambda I - A^{*-1})^{-1}A^{*-1}C^*$$

is a minimal realization for $W(\lambda)$. Hence there exists a unique invertible matrix S such that

$$SA = A^{*-1}S, \quad A^{*-1}C^* = SB, \quad -B^*A^{*-1}S = C. \qquad (6.6.5)$$

Taking adjoints in (6.6.5) we find from the uniqueness of S that $S = -S^*$. It follows that the matrix iS is self-adjoint and invertible, and A is iS-unitary, i.e. introducing the indefinite scalar product $[x,y] = \langle iSx, y \rangle$ we have

$$[Ax, Ay] = [x, y]$$

for all $x, y \in \mathbb{C}^m$, or, equivalently, $A^*(iS)A = iS$.

It turns out that the matrix $A^\times := A - BD^{-1}C$ is iS-unitary as well. Indeed, in view of (6.6.5) we have

$$SA^\times = SA - SBD^{-1}C = (A^{*-1} + A^{*-1}C^*D^{-1}B^*A^{*-1})S.$$

A straightforward computation using (6.6.4) yields

$$(A^\times)^{-1} = A^{-1} + A^{-1}BD^{*-1}CA^{-1}.$$

Hence

$$SA^\times = (A^\times)^{*-1}S.$$

Recall from Section 6.4 that a projector $\pi : \mathbb{C}^m \to \mathbb{C}^m$, where m is the size of A, is called *supporting* if Ker π is A-invariant and Im π is $A^\times$-invariant. By Theorem 6.4.2 (suitably modified to account for the value of $W(\lambda)$ at infinity which need not be I) every supporting projector gives rise to a minimal factorization

$$W(\lambda) = K(\lambda)D_0 L(\lambda) \qquad (6.6.6)$$

where D_0 is some invertible matrix, and where

$$K(\lambda) = DD_0^{-1} + C(\lambda - A)^{-1}(I - \pi)BD_0^{-1}, \quad L(\lambda) = I + D^{-1}C\pi(\lambda - A)^{-1}B.$$

We say that this factorization is *associated* with π. Note that $L(\infty) = I$.

The following proposition will be useful in what follows.

Proposition 6.6.1. *Assume that $W(\lambda)$ satisfies the condition (6.5.2), and let $W(\lambda) = K(\lambda)D_0 L(\lambda)$ be the minimal factorization associated with a supporting projector π, where the invertible matrix D_0 is chosen so that $K(0) = I$. Then the minimal factorization*

$$W(\lambda) = L(\bar{\lambda}^{-1})^* D_0^* K(\bar{\lambda}^{-1})^* \qquad (6.6.7)$$

is associated with the supporting projection $\tau = S^{-1}(I - \pi^*)S$.

Note that in view of (6.6.2), the formula (6.6.6) implies (6.6.7).

Proof. First we compute a formula for D_0. Since $W(\infty) = D$ we have
$$D = L(0)^* D_0^* = (I - B^* A^{*-1}\pi^* C^* D^{*-1})D_0^*.$$
So
$$D_0^{*-1} = D^{-1}(I - B^* A^{*-1}\pi^* C^* D^{*-1}). \tag{6.6.8}$$
Using (6.6.5) and the equality
$$(\lambda^{-1}I - A^*)^{-1} = -A^{*-1} - A^{*-1}(\lambda I - A^{*-1})^{-1}A^{*-1}$$
computation of $L(\bar{\lambda}^{-1})^*$ gives
$$\begin{aligned}L(\bar{\lambda}^{-1})^* &= I - B^* A^{*-1}\pi^* C^* D^{*-1} - B^* A^{*-1}(\lambda I - A^{*-1})^{-1} A^{*-1} \pi^* C^* D^{*-1}\\ &= DD_0^{*-1} + C(\lambda I - A)^{-1}AS^{-1}\pi^* SA^{-1}BD^{*-1}.\end{aligned}$$
Now we claim that $AS^{-1}\pi^* SA^{-1}BD^{*-1} = S^{-1}\pi^* SBD_0^{*-1}$. Indeed, using (6.6.5) and (6.6.8) we have
$$\begin{aligned}S^{-1}\pi^* SBD_0^{*-1}D^* &= S^{-1}\pi^* SB(D^{-1}D^* - D^{-1}B^* A^{*-1}\pi^* C^*)\\ &= S^{-1}\pi^* SB(I - D^{-1}CA^{-1}B + D^{-1}CS^{-1}\pi^* SA^{-1}B)\\ &= S^{-1}\pi^* SB - S^{-1}\pi^* SBD^{-1}CS^{-1}(I - \pi^*)SA^{-1}B.\end{aligned}$$
Now $SBD^{-1}CS^{-1} = S(A - A^\times)S^{-1} = A^{*-1} - (A^\times)^{*-1}$; and since $\operatorname{Im}\pi$ is $(A^\times)^{-1}$-invariant, which gives $(I - \pi)(A^\times)^{-1}\pi = 0$, we obtain
$$\begin{aligned}S^{-1}\pi^* SBD_0^{*-1}D^* &= S^{-1}\pi^* SB - S^{-1}\pi^* A^{*-1}(I - \pi^*)SA^{-1}B\\ &= S^{-1}\pi^* SB - S^{-1}\pi^* A^{*-1}(I - \pi^*)A^* SB\\ &= S^{-1}\pi^* A^{*-1}\pi^* A^* SB = S^{-1}\pi^* A^{*-1}\pi^* SA^{-1}B\\ &= S^{-1}A^{*-1}\pi^* SA^{-1}B,\end{aligned}$$
the last equality following easily from the relation $(I-\pi)A^{-1}(I-\pi) = A^{-1}(I-\pi)$. Hence
$$S^{-1}\pi^* SBD_0^{*-1}D^* = AS^{-1}\pi^* SA^{-1}B.$$
Our claim is established.

From this one sees that
$$L(\bar{\lambda}^{-1})^* = DD_0^{*-1} + C(\lambda I - A)^{-1}S^{-1}\pi^* SBD_0^{*-1}.$$
Similarly one shows
$$K(\bar{\lambda}^{-1})^* = I + D^{-1}CS^{-1}(I - \pi^*)S(\lambda I - A)^{-1}B.$$
Since $S^{-1}(I - \pi^*)S$ is easily seen to be a supporting projector it follows that the factorization associated with $S^{-1}(I - \pi^*)S$ is $W(\lambda) = L(\bar{\lambda}^{-1})^* D_0^* K(\bar{\lambda}^{-1})^*$. □

The case when $W(\lambda)$ is positive semidefinite on the unit circle is of particular interest in the applications to discrete Riccati equations:

Corollary 6.6.2. *Let $W(\lambda)$ be a rational matrix function such that $W(\lambda)$ is analytic and invertible at infinity and such that $W(\lambda) \geq 0$ for all λ on the unit circle which are not poles of $W(\lambda)$. Then $W(\lambda)$ admits a minimal factorization of the form*

$$W(\lambda) = L(\bar{\lambda}^{-1})^* D L(\lambda), \tag{6.6.9}$$

where $L(\lambda)$ is a rational matrix function having value I at infinity.

More precisely, fix a minimal realization (6.6.3). Then the set of minimal factorizations of the form (6.6.9) is in one-to-one correspondence with the set of supporting projectors π satisfying $S\pi = (I - \pi^*)S$; this correspondence is given by the formula

$$L(\lambda) = I + D^{-1} C \pi (\lambda I - A)^{-1} B.$$

Proof. We only have to prove existence of minimal factorizations (6.6.9); all other parts of the corollary follow from Proposition 6.6.1.

Choose complex numbers $\omega \neq \bar{\omega}$ and $|\eta| = 1$ such that $W(\eta)$ is invertible. Let

$$\tilde{W}(\lambda) = W(\eta(\lambda - \bar{\omega})(\lambda - \omega)^{-1}).$$

Then $\tilde{W}(\lambda)$ is nonnegative on the real axis and $\tilde{W}(\infty) = W(\eta)$ is invertible. By Theorem 6.5.3 $\tilde{W}(\lambda)$ admits a minimal factorization

$$\tilde{W}(\lambda) = (\tilde{L}(\bar{\lambda}))^* D \tilde{L}(\lambda), \tag{6.6.10}$$

for some rational matrix function $\tilde{L}(\lambda)$ such that $\tilde{L}(\infty) = I$. Let

$$L(\lambda) = \tilde{L}((\eta\bar{\omega} - \lambda\omega)(\eta - \lambda)^{-1}).$$

Since the functions $f(\lambda) = \eta(\lambda - \bar{\omega})(\lambda - \omega)^{-1}$ and $g(\lambda) = (\eta\bar{\omega} - \lambda\omega)(\eta - \lambda)^{-1}$ are inverses to each other, (6.6.10) implies the factorization

$$W(\lambda) = L(\bar{\lambda}^{-1})^* D L(\lambda) \tag{6.6.11}$$

It is easy to check that (6.6.11) is minimal (because (6.6.10) is) using the formulas for the minimal realization of $L(\lambda)$ in terms of the minimal realization for $\tilde{L}(\lambda)$, for example. Thus, if

$$\tilde{L}(\lambda) = I + \tilde{C}(\lambda I - \tilde{A})^{-1} \tilde{B}$$

is a minimal realization of $\tilde{L}(\lambda)$, then

$$L(\lambda) = I - C(-\eta I + A)^{-1} B$$
$$+ (-\eta\omega + \eta\bar{\omega}) C(-\eta I + A)^{-1} (\lambda I + (\eta\bar{\omega} I - \omega A)(-\eta I + A)^{-1})(-\eta I + A)^{-1} B \tag{6.6.12}$$

is a minimal realization of $L(\lambda)$. The formula (6.6.12) can be established using simple algebraic manipulations (for details see Section 1.5 of Bart et al. (1979), for example). □

A result analogous to Theorem 6.5.4 is valid for minimal factorizations (6.6.9) as well. Namely, let A and $A^\times := A - BD^{-1}C$ be taken from a minimal realization (6.6.3), where $W(\lambda) \geq 0$ for all λ with $|\lambda| = 1$ which are not poles of $W(\lambda)$. A set $\mathcal{C}$ of nonunimodular eigenvalues of A will be called a *d-set* for A if $\lambda \in \mathcal{C} \Rightarrow \bar{\lambda}^{-1} \notin \mathcal{C}$ and $\mathcal{C}$ is maximal with respect to this property. Then for any choice of a *d*-set $\mathcal{C}_A$ for A and of a *d*-set $\mathcal{C}_{A^\times}$ for $A^\times$ there exists a minimal factorization (6.6.9), where the set of nonunimodular poles (resp. zeros) of $L(\lambda)$ coincides with $\mathcal{C}_A$ (resp. $\mathcal{C}_{A^\times}$). The proof of this statement can be reduced to Theorem 6.5.4, as in the proof of Corollary 6.6.2.

6.7 The real case

In this section we consider briefly real rational matrix functions, i.e. functions of the form

$$W(\lambda) = \left[\frac{p_{ij}(\lambda)}{q_{ij}(\lambda)}\right]_{i,j=1}^{r,n},$$

which are finite at infinity, and where $p_{ij}(\lambda)$ and $q_{ij}(\lambda)$ are scalar polynomials of the complex variable λ with *real* coefficients.

All the results and proofs of Section 6.1 remain valid for such functions if we admit only realizations

$$W(\lambda) = D + C(\lambda I - A)^{-1}B \qquad (6.7.1)$$

with real matrices A, B, C, D. For example, if (6.7.1) and

$$W(\lambda) = D + C'(\lambda I - A')^{-1}B' \qquad (6.7.2)$$

are two minimal realizations of the same $W(\lambda)$ and all the matrices appearing in the realizations (6.7.1) and (6.7.2) are real, then there exists a unique invertible real matrix S such that

$$C = C'S, \quad A = S^{-1}A'S, \quad B = S^{-1}B'.$$

The results of Section 6.3 can be applied, in particular, for a real rational matrix function $W(\lambda)$. As a byproduct of the existence of a real minimal realization (6.7.1) we obtain the following fact.

Proposition 6.7.1. *Let $W(\lambda)$ be an $n \times n$ regular real rational matrix function. Then the poles, as well as the zeros, of $W(\lambda)$ are symmetric relative to the real axis, and this symmetry extends to partial multiplicities. In other words, if λ_0 is a pole of $W(\lambda)$, then $\bar{\lambda}_0$ is also a pole of $W(\lambda)$ with the same partial pole multiplicities as λ_0. A similar statement applies to the zeros of $W(\lambda)$.*

The results and proofs of Section 6.4 extend to the real case as well. Here, we use minimal realizations of the form (6.4.1) with real matrices A, B, C, and the supporting decompositions are considered in $\mathbb{R}^m$ (rather than in $\mathbb{C}^m$, see equation (6.4.7)). We will not reformulate these results for the real case, but consider an instructive example instead.

Example 6.7.1. Let

$$W(\lambda) = \begin{bmatrix} 1 + \lambda^{-1}(\lambda - 1)^{-1} & \lambda^{-1} \\ 0 & 1 + \lambda^{-1} \end{bmatrix},$$

as in Example 6.4.1. The minimal realization (6.4.16) is obviously real. The minimal factorizations $W(\lambda) = W_1(\lambda)W_2(\lambda)$ with real rational matrix functions $W_1(\lambda)$ and $W_2(\lambda)$ such that $W_1(\infty) = W_2(\infty) = I$, correspond to supporting decompositions

$$\mathbb{R}^3 = \mathcal{L} + \mathcal{N}$$

(i.e. $\mathcal{L}$ and $\mathcal{N}$ are real subspaces). The supporting decompositions with $\dim \mathcal{L} = 1$, $\dim \mathcal{N} = 2$ are given by the formula

$$\mathcal{L} = \operatorname{span}\{\langle 0, \alpha, 1\rangle\}, \quad \mathcal{N} = \operatorname{span}\{\langle 1, 0, 0\rangle, \langle 0, 1, 0\rangle\}, \qquad (6.7.3)$$

where $\alpha \in \mathbb{R}$ (cf. the list (1)–(7) in Example 6.4.1 of supporting decompositions when $W_1(\lambda)$ and $W_2(\lambda)$ are not required to be real). Write down the matrices

$$A = \begin{bmatrix} 1 & 0 & 0 \\ 1 & 0 & 0 \\ 0 & 0 & 0 \end{bmatrix}, \quad B = \begin{bmatrix} 1 & 0 \\ 0 & 1 \\ 0 & 1 \end{bmatrix},$$

$$C = \begin{bmatrix} 0 & 1 & 0 \\ 0 & 0 & 1 \end{bmatrix}, \quad A^\times = \begin{bmatrix} 1 & -1 & 0 \\ 1 & 0 & -1 \\ 0 & 0 & -1 \end{bmatrix}$$

(understood as linear transformations with respect to the standard bases in $\mathbb{R}^2$ and $\mathbb{R}^3$) with respect to the ordered basis $\langle 0, \alpha, 1\rangle$, $\langle 1, 0, 0\rangle$, $\langle 0, 1, 0\rangle$ in $\mathbb{R}^3$ and the standard basis in $\mathbb{R}^2$. We obtain the following forms:

$$A = \begin{bmatrix} 0 & 0 & 0 \\ 0 & 1 & 0 \\ 0 & 1 & 0 \end{bmatrix}, \quad B = \begin{bmatrix} 0 & 1 \\ 1 & 0 \\ 0 & 1-\alpha \end{bmatrix}, \quad C = \begin{bmatrix} \alpha & 0 & 1 \\ 1 & 0 & 0 \end{bmatrix},$$

$$A^\times = \begin{bmatrix} -1 & 0 & 0 \\ -\alpha & 1 & -1 \\ -1+\alpha & 1 & 0 \end{bmatrix}.$$

By formula (6.4.8) the minimal factorization of $W(\lambda)$ that corresponds to the supporting decomposition (6.7.3) is given by $W(\lambda) = W_1(\lambda)W_2(\lambda)$, where

$$W_1(\lambda) = I + \begin{bmatrix} \alpha \\ 1 \end{bmatrix} \lambda^{-1}[0 \ 1] = \begin{bmatrix} 1 & \alpha\lambda^{-1} \\ 0 & 1+\lambda^{-1} \end{bmatrix};$$

$$W_2(\lambda) = I + \begin{bmatrix} 0 & 1 \\ 0 & 0 \end{bmatrix} \left(\lambda I - \begin{bmatrix} 1 & 0 \\ 1 & 0 \end{bmatrix} \right)^{-1} \begin{bmatrix} 1 & 0 \\ 0 & 1-\alpha \end{bmatrix}$$
$$= \begin{bmatrix} 1 + \lambda^{-1}(\lambda-1)^{-1} & \lambda^{-1}(1-\alpha) \\ 0 & 0 \end{bmatrix}. \quad \square$$

Theorem 6.5.3 in its entirety is not valid in the real case, as it is easy to find real nonnegative rational matrix functions $W(\lambda)$ which do not admit minimal factorizations of the form $W(\lambda) = (L(\bar{\lambda}))^*L(\lambda)$, where $L(\lambda)$ is real. Take $W(\lambda) = 1 + \lambda^{-2}$, a scalar function, for example. We state the real analogue of Theorem 6.5.3.

Theorem 6.7.2. *Let $W(\lambda)$ be a real nonnegative rational matrix function having value I at infinity. Then $W(\lambda)$ admits a minimal factorization of the form*

$$W(\lambda) = (L(\bar{\lambda}))^*L(\lambda) \qquad (6.7.4)$$

where $L(\lambda)$ is a real rational matrix function having value I at infinity if and only if the following condition holds:

(i) *for every non-real zero (resp. pole) λ_0 of $W(\lambda)$ the sum of positive (resp. negative) partial multiplicities of $W(\lambda)$ at λ_0 is even.*

If the condition (i) holds, then the set of all minimal factorizations of the form (6.7.4) is in one-to-one correspondence with the set of all ordered pairs $\{\mathcal{L}, \mathcal{N}\}$, where $\mathcal{L}$ is A-invariant, $\mathcal{N}$ is $(A - BB^*H)$-invariant, and both $\mathcal{L}$ and $\mathcal{N}$ are $\frac{m}{2}$-dimensional H-neutral subspaces in $\mathbb{R}^m$. This correspondence is given by the same formula (6.5.13) as in Theorem 6.5.3.

The proof of Theorem 6.7.2 is analogous to that of Theorem 6.5.3 (proof of necessity and sufficiency of condition (i) is based on Theorem 2.6.3). We omit further details.

A special situation when all poles and zeros of $W(\lambda)$ are real warrants a separate statement.

Theorem 6.7.3. *Let $W(\lambda)$ be a real nonnegative rational matrix function with value I at infinity, and having all poles and zeros on the real axis. Then $W(\lambda)$ admits a unique minimal factorization of the form (6.7.4).*

Proof. The existence of such factorization of $W(\lambda)$ follows from Theorem 6.7.2, and its uniqueness is a consequence of Theorem 6.5.4. $\square$

6.8 Exercises

6.8.1. Find minimal realizations for the following rational matrix functions:

(a) $\begin{bmatrix} 1 & 2\lambda^{-1} \\ 2\lambda^{-1} & 1 \end{bmatrix}$;

(b) $\begin{bmatrix} 1 & 2\lambda^{-1} \\ (\lambda+1)^{-1} & 1-\lambda^{-1} \end{bmatrix}$;

(c) $\begin{bmatrix} 1+(\lambda-1)^{-1} & 0 \\ \lambda^{-1} & 1+(\lambda+1)^{-1} \end{bmatrix}$.

6.8.2. Let $W(\lambda)$ be an $m \times n$ rational matrix function (not necessarily analytic at infinity). Show that $W(\lambda)$ admits realizations of the form

$$W(\lambda) = D + C(\lambda I - A)^{-1}B + E(I - \lambda F)^{-1}G, \qquad (6.8.1)$$

where the matrix F is nilpotent. [Hint: Write $W(\lambda) = W_1(\lambda) + W_2(\lambda)$, where $W_1(\lambda)$ is analytic at infinity and $W_2(\lambda)$ is a polynomial, and apply Theorem 6.1.2 to $W_1(\lambda)$ and $W_2(\lambda^{-1})$].

6.8.3. State and prove the analogues of Theorems 6.1.4 and 6.1.5 for the realizations (6.8.1).

6.8.4. Prove part (I) of Proposition 6.3.1.

6.8.5. Show that for rational matrix functions $W_1(\lambda)$ and $W_2(\lambda)$ with values I at infinity the inequalities

$$|\delta(W_1) - \delta(W_2)| \leq \delta(W_1 W_2) \leq \delta(W_1) + \delta(W_2)$$

hold.

6.8.6. Compute the minimal factorizations of the rational matrix function (6.4.16) corresponding to the supporting decompositions $\mathbb{C}^3 = \mathcal{L} \dotplus \mathcal{N}$, where $\mathcal{L}$ is two-dimensional and $\mathcal{N}$ is one-dimensional.

6.8.7. Show that $W(\lambda) = \begin{bmatrix} 1 & \lambda^{-2} \\ 0 & 1 \end{bmatrix}$ does not admit nontrivial minimal factorizations.

6.8.8. (Generalization of (6.8.7).) Let $W(\lambda) = D + C(\lambda I - A)^{-1}B$ be a minimal realization, where $BD^{-1}C = 0$. Show that $W(\lambda)$ admits a nontrivial minimal factorization if and only if A is not similar to one Jordan block.

6.8.9. Find minimal realizations (6.5.3) and the matrix H (determined by (6.5.5)) for the following hermitian rational matrix functions:

(a) $\begin{bmatrix} 1 & 2\lambda^{-1} \\ 2\lambda^{-1} & 1 \end{bmatrix}$;

(b) $\begin{bmatrix} 1+(\lambda-i)^{-1}+(\lambda+i)^{-1}+(\lambda^2+1)^{-1} & \lambda^{-1}+\lambda^{-1}(\lambda-i)^{-1} \\ \lambda^{-1}+\lambda^{-1}(\lambda+i)^{-1} & 1+\lambda^{-2}+(\lambda-i)^{-1}+(\lambda+i)^{-1}+(\lambda^2+1)^{-1} \end{bmatrix}.$

6.8.10. Prove that the rational matrix function of 6.8.9(b) is nonnegative and find for this function all minimal factorizations (6.5.10).

6.8.11. Find all minimal factorizations (6.6.9) of the following rational matrix function:

$$W(\lambda) = \begin{bmatrix} 1 & (\lambda+1)^{-2} \\ (\lambda^{-1}+1)^{-2} & 1+(2+\lambda+\lambda^{-1})^{-2} \end{bmatrix}.$$

6.8.12. Prove that for a given $W(\lambda)$ as in Corollary 6.6.2 the minimal factorization (6.6.9) is unique if and only if all poles and zeros of $W(\lambda)$ are on the unit circle.

6.9 Notes

Rational matrix functions appear as transfer functions in the analysis of time-invariant linear systems and in signal processing. The associated realization problems are fundamental and development of the theory can be found in the following sources (among many others): Anderson and Vongpanitlerd (1973), Belevitch (1968), Brockett (1970), Gohberg et al. (1986a), Kailath (1980), Kalman et al. (1969), Rosenbrock (1970).

Factorization of rational matrix functions, and especially minimal factorization, have emerged more recently as important concepts in both systems theory and mathematical analysis. The first results concerning minimal factorization in special cases can be found in the books of Anderson and Vongpanitlerd (1973) and Belevitch (1968). An important part in the further development of the theory was played by the papers of Dewilde and Vandewalle (1975) and Vandewalle and Dewilde (1977). A comprehensive theory of minimal factorization of rational matrix functions emerged in the works of Bart et al. (1979 and 1980). For continued development and applications see the books of Clancey and Gohberg (1981), Gohberg and Kaashoek (1986), and Gohberg et al. (1986a).

The concept of locally minimal realizations, Proposition 6.3.4 and its proof appeared first in a paper of Gohberg and Kaashoek (1987). Other material on locally minimal realizations in Section 6.3, as well as Theorem 6.5.5, seem to be new.

Theorem 6.5.3, the main result of Section 6.5, was obtained by Ran (1982). Factorization results of this type for other classes of matrix functions with symmetries are known; see Gohberg *et al.* (1980, 1982a, 1983) for nonnegative matrix polynomials; Ran and Rodman (1990) for real rational matrix functions with various symmetries.

Part II

CONTINUOUS ALGEBRAIC RICCATI EQUATIONS

In this part we develop the analysis of one of our major problems: to describe, as far as possible, the solution set of Riccati equations of the form

$$XDX + XA + A^*X - C = 0$$

where A, D, C are given matrices in $\mathbb{C}^{n \times n}$, or in $\mathbb{R}^{n \times n}$, with $D^* = D$, $C^* = C$, and additional hypotheses, as required. The solution set is defined in $\mathbb{C}^{n \times n}$, or in $\mathbb{R}^{n \times n}$, in the two cases, and we are particularly interested in, respectively, hermitian solutions, and real symmetric solutions.

The Part consists of five chapters, numbers 7 through 11. In Chapter 7 we develop a geometric approach based on the connection of solutions with certain invariant subspaces of the (Hamiltonian) matrix

$$i \begin{bmatrix} A & D \\ C & -A^* \end{bmatrix},$$

and depends on the fact that this matrix is self-adjoint in a suitable *indefinite* scalar product. Canonical reductions developed in Chapter 2 play an important part here. In Chapter 8 a similar program is followed in the "real case". However, there are significant differences in the results depending on the properties of real (rather than complex) canonical forms (see Chapters 2 and 3). Chapters 7 and 8 form the central core of the whole book.

Another approach to this Riccati equation is based on comparisons between two equations, and a constructive iterative approach to their solution. This is the subject matter of Chapter 9. Connections with rational matrix functions, using materials developed in Chapter 6, are the topic of Chapter 10 and, finally, Chapter 11 is devoted to perturbation theory.

7
GEOMETRIC THEORY: THE COMPLEX CASE

Using some of the material developed in Part I we can now begin with the first central chapter on solution of Riccati equations. The "geometric theory" of the title refers to the geometry of subspaces in a finite dimensional complex linear space. It is shown that solutions of Riccati equations, and hermitian solutions in particular, are strongly connected with invariant subspaces of a certain operator defined by the coefficients of the equation. By examining these invariant subspaces in the context of a suitable indefinite scalar product (see Chapter 2) considerable information on the structure of the solution set of the Riccati equation is obtained. Of course, the "solution set" here admits both real and complex matrices.

7.1 Solutions and invariant subspaces

We start with some general results concerning solutions of matrix quadratic equations; (not necessarily with the symmetries prescribed in the introduction to Part II). Thus, consider the equation

$$XBX + XA - DX - C = 0, \qquad (7.1.1)$$

where A, B, C, D have sizes $n \times n$, $n \times m$, $m \times n$, and $m \times m$, respectively, and $m \times n$ matrix solutions X are to be found.

For any $m \times n$ matrix X and $n \times n$ identity I, we call the n-dimensional subspace

$$G(X) = \operatorname{Im} \begin{bmatrix} I \\ X \end{bmatrix} \subseteq \mathbb{C}^{m+n}$$

the *graph* of X. Also a subspace of $\mathbb{C}^{m+n}$ is called a *graph subspace* if it has the form $G(X)$ for some X.

The first simple proposition connects solutions of (7.1.1) with invariant subspaces of the $(m+n) \times (m+n)$ matrix

$$T = \begin{bmatrix} A & B \\ C & D \end{bmatrix}. \qquad (7.1.2)$$

Proposition 7.1.1. *For any $m \times n$ matrix X, the graph of X is T-invariant if and only if X is a solution of (7.1.1).*

Proof. If $G(X)$ if T-invariant, then

$$\begin{bmatrix} A & B \\ C & D \end{bmatrix} \begin{bmatrix} I \\ X \end{bmatrix} = \begin{bmatrix} I \\ X \end{bmatrix} Z \qquad (7.1.3)$$

for a suitable matrix Z. The first block row in this equality gives $Z = A + BX$, and then the second block row gives

$$C + DX = X(A + BX).$$

In other words, X solves (7.1.1).
  Conversely, if X solves (7.1.1), then (7.1.3) holds with $Z = A + BX$. □

Representing the T-invariant subspace $G(X)$ as the linear span of a set of Jordan chains of T, we obtain the following result.

Theorem 7.1.2. *Equation (7.1.1) has a solution $X \in \mathbb{C}^{m \times n}$ if and only if there is a set of vectors $v_1, \ldots, v_n$ in $\mathbb{C}^{m+n}$ forming a set of Jordan chains for T (of equation (7.1.2)), and if*

$$v_j = \begin{bmatrix} y_j \\ z_j \end{bmatrix}, \quad j = 1, 2, \ldots,$$

where $y_j \in \mathbb{C}^n$, then $y_1, y_2, \ldots, y_n$ form a basis for $\mathbb{C}^n$.
  Furthermore, if

$$Y = [y_1 \ y_2 \ \cdots \ y_n] \in \mathbb{C}^{n \times n}, \quad Z = [z_1 \ z_2 \ \cdots \ z_m] \in \mathbb{C}^{m \times n},$$

every solution of (7.1.1) has the form $X = ZY^{-1}$ for some set of Jordan chains $v_1, v_2, \ldots, v_n$ for T such that Y is invertible.

The invertibility of Y is the condition ensuring that $\text{span}\{v_1, v_2, \ldots, v_n\}$ is a graph subspace of $\mathbb{C}^{m+n}$.
  Clearly, $G(X_1) = G(X_2)$ if and only if $X_1 = X_2$. Thus, Proposition 7.1.1 (or Theorem 7.1.2) establishes a one-to-one correspondence between the set of solutions of (7.1.1) and a subset of the set of n-dimensional T-invariant subspaces. This subset may be a proper subset of the set of n-dimensional T-invariant subspaces, or it may even be empty as, in general, not every n-dimensional T-invariant subspace is a graph subspace. A well-known example of a 2×2 matrix without a square root illustrates this point:

Example 7.1.1. Let $B = I$, $A = D = 0$, $C = \begin{bmatrix} 0 & 1 \\ 0 & 0 \end{bmatrix}$. Then

$$T = \begin{bmatrix} 0 & 0 & 1 & 0 \\ 0 & 0 & 0 & 1 \\ 0 & 1 & 0 & 0 \\ 0 & 0 & 0 & 0 \end{bmatrix}.$$

It is easy to see that the Jordan form of T consists of one 4×4 nilpotent Jordan block. Thus, there is a unique 2-dimensional T-invariant subspace $\mathcal{M}$, and it is spanned by $(1\ 0\ 0\ 0)^T$ and $(0\ 0\ 1\ 0)^T$. Clearly, $\mathcal{M}$ is not a graph subspace for any 2×2 matrix X, and we conclude that the equation

$$X^2 - \begin{bmatrix} 0 & 1 \\ 0 & 0 \end{bmatrix} = 0$$

has no solutions X. □

We remark also that generically (for example, when the matrix T is nonderogatory) the number of solutions of (7.1.1) is finite and bounded above by the binomial coefficient $\binom{m+n}{n}$, the number of ways in which n vectors can be chosen from a basis of $m + n$ eigenvectors and generalized eigenvectors of T.

Finally, observe that all results and observations of this section are valid for real solutions X of the equation (7.1.1) with real coefficients A, B, C and D.

7.2 Symmetric equations and invariant semidefinite subspaces

The equation we consider now has the form

$$XDX + XA + A^*X - C = 0, \qquad (7.2.1)$$

where $D = D^*$, $C = C^*$ and all matrices are in $\mathbb{C}^{n \times n}$. For reasons noted in our Preface, this is known as the *continuous algebraic Riccati equation*, or CARE. Define $2n \times 2n$ matrices M, H and $\hat{H}$ by

$$M = i \begin{bmatrix} A & D \\ C & -A^* \end{bmatrix}, \quad H = \begin{bmatrix} -C & A^* \\ A & D \end{bmatrix}, \quad \hat{H} = i \begin{bmatrix} 0 & I \\ -I & 0 \end{bmatrix}.$$

Observe that (using the notation of (7.1.2)) $M = iT$, and therefore Proposition 7.1.1 and Theorem 7.1.2 are valid with T replaced by M. Thus, there is a one-to-one correspondence between the invariant subspaces $\mathcal{M}$ of M which are graph subspaces and solutions X of (7.2.1), via

$$\mathcal{M} = \text{Im} \begin{bmatrix} I \\ X \end{bmatrix}.$$

But, now, because of the symmetry of the equation (7.2.1), we can say more about the $\mathcal{M}$-invariant subspaces.

A direct computation serves to verify that

$$H = H^*, \quad \hat{H} = \hat{H}^*, \quad HM = M^*H, \quad \hat{H}M = M^*\hat{H}, \quad H = \hat{H}M.$$

It may be assumed that H is invertible with n positive eigenvalues and n negative eigenvalues (counted with algebraic multiplicities in both cases). This incurs no loss of generality because, if necessary, A can be replaced by $A + i\alpha I$ where α is a sufficiently large real number, without altering the solution set of (7.2.1).

In what follows we use the notation

$$\text{Im}\, X = \frac{1}{2i}(X - X^*).$$

Clearly, $\text{Im}\, X$ is a hermitian matrix, and $\text{Im}\, X = 0$ if and only if X is hermitian.

Proposition 7.2.1. *Let X be a solution of the CARE (7.2.1) with the corresponding graph subspace $G(X)$. Then:*

(α) $\text{Im}\, X \geq 0$ (resp. $\text{Im}\, X \leq 0$) if and only if $G(X)$ is $\hat{H}$-nonnegative (resp. $\hat{H}$-nonpositive).

(β) X is hermitian if and only if $G(X)$ is $\hat{H}$-neutral.

(γ) The graph $G(X)$ is H-nonpositive (resp. H-nonnegative) if and only if

$$(X^* - X)(A + DX) \leq 0$$

(resp. $(X^* - X)(A + DX) \geq 0$).

(δ) The graph $G(X)$ is H-neutral if and only if

$$(X^* - X)(A + DX) = 0.$$

All parts of this proposition follow by straightforward computations. In particular, the following equality is used:

$$\begin{bmatrix} I \\ X \end{bmatrix}^* H \begin{bmatrix} I \\ X \end{bmatrix} = (X^* - X)(A + DX) + XDX + XA + A^*X - C.$$

It turns out that, under the hypotheses of Proposition 7.2.1, the graph subspace $G(X)$ is $\hat{H}$-neutral if and only if it is H-neutral. Indeed, if $G(X)$ is $\hat{H}$-neutral then, by Proposition 7.2.1, parts (β) and (δ), it follows that $G(X)$ is H-neutral. Conversely, assume $G(X)$ is H-neutral, i.e. $x^*Hy = 0$ for all $x, y \in G(X)$. Because of the equality $H = \hat{H}M$ and the invertibility of H, the matrix M is invertible as well, and the subspace $G(X)$ is M^{-1}-invariant. Now for any $x, z \in G(X)$ we have

$$x^*\hat{H}z = x^*H(M^{-1}z) = 0$$

because $M^{-1}z \in G(X)$.

SYMMETRIC EQUATIONS 153

In particular, for a solution X of (7.2.1) the conditions $X = X^*$ and $(X^* - X)(A + DX) = 0$ are equivalent.

The result of Proposition 7.2.1 suggests that one can find solutions to (7.2.1) with various additional properties (e.g., hermitian solutions) by finding corresponding M-invariant subspaces with additional properties. Of course, when using this approach we need extra hypotheses to make sure that the corresponding subspace is a graph subspace. A first important result in this direction is the following:

Lemma 7.2.2. *Assume that $D \geq 0$ and the pair (A, D) is controllable. Let $\mathcal{L}$ be an n-dimensional M-invariant H-nonpositive subspace of $\mathbb{C}^{2n}$. Then $\mathcal{L}$ is a graph subspace.*

Proof. For a subspace $\mathcal{L}$ defined in the statement, write

$$\mathcal{L} = \mathrm{Im} \begin{bmatrix} X_1 \\ X_2 \end{bmatrix} \tag{7.2.2}$$

for some $n \times n$ matrices X_1 and X_2. We are going to prove that X_1 is invertible.

First, observe that M-invariance of $\mathcal{L}$ means that

$$\begin{bmatrix} A & D \\ C & -A^* \end{bmatrix} \begin{bmatrix} X_1 \\ X_2 \end{bmatrix} = \begin{bmatrix} X_1 \\ X_2 \end{bmatrix} T$$

for some $n \times n$ matrix T. In other words,

$$AX_1 + DX_2 = X_1 T; \tag{7.2.3}$$

$$CX_1 - A^* X_2 = X_2 T. \tag{7.2.4}$$

Then H-nonpositivity of $\mathcal{L}$ means that the matrix

$$[X_1^* \ X_2^*] \begin{bmatrix} -C & A^* \\ A & D \end{bmatrix} \begin{bmatrix} X_1 \\ X_2 \end{bmatrix} = X_2^* D X_2 + X_1^* A^* X_2 + X_2^* A X_1 - X_1^* C X_1 \tag{7.2.5}$$

is negative semidefinite.

Let $\mathcal{K} = \mathrm{Ker}\, X_1$. Since (7.2.5) is negative semidefinite, we have for every $x \in \mathcal{K}$:

$$0 \geq x^* X_2^* D X_2 + x^* X_1 A^* X_2 x + x^* X_2 A X_1 x - x X_1^* C X_1 x = x^* X_2^* D X_2 x,$$

and since $D \geq 0$, $X_2 x \in \mathrm{Ker}\, D$, i.e.

$$X_2 \mathcal{K} \subseteq \mathrm{Ker}\, D. \tag{7.2.6}$$

Further, equation (7.2.3) implies that

$$TK \subseteq K. \qquad (7.2.7)$$

Indeed, for $x \in K$ we have in view of (7.2.3) and (7.2.6):

$$X_1 T x = A X_1 x + D X_2 x = 0.$$

Now equation (7.2.4) gives for every $x \in K$:

$$A^* X_2 x = -C X_1 x + A^* X_2 x = -X_2 T x \in X_2 K$$

and so

$$A^* X_2 K \subseteq X_2 K. \qquad (7.2.8)$$

We see from (7.2.3) that $A^* X_2 K \subseteq \operatorname{Ker} D$ and we now claim, more generally, that

$$A^{*r} X_2 K \subseteq \operatorname{Ker} D, \quad r = 0, 1, 2, \ldots. \qquad (7.2.9)$$

We have already proved this inclusion for $r = 0$ and $r = 1$. Assuming, inductively, that (7.2.9) holds for $r - 1$, and using (7.2.8) it is found that

$$A^{*r}(X_2 K) = A^{*r-1}(A^* X_2 K) \subseteq A^{*r-1} X_2 K \subseteq \operatorname{Ker} D;$$

so (7.2.9) holds. Now for every $x \in K$:

$$\begin{bmatrix} D \\ DA^* \\ \vdots \\ DA^{*n-1} \end{bmatrix} (X_2 x) = 0,$$

or $(X_2 x)^* [D, AD, \ldots, A^{n-1} D] = 0$. But $\operatorname{rank}[D, AD, \ldots, A^{n-1} D] = n$, so $X_2 x = 0$. But the only n-dimensional vector x for which $X_1 x = X_2 x = 0$ is the zero vector; otherwise $\dim \mathcal{L} < n$, which contradicts our assumptions. So $K = \{0\}$ and X_1 is invertible. Then we can write

$$\mathcal{L} = \operatorname{Im} \begin{bmatrix} I \\ X \end{bmatrix},$$

where $X = X_2 X_1^{-1}$, and so $\mathcal{L}$ is indeed a graph subspace. $\square$

Recall that M is H-self-adjoint and that H is assumed to have n positive eigenvalues and n negative eigenvalues. Thus, the existence of n-dimensional M-invariant H-nonpositive subspaces (or, what is the same, M-invariant maximal H-nonpositive subspaces) is ensured by Theorem 2.4.2. So, combining Proposition 7.2.1 and Lemma 7.2.2, we arrive at a complete description of a class of solutions of (7.2.1) in terms of invariant subspaces:

Theorem 7.2.3. *Assume that $D \geq 0$ and the pair (A, D) is controllable. Then the CARE (7.2.1) admits solutions X such that*

$$(X^* - X)(A + DX) \leq 0. \tag{7.2.10}$$

Moreover, the formula

$$\mathcal{M} = \operatorname{Im} \begin{bmatrix} I \\ X \end{bmatrix} \tag{7.2.11}$$

establishes a one-to-one correspondence between the set of solutions X with the property (7.2.10) and the set of M-invariant maximal H-nonpositive subspaces $\mathcal{M}$.

Specializing to hermitian solutions the following theorem is obtained. The nonhermitian solutions will be considered in Section 7.4.

Theorem 7.2.4. *Assume that $D \geq 0$ and the pair (A, D) is controllable. Then the CARE admits hermitian solutions if and only if there exists an n-dimensional M-invariant H-neutral subspace. Moreover, the formula (7.2.11) establishes a one-to-one correspondence between the set of hermitian solutions X and the set of n-dimensional M-invariant H-neutral subspaces $\mathcal{M}$.*

As in the remark after Proposition 7.2.1 we verify that a subspace is M-invariant H-neutral if and only if it is M-invariant $\hat{H}$-neutral. Thus, in the formulation of Theorem 7.2.4 we can replace H by $\hat{H}$.

Later in this chapter we will study the situations in which there exist subspaces $\mathcal{M}$ as described in the theorem.

If one is concerned only with the *existence* of solutions X in a certain class, and leaves aside the problem of describing *all* such solutions, then the controllability hypothesis in Theorems 7.2.3 and 7.2.4 can be relaxed considerably.

Let (A, B) be a pair of matrices where A is $n \times n$ and B is $n \times m$, and let $\mathcal{C}_{A,B}$ be their controllable subspace (see Section 4.1). The pair (A, B) is called *sign controllable* if either

$$\mathcal{R}_{\alpha+i\beta}(A) \subseteq \mathcal{C}_{A,B}, \quad \text{or} \quad \mathcal{R}_{-\alpha+i\beta}(A) \subseteq \mathcal{C}_{A,B} \tag{7.2.12}$$

(or both) for every complex number $\alpha + i\beta$, $\alpha, \beta \in \mathbb{R}$, and we define $\mathcal{R}_{\alpha+i\beta}(A)$ to be the zero subspace if $\alpha + i\beta$ is not an eigenvalue of A. Clearly, if $\alpha \neq 0$ and one of $\pm \alpha + i\beta$ is not an eigenvalue of A then this condition is automatically satisfied. The cases of interest are when $\alpha \neq 0$ and both $\pm \alpha + i\beta$ are eigenvalues of A, and also when A has pure imaginary eigenvalues. Evidently, any controllable pair is sign controllable. More generally, *any c-stabilizable pair is also sign controllable* (see Section 4.5) and note that this is not necessarily the case for d-stabilizable pairs.

Theorem 7.2.5. *Assume that $D \geq 0$ and the pair (A, D) is sign controllable. Then (7.2.1) admits a solution X such that*

$$(X^* - X)(A + DX) \leq 0.$$

Furthermore, the equation (7.2.1) admits a hermitian solution if and only if there exists an n-dimensional M-invariant H-neutral subspace.

For the proof of Theorem 7.2.5 we need a suitable variation of Lemma 7.2.2.

Lemma 7.2.6. *Assume that $D \geq 0$ and the pair (A, D) is sign controllable. Then there exists an n-dimensional M-invariant H-nonpositive subspace which is a graph subspace.*

Proof. Let

$$\mathcal{L} = \operatorname{Im} \begin{bmatrix} X_1 \\ X_2 \end{bmatrix}$$

be an n-dimensional M-invariant H-nonpositive subspace. Then, as in the proof of Lemma 7.2.2,

$$\begin{bmatrix} A & D \\ C & -A^* \end{bmatrix} \begin{bmatrix} X_1 \\ X_2 \end{bmatrix} = \begin{bmatrix} X_1 \\ X_2 \end{bmatrix} T, \qquad (7.2.13)$$

$$T\mathcal{K} \subseteq \mathcal{K},$$

where $\mathcal{K} = \operatorname{Ker} X_1$, and

$$A^* X_2 x = -X_2 T x \qquad (7.2.14)$$

for every $x \in \mathcal{K}$. As $\mathcal{L}$ is n-dimensional, it follows that $X_2 x = 0$, $x \in \mathcal{K}$, is only possible when $x = 0$. Therefore, (7.2.14) implies that $A^*|_{X_2\mathcal{K}}$ is similar to $(-T)|_\mathcal{K}$. In turn, (7.2.13) shows that $-T$ is the restriction of iM to its invariant subspace $\mathcal{L}$. Thus, $A^*|_{X_2\mathcal{K}}$ is similar to a restriction of iM to its invariant subspace which is contained in $\mathcal{L}$. Therefore the spectrum of $A^*|_{X_2\mathcal{K}}$ is contained in the spectrum of $iM|_\mathcal{L}$.

We will now make a special choice of $\mathcal{L}$ (which is admitted by Theorem 2.4.1). Namely, let S be a maximal subset of the complex plane with the properties that

$$\mathcal{R}_\lambda(A) \subseteq \operatorname{Im} [D, AD, \ldots, A^{n-1}D] = \mathcal{C}_{A,D} \qquad (7.2.15)$$

for every $\lambda \in S$, and if $\alpha + i\beta \in S$, $\alpha, \beta \in \mathcal{R}$, $\alpha \neq 0$, then $-\alpha + i\beta \notin S$. (In particular, all the pure imaginary eigenvalues of iM are in S.) Choose $\mathcal{L}$ so that all the eigenvalues of $iM|_\mathcal{L}$ with nonzero real part are in

$$\bar{S} = \{\bar{\lambda} | \lambda \in S\}.$$

Let $X_2 x$, $x \in \mathcal{K}$ be an eigenvector of $A^*|_{X_2\mathcal{K}}$. As in the proof of Lemma 7.2.2 we obtain

$$X_2 x \in \operatorname{Ker} \begin{bmatrix} D \\ DA^* \\ \vdots \\ DA^{*n-1} \end{bmatrix} = (\mathcal{C}_{A,D})^\perp \subseteq (\sum_{\lambda \in S} \mathcal{R}_\lambda(A))^\perp.$$

The inclusion here follows from (7.2.15). Using a block representation

$$A = \begin{bmatrix} A_{11} & A_{12} \\ 0 & A_{22} \end{bmatrix}$$

with respect to the orthogonal decomposition $\mathbb{C}^n = \mathcal{N} \oplus \mathcal{N}^\perp$, where $\mathcal{N} = \sum_{\lambda \in S} \mathcal{R}_\lambda(A)$, it is easy to verify that

$$\left(\sum_{\lambda \in S} \mathcal{R}_\lambda(A) \right)^\perp = \sum_{\lambda \notin \bar{S}} \mathcal{R}_\lambda(A^*).$$

Thus, $X_2 x \in \sum_{\lambda \notin \bar{S}} \mathcal{R}_\lambda(A^*)$, and therefore $X_2 x$ is an eigenvector of $A^*|X_2\mathcal{K}$ corresponding to an eigenvalue $\lambda_0 \notin \bar{S}$. As S contains the imaginary axis, so does $\bar{S}$, and therefore λ_0 has nonzero real part. But we have seen previously that $\sigma(A^*|X_2\mathcal{K}) \subseteq \sigma(iM|\mathcal{L})$; so $\lambda_0 \in \sigma(iM|\mathcal{L})$, and by the choice of $\mathcal{L}$, $\lambda_0 \in \bar{S}$. We have obtained a contradiction unless $X_2\mathcal{K} = \{0\}$. But the only vector x for which $X_1 x = X_2 x = 0$ is the zero vector (because $\dim \mathcal{L} = n$), and therefore we must conclude that $\mathcal{K} = \operatorname{Ker} X_1 = \{0\}$. Thus,

$$\mathcal{L} = \operatorname{Im} \begin{bmatrix} I \\ X_2 X_1^{-1} \end{bmatrix}$$

is a graph subspace. □

Proof of Theorem 7.2.5. The first part follows by combining Proposition 7.2.1 and Lemma 7.2.6, just as in the proof of Theorem 7.2.3. The second part of Theorem 7.2.5 follows by combining Proposition 7.2.1 and Lemma 7.2.7 to follow. □

Lemma 7.2.7. *Assume that $D \geq 0$ and that the pair (A, D) is sign controllable. If there exists an n-dimensional M-invariant H-neutral subspace, then there exists such a subspace which is also a graph subspace.*

Proof. Observe that, by Theorem 2.5.4, if there is an n-dimensional M-invariant H-neutral subspace, then there is such a subspace $\mathcal{L}$ with the non-real eigenvalues of $M|\mathcal{L}$ lying in any prescribed c-set. Using this observation, Lemma 7.2.7 is proved in exactly the same way as Lemma 7.2.6.

Going back to Theorem 7.2.5, one situation in which the existence of an n-dimensional M-invariant H-neutral subspace is ensured is when M has no real eigenvalues (see Theorem 2.5.1). The following proposition shows that this situation is realized in some cases of interest. Note that the concept of stability used here (and throughout this chapter) is the "c-stability" of Section 4.4.

Theorem 7.2.8. *If $D \geq 0$, $C \geq 0$, (A, D) is stabilizable and (C, A) detectable, then M has no real eigenvalues.*

Proof. Suppose $\lambda \in \mathbb{R}$ and $M \begin{bmatrix} x_1 \\ x_2 \end{bmatrix} = \lambda \begin{bmatrix} x_1 \\ x_2 \end{bmatrix}$, $\begin{bmatrix} x_1 \\ x_2 \end{bmatrix} \neq \begin{bmatrix} 0 \\ 0 \end{bmatrix}$. Thus

$$-Ax_1 + Dx_2 = -i\lambda x_1, \quad Cx_1 + A^* x_2 = -i\lambda x_2. \tag{7.2.16}$$

The first equation implies

$$x_2^* D x_2 = -i\lambda x_2^* x_1 + x_2^* A x_1 = -i\lambda x_1^* x_2 + x_1^* A^* x_2,$$

and using the second equation, $x_2^* D x_2 = -x_1^* C x_1$. The hypotheses $D \geq 0$ and $C \geq 0$ imply $Cx_1 = Dx_2 = 0$.

Using the first equation of (7.2.16) we now have

$$x_1^*(A^* + i\lambda I) = 0, \quad x_1^* C = 0.$$

But (A^*, C) is stabilizable and, since $\lambda \in \mathbb{R}$, Theorem 4.5.6(a) implies

$$\text{rank}\,[-(i\lambda I + A^*), C] = n.$$

Consequently, $x_1 = 0$. Similarly, the stabilizability of (A, D) can be used to show that $x_2 = 0$ as well. Hence a contradiction. □

Note that, when M has no real eigenvalues, as in this theorem, the existence of a solution of (7.2.1) follows readily from the construction used in the proof of Lemma 7.2.6. This will be a special case of the general existence theorem to be proved in the next section. Other existence theorems will be established in Section 9.1 using completely different methods.

7.3 Existence of hermitian solutions and partial multiplicities

At the end of the last section it is observed that when the Hamiltonian matrix M has no real eigenvalues, the existence of a solution of the CARE (7.2.1) is assured. This section is devoted to resolution of the more delicate case in which real eigenvalues for M are admitted. The first theorem is a major technical step toward the proof of the main result of the section, Theorem 7.3.7, in which the controllable or stabilizable hypothesis on (A, D) is replaced by sign-controllability.

EXISTENCE OF HERMITIAN SOLUTIONS

Theorem 7.3.1. *Assume that $D \geq 0$ and $C^* = C$. If the CARE has a hermitian solution X, and*

$$\mathcal{R}_{\lambda_0}(A + DX) \subseteq \mathcal{C}_{A,D} \tag{7.3.1}$$

for every pure imaginary or zero eigenvalue λ_0 of $A + DX$, then all the partial multiplicities corresponding to the real eigenvalues of M (if any) are even. In fact, the partial multiplicities of a real eigenvalue λ_0 of M are twice the partial multiplicities of $-i\lambda_0$ as an eigenvalue of $A + DX$.

For the proof of Theorem 7.3.1 we need two technical lemmas which are, however, of independent interest.

Lemma 7.3.2. *Let $J_1, \ldots, J_k$ be the nilpotent Jordan blocks of sizes $\alpha_1 \geq \cdots \geq \alpha_k$ respectively. Let*

$$J_0 = \begin{bmatrix} J_1 & 0 & \cdots & 0 \\ 0 & J_2 & \cdots & 0 \\ \vdots & & \cdots & \vdots \\ 0 & 0 & \cdots & J_k \end{bmatrix}$$

and

$$\Phi = \begin{bmatrix} J_0 & \Phi_0 \\ 0 & J_0^T \end{bmatrix}$$

where $\Phi_0 = [\phi_{ij}]_{i,j=1}^{\alpha}$ is a matrix of size $\alpha \times \alpha$ ($\alpha = \alpha_1 + \cdots + \alpha_k$). Let

$$\beta_i = \alpha_1 + \cdots + \alpha_i, \quad i = 1, \ldots, k$$

and suppose that the $k \times k$ submatrix $\Psi = [\phi_{\beta_i \beta_j}]_{i,j=1}^{k}$ of Φ_0 is invertible and $\Psi = \Psi_1 \Psi_2$, where Ψ_1 is a lower triangular matrix and Ψ_2 is an upper triangular matrix. Then the sizes of Jordan blocks in the Jordan form of Φ are $2\alpha_1, \ldots, 2\alpha_k$.

Proof. We shall apply similarity transformations to the matrix Φ step by step, and eventually obtain a matrix similar to Φ for which the elementary divisors obviously are $2\alpha_1, \ldots, 2\alpha_k$. For convenience we shall denote block triangular matrices of the form $\begin{bmatrix} K & M \\ 0 & N \end{bmatrix}$ by triang $[K, M, N]$.

Let $i \notin \{\beta_1, \beta_2, \ldots, \beta_k\}$ be an integer; $1 \leq i \leq \alpha$. Let U_{i1} be the $\alpha \times \alpha$ matrix such that all its rows (except for the $(i+1)$-th row) are zeros, and its $(i+1)$-th row is $[\phi_{i1}\phi_{i2}\ldots\phi_{i\alpha}]$. Put $S_{i1} = $ triang $[I, U_{i1}, I]$; then $S_{i1}\Phi S_{i1}^{-1} = $ triang $[J, \Phi_{i0}, J^T]$, where the i-th row of Φ_{i0} is zero, and all other rows (except for the $(i+1)$-th) of Φ_{i0} are the same as in Φ_0. Note also that the submatrix Ψ is the same in Φ_{i0} and in Φ_0. Applying this transformation sequentially for every $i \in \{1, \ldots, \alpha\} \setminus \{\beta_1, \ldots, \beta_k\}$, we find that Φ is similar to a matrix of the form $\Phi_1 = $ triang $[J, V, J^T]$, where the i-th row of V is zero for $i \notin \{\beta_1, \ldots, \beta_k\}$, and $v_{\beta_p \beta_1} = \phi_{\beta_p \beta_q}$, $p, q = 1, \ldots, k$, where $V = (v_{ij})_{i,j=1}^{\alpha}$.

We show now that, by applying a similarity transformation to Φ_1, it is possible to make $v_{ij} = 0$ for $i \notin \{\beta_1, \ldots \beta_k\}$ or $j \notin \{\beta_1, \ldots, \beta_k\}$. Let $V_j = $ col $[v_{ij}]_{i=1}^{\alpha}$

be the j-th column of V. For fixed $j \notin \{\beta_1, \ldots, \beta_k\}$ we have $v_{ij} = 0$ for $i \notin \{\beta_1, \ldots, \beta_k\}$, and, since Ψ is invertible, there exist $\sigma_1, \ldots, \sigma_k \in \mathbb{C}$ such that

$$V_i + \sum_{i=1}^{k} \sigma_i V_{\beta_i} = 0.$$

Let $S_{2j} = \text{diag}[I, I + U_{2j}]$ where, but for the j-th column, U_{2j} consists of zeros, and the i-th entry in the j-th column of U_{2j} is σ_{β_m} if $i = \beta_m$ for some m and zero otherwise. Then $S_{2j}^{-1} \Phi_1 S_{2j} = \text{triang}[J, V - Z_1, J^T - Z_2]$ with the following structure of the matrices Z_1 and Z_2: $Z_1 = [0 \ldots 0 V_j 0 \ldots 0]$, where V_j is in the j-th place; $Z_2 = [0 \ldots 0 Z_{2,j-1} 0 \ldots 0]$, where $Z_{2,j-1} = \text{col}\,[\text{col}\,[-\delta_{p,\alpha_q}\sigma_q]_{p=1}^{\alpha_q}]_{q=1}^{k}$ is an $\alpha \times 1$ column in the $(j-1)$-th place in Z_2 (δ_{p,α_q} denotes the Kronecker symbol). If $j = \beta_m + 1$ for some m, then we put $Z_2 = 0$. It is easy to see that $S_{2j}^{-1} \Phi_1 S_{2j}$ can be reduced by a similarity transformation with a matrix of type $\text{diag}[I, U_{3j}]$ to the form $\text{triang}[J, W_1, J^T]$, where the m-th column of W_1 coincides with the m-th column of $V - Z_1$ for $m \geq j$ and for $m \in \{\beta_1, \ldots, \beta_k\}$. Applying this similarity sequentially for every $j \notin \{\beta_1, \ldots, \beta_k\}$ (starting with $j = \beta_k - 1$ and finishing with $j = 1$), we find that Φ_1 is similar to the matrix $\Phi_2 = \text{triang}[J, W_2, J^T]$, where the (β_i, β_j)-entries of W_2 ($i, j = 1, \ldots, k$) form the matrix Ψ, and all other entries of W_2 are zeros.

The next step is to replace the invertible submatrix Ψ in W_2 by I. We have $\Psi_1^{-1}\Psi = \Psi_2$, where $\Psi_1^{-1} = (b_{ij})_{i,j=1}^{k}$, $b_{ij} = 0$ for $i < j$, is a lower triangular matrix; $\Psi_2 = (c_{ij})_{i,j=1}^{k}$, $c_{ij} = 0$ for $i > j$ is an upper triangular matrix. Define the $2\alpha \times 2\alpha$ invertible matrix $S_3 = (s_{ij}^{(3)})_{i,j=1}^{2\alpha}$ as follows: $s_{\beta_i \beta_j}^{(3)} = b_{ij}$ for $i, j = 1, \ldots, k$; $s_{mm}^{(3)} = 1$ for $m \notin \{\beta_1, \ldots, \beta_k\}$; $s_{ij}^{(3)} = 0$ otherwise. Then $S_3 \Phi_2 S_3^{-1} = \text{triang}[J + Z_3, W_3, J^T]$, where the (β_i, β_j) entries of W_3 ($i, j = 1, \ldots, k$) form the upper triangular matrix Ψ_2, and all other entries of W_3 are zeros; the $\alpha \times \alpha$ matrix Z_3 can contain nonzero entries only on the places $(\beta_i - 1, \beta_j)$ for $i < j$ and such that $\alpha_i > 1$. It is easy to see that $S_3 \Phi_2 S_3^{-1}$ is similar to $\Phi_3 = \text{triang}[J, W_3, J^T]$. Define the $2\alpha \times 2\alpha$ invertible matrix $S_4 = (s_{ij}^{(4)})_{i,j=1}^{2\alpha}$ by the following equalities: $s_{\beta_i \beta_j}^{(4)} - c_{ij}$ for $i, j = 1, \ldots, k$; $s_{mm}^{(4)} = 1$ for $m \notin \{\beta_1, \ldots, \beta_k\}$; $s_{ij}^{(4)} = 0$ otherwise. Then $S_4 \Phi_3 S_4^{-1} = \text{triang}[J, W_4, J^T + Z_4]$, where the (β_i, β_j) entries of W_4 form the unit matrix I, and all other entries are zeros; the $\alpha \times \alpha$ matrix Z_4 can contain nonzero entries only in the positions $(\beta_i, \beta_j - 1)$ for $i < j$ and such that $\alpha_j > 1$. Again, it is easy to see that $S_4 \Phi_3 S_4^{-1}$ is similar to $\Phi_4 = \text{triang}[J, W_4, J^T]$.

Evidently, the degrees of the elementary divisors of Φ_4 are $2\alpha_1, \ldots, 2\alpha_k$. So the same is true for Φ. □

Lemma 7.3.3. *Let Y, D be $k \times k$ matrices with $D \geq 0$ and let X_+, X_- be matrices of the form*

$$X_{\pm} = \begin{bmatrix} Y & D \\ 0 & \pm Y^* \end{bmatrix}.$$

EXISTENCE OF HERMITIAN SOLUTIONS

Assume also that with the minus signs (with the plus signs)

$$\mathcal{R}_{\lambda_0}(Y) \subseteq \mathcal{C}_{Y,D} \tag{7.3.2}$$

for every eigenvalue λ_0 of Y on the imaginary axis (on the real axis, respectively).

Then the partial multiplicities of any eigenvalue λ_0 of X_- (or X_+) on the imaginary axis (on the real axis) are all even. Indeed, they are twice the partial multiplicities of λ_0 as an eigenvalue of Y.

Proof. We consider the case of X_-; the argument for X_+ is similar. Let Z be a Jordan form for Y and $Z = SYS^{-1}$. Then

$$\begin{bmatrix} Y & D \\ 0 & -Y^* \end{bmatrix} = \begin{bmatrix} S^{-1} & 0 \\ 0 & S^* \end{bmatrix} \begin{bmatrix} Z & D_0 \\ 0 & -Z^* \end{bmatrix} \begin{bmatrix} S & 0 \\ 0 & (S^{-1})^* \end{bmatrix}$$

where $D_0 = SDS^* \geq 0$, and it is sufficient to prove that all eigenvalues of $\begin{bmatrix} Z & D_0 \\ 0 & -Z^* \end{bmatrix}$ on the imaginary axis (if any) have the required properties.

Let λ_0 be such a pure imaginary or zero eigenvalue. Let $Z_1, \ldots, Z_k$ be the Jordan blocks of Z corresponding to λ_0, let their sizes be $\alpha_1, \ldots, \alpha_k$, and denote $\alpha = \alpha_1 + \cdots + \alpha_k$. Without loss of generality we can suppose that these blocks are in the northwest corner of Z. So we can write

$$\begin{bmatrix} Z & D_0 \\ 0 & -Z^* \end{bmatrix} = \begin{bmatrix} Z' & 0 & D_1 & D_2 \\ 0 & Z'' & D_2 & D_3 \\ 0 & 0 & -Z'^* & 0 \\ 0 & 0 & 0 & -Z''^* \end{bmatrix}$$

where $Z' = Z_1 \oplus \cdots \oplus Z_k$, Z'' is the 'rest' of Z, and

$$D_0 = \begin{bmatrix} D_1 & D_2 \\ D_2^* & D_3 \end{bmatrix}$$

is the corresponding partition of D_0. The condition (7.3.2) now takes the form

$$\operatorname{Im}\left(\begin{bmatrix} D_1 & D_2 \\ D_2^* & D_3 \end{bmatrix}, \begin{bmatrix} Z' & 0 \\ 0 & Z'' \end{bmatrix}\begin{bmatrix} D_1 & D_2 \\ D_2^* & D_3 \end{bmatrix}, \ldots, \right.$$
$$\left.\begin{bmatrix} (Z')^{n-1} & 0 \\ 0 & (Z'')^{n-1} \end{bmatrix}\begin{bmatrix} D_1 & D_2 \\ D_2^* & D_3 \end{bmatrix}\right) \supseteq \begin{bmatrix} \mathbb{C}^{\alpha} \\ 0 \end{bmatrix}. \tag{7.3.3}$$

Let $D_1 = (D_{1ij})_{i,j=1}^k$ be the partition of D_1 consistent with the partitioning of Z'. It is enough to prove that in the Jordan form of the matrix

$$\begin{bmatrix} Z_1 - \lambda_0 I & 0 & \cdot & 0 & D_{111} & D_{112} & \cdot & D_{11k} \\ 0 & Z_2 - \lambda_0 I & \cdot & 0 & D_{121} & D_{122} & \cdot & D_{12k} \\ \cdot & \cdot & & \cdot & \cdot & \cdot & & \cdot \\ 0 & 0 & \cdot & Z_k - \lambda_0 I & D_{1k1} & D_{1k2} & \cdot & D_{1kk} \\ 0 & 0 & \cdot & 0 & -Z_1^* - \lambda_0 I & 0 & \cdot & 0 \\ 0 & 0 & \cdot & 0 & 0 & -Z_2^* - \lambda_0 I & \cdot & 0 \\ \cdot & \cdot & & \cdot & \cdot & \cdot & & \cdot \\ 0 & 0 & \cdot & 0 & 0 & 0 & \cdot & -Z_k^* - \lambda_0 I \end{bmatrix}$$

the blocks with eigenvalue 0 have sizes $2\alpha_1, \ldots, 2\alpha_k$. Let f_{ij} $(i, j = 1, \ldots, k)$ be the entry in the southeast corner of D_{1ij}; consider the matrix $F = [f_{ij}]_{i,j=1}^k$ formed by all these entries. Since F is a principal submatrix of D_1, and hence of D_0, $F \geq 0$. Let us show that F is invertible. Suppose not; then there exists an invertible matrix $U = [u_{ij}]_{i,j=1}^k$ such that UFU^* has a zero in the southeast corner. Let $G = [g_{ij}]_{i,j=1}^\alpha$ be an $\alpha \times \alpha$ invertible matrix of the following structure

$$\begin{aligned} g_{\beta_i \beta_j} &= u_{ij} \quad \text{where} \quad \beta_i = \alpha_1 + \cdots + \alpha_i; \quad i, j = 1, \ldots, k \\ g_{qq} &= 1 \quad \text{for} \quad q \notin \{\beta_1, \ldots, \beta_k\} \\ g_{pq} &= 0 \quad \text{for} \quad p \neq q \text{ and } \{p, q\} \not\subset \{\beta_1, \ldots, \beta_k\}. \end{aligned}$$

Then the matrix GD_1G^* has a zero in the southeast corner, and since $GD_1G^* \geq 0$, the last column and last row of GD_1G^* are also zeros. On the other hand, from the structure of G it is seen (bearing in mind that the β_1-th, ..., β_k-th rows of $Z' - \lambda_0 I$ are zeros) that the last row of $G(Z' - \lambda_0 I)G^{-1}$ is also zero. Now let

$$\tilde{G} = \begin{bmatrix} G & 0 \\ 0 & I \end{bmatrix}$$

be an $n \times n$ matrix where I is the $(n - \alpha) \times (n - \alpha)$ unit matrix. It is clear that the β_k-th row of $\tilde{G}D_0\tilde{G}^*$ is zero, as well as the β_k-th row of $\tilde{G}(Z - \lambda_0 I)\tilde{G}^{-1}$. However, this contradicts (7.3.3).

So the matrix F is invertible, and since $F > 0$, F can be represented as a product of upper and lower triangular matrices. (The Cholesky factorization; see Chapter 4 of Golub and Van Loan (1989), for example.) Thus, Lemma 7.3.2 applies and the proof is complete. □

Proof of Theorem 7.3.1. Observe that (by Lemma 4.4.1) the right-hand side of (7.3.1) can be written in the form

$$\mathcal{C}_{A,D} = \text{Im } [D, (A + DX)D, \ldots, (A + DX)^{n-1}D]$$

for any $n \times n$ matrix X. Let $X = X^*$ be a solution of the CARE. Then the similarity

$$\begin{bmatrix} I & 0 \\ -X & I \end{bmatrix} \begin{bmatrix} A & D \\ C & -A^* \end{bmatrix} \begin{bmatrix} I & 0 \\ X & I \end{bmatrix} = \begin{bmatrix} A + DX & D \\ 0 & -(A^* + XD) \end{bmatrix}$$

is easily verified. So we can consider the matrix

$$i \begin{bmatrix} A + DX & D \\ 0 & -(A^* + XD) \end{bmatrix}$$

in place of M. It remains to apply Lemma 7.3.3. □

EXISTENCE OF HERMITIAN SOLUTIONS

Of particular interest in Theorem 7.3.1 is the case when the pair (A, D) is controllable. In this case condition (7.3.1) is satisfied always, and we obtain the following corollary.

Corollary 7.3.4. *Assume that $D \geq 0$ and the pair (A, D) is controllable. If the CARE admits a hermitian solution, then all the partial multiplicities of M corresponding to real eigenvalues (if any) are even.*

We now return to Theorem 7.3.1. As we have observed, the matrix M is $\hat{H}$-self-adjoint, where $\hat{H} = i \begin{bmatrix} 0 & I \\ -I & 0 \end{bmatrix}$. It turns out that in this context the sign characteristic of the pair $(M, \hat{H})$ is completely determined.

Theorem 7.3.5. *Under the hypotheses of Theorem 7.3.1, and assuming that the CARE has a hermitian solution X satisfying (7.3.1), the sign characteristic of $(M, \hat{H})$ consists of $+1$'s only.*

We need some preparation for the proof of Theorem 7.3.5, and we start with a general result concerning the sign characteristic of pairs of matrices depending continuously on a parameter.

Lemma 7.3.6. *Let $H = H^*$ be an $n \times n$ invertible matrix, and let $A(t)$ be an $n \times n$ matrix depending continuously on a parameter $t \in [0, 1]$, and such that*

(i) $A(t)$ is H-self-adjoint for all $t \in [0, 1]$,

(ii) $A(t)$ is similar to $A(0)$ for all $t \in [0, 1]$.

Then the sign characteristic of the pair $(A(t), H)$ is independent of t.

More exactly, it means the following: Let λ_0 be a real eigenvalue of $A(t)$ having partial multiplicities $m_1, \ldots, m_r$ (because of condition (ii) λ_0 and $m_1, \ldots, m_r$ are independent of t in $[0, 1]$). Then the signs $\epsilon_1, \ldots, \epsilon_r$ in the sign characteristic of $(A(t), H)$ do not depend on $t \in [0, 1]$ (up to permutation of signs corresponding to equal partial multiplicities).

The full proof of Lemma 7.3.6 is beyond the scope of this book. We only indicate that it can be easily deduced from the following perturbation theorem, which is a particular case of Theorem II.1.1 in Gohberg et al (1982b). Let A be an H-self-adjoint matrix. Then there exists an $\varepsilon > 0$ such that every H-self-adjoint matrix B which is similar to A and satisfies $\|B - A\| < \varepsilon$ has the same sign characteristic at every real $\lambda_0 \in \sigma(A) = \sigma(B)$ as A does.

Proof of Theorem 7.3.5. Let X be a hermitian solution of the CARE (7.2.1) which also satisfies (7.3.1). As in the proof of Theorem 7.3.1, we apply the transformation

$$T^{-1} i \begin{bmatrix} A & D \\ C & -A^* \end{bmatrix} T = i \begin{bmatrix} Z & D_0 \\ 0 & -Z^* \end{bmatrix},$$

where $T = \begin{bmatrix} I & 0 \\ X & I \end{bmatrix} \begin{bmatrix} S & 0 \\ 0 & S^{-1*} \end{bmatrix}$. Here Z is a matrix in Jordan form, and D_0 is positive semidefinite. A straightforward computation shows that

$$T^* \hat{H} T = \hat{H}.$$

Thus, we can consider the sign characteristic of $\left(i \begin{bmatrix} Z & D_0 \\ 0 & -Z^* \end{bmatrix}, \hat{H} \right)$ instead of the original pair (M, H).

Let λ_0 be a real eigenvalue of $i \begin{bmatrix} Z & D_0 \\ 0 & -Z^* \end{bmatrix}$. As in the proof of Lemma 7.3.3, we can assume that

$$iZ = \begin{bmatrix} iZ' & 0 \\ 0 & iZ'' \end{bmatrix},$$

where $Z' = Z_1 \oplus \cdots \oplus Z_k$ is the Jordan matrix having the eigenvalue $-i\lambda_0$ and partitioned into Jordan blocks $Z_1, \ldots, Z_k$, and where Z'' is the Jordan matrix having eigenvalues different from $-i\lambda_0$. Partition D_0 accordingly

$$D_0 = \begin{bmatrix} D_1 & D_2 \\ D_2^* & D_3 \end{bmatrix},$$

and partition D_1 further conformally with the partition of Z' into Jordan blocks:

$$D_1 = \begin{bmatrix} D_{111} & D_{112} & \cdots & D_{11k} \\ D_{121} & D_{122} & \cdots & D_{12k} \\ \vdots & \vdots & & \vdots \\ D_{1k1} & D_{1k2} & \cdots & D_{1kk} \end{bmatrix}. \quad (7.3.4)$$

Let f_{ij} ($i, j = 1, \ldots, k$) be the entry in the southeast corner of D_{1ij}. As shown in the proof of Lemma 7.3.3 the matrix $[f_{ij}]_{i,j=1}^k$ is positive definite. In particular, $f_{ii} > 0$ for $i = 1, \ldots, k$.

Let $E_1 = [E_{ij}]_{i,j=1}^k$ be the matrix of the same size as D_1 and partitioned conformally with the partition (7.3.4) (so E_{ij} has same size of D_{1ij}), where $E_{ij} = 0$ for $i \neq j$, and E_{ii} has all entries zero except for the southeast corner whose value is 1. Clearly, E_1 is positive semidefinite. Let

$$M(t) = i \begin{bmatrix} Z' & 0 & tD_1 + (1-t)E_1 & tD_2 \\ 0 & Z'' & tD_2^* & D_3 \\ 0 & 0 & -Z'^* & 0 \\ 0 & 0 & 0 & -Z''^* \end{bmatrix}, \quad 0 \leq t \leq 1.$$

We have

$$M(1) = i \begin{bmatrix} Z & D_0 \\ 0 & -Z^* \end{bmatrix}, \quad M(0) = i \begin{bmatrix} Z'' & 0 & E_1 & 0 \\ 0 & Z''' & 0 & D_3 \\ 0 & 0 & -Z'^* & 0 \\ 0 & 0 & 0 & -Z'''^* \end{bmatrix},$$

and since the matrix

$$\begin{bmatrix} tD_1 + (1-t)E_1 & tD_2 \\ tD_2^* & D_3 \end{bmatrix}$$

is hermitian for $t \in [0,1]$, the matrix $M(t)$ is $\hat{H}$-self-adjoint for all such t. Since $\sigma(Z') \cap \sigma(-Z'''^*) = \emptyset$, we can apply a similarity transformation to $M(t)$ with a similarity matrix of the type

$$\begin{bmatrix} I & 0 & 0 & T \\ 0 & I & 0 & 0 \\ 0 & 0 & I & 0 \\ 0 & 0 & 0 & I \end{bmatrix}$$

for a suitable T, in order to transform $M(t)$ to the block diagonal form

$$i \begin{bmatrix} Z' & tD_1 + (1-t)E_1 \\ 0 & -Z'^* \end{bmatrix} \oplus i \begin{bmatrix} Z'' & D_3 \\ 0 & -Z'''^* \end{bmatrix}.$$

This form, together with Lemma 7.3.2, show that the Jordan form of $M(t)$ is independent of t on the interval $[0,1]$.

We now appeal to Lemma 7.3.6. According to this lemma, the sign characteristic of $(M(t), \hat{H})$ is independent of $t \in [0,1]$. Therefore, we can compute the sign characterisitc of $(M(0), \hat{H})$ at its real eigenvalue λ_0, which is the same as the sign characteristic of $(M(1), \hat{H})$ at λ_0 which we were originally interested in. But $M(0)$ is a much simpler matrix, and therefore the sign characteristic of $(M(0), \hat{H})$ can be computed in a straightforward way, as follows.

Since

$$Z' = Z_1 \oplus \cdots \oplus Z_k, \quad E_1 = E_{11} \oplus \cdots \oplus E_{kk},$$

it is easily seen that it is sufficient to consider the situation where Z' has just one Jordan block. In other words, the problem of computing the sign characteristic of $(M(0), \hat{H})$ is reduced to computing the sign characteristic of the pair (N_0, H_0), where

$$N_0 = i[J_p(-i\lambda_0) \oplus -J_p(i\lambda_0)^T]$$

and $H_0 = i \begin{bmatrix} 0 & I \\ -I & 0 \end{bmatrix}$ are $2p \times 2p$ matrices. Without loss of generality we can assume $\lambda_0 = 0$ (this amounts to subtracting λ_0 from the main diagonal in N_0 which does not affect the sign characteristic). A Jordan chain of N_0 of length $2p$ is easily found; indeed, letting $e_j \in \mathbb{C}^{2p}$ be the j-th unit coordinate vector, it turns out that with y_{2k} defined by $y_{2k} = -ie_{p+1}$ we have $Ny_q = y_{q-1}$ ($q = 2k, 2k-1, \ldots, 2$) and $y_1 = e_1$, $Ny_1 = 0$. Now the canonical form of the pair (N_0, H_0) shows that the sign in the sign characteristic of (N_0, H_0), corresponding to the sole partial multiplicity $2p$, coincides with the sign of the nonzero real number

$$\langle H_0 y_1, y_{2k} \rangle = \langle -ie_{p+1}, -ie_{p+1} \rangle = 1.$$

Thus, the sign characteristic does indeed consist of $+1$'s. □

Combining Theorems 7.2.5, 7.3.1 and 7.3.5 we obtain the following general characterizations of solvability of the CARE under the sign controllability condition.

Theorem 7.3.7. *Assume that $D \geq 0$, $C^* = C$, and that the pair (A, D) is sign controllable. Then the following statements are equivalent:*

(i) *The CARE (7.2.1) has hermitian solutions;*

(ii) *There exists an n-dimensional M-invariant $\hat{H}$-neutral subspace.*

(iii) *M has only even multiplicities corresponding to the real eigenvalues (if any) and the sign characteristic of $(M, \hat{H})$ consists of $+1$'s only.*

Proof. The equivalence of (i) and (ii) is established in Theorem 7.2.5, while (iii) $\Rightarrow$ (ii) follows from Theorem 2.5.2. To show that (i) $\Rightarrow$ (iii), Theorems 7.3.1 and 7.3.5 are used. Indeed, they immediately yield the required result if hypothesis (7.3.2) can be established, i.e. that for every $n \times n$ matrix X we have

$$\mathcal{R}_{\lambda_0}(A + DX) \subseteq \mathcal{C}_{A,D} \tag{7.3.5}$$

for every pure imaginary or zero eigenvalue λ_0 of $A + DX$. Write $\mathcal{C} = \mathcal{C}_{A,D}$ for the controllable subspace of (A, D). With respect to the orthogonal decomposition $\mathbb{C}^n = \mathcal{C} \oplus \mathcal{C}^\perp$ we have

$$A = \begin{bmatrix} A_1 & A_{12} \\ 0 & A_2 \end{bmatrix}, \quad D = \begin{bmatrix} D_1 \\ 0 \end{bmatrix}.$$

The sign controllability condition implies that A_2 has no pure imaginary or zero eigenvalues. Indeed, for every pure imaginary or zero eigenvalue λ_0 of A, it follows from the defining property (7.2.12) that $\mathcal{R}_{\lambda_0}(A) \subseteq \mathcal{C}$ and, therefore, $\dim \mathcal{R}_{\lambda_0}(A) = \dim(\mathcal{R}_{\lambda_0}(A) \cap \mathcal{C}) = \dim \mathcal{R}_{\lambda_0}(A|\mathcal{C}) = \dim \mathcal{R}_{\lambda_0}(A_1)$. On the other hand, $\dim \mathcal{R}_{\lambda_0}(A)$ (resp. $\dim \mathcal{R}_{\lambda_0}(A_1)$) is just the multiplicity of λ_0 as a zero of $\det(\lambda I - A)$ (resp. of $\det(\lambda I - A_1)$). Since $\det(\lambda I - A) = \det(\lambda I - A_1) \det(\lambda I - A_2)$, it follows that λ_0 is not a zero of $\det(\lambda I - A_2)$.

Since

$$A + DX = \begin{bmatrix} A_1 + D_1 X_1 & A_{12} + D_1 X_2 \\ 0 & A_2 \end{bmatrix},$$

where $X = [X_1 X_2]$, (7.3.5) follows immediately. □

7.4 Special and hermitian solutions

In this section we consider the CARE of equation (7.2.1);

$$XDX + XA + A^*X - C = 0 \qquad (7.4.1)$$

under the assumptions that C is hermitian, D positive semidefinite, and the pair (A, D) is controllable. A solution X is said to be *special* if

$$(X^* - X)(A + DX) \leq 0.$$

By Theorem 7.2.3, the special solutions X are in one-to-one correspondence with the M-invariant maximal H-nonpositive subspaces $\mathcal{M}$ via the formula

$$\mathcal{M} = \operatorname{Im}\begin{bmatrix} I \\ X \end{bmatrix}.$$

Therefore, many results concerning M-invariant maximal H-nonpositive subspaces can be interpreted as results concerning special solutions of (7.4.1). The next results are of this kind. First recall the notion of the *sign condition* given in Section 2.4.

Theorem 7.4.1. *Let S be a set of eigenvalues of $\begin{bmatrix} A & D \\ C & -A^* \end{bmatrix}$ with nonzero real parts, such that $\lambda \in S$ implies $-\bar{\lambda} \notin S$ and which is maximal (subject to these properties). Then there exists a special solution X of (7.4.1) such that S is exactly the set of eigenvalues of $A + DX$ having nonzero real parts.*

If, in addition, the pair (M, H) satisfies the sign condition, then this solution X is unique (for a given S).

In the terminology of Theorem 2.4.1, iS is a c-set for M and $k = n$, and so the first part of Theorem 7.4.1 follows from Theorem 2.4.1. Then the second part follows from Theorem 2.4.2.

We now move on to the hermitian solutions. By Theorem 7.2.4, hermitian solutions X are in one-to-one correspondence with n-dimensional M-invariant H-neutral subspaces. Furthermore, existence of a hermitian solution implies that the partial multiplicities of M corresponding to real eigenvalues are all even (Theorem 7.3.1). Appealing to Theorem 2.5.1, we obtain the existence part of the following result.

Theorem 7.4.2. *Let S be as in Theorem 7.4.1, and assume that the CARE of equation (7.4.1) has hermitian solutions. Then there exists a unique hermitian solution X of (7.4.1) such that S is exactly the set of eigenvalues of $A + DX$ having nonzero real parts.*

To prove the uniqueness part, appeal to Theorems 2.4.2 and 7.3.5 (the latter implies that under the hypotheses of this section the pair (M, H) satisfies the sign condition). Note in particular, that Theorem 7.4.2 implies the existence of a unique hermitian solution X_0 such that $\sigma(A + DX_0)$ is in the closed left half-plane. Similarly, there is a unique hermitian solution X_1 such that $\sigma(A + DX_1)$ is in the closed right half-plane.

7.5 Extremal solutions

In this and the next section we continue our study of the CARE (7.4.1) under the assumptions that $C = C^*$, $D \geq 0$ and (A, D) is controllable. It will also be assumed throughout this section that the CARE has at least one hermitian solution. We shall prove the existence of maximal and minimal solutions under these hypotheses.

There is a natural order relation in the set of all hermitian matrices. Namely, $X_1 \leq X_2$ for hermitian matrices X_1 and X_2 means that $X_2 - X_1$ is nonnegative definite. A hermitian solution X_+ (resp. X_-) of the CARE is called *maximal* (resp. *minimal*) if $X_- \leq X \leq X_+$ for every hermitian solution X. Obviously, if a maximal (resp. minimal) solution exists, it is unique. The following theorem establishes the existence of extremal hermitian solutions and characterizes them in spectral terms.

Theorem 7.5.1. *If $C^* = C$, $D \geq 0$, (A, D) is controllable, and the CARE has a hermitian solution then there exist a maximal hermitian solution X_+, and a minimal hermitian solution, X_-. The solution X_+ is the unique hermitian solution for which $\sigma(A+DX_+)$ lies in the closed right half-plane, and is obtained from Theorem 7.4.2 by taking S to be the set of eigenvalues of $\begin{bmatrix} A & D \\ C & -A^* \end{bmatrix}$ having positive real parts. The solution X_- is the unique hermitian solution with $\sigma(A + DX_-)$ in the closed left half-plane, and is obtained from Theorem 7.4.2 by taking S to be the set of eigenvalues of $\begin{bmatrix} A & D \\ C & -A^* \end{bmatrix}$ having negative real parts.*

For convenience, we state and prove a lemma which will be used in the proof of Theorem 7.5.1.

Lemma 7.5.2. *Let Q, R be $n \times n$ complex matrices such that R is positive semidefinite and (Q, R) is a controllable pair. Then the matrix*

$$\Omega(t) = -\int_0^t e^{-\tau Q} R e^{-\tau Q^*} \, d\tau \tag{7.5.1}$$

is negative (resp. positive) definite for all $t > 0$ (resp. $t < 0$) and the matrix $\hat{Q} = Q + R\Omega(t)^{-1}$, $t > 0$ is stable (i.e. $\sigma(\hat{Q})$ lies in the open left half-plane).

EXTREMAL SOLUTIONS

Proof. Let $R^{1/2}$ be the unique positive semidefinite square root of R and observe that Im $R^{1/2}$ = Im R. Then, from Proposition 4.1.2, (Q, R) is controllable if and only if $(Q, R^{1/2})$ is controllable. Now the lemma follows from Theorem 4.4.2(b). □

Proof of Theorem 7.5.1. Let X be a hermitian solution of the CARE. Let $\tilde{A} = A + DX$ and

$$U(t) = -\int_0^t e^{-\tau \tilde{A}} D e^{-\tau \tilde{A}^*} \, d\tau, \quad t \in \mathbb{R}.$$

Since (A, D) is controllable, so is $(\tilde{A}, D)$ (by Lemma 4.4.1). So Lemma 7.5.2 is applicable, and $U(t)$ is positive (resp. negative) definite for $t < 0$ (resp. $t > 0$). Also,

$$U(t_1) \leq U(t_2) \quad \text{for} \quad t_1 \geq t_2,$$

and therefore (using the easily verified property that $X \geq Y > 0$ implies $0 < X^{-1} \leq Y^{-1}$)

$$U(t_1)^{-1} \geq U(t_2)^{-1} > 0 \quad \text{for} \quad 0 > t_1 \geq t_2. \tag{7.5.2}$$

It follows that the limit $\lim_{t \to -\infty} U(t)^{-1}$ exists (indeed, this limit can be uniquely determined by

$$\langle \lim_{t \to -\infty} U(t)^{-1} x, x \rangle = \lim_{t \to -\infty} \langle U(t)^{-1} x, x \rangle$$

for any $x \in \mathbb{C}^n$, where the limit on the right-hand side exists by (7.5.2)). Moreover, $\lim_{t \to -\infty} U(t)^{-1}$ is positive semidefinite.

For any positive real T and $t < T$ define

$$X_T(t) = X + (U(t - T))^{-1}. \tag{7.5.3}$$

Since $X_T(t) = [XU(t - T) + I]U(t - T)^{-1}$, it follows that $X_T(t)$ is invertible for $t \in [T - \delta, T)$, where $\delta > 0$ is small enough, and (because $U(0) = 0$)

$$\lim_{t \to T} X_T(t)^{-1} = 0.$$

By a direct computation, using the facts that $X = X^*$ is a solution of the CARE (7.4.1) and that $U(t)$ satisfies the differential equation

$$U(t)' = -\tilde{A}U - U\tilde{A}^* - D,$$

one checks easily that $X_T(t)$ satisfies the differential Riccati equation

$$X_T(t)' + C - X_T(t)A - A^* X_T(t) - X_T(t) D X_T(t) = 0, \quad t < T. \tag{7.5.4}$$

Equation (7.5.4), together with the boundary condition $\lim_{t \to T} X_T(t)^{-1} = 0$, determine $X_T(t)$ uniquely as a quantity independent of the choice of the hermitian solution X of (7.4.1). Indeed, let $V_T(t) = X_T(t)^{-1}$. Then, by differentiating the equation $X_T(t) \cdot X_T(t)^{-1} = I$, it is found that

$$V_T(t)' = -V_T(t)X_T(t)'V_T(t),$$

and, using (7.5.4), $V_T(t)$ satisfies the differential equation

$$-V_T(t)' + V_T(t)CV_T(t) - AV_T(t) - V_T(t)A^* - D = 0$$

for $T - \delta < t < T$ (this is the interval where the invertibility of $X_T(t)$ is guaranteed). Also, $V_T(t)$ satisfies the initial condition

$$\lim_{t \to T} V_T(t) = 0.$$

Hence, by the uniqueness theorem for solutions of the initial value problem we find that $V_T(t)$ and, consequently, also $X_T(t)$ are uniquely determined (i.e. independent of the choice of X) for $T - \delta_0 < t < T$. Here $\delta_0 \leq \delta$ is a suitable positive number. But then $X_T(t)$ is uniquely determined for all $t < T$, as is easily seen from the formula (7.5.3).

As $T \to \infty$, the matrix function $X_T(t)$ converges uniformly on compact intervals to the constant matrix

$$X_+ = X + \lim_{t \to -\infty} U(t)^{-1}. \tag{7.5.5}$$

Of course, X_+ is independent of X. Now (7.5.5) implies that $X_+ = X_+^*$ and $X \leq X_+$. Further,

$$\begin{aligned}
X_T'(t) &= \frac{dU(t-T)^{-1}}{dt} = -U(t-T)^{-1} \cdot \frac{dU(t-T)}{dt} U(t-T)^{-1} \\
&= U(t-T)^{-1}[\tilde{A}U(t-T) + U(t-T)\tilde{A}^* + D]U(t-T)^{-1} \\
&= U(t-T)^{-1}\tilde{A} + \tilde{A}^*U(t-T)^{-1} + U(t-T)^{-1}DU(t-T)^{-1}
\end{aligned}$$

has a limit when $T \to \infty$ (keeping t fixed). As the function $U(t-T)^{-1}$ itself has a limit when $T \to \infty$ (and t is fixed), it follows that

$$\lim_{T \to \infty} X_T'(t) = 0.$$

For a fixed t, passing to the limit when $T \to \infty$ in (7.5.4), we find that X_+ is a solution of the CARE. Since $X \leq X_+$ for any hermitian solution X, X_+ is maximal.

Furthermore,
$$A + DX_+ = \lim_{T \to \infty} (A + DX_T(0)), \tag{7.5.6}$$

and

$$A + DX_T(0) = \tilde{A} + D(U(-T))^{-1} = \tilde{A} - D\left[\int_0^{-T} e^{-\tau \tilde{A}} D e^{-\tau \tilde{A}^*} d\tau\right]^{-1}$$

$$= -\left[(-\tilde{A}) - D\left(\int_0^T e^{-\tau(-\tilde{A})} D e^{-\tau(-\tilde{A}^*)} d\tau\right)^{-1}\right].$$

Since rank $[D, -\tilde{A}D, \ldots, (-1)^{n-1}\tilde{A}^{n-1}D] = n$, the second part of Lemma 7.5.2 ensures that $\sigma(A + DX_T(0))$ lies in the open right half-plane. Using (7.5.6) and the continuity of eigenvalues of $A + DX_T(0)$ as functions of T, we find that $\operatorname{Re} \lambda_0 \geq 0$ for every $\lambda_0 \in \sigma(A + DX_+)$. Uniqueness of a hermitian solution X of (7.4.1) with the additional property that $\operatorname{Re} \sigma(A + DX) \geq 0$ follows from Theorem 7.4.2.

Applying the results we have obtained concerning maximal solutions of the CARE to the equation

$$XDX - XA - A^*X - C = 0 \tag{7.5.7}$$

and noting that X is a solution of (7.5.7) if and only if $-X$ is a solution of (7.4.1), we obtain the corresponding results for minimal solutions. □

Although Theorem 7.5.1 refers only to extremal solutions, the idea emerging here is that the choice of S in the spectrum of $\begin{bmatrix} A & D \\ C & -A^* \end{bmatrix}$ determines a hermitian solution X (by Theorem 7.4.2) and the set S reappears as that part of the spectrum of the resulting modified state matrix $A + DX$ which is not on the imaginary axis.

Another interesting and closely related idea is that, in a certain sense, every hermitian solution of (7.4.1) can be obtained as a combination of the maximal and the minimal solutions. We will make this statement precise in Theorem 7.5.4 below. First define the hermitian matrix $X_+ - X_-$ to be the "gap" between the extremal solutions, and let $\mathcal{N} = \operatorname{Ker}(X_+ - X_-)$.

Theorem 7.5.3. *Suppose $D \geq 0$, $C^* = C$, (A, D) is controllable, and there is a hermitian solution of the CARE (7.4.1). Then for any $x \in \mathcal{N}$ and any hermitian solution X,*
$$(A + DX)x = (A + DX_+)x = (A + DX_-)x. \tag{7.5.8}$$

Moreover, for each such X, $\mathcal{N}$ is the spectral subspace of $A + DX$ corresponding to all of its pure-imaginary or zero eigenvalues.

Proof. Let $X = X^*$ be a solution of (7.4.1) and write $\mathcal{N}_1 = \mathrm{Ker}\,(X - X_-)$, $\mathcal{N}_2 = \mathrm{Ker}\,(X_+ - X)$. Observe first that $\mathcal{N} = \mathcal{N}_1 \cap \mathcal{N}_2$. The inclusion $\mathcal{N}_1 \cap \mathcal{N}_2 \subseteq \mathcal{N}$ is clear. Also, if $x \in \mathcal{N}$, then $(X - X_+)x = (X - X_-)x$ and we have

$$\langle (X - X_+)x, x \rangle = \langle (X - X_-)x, x \rangle.$$

But $(X - X_+) \leq 0$ and $X - X_- \geq 0$, so $(X - X_+)x = (X - X_-)x = 0$, and so $\mathcal{N} \subseteq \mathcal{N}_1 \cap \mathcal{N}_2$.

We obviously have

$$(A + DX_-)x = (A + DX)x, \quad x \in \mathcal{N}_1;$$

$$(A + DX_+)x = (A + DX)x, \quad x \in \mathcal{N}_2,$$

and so (7.5.8) follows from $\mathcal{N} = \mathcal{N}_1 \cap \mathcal{N}_2$.

A direct computation using the fact that both X_+ and X_- satisfy (7.4.1) shows that

$$\begin{aligned}-(X_+ - X_-)D(X_+ - X_-)+ \\ (X_+ - X_-)(A + DX_+) + (A^* + X_+ D)(X_+ - X_-) = 0 \end{aligned} \quad (7.5.9)$$

Using this, it is easily verified that $\mathcal{N}$ is $(A + DX_+)$-invariant and hence, using (7.5.8), that $\mathcal{N}$ is $(A + DX)$-invariant for any hermitian solution X of (7.4.1). Furthermore, $\sigma((A + DX)|\mathcal{N})$ does not depend on X. As the nonpure imaginary parts of $\sigma(A + DX_-)$ and $\sigma(A + DX_+)$ are disjoint, it follows that all the eigenvalues of the restriction $(A + DX_+)|\mathcal{N}$ are pure imaginary, and similarly for $\sigma(A + DX)|\mathcal{N}$.

Next, it is shown that the subspace $\mathcal{N}$ contains the spectral subspace of $A + DX_+$ corresponding to *all* of its pure imaginary or zero eigenvalues. Suppose not. Then there exists a pure imaginary eigenvalue $-i\omega$ of $A + DX_+$ and a vector $z \notin \mathcal{N}$ such that

$$[(A + DX_+) + i\omega]z \in \mathcal{N}.$$

Let $z_0 = [(A + DX_+) + i\omega]z$ and $x = (X_+ - X_-)z$; then $(X_+ - X_-)z_0 = 0$ and $x \neq 0$. Now (7.5.9) implies

$$\begin{aligned} x^* Dx &= x^*(A + DX_+)z + z^*(A^* + X_+ D)x \\ &= x^*(z_0 - i\omega z) + (z_0^* + i\omega z^*)x = x^* z_0 + z_0^* x, \end{aligned} \quad (7.5.10)$$

since $x^* z = z^*(X_+ - X_-)z$ is a real number. On the other hand, again by (7.5.9),

$$
\begin{aligned}
(X_+ - X_-)Dx &= (X_+ - X_-)D(X_+ - X_-)z \\
&= (X_+ - X_-)(z_0 - i\omega z) + (A^* + X_+ D)(X_+ - X_-)z \\
&= [(A^* + X_+ D) - i\omega](X_+ - X_-)z = (A^* + X_+ D - i\omega)x.
\end{aligned}
$$
(7.5.11)

Premultiply by z^* to get $x^* Dx = z_0^* x$. Comparing with (7.5.10) we easily deduce that $x^* Dx = 0$ and (because $D \geq 0$) $Dx = 0$. Using (7.5.11) we now obtain

$$(A + DX_+)^* x = i\omega x.$$

Consequently, for $m = 1, 2, \ldots$,

$$D(A + DX_+)^{*m} x = (i\omega)^m Dx = 0, \quad x \neq 0.$$

This contradicts the controllability of $(A + DX_+, D)$ (which, by Lemma 4.4.1, follows from the controllability of (A, D)).

It has been proved that $\mathcal{N}$ is the spectral subspace of $A + DX_+$ corresponding to all of its pure imaginary eigenvalues.

The similarity

$$T^{-1} M_0 T = \begin{bmatrix} A + DX & D \\ 0 & -(A + DX)^* \end{bmatrix},$$

where $T = \begin{bmatrix} I & 0 \\ X & I \end{bmatrix}$ and where X is a hermitian solution of the CARE implies that, for any such X, the eigenvalues of $A + DX$ on the imaginary axis are just the eigenvalues of M_0 on the imaginary axis. Moreover, the dimension of the corresponding spectral subspace of $A + DX$ does not depend on X. Since (7.5.8) implies that $\mathcal{N}$ is also $(A + DX)$-invariant for any hermitian solution X, it follows that $\mathcal{N}$ is, indeed, the spectral subspace of $A + DX$ corresponding to all of its pure imaginary or zero eigenvalues. □

We can now set up the description of all hermitian solutions of the CARE as combinations of the maximal and minimal solutions. Let $\mathcal{N} = \text{Ker}\,(X_+ - X_-)$, as before, and let $\mathcal{N}_+$ be the spectral subspace of $A + DX_+$ that is complementary to $\mathcal{N}$. Since $\sigma(A + DX_+)$ is in the closed right half-plane, $\mathcal{N}_+$ corresponds to all the eigenvalues of $A + DX_+$ with positive real parts. Then we have:

Theorem 7.5.4. *Suppose $D \geq 0$, $C^* = C$, (A, D) is controllable, and that (7.4.1) has at least one hermitian solution. Let S be an $(A + DX_+)$-invariant subspace with $S \subseteq \mathcal{N}_+$ and define*

$$T = ((X_+ - X_-)S)^\perp. \tag{7.5.12}$$

Then T is $(A + DX_-)$-invariant, $\mathbb{C}^n = S \dotplus T$, and if P is the projection onto S along T, then

$$X = X_+ P + X_-(I - P) \tag{7.5.13}$$

is a hermitian solution of the CARE (7.4.1). Conversely, if X is a hermitian solution of (7.4.1), then there is an $(A + DX_+)$-invariant subspace $S \subseteq \mathcal{N}_+$ for which X is given by (7.5.13) and this correspondence is one-to-one.

Proof. Let $x \in S \cap T$. Then $x^*(X_+ - X_-)x = 0$. Since $X_+ \geq X_-$ it follows that $(X_+ - X_-)x = 0$. Thus, $x \in \mathcal{N}$. But $\mathcal{N} \cap \mathcal{N}_+ = \{0\}$ since $\mathcal{N}$ and $\mathcal{N}_+$ are spectral $(A + DX_+)$-invariant subspaces corresponding to disjoint sets of eigenvalues. So $x = 0$.

The above argument also shows that $(X_+ - X_-)$ is one-to-one on S, and therefore

$$\dim T = n - \dim((X_+ - X_-)S) = n - \dim S,$$

and so $\mathbb{C}^n = S \dotplus T$ is indeed a direct sum.

Since both X_+ and X_- are hermitian solutions of (7.4.1) a straightforward computation shows that

$$(A + DX_+)^*(X_+ - X_-) = -(X_+ - X_-)(A + DX_-). \tag{7.5.14}$$

So if $x \in S$, $y \in T$, then

$$x^*(X_+ - X_-)(A + DX_-)y = -x^*(A + DX_+)^*(X_+ - X_-)y = 0$$

because S is $(A + DX_+)$-invariant. This shows that T is $(A + DX_-)$-invariant. Hence the projection P satisfies

$$P(A + DX_+)P = (A + DX_+)P \tag{7.5.15}$$

and

$$(I - P)(A + DX_-)(I - P) = (A + DX_-)(I - P);$$

that is,

$$P(A + DX_-)P = P(A + DX_-). \tag{7.5.16}$$

Since $(X_+ - X_-)T$ is orthogonal to S we have also $P^*(X_+ - X_-)(I - P) = 0$. That is,

$$P^*(X_+ - X_-) = P^*(X_+ - X_-)P.$$

Since the right side is hermitian this gives

$$P^*(X_+ - X_-) = (X_+ - X_-)P. \tag{7.5.17}$$

By (7.5.15) and (7.5.16) we have

$$(I - P)(A + DX_+) = (I - P)(A + DX_+)(I - P)$$
$$= (I - P)(A + DX_- + D(X_+ - X_-))(I - P)$$
$$= (A + DX_-)(I - P) + (I - P)D(X_+ - X_-)(I - P).$$

Therefore, by (7.5.14)

$$(X_+ - X_-)(I - P)D(X_+ - X_-)(I - P)$$
$$-(A + DX_+)^*(X_+ - X_-)(I - P) - (X_+ - X_-)(I - P)(A + DX_+) = 0.$$
(7.5.18)

Define X by (7.5.13), and let

$$F = X_+ - X = (X_+ - X_-)(I - P).$$

By (7.5.17) $(X_+ - X_-)(I - P)$ is hermitian, and therefore so is X. Equality (7.5.18) can be rewritten in the form

$$FDF - (A + DX_+)^*F - F(A + DX_+) = 0.$$

Now

$$XDX + XA + A^*X - C = (X_+ - F)D(X_+ - F) + (X_+ - F)A$$
$$+ A^*(X_+ - F) - C$$
$$= (X_+DX_+ + X_+A + A^*X_+ - C) - FDX_+ -$$
$$X_+DF + FDF - FA - A^*F$$
$$= FDF - F(A + DX_+) - (A^* + X_+D)F = 0,$$

so X is indeed a solution of (7.4.1). Furthermore, (by (7.5.13))

$$Xx = X_+x \quad \text{for} \quad x \in \mathcal{S}$$

$$Xy = X_-y \quad \text{for} \quad y \in \mathcal{T}.$$

Therefore, $\mathcal{S}$ and $\mathcal{T}$ are $(A+DX)$-invariant and all the eigenvalues of $(A+DX)|\mathcal{S}$ have positive real parts, while all the eigenvalues of $(A+DX)|\mathcal{T}$ have nonpositive real parts (indeed, $(A+DX)|\mathcal{T} = (A+DX_-)|\mathcal{T}$, and all eigenvalues of $A+DX_-$ have nonpositive real parts by Theorem 7.5.1). This shows that $\mathcal{S}$ is uniquely determined by X.

Conversely, let X be a hermitian solution of the CARE. Write $\mathbb{C}^n = \mathcal{S} \dotplus \mathcal{T}$, where $\mathcal{S}$ (resp. $\mathcal{T}$) is the spectral $(A + DX)$-invariant subspace corresponding to the eigenvalues of $A + DX$ in the open right (resp. closed left) half-plane.

We next prove that

$$Xx = X_+ x \quad \text{for} \quad x \in \mathcal{S}; \quad Xy = X_- y \quad \text{for} \quad y \in \mathcal{T}. \tag{7.5.19}$$

In fact, only the second equality in (7.5.19) will be proved; the proof of the first one is similar. In view of Theorem 7.5.3 we only have to verify that

$$Xy = X_- y \quad \text{for} \quad y \in \mathcal{T}_0, \tag{7.5.20}$$

where $\mathcal{T}_0$ is the spectral $(A+DX)$-invariant subspace corresponding to the eigenvalues of $A + DX$ in the open left half-plane. As in (7.5.9) we have

$$-(X-X_-)D(X-X_-)+(X-X_-)(A+DX)+(A^*+XD)(X-X_-) = 0. \tag{7.5.21}$$

Let x_0 be an eigenvector of $A + DX$ corresponding to an eigenvalue λ_0 with $\operatorname{Re}\lambda_0 < 0$. Then (7.5.21) leads to the equality

$$-x_0^*(X - X_-)D(X - X_-)x_0 + 2\operatorname{Re}\lambda_0 x_0^*(X - X_-)x_0 = 0.$$

Since both matrices $(X - X_-)D(X - X_-)$ and $X - X_-$ are positive semidefinite, we must have $(X - X_-)x_0 = 0$. Now we apply an inductive argument, and suppose that for all x_0 such that $(A+DX-\lambda_0 I)^k x_0 = 0$, $\operatorname{Re}\lambda_0 < 0$ it is already proved that $(X - X_-)x_0 = 0$. Let y be such that $(A + DX - \lambda_0 I)^{k+1} y = 0$, $\operatorname{Re}\lambda_0 < 0$. Letting $x_0 = (A + DX - \lambda_0 I)y$, we have (using (7.5.21))

$$-y^*(X - X_-)D(X - X_-) + y^*(X - X_-)(A + DX - \lambda_0 I)y$$
$$+ y^*(A^* + XD - \bar{\lambda}_0 I)(X - X_-) + 2\operatorname{Re}\lambda_0 y^*(X - X_-)y = 0.$$

By the induction hypothesis, $(X - X_-)(A + DX - \lambda_0 I)y = 0$, and we conclude that $(X - X_-)y = 0$. Thus, $(X - X_-)y = 0$ for every $y \in \operatorname{Ker}(A+DX-\lambda_0 I)^k$; $k = 1, 2, \ldots$, $\operatorname{Re}\lambda_0 < 0$. Since $\mathcal{T}_0$ is the sum of all such subspaces $\operatorname{Ker}(A + DX - \lambda_0 I)^k$, the equality (7.5.20) follows.

Now, because of (7.5.19), if P is the projection onto $\mathcal{S}$ along $\mathcal{T}$, then X is given by (7.5.13). Moreover, $\mathcal{S}$ is invariant under $A + DX_+$ and $\mathcal{S} \cap \mathcal{N} = \{0\}$. Then in fact $\mathcal{S} \subseteq \mathcal{N}_+$ (indeed, $(A + DX)|\mathcal{S} = (A + DX_+)|\mathcal{S}$, and all the eigenvalues of $(A + DX_+)|\mathcal{S}$ have positive real parts in view of the definition of $\mathcal{S}$). Since $X_+ - X = (X_+ - X_-)(I - P)$ is hermitian, the equality (7.5.17) holds. Therefore, since P is a projection, $P^*(X_+ - X_-) = P^*(X_+ - X_-)P$, that is, $P^*(X_+ - X_-)(I - P) = 0$. Thus $(X_+ - X_-)\mathcal{T}$ is orthogonal to $\mathcal{S}$. Counting dimensions (as in the first part of the proof) now shows that in fact $\mathcal{T} = ((X_+ - X_-)\mathcal{S})^\perp$. This completes the proof of Theorem 7.5.4. □

7.6 Partial order

To summarize some of the results obtained earlier concerning the existence of hermitian solutions of the CARE, we have

Theorem 7.6.1. *Assume that $D \geq 0$, $C = C^*$ and (A, D) is a controllable pair. Let*

$$M = i \begin{bmatrix} A & D \\ C & -A^* \end{bmatrix}, \quad \hat{H} = i \begin{bmatrix} 0 & I \\ -I & 0 \end{bmatrix}.$$

Then the following statements are equivalent:

(i) *There exists a hermitian solution X of the CARE (7.4.1).*

(ii) *There exists an n-dimensional M-invariant $\hat{H}$-neutral subspace $\mathcal{M}$.*

(iii) *The partial multiplicities of M corresponding to the real eigenvalues are all even.*

(iv) *The partial multiplicities of M corresponding to the real eigenvalues are all even, and the sign characteristic of $(M, \hat{H})$ consists of $+1$'s only.*

Moreover, the formula

$$\mathcal{M} = \text{Im} \begin{bmatrix} I \\ X \end{bmatrix} \tag{7.6.1}$$

establishes a one-to-one correspondence between the hermitian solutions X and the n-dimensional M-invariant $\hat{H}$-neutral subspaces $\mathcal{M}$.

We now study the latter correspondence in more detail. We will be interested in the partial order on the set of all hermitian solutions introduced in the preceding section. Problems concerning existence and identification of maximal and minimal solutions have been resolved in Theorem 7.5.1, and that theorem will play an important role here.

In the rest of this section we assume that the hypotheses and the equivalent conditions (i)–(iv) of Theorem 7.6.1 are satisfied. It will be more convenient to work with the matrix $M_0 = \begin{bmatrix} A & D \\ C & -A^* \end{bmatrix}$ rather than with M.

Let $\mathbb{C}_+ = \{\lambda \in \mathbb{C} : \text{Re}\,\lambda > 0\}$ be the open right half-plane, and let $\mathcal{M}_+$ be the spectral invariant subspace of M_0 corresponding to the eigenvalues in $\mathbb{C}_+$. By Theorem 2.5.2 (using also Theorem 7.3.5) we know that, for every M_0-invariant subspace $\mathcal{N} \subseteq \mathcal{M}_+$, there exists a unique n-dimensional M_0-invariant $\hat{H}$-neutral subspace $\mathcal{M}$ such that

$$\mathcal{M} \cap \mathcal{M}_+ = \mathcal{N}. \tag{7.6.2}$$

In other words, this relation establishes a one-to-one correspondence between the invariant subspaces $\mathcal{N}$ of the linear transformation $M_0|\mathcal{M}_+$ and n-dimensional

M_0-invariant $\hat{H}$-neutral subspaces $\mathcal{M}$. Using (7.6.1), we extend this correspondence to hermitian solutions X of the CARE (7.4.1):

$$\operatorname{Im} \begin{bmatrix} I \\ X \end{bmatrix} \cap \mathcal{M}_+ = \mathcal{N}. \tag{7.6.3}$$

It will be convenient to restate this correspondence more formally:

Theorem 7.6.2. *Consider the algebraic Riccati equation*

$$XDX + XA + A^*X - C = 0, \tag{7.6.4}$$

where $D \geq 0$, $C = C^*$, and (A, D) is a controllable pair. Assume that there exists a hermitian solution of (7.6.4). Let $\mathcal{M}_+$ be the spectral invariant subspace of $M_0 = \begin{bmatrix} A & D \\ C & -A^* \end{bmatrix}$ corresponding to the eigenvalues of M_0 in the open right half-plane. Then the formula (7.6.3) gives a one-to-one correspondence between hermitian solutions X of (7.6.4) and M_0-invariant subspaces $\mathcal{N}$ such that $\mathcal{N} \subseteq \mathcal{M}_+$.

Using Theorems 2.4.2 and 7.3.5, it is possible to generalize this result. Thus, replace $\mathcal{M}_+$ by the spectral subspace of M_0 corresponding to a set of eigenvalues $\mathcal{S}$ with the following properties:

(i) $\lambda \in \mathcal{S} \to \operatorname{Re} \lambda \neq 0$.

(ii) $\lambda \in \mathcal{S} \to -\bar{\lambda} \notin \mathcal{S}$.

(iii) $\mathcal{S}$ is a maximal set of eigenvalues of M_0 subject to the conditions (i) and (ii), namely, if $\mathcal{S} \subseteq \mathcal{S}_1 \subseteq \sigma(M_0)$, and $\mathcal{S}_1$ has the properties (i) and (ii), then in fact $\mathcal{S}_1 = \mathcal{S}$.

Such a set of eigenvalues $\mathcal{S}$ will be called a *c-set*. In the rest of this section we will use Theorem 7.6.2 as stated, i.e. with $\mathcal{M}_+$. In the next section, however, we will find it convenient to use this generalization.

We express the correspondence given by (7.6.3) by introducing a function "Ric" from subspaces to hermitian matrices, and write

$$X = \operatorname{Ric}(\mathcal{N}), \quad \text{or} \quad X = \operatorname{Ric}(\mathcal{N}; A, D, C),$$

if we wish to emphasize explicitly that X is a solution of the equation (7.6.4) determined by its coefficients A, D and C.

Before we go on to study the partial order, let us observe that, because of the one-to-one correspondence described in Theorem 7.6.2, we can count the hermitian solutions:

Corollary 7.6.3. *The number of hermitian solutions of (7.6.4) is finite if and only if*

$$\dim \text{Ker}\,(M_0 - \lambda_0 I) = 1 \qquad (7.6.5)$$

for every eigenvalue λ_0 of M_0 in $\mathbb{C}_+$. In this case, the number of hermitian solutions of (7.6.4) is given by

$$\prod_{i=1}^{p} (1 + \dim \mathcal{R}_{\lambda_0}(M_0)) \qquad (7.6.6)$$

where $\lambda_1, \ldots, \lambda_p$ are all the distinct eigenvalues of M_0 in $\mathbb{C}_+$. If

$$\dim \text{Ker}\,(M_0 - \lambda_0 I) > 1 \qquad (7.6.7)$$

for at least one eigenvalue $\lambda_0 \in \mathbb{C}_+$ of M_0, then (7.6.4) has a continuum of hermitian solutions X.

Proof. By Theorem 7.6.2 we may count the invariant subspaces of $M_0|_{\mathcal{M}_+}$ instead of the hermitian solutions of (7.6.4).

If (7.6.7) holds, let x and y be linearly independent eigenvectors of M_0 corresponding to the eigenvalue λ_0. Then span $\{x + \alpha y\}$, $\alpha \in \mathbb{C}$ provides a continuum of one-dimensional $M_0|_{\mathcal{M}_+}$-invariant subspaces.

If (7.6.5) holds for every $\lambda_0 \in \sigma(M_0) \cap \mathbb{C}_+$, then the Jordan form of the restriction $M_0|_{\mathcal{M}_+}$ is

$$J = J_{m_1}(\lambda_1) \oplus \cdots \oplus J_{m_p}(\lambda_p),$$

where $m_i = \dim \mathcal{R}_{\lambda_i}(M_0)$. By Exercise 1.8.2, each Jordan block $J_{m_i}(\lambda_i)$ has exactly $(m_i + 1)$ invariant subspaces, and by Theorem 1.2.5 any J-invariant subspace $\mathcal{M}$ has the form $\mathcal{M} = \mathcal{M}_1 \oplus \cdots \oplus \mathcal{M}_p$, where $\mathcal{M}_i$ is $J_{m_i}(\lambda_i)$-invariant. The formula (7.6.6) follows. □

There is a natural partial order (by inclusion) in the set of $M_0|_{\mathcal{M}_+}$-invariant subspaces. There is also a natural partial order on the set of hermitian solutions X of (7.6.4): namely, $X_1 \leq X_2$ if and only if the difference $X_2 - X_1$ is positive semidefinite. It turns out that the corespondence described above is order preserving:

Theorem 7.6.4. *Let $X = \text{Ric}\,(\mathcal{N})$ and $Y = \text{Ric}\,(\mathcal{L})$ be hermitian solutions of the CARE (7.6.4). Then $X \leq Y$ if and only if $\mathcal{N} \subseteq \mathcal{L}$.*

Proof. First we show that the theorem can be reduced to the special case in which $C = 0$ and $\sigma(A)$ lies in the closed right half-plane. Let X_0 be the maximal solution of (7.6.4) which exists by Theorem 7.5.1. Then $\sigma(A + DX_0)$ lies in the closed right half-plane. Consider the Riccati equation

$$YDY + (A + DX_0)Y + Y(A + DX_0) = 0. \qquad (7.6.8)$$

A straightforward calculation shows that X is a hermitian solution of (7.6.4) if and only if $Y = X - X_0$ is a hermitian solution of (7.6.8). Put

$$\hat{M}_0 = \begin{bmatrix} A + DX_0 & D \\ 0 & -(A + DX_0)^* \end{bmatrix}.$$

Note that $\hat{M}_0 = EM_0E^{-1}$, where $E = \begin{bmatrix} I & 0 \\ -X_0 & I \end{bmatrix}$. By Theorem 7.6.2 there is a one-to-one correspondence between hermitian solutions Y of (7.6.8) and $\hat{M}_0$-invariant subspaces $\hat{\mathcal{N}}$ contained in $\hat{\mathcal{M}}_+$, the spectral subspace of $\hat{M}_0$ corresponding to its eigenvalues in $\mathbb{C}_+$. This correspondence is given by the formula

$$\hat{\mathcal{N}} = \operatorname{Im} \begin{bmatrix} I \\ Y \end{bmatrix} \cap \hat{\mathcal{M}}_+.$$

It is easy to see that $X = \operatorname{Ric}(\mathcal{N}; A, D, C)$ if and only if $X - X_0 = \operatorname{Ric}(E\mathcal{N}; A + DX_0, D, 0)$. Thus, if we have already proved Theorem 7.6.2 for the equation (7.6.8), the result for the original equation (7.6.4) will follow.

We assume for the rest of this proof that $C = 0$ and $\sigma(A)$ lies in the closed right half-plane. Then

$$\mathcal{M}_+ = \operatorname{Im} \begin{bmatrix} I \\ 0 \end{bmatrix} \cap \begin{bmatrix} \mathcal{A}_+ \\ 0 \end{bmatrix},$$

where $\mathcal{A}_+$ is the spectral subspace of A corresponding to its eigenvalues in the open right half-plane. By Theorem 7.5.1 the zero matrix is the maximal hermitian solution of the equation

$$XDX + A^*X + XA = 0. \qquad (7.6.9)$$

Now suppose X_1 and X_2 are two hermitian solutions of (7.6.9) with $X_1 \leq X_2 \leq 0$. Then for $i = 1, 2$:

$$\mathcal{N}_i \stackrel{\text{def}}{=} \operatorname{Im} \begin{bmatrix} I \\ X_i \end{bmatrix} \cap \mathcal{M}_+ = \left\{ \begin{bmatrix} x \\ 0 \end{bmatrix} \mid x \in \operatorname{Ker} X_i \right\} \cap \begin{bmatrix} \mathcal{A}_+ \\ 0 \end{bmatrix}.$$

So in order to prove that $\mathcal{N}_1 \subseteq \mathcal{N}_2$ it suffices to show that $\operatorname{Ker} X_1 \subseteq \operatorname{Ker} X_2$. But this follows easily from the inequalities $X_1 \leq X_2 \leq 0$.

Conversely, assume $\mathcal{N}_1 \subseteq \mathcal{N}_2$, i.e., $\operatorname{Ker} X_1 \subseteq \operatorname{Ker} X_2$. With respect to the orthogonal decomposition $\mathbb{C}^n = \operatorname{Ker} X_1 \oplus \operatorname{Im} X_1$ write

$$X_1 = \begin{bmatrix} 0 & 0 \\ 0 & \tilde{X}_1 \end{bmatrix}, \quad X_2 = \begin{bmatrix} 0 & 0 \\ 0 & \tilde{X}_2 \end{bmatrix}, \quad D = \begin{bmatrix} D_1 & D_2 \\ D_2^* & D_3 \end{bmatrix}.$$

As the subspace $\left\{ \begin{bmatrix} x \\ 0 \end{bmatrix} \mid x \in \operatorname{Ker} X_1 \right\}$ is M_0-invariant, the subspace $\operatorname{Ker} X_1$ is A-invariant, so with respect to the same decomposition we have

$$A = \begin{bmatrix} A_{11} & A_{12} \\ 0 & \tilde{A} \end{bmatrix}.$$

Now the equalities

$$X_i D X_i + A^* X_i + X_i A = 0, \quad i = 1, 2$$

imply

$$\tilde{X}_i D_3 \tilde{X}_i + \tilde{A}^* \tilde{X}_i + \tilde{X}_i \tilde{A} = 0, \quad i = 1, 2. \tag{7.6.10}$$

First we show that $\tilde{A}$ has no pure imaginary eigenvalues. Indeed, we have

$$M_0 = \begin{bmatrix} A_{11} & A_{12} & D_1 & D_2 \\ 0 & \tilde{A} & D_2^* & D_3 \\ 0 & 0 & -A_{11}^* & 0 \\ 0 & 0 & -A_{12}^* & -\tilde{A}^* \end{bmatrix}.$$

Let $X_+ = 0$ and X_- be the maximal and minimal hermitian solutions of (7.6.9) respectively. Let $\mathcal{S}$ be the spectral M_0-invariant subspace corresponding to the eigenvalues of M_0 on the imaginary axis, and let $\mathcal{S}_0$ be the unique $\frac{1}{2}\dim \mathcal{S}$-dimensional $M_0|\mathcal{S}$-invariant $\hat{H}$-neutral subspace (the uniqueness of such an $\mathcal{S}_0$ follows from Theorem 7.3.5 and Corollary 2.5.3). Theorem 7.5.1 now implies that

$$\mathcal{S}_0 = \operatorname{Im} \begin{bmatrix} I \\ X_+ \end{bmatrix} \cap \operatorname{Im} \begin{bmatrix} I \\ X_- \end{bmatrix} = \left\{ \begin{bmatrix} x \\ 0 \end{bmatrix} : x \in \operatorname{Ker} X_- \right\}$$

where the last equality follows because $X_+ = 0$.

Since $\mathcal{S}_0 \subseteq \operatorname{Im} \begin{bmatrix} I \\ X \end{bmatrix}$ for every hermitian solution X of (7.6.9) it follows that $\operatorname{Ker} X_- \subseteq \operatorname{Ker} X_1$. So $M_0|\mathcal{S}_0$ is similar to $A_{11}|\operatorname{Ker} X_-$. Hence A_{11} has the same pure imaginary or zero eigenvalues as M_0, with multiplicities precisely half of the multiplicities of M corresponding to pure imaginary or zero eigenvalues (see Theorem 2.5.2). As

$$\det(\lambda - M_0) = \det(\lambda - A_{11}) \det(\lambda - \tilde{A}) \det(\lambda + A_{11}^*) \det(\lambda + \tilde{A}^*),$$

by our preceding remark we find that $\det(\lambda - \tilde{A}) \det(\lambda + \tilde{A}^*)$ has no zeros on the imaginary line. Hence $\tilde{A}$ has no pure imaginary or zero eigenvalues, so that $\sigma(\tilde{A})$ is in the *open* right half-plane.

Let us check that the pair $(\tilde{A}, D_3)$ is controllable. Indeed, applying a congruence transformation to D_3 (and the corresponding similarity to $\tilde{A}$), and taking into account that D, and therefore D_3, are positive semidefinite, we can assume for the purpose of this check that $D_3 = \begin{bmatrix} I & 0 \\ 0 & 0 \end{bmatrix}$ with respect to some orthogonal decomposition of Im X_1. But then D_2^* must be of the form $\begin{bmatrix} Z \\ 0 \end{bmatrix}$ for some matrix Z, so $D_2^* = D_3 D_2^*$. Now

$$[D, AD, \ldots, A^{n-1}D] = \begin{bmatrix} 0 & D_3 & 0 & \tilde{A}D_3 & \ldots & 0 & \tilde{A}^{n-1}D_3 \end{bmatrix}^*$$
$$\left(\begin{bmatrix} I & 0 \\ D_2^* & I \end{bmatrix} \oplus \begin{bmatrix} I & 0 \\ D_2^* & I \end{bmatrix} \oplus \cdots \oplus \begin{bmatrix} I & 0 \\ D_2^* & I \end{bmatrix}\right),$$

where $*$ denotes a part of the matrix which is not of interest to us. As (A, D) is controllable, the rows of $[D_3, \tilde{A}D_3, \ldots, \tilde{A}^{n-1}D_3]$ are linearly independent, and the controllability of $(\tilde{A}, D_3)$ follows.

Since $\tilde{X}_1$ is invertible, we can rewrite (7.6.10) in the form $\tilde{X}_1^{-1}\tilde{A}^* + \tilde{A}\tilde{X}_1^{-1} = -D_3$. Since $\sigma(\tilde{A})$ is in the open right half-plane the equation $Y\tilde{A}^* + \tilde{A}Y = -D_3$ has a unique solution and it is negative definite (see Theorem 5.3.1). It follows that the equation

$$XD_3X + \tilde{A}^*X + X\tilde{A} = 0 \tag{7.6.11}$$

has only one negative definite solution, namely $\tilde{X}_1$. This implies that $\tilde{X}_1$ is the minimal solution of the Riccati equation (7.6.11). As $\tilde{X}_2$ is also a hermitian solution of this equation, we get $\tilde{X}_1 \leq \tilde{X}_2$ and consequently $X_1 \leq X_2$. □

Theorem 7.6.4 has several interesting consequences. Let us start with a corollary:

Corollary 7.6.5. *Let X_+ and X_- be the maximal and minimal solutions of the CARE (7.6.4), respectively. Then $X_+ - X_-$ is invertible if and only if M_0 has no pure imaginary or zero eigenvalues.*

Proof. As in the proof of Theorem 7.6.4 we can assume that $C = 0$ and A has all its eigenvalues in the closed right half-plane. Then $X_+ = 0$. Note that X_- is invertible if and only if the subspace $\mathcal{S}_0$ introduced in the proof of Theorem 7.6.4 is the zero subspace. By definition, $\mathcal{S}_0 = \{0\}$ if and only if M_0 has no eigenvalues on the imaginary axis. □

Let $\text{Her}(A, D, C)$ be the set of hermitian solutions of the CARE (7.6.4). By Theorem 7.6.4 the set $\text{Her}(A, D, C)$ as a partially ordered set looks exactly the same as the set of invariant subspaces of $M_0|\mathcal{M}_+$. For any set $\mathcal{N} = \{\mathcal{N}_\alpha\}_{\alpha \in \ell}$

of $M_0|\mathcal{M}_+$-invariant subspaces there exists the least upper bound $\mathcal{N}_u$ and the greatest lower bound $\mathcal{N}_\ell$. Indeed,

$$\mathcal{N}_u = \sum_{\alpha \in \Omega} \mathcal{N}_\alpha$$

has the property that $\mathcal{N}_\alpha \subseteq \mathcal{N}_u$ for all $\alpha \in \Omega$, and if $\mathcal{N}_\alpha \subseteq \mathcal{M}$ for all $\alpha \in \Omega$, where $\mathcal{M}$ is an $M_0|\mathcal{M}_+$-invariant subspace, then $\mathcal{N}_u \subseteq \mathcal{M}$. The subspace $\mathcal{N}_\ell = \cap_{\alpha \in \Omega} \mathcal{N}_\alpha$ has the dual property: $\mathcal{N}_\ell \subseteq \mathcal{N}_\alpha$ for all $\alpha \in \Omega$, and if $\mathcal{M} \subseteq \mathcal{N}_\alpha$ for all $\alpha \in \Omega$ and some $M_0|\mathcal{M}_+$-invariant subspace $\mathcal{M}$, then $\mathcal{M} \subseteq \mathcal{N}_\ell$. Therefore, the set Her (A, D, C) enjoys the analogous properties:

Corollary 7.6.6. *Given the hypotheses of Theorem 7.6.2, any set $Q \subseteq$ Her (A, D, C) has a least upper bound X_u and a greatest lower bound X_ℓ. Here X_u and X_ℓ are defined by the properties that $X_u \geq X$ (resp. $X_\ell \leq X$) for all $X \in Q$ and if for some $Y \in$ Her (A, D, C) the inequalities $Y \geq X$ (resp. $Y \leq X$) hold for all $X \in Q$, then $Y \geq X_u$ (resp. $Y \leq X_\ell$).*

In contrast, there need not be a least upper bound (or a greatest lower bound) for a set Q of hermitian matrices *in the set of all hermitian matrices*. The following example illustrates this point.

Example 7.6.1. Let

$$Q = \left\{ \begin{bmatrix} 1 & 0 \\ 0 & 0 \end{bmatrix}, \begin{bmatrix} 0 & 1 \\ 1 & 1 \end{bmatrix} \right\}.$$

The matrices $X_1 = \begin{bmatrix} 2 & 1 \\ 1 & 1 \end{bmatrix}$ and $X_2 = \begin{bmatrix} 1 & 0 \\ 0 & 2 \end{bmatrix}$ have the property that $X \leq X_1$ and $X \leq X_2$ for all $X \in Q$. If there were a least upper bound $X_0 = \begin{bmatrix} a & b \\ \bar{b} & c \end{bmatrix}$ for the set Q, then we would have $X_0 \leq X_1$ and $X_0 \leq X_2$. This implies $a \leq 1$ and $c \leq 1$. But also

$$\begin{bmatrix} a & b \\ \bar{b} & c \end{bmatrix} \geq \begin{bmatrix} 0 & 1 \\ 1 & 1 \end{bmatrix}, \begin{bmatrix} a & b \\ \bar{b} & c \end{bmatrix} \geq \begin{bmatrix} 1 & 0 \\ 0 & 0 \end{bmatrix}$$

which implies $a = c = 1$. But now the second inequality implies $b = 0$, while the first inequality implies the contradictory statement $b = 1$. □

7.7 Continuity

We continue to study the hermitian solutions X of the CARE (7.6.4) in terms of M_0-invariant subspaces $\mathcal{N}$ such that $\mathcal{N} \subseteq \mathcal{M}_+$, as expressed in Theorem 7.6.2:

$$\operatorname{Im} \begin{bmatrix} I \\ X \end{bmatrix} \cap \mathcal{M}_+ = \mathcal{N}, \tag{7.7.1}$$

or

$$X = \operatorname{Ric}(\mathcal{N}) = \operatorname{Ric}(\mathcal{N}; A, D, C). \tag{7.7.2}$$

In this section we retain the hypotheses of Sections 7.5 and 7.6: $D \geq 0$, $C = C^*$ and (A, D) controllable, and study the continuity and analyticity properties of the function $X = \operatorname{Ric}(\mathcal{N})$ and its inverse $\mathcal{N} = \operatorname{Ric}^{-1}(X)$, defined by (7.7.1) and (7.7.2). The $n \times n$ matrices A, D and C will be kept fixed throughout this section. The topology in the set of hermitian solutions of the CARE is the natural one induced by the operator norm:

$$\|X\| = \max_{\|x\|=1} \|Xx\|, \tag{7.7.3}$$

with the usual concept of convergence $X = \lim_{m \to \infty} X_m$. The topology in the set of $M_0 | \mathcal{M}_+$-invariant subspaces is the metric topology induced by the gap metric (see the Appendix 7.A to this chapter), with the convergence $\lim_{m \to \infty} \mathcal{N}_m = \mathcal{N}$ defined by $\lim_{m \to \infty} \theta(\mathcal{N}_m, \mathcal{N}) = 0$, where $\theta(\mathcal{H}, \mathcal{Y})$ is the gap between subspaces $\mathcal{X}, \mathcal{Y} \subseteq \mathbb{C}^{2n}$.

Theorem 7.7.1. *The functions* Ric *and its inverse* Ric^{-1} *are continuous. In other words, if* $\{\mathcal{N}_m\}_{m=1}^{\infty}$ *are* $M_0 | \mathcal{M}_+$-*invariant subspaces such that* $\mathcal{N} = \lim_{m \to \infty} \mathcal{N}_m$, *then*

$$\lim_{m \to \infty} \operatorname{Ric}(\mathcal{N}_m) = \operatorname{Ric}(\mathcal{N}), \tag{7.7.4}$$

and conversely, if $\{X_m\}_{m=1}^{\infty}$ *are hermitian solutions of the CARE and* $X = \lim_{m \to \infty} X_m$, *then* $\lim_{m \to \infty} \mathcal{N}_m = \mathcal{N}$, *where* $\mathcal{N}_m$ *and* $\mathcal{N}$ *are defined by*

$$\operatorname{Im} \begin{bmatrix} I \\ X_m \end{bmatrix} \cap \mathcal{M}_+ = \mathcal{N}_m, \quad \operatorname{Im} \begin{bmatrix} I \\ X \end{bmatrix} \cap \mathcal{M}_+ = \mathcal{N}. \tag{7.7.5}$$

Observe that by Corollary 7.A.6 in the Appendix the subspace $\mathcal{N} = \lim_{m \to \infty} \mathcal{N}_m$ is M_0-invariant and since all $\mathcal{N}_m$ are contained in $\mathcal{M}_+$, so is $\mathcal{N}$ (one easily verifies this using Theorem 7.A.5). Therefore $\operatorname{Ric}(\mathcal{N})$ is well-defined and (7.7.4) makes sense. Similarly, if $\{X_m\}_{m=1}^{\infty}$ are hermitian solutions of the CARE, so is $X = \lim_{m \to \infty} X_m$ (just pass to the limit in the equalities

$$X_m D X_m + X_m A + A^* X_m - C = 0, \quad m = 1, 2, \ldots)$$

and therefore $\mathcal{N}$ defined by (7.7.5) is M_0-invariant (and clearly contained in $\mathcal{M}_+$).

Proof. It is easily verified that the orthogonal projector P_m on the subspace $\text{Im}\begin{bmatrix} I \\ X_m \end{bmatrix}$ is given by

$$\begin{bmatrix} Z_m & Z_m X_m^* \\ X_m Z_m^* & X_m Z_m X_m^* \end{bmatrix},$$

where $Z_m = (I + X_m^* X_m)^{-1/2}$. So if $\{X_m\}_{m=1}^\infty$ are hermitian solutions of the CARE (7.6.4) and $X_m \to X$ as $m \to \infty$, then $\lim_{m \to \infty} P_m = P$, where

$$P = \begin{bmatrix} Z & ZX^* \\ XZ & XZX^* \end{bmatrix}, \quad Z = (I + X^*X)^{-1/2}, \tag{7.7.6}$$

is the orthogonal projector on $\text{Im}\begin{bmatrix} I \\ X \end{bmatrix}$. Thus (in terms of the "gap" introduced in Appencix 7.A),

$$\lim_{m \to \infty} \theta\left(\text{Im}\begin{bmatrix} I \\ X_m \end{bmatrix}, \text{Im}\begin{bmatrix} I \\ X \end{bmatrix}\right) = 0$$

and by formula (7.7.5) and Corollary 7.A.10 we have

$$\lim_{m \to \infty} \mathcal{N}_m = \mathcal{N}. \tag{7.7.7}$$

Conversely, assume (7.7.7) holds, and let X_m, X be hermitian solutions of the CARE defined by (7.7.5). We need some preliminary observations. Recall that $M = iM_0$ is $\hat{H}$-self-adjoint. Using the canonical form of Theorem 2.3.2 for the pair $(iM_0, \hat{H})$, we can choose S to be an invertible $2n \times 2n$ matrix in such a way that

$$S^{-1}(iM_0)S = \begin{bmatrix} B_1 & 0 & 0 \\ 0 & B_0 & 0 \\ 0 & 0 & B_1^* \end{bmatrix}, \quad S^*\hat{H}S = \begin{bmatrix} 0 & 0 & I_k \\ 0 & G_0 & 0 \\ I_k & 0 & 0 \end{bmatrix},$$

where $\sigma(B_1)$ lies in the open upper half-plane, $\sigma(B_0)$ lies on the real line, and B_1 is of size $k \times k$. Clearly, B_0 is G_0-self-adjoint, the partial multiplicities of B_0 are all even, and the sign characteristic of (B_0, G_0) consists of $+1$'s only. Let

$$\tilde{M}_0 = \begin{bmatrix} -iB_1 & 0 & 0 \\ 0 & -iB_0 & 0 \\ 0 & 0 & -iB_1^* \end{bmatrix}, \quad \tilde{G} = \begin{bmatrix} 0 & 0 & I_k \\ 0 & G_0 & 0 \\ I_k & 0 & 0 \end{bmatrix}.$$

The spectral subspace of $\tilde{M}_0$ corresponding to the eigenvalues in the open right (resp. left) half-plane is $\tilde{\mathcal{M}}_+ = [\mathbb{C}^k \oplus 0 \oplus 0]$ (resp. $\tilde{\mathcal{M}}_- = [0 \oplus 0 \oplus \mathbb{C}^k]$), and the spectral subspace of $\tilde{M}_0$ corresponding to its eigenvalues on the imaginary axis is $\tilde{\mathcal{M}}_0 = [0 \oplus \mathbb{C}^{2n-2k} \oplus 0]$. A straightforward verification shows that for any $\tilde{M}_0|\tilde{\mathcal{M}}_+$-invariant subspace $\tilde{\mathcal{N}}$ the unique n-dimensional $\tilde{M}_0$-invariant $\tilde{G}$-neutral subspace $\tilde{\mathcal{Y}}$ such that

$$\tilde{\mathcal{Y}} \cap \tilde{\mathcal{M}}_+ = \tilde{\mathcal{N}} \tag{7.7.8}$$

is given by

$$\tilde{\mathcal{Y}} = \begin{bmatrix} \tilde{\mathcal{N}} \\ 0 \\ 0 \end{bmatrix} + \begin{bmatrix} 0 \\ \tilde{\mathcal{Y}}_0 \\ 0 \end{bmatrix} + \begin{bmatrix} 0 \\ 0 \\ \tilde{\mathcal{N}}^\perp \end{bmatrix}, \tag{7.7.9}$$

where $\tilde{\mathcal{Y}}_0$ is the unique $(n-k)$-dimensional B_0-invariant G_0-neutral subspace, and $\tilde{\mathcal{N}}^\perp$ is the orthogonal complement to $\tilde{\mathcal{N}}$ in $\mathbb{C}^k$. The formula (7.7.9) shows that for $M_0|\tilde{\mathcal{M}}_+$-invariant subspaces $\tilde{\mathcal{N}}_1$ and $\tilde{\mathcal{N}}_2$ we have

$$\theta(\tilde{\mathcal{N}}_1, \tilde{\mathcal{N}}_2) = \theta(\tilde{\mathcal{Y}}_1, \tilde{\mathcal{Y}}_2), \tag{7.7.10}$$

where $\tilde{\mathcal{Y}}_i$ is the n-dimensional $\tilde{M}_0$-invariant $\tilde{G}$-neutral subspace defined by (7.7.8) with $\tilde{\mathcal{N}}$ replaced by $\tilde{\mathcal{N}}_i$. We now apply formula (7.7.10) with

$$\tilde{\mathcal{Y}}_1 = S^{-1}\left(\operatorname{Im}\begin{bmatrix} I \\ X \end{bmatrix}\right), \quad \tilde{\mathcal{Y}}_2 = S^{-1}\left(\operatorname{Im}\begin{bmatrix} I \\ X_m \end{bmatrix}\right),$$

$$\tilde{\mathcal{N}}_1 = S^{-1}\mathcal{N}, \quad \tilde{\mathcal{N}}_2 = S^{-1}\mathcal{N}_m.$$

So, denoting $\mathcal{Y} = \operatorname{Im}\begin{bmatrix} I \\ X \end{bmatrix}$, $\mathcal{Y}_m = \operatorname{Im}\begin{bmatrix} I \\ X_m \end{bmatrix}$ and letting P_χ be the orthogonal projector on the subspace χ, we have (in this calculation the first and the last but one inequalities follow from Theorem 7.A.2):

$$\theta(\mathcal{Y}, \mathcal{Y}_m) \leq \|SP_{\tilde{\mathcal{Y}}_1}S^{-1} - SP_{\tilde{\mathcal{Y}}_2}S^{-1}\|$$
$$\leq \|S\| \cdot \|S^{-1}\| \cdot \|P_{\tilde{\mathcal{Y}}_1} - P_{\tilde{\mathcal{Y}}_2}\| = \|S\| \cdot \|S^{-1}\| \cdot \theta(\tilde{\mathcal{Y}}_1, \tilde{\mathcal{Y}}_2)$$
$$= \|S\| \, \|S^{-1}\| \theta(\tilde{\mathcal{N}}_1, \tilde{\mathcal{N}}_2)$$
$$\leq \|S\| \cdot \|S^{-1}\| \, \|S^{-1}P_\mathcal{N}S - S^{-1}P_{\mathcal{N}_m}S\|$$
$$\leq \|S\|^2 \, \|S^{-1}\|^2 \, \|P_\mathcal{N} - P_{\mathcal{N}_m}\| = \|S\|^2 \, \|S^{-1}\|^2 \, \theta(\mathcal{N}, \mathcal{N}_m).$$

In view of (7.7.7) we have

$$\lim_{m \to \infty} \left(\operatorname{Im}\begin{bmatrix} I \\ X_m \end{bmatrix}\right) = \operatorname{Im}\begin{bmatrix} I \\ X \end{bmatrix},$$

or (using the formula (7.7.6) for the orthogonal projector on $\operatorname{Im}\begin{bmatrix} I \\ X \end{bmatrix}$),

$$(I + X_m^* X_m)^{-1/2} \to (I + X^* X)^{-1/2}, \quad X_m(I + X_m^* X_m)^{-1/2} \to X(I + X^* X)^{-1/2}$$

as $m \to \infty$. This clearly implies $X_m \to X$ (as $m \to \infty$), and (7.7.4) is proved. □

A more detailed analysis of the proof of Theorem 7.7.1 reveals that the functions Ric and Ric^{-1} are not only continuous but they are also Lipschitz continuous. Namely, there exist positive constants K_1, K_2 depending on A, D and C only such that $\|\operatorname{Ric}(\mathcal{N}_1) - \operatorname{Ric}(\mathcal{N}_2)\| \leq K_1 \theta(\mathcal{N}_1, \mathcal{N}_2)$, and

$$\theta(\operatorname{Ric}^{-1}(X_1), \operatorname{Ric}^{-1}(X_2)) \leq K_2 \|X_1 - X_2\|.$$

The result of Theorem 7.7.1 is valid also for any c-set S of eigenvalues of M_0 (the definition of a c-set is given after Theorem 7.6.2). By Theorems 7.3.5 and 2.5.2, given a c-set S and the spectral M_0-invariant subspace $\mathcal{M}_S$ corresponding to the eigenvalues in S, there is a one-to-one correspondence between the set of hermitian solutions X of the CARE (7.6.4) and the set of M_0-invariant subspaces $\mathcal{N}$ such that $\mathcal{N} \subseteq \mathcal{M}_S$ given by

$$\operatorname{Im} \begin{bmatrix} I \\ X \end{bmatrix} \cap \mathcal{M}_S = \mathcal{N}.$$

This one-to-one correspondence is continuous in both directions, in the same sense as stated in Theorem 7.7.1. The proof of this statement is completely analogous to the proof of Theorem 7.7.1 and is therefore omitted.

Using the continuity of Ric and Ric^{-1}, we now characterize isolated solutions. A hermitian solution X of the CARE is called *isolated* if there is an open neighborhood of X contaning no other hermitian solution.

Theorem 7.7.2. *The following statements are equivalent for a hermitian solution X_0 of the CARE (7.6.4):*

(i) *X_0 is isolated.*

(ii) *each common eigenvalue λ with nonzero real part of $A + DX_0$ and of $-(A^* + X_0 D)$ is an eigenvalue of $M_0 = \begin{bmatrix} A & D \\ C & -A^* \end{bmatrix}$ with $\dim \operatorname{Ker}(\lambda I - M_0) = 1$.*

(iii) *for every eigenvalue λ of M_0 with $\operatorname{Re}\lambda \neq 0$ and with $\dim \operatorname{Ker}(\lambda I - M_0) > 1$, we have either*

$$\mathcal{R}_\lambda(M) \subseteq \operatorname{Im} \begin{bmatrix} I \\ X_0 \end{bmatrix} \quad or \quad \mathcal{R}_\lambda(M) \cap \operatorname{Im} \begin{bmatrix} I \\ X_0 \end{bmatrix} = \{0\}.$$

(iv) *Statement (iii) with $\operatorname{Re}\lambda \neq 0$ replaced by $\operatorname{Re}\lambda > 0$.*

(v) *Statement (iii) with $\operatorname{Re}\lambda \neq 0$ replaced by $\operatorname{Re}\lambda < 0$.*

In particular, Theorem 7.7.2 implies that the minimal and maximal solutions of (7.6.4) are isolated.

The proof of Theorem 7.7.2 depends on the following description of isolated invariant subspaces of a given $n \times n$ matrix Z. A Z-invariant subspace $\mathcal{Y}_0 \subseteq \mathbb{C}^n$ is called *isolated* if there is $\epsilon > 0$ such that $\theta(\mathcal{Y}_0, \mathcal{Y}) \geq \epsilon$ for every other Z-invariant subspace $\mathcal{Y}$. The trivial Z-invariant subspaces $\{0\}$ and $\mathbb{C}^n$ are always isolated.

Lemma 7.7.3. *A Z-invariant subspace $\mathcal{Y}_0$ is isolated if and only if for every eigenvalue λ_0 of Z such that $\dim \mathrm{Ker}\,(\lambda_0 I - Z) \geq 2$ either $\mathcal{Y}_0 \supseteq \mathcal{R}_{\lambda_0}(Z)$ or $\mathcal{Y}_0 \cap \mathcal{R}_{\lambda_0}(Z) = \{0\}$ holds.*

A full proof of Lemma 7.7.3 is found in Gohberg *et al.* (1986a) (Theorem 14.3.1) and will not be reproduced here. It follows, for example, that every spectral subspace Z-invariant subspace is isolated, and that the nonderogatory matrices are the only ones with the property that every invariant subspace is isolated.

Proof of Theorem 7.7.2. In view of Theorem 7.7.1 the isolatedness of X_0 means that the $M_0|\mathcal{M}_+$-invariant subspace

$$\mathrm{Im}\,\begin{bmatrix} I \\ X_0 \end{bmatrix} \cap \mathcal{M}_+ \tag{7.7.11}$$

is isolated (in the set of all $M_0|\mathcal{M}_+$-invariant subspaces). By Lemma 7.7.3 the isolatedness of (7.7.11) is equivalent to condition (iii) applied to the eigenvalues λ of M_0 with positive real parts. Apply the analogue of Theorem 7.7.1 with $\mathcal{M}_+$ replaced by the spectral M_0-invariant subspace $\mathcal{M}_-$ corresponding to the eigenvalues of M_0 with negative real parts. (The validity of such an analogue is explained in a remark after the proof of Theorem 7.7.1.) We find that the isolatedness of X is equivalent to condition (iii) applied for $\lambda \in \sigma(M_0)$ with $\mathrm{Re}\,\lambda < 0$. Thus, (i)$\Leftrightarrow$(iii)$\Leftrightarrow$(iv)$\Leftrightarrow$(v).

To prove the equivalence of (i) and (ii) we consider the equation

$$YDY + (A + DX_0)Y + Y(A + DX_0) = 0 \tag{7.7.12}$$

and observe that $X = X^*$ solves the CARE (7.6.4) if and only if $Y = Y^* = X - X_0$ solves (7.7.12) (cf. the proof of Theorem 7.6.4). Thus (i) is equivalent to the isolatedness of the solution $Y_0 = 0$ of (7.7.12) which, by the already proved part of Theorem 7.7.2 is equivalent to the condition (iii) applied to equation (7.7.12), namely:

(iii') *for every eigenvalue λ of*

$$\tilde{M}_0 = \begin{bmatrix} A + DX_0 & D \\ 0 & -(A + DX_0)^* \end{bmatrix}$$

with $\mathrm{Re}\,\lambda \neq 0$ and with $\dim \mathrm{Ker}\,(\lambda I - \tilde{M}_0) > 1$, we have either

ANALYTICITY

$$R_\lambda(\tilde{M}_0) \subseteq \operatorname{Im} \begin{bmatrix} I \\ 0 \end{bmatrix} = \left\{ \begin{bmatrix} x \\ 0 \end{bmatrix} : x \in \mathbb{C}^n \right\}, \qquad (7.7.13)$$

or

$$R_\lambda(\tilde{M}_0) \cap \left\{ \begin{bmatrix} x \\ 0 \end{bmatrix} : x \in \mathbb{C}^n \right\} = \{0\}. \qquad (7.7.14)$$

If neither (7.7.13) nor (7.7.14) holds, then clearly λ is a common eigenvalue of $A + DX_0$ and of $-(A^* + X_0 D)$, and so (iii') is equivalent to (ii). □

7.8 Analyticity

We consider here the analytic properties of the maps Ric and Ric^{-1} defined in Section 7.6. Let Ω be an open connected set in $\mathbb{R}^k$. A $p \times q$ matrix function $X(t)$, $t \in \Omega$, is said to be *analytic* (more precisely, "real analytic") on Ω if in a neighborhood of each point $t_0 = (t_{01}, t_{02}, \ldots, t_{0k}) \in \Omega$, $X(t)$ is represented by a power series

$$X(t) = \sum \gamma(\alpha_1, \alpha_2, \ldots, \alpha_k)(t_1 - t_{01})^{\alpha_1}(t_2 - t_{02})^{\alpha_2} \ldots (t_k - t_{0k})^{\alpha_k},$$

$$t = (t_1, t_2, \ldots, t_k),$$

where the summation is taken over all k-tuples of nonnegative integers $(\alpha_1, \ldots, \alpha_k)$, and $\gamma(\alpha_1, \alpha_2, \ldots, \alpha_k)$ is a $p \times q$ matrix depending on α_i, $i = 1, \ldots, k$. A subspace valued function $\mathcal{L}(t)$, $t \in \Omega$, so that $\mathcal{L}(t)$ is a subspace in $\mathbb{C}^m$ for each $t \in \Omega$, is said to be *analytic* in Ω if the orthogonal projector on $\mathcal{L}(t)$ is an analytic matrix function on Ω.

Observe that in view of Theorem 7.A.3 the dimension of an analytic subspace valued function $\mathcal{L}(t)$ on Ω is constant throughout Ω (here it is essential that Ω be connected).

For future reference we record the following fact.

Lemma 7.8.1. *If the subspace valued function $\mathcal{L}(t) \subseteq \mathbb{C}^m$ is analytic on Ω, then the subspace valued function $S\mathcal{L}(t)$ is analytic as well, where S is any fixed invertible $m \times m$ matrix.*

Proof. Let $P_{\mathcal{L}(t)}$ (resp. $P_{S\mathcal{L}(t)}$) be the orthogonal projector on $\mathcal{L}(t)$ (resp. $S\mathcal{L}(t)$). We have

$$P_{S\mathcal{L}(t)} = V(t) S P_{\mathcal{L}(t)} S^{-1} V(t)^{-1},$$

where $V(t)$ is any invertible matrix with the properties that

$$V(t)(S\mathcal{L}(t)) = S\mathcal{L}(t) \qquad (7.8.1)$$

and

$$V(t)(\operatorname{Ker}(P_{\mathcal{L}(t)} S^{-1})) = (S\mathcal{L}(t))^\perp. \qquad (7.8.2)$$

Fix $t_0 \in \Omega$ and we will show that $V(t)$ can be chosen analytic on a neighborhood of t_0, thereby proving Lemma 7.8.1. Indeed, let $x_1(t), \ldots, x_p(t)$ be the columns of the matrix $SP_{\mathcal{L}(t)}$ such that $x_1(t_0), \ldots, x_p(t_0)$ form a basis in the column space of $SP_{\mathcal{L}(t_0)}$. As the property of being linearly independent is preserved under sufficiently small perturbations, and because $x_1(t), \ldots, x_p(t)$ are analytic (in particular, continuous) in $t \in \Omega$, there exists an $\epsilon_1 > 0$ such that $x_1(t), \ldots, x_p(t)$ are linearly independent for $\|t - t_0\| < \epsilon_1$, $t \in \Omega$. But the dimension of $\operatorname{Im}(SP_{\mathcal{L}(t)}) = S\mathcal{L}(t)$ is constant throughout Ω (it is equal to the dimension of $\mathcal{L}(t)$), so in fact $x_1(t), \ldots, x_p(t)$ form a basis in $S\mathcal{L}(t)$ for t sufficiently close to t_0. Similarly, choose analytic vector functions $x_{p+1}(t), \ldots, x_m(t)$ which form a basis in $\operatorname{Im}(S(I - P_{\mathcal{L}(t)}))$ for t sufficiently close to t_0. The subspaces $\operatorname{Im}(SP_{\mathcal{L}(t)})$ and $\operatorname{Im}(S(I - P_{\mathcal{L}(t)})) = \operatorname{Ker}(P_{\mathcal{L}(t)}S^{-1})$ are direct complements of one another, so the $m \times m$ matrix $[x_1(t) \ldots x_p(t) x_{p+1}(t) \ldots x_m(t)]$ is invertible and its inverse is analytic as well (for t close to t_0). The vectors $y_{p+1}(t), \ldots, y_m(t)$ defined by

$$\begin{bmatrix} (y_1(t))^* \\ (y_2(t))^* \\ \vdots \\ (y_m(t))^* \end{bmatrix} = [x_1(t) \ldots x_m(t)]^{-1}$$

form a basis in $(S\mathcal{L}(t))^\perp$ and are clearly analytic (again, for t sufficiently close to t_0). Now $V(t)$ defined by

$$V(t) x_i(t) = x_i(t); \quad i = 1, \ldots, p$$

$$V(t) x_i(t) = y_i(t); \quad i = p+1, \ldots, m$$

satisfies the required properties (7.8.1) and (7.8.2), and is invertible and analytic in a neighborhood of t_0. □

We now state and prove the main result of this section.

Theorem 7.8.2. *Assume that $D \geq 0$, $C^* = C$ and (A, D) is controllable. Then the maps* Ric *and* Ric^{-1} *are analytic in the following sense. Given an analytic hermitian valued function $X(t)$ on Ω, where $X(t)$ is a solution of the CARE (7.6.4) for each $t \in \Omega$, the subspace valued function*

$$\mathcal{N}(t) = \operatorname{Im} \begin{bmatrix} I \\ X(t) \end{bmatrix} \cap \mathcal{M}_+ \tag{7.8.3}$$

is analytic as well. Conversely, let $\mathcal{N}(t) \subseteq \mathbf{C}^{2n}$ be an analytic subspace valued function defined on Ω such that for every $t \in \Omega$ the subspace $\mathcal{N}(t)$ is M_0-invariant and contained in $\mathcal{M}_+$. Then the hermitian valued function $X(t)$, defined on Ω by (7.8.3), where $X(t)$ is a solution of the CARE for every $t \in \Omega$, is also analytic.

As for the continuity properties of Ric and Ric^{-1}, here also one can replace $\mathcal{M}_+$ by the spectral M_0-invariant subspace corresponding to any given c-set of eigenvalues.

Proof. Assume that $X(t)$ is an analytic hermitian solution of (7.6.4) on Ω. The subspace $\mathcal{N}(t)$, which is by definition $\operatorname{Im} \begin{bmatrix} I \\ X(t) \end{bmatrix} \cap \mathcal{M}_+$, is equal to $\operatorname{Im} \left(P_+ \begin{bmatrix} I \\ X(t) \end{bmatrix} \right)$, where P_+ is the projection on $\mathcal{M}_+$ along the spectral subspace of M_0 corresponding to its eigenvalues in the set $\{\lambda \in \mathbb{C} | \operatorname{Re}\lambda \leq 0\}$. Indeed, the inclusion

$$\operatorname{Im} \begin{bmatrix} I \\ X(t) \end{bmatrix} \cap \mathcal{M}_+ \subseteq \operatorname{Im} \left(P_+ \begin{bmatrix} I \\ X(t) \end{bmatrix} \right)$$

is immediate. The opposite inclusion follows easily since the subspace $\operatorname{Im} \begin{bmatrix} I \\ X(t) \end{bmatrix}$ is M_0-invariant, and therefore also P_+-invariant. From the continuity of Ric^{-1} and from the fact that $\dim V_1 = \dim V_2$ as long as $\theta(V_1, V_2) < 1$ for subspaces V_1, V_2 in $\mathbb{C}^{2n}$ (Theorem 7.A.3), it follows that the rank of the $2n \times n$ matrix $P_+ \begin{bmatrix} I \\ X(t) \end{bmatrix}$ does not depend on t (here also the connectedness of Ω is necessary).

Therefore, for every $t_0 \in \Omega$ there exists a basis $x_1(t), \ldots, x_k(t)$ in $\operatorname{Im} \left[P_+ \begin{bmatrix} I \\ X(t) \end{bmatrix} \right]$ such that the vectors $x_1(t), \ldots, x_k(t)$ are analytic in some open neighborhood $U \subseteq \Omega$ of t_0 (indeed, take $x_1(t), \ldots, x_k(t)$ to be the columns in $P_+ \begin{bmatrix} I \\ X(t) \end{bmatrix}$ which, when evaluated at t_0 form a basis in the column space of $P_+ \begin{bmatrix} I \\ X(t_0) \end{bmatrix}$). Performing the Gram–Schmidt orthogonalization (which does not effect the analyticity of $x_i(t)$ because t represents *real* variables), we can assume that $x_i(t)$ are orthogonal. By a similar argument, choose an orthogonal basis $x_{k+1}(t), \ldots, x_{2n}(t)$ in $\left\{ \operatorname{Im} \left(P_+ \begin{bmatrix} I \\ X(t) \end{bmatrix} \right) \right\}^\perp = \operatorname{Ker}\left([I\ X(t)]P_+^*\right)$ in such a way that $x_j(t)$ is analytic on U, $j = k+1, \ldots, 2n$. Now the matrix

$$[x_1(t), x_2(t), \ldots, x_{2n}(t)] \begin{bmatrix} I_k & 0 \\ 0 & 0 \end{bmatrix} [x_1(t), x_2(t), \ldots, x_{2n}(t)]^{-1}$$

is the orthogonal projection on $\mathcal{N}(t)$ for each $t \in U$. This matrix is clearly analytic on U. Since $t_0 \in \Omega$ was arbitrary, the subspaces $\mathcal{N}(t)$ are analytic on Ω.

Conversely, assume that $\mathcal{N}(t)$ is analytic on Ω. Let $\hat{H} = i \begin{bmatrix} 0 & I \\ -I & 0 \end{bmatrix}$. By applying a simultaneous similarity-congruence $(iM_0, \hat{H}) \to (S^{-1}(iM_0)S, S^*\hat{H}S)$ for some invertible matrix S, we can assume, as in the proof of Theorem 7.7.1, that

$$S^{-1}(iM_0)S = \begin{bmatrix} B_1 & 0 & 0 \\ 0 & B_0 & 0 \\ 0 & 0 & B_1^* \end{bmatrix}, \quad S^*\hat{H}S = \begin{bmatrix} 0 & 0 & I_k \\ 0 & G_0 & 0 \\ I_k & 0 & 0 \end{bmatrix},$$

where $\sigma(B_1)$ lies in the open upper half-plane and $\sigma(B_0)$ lies on the real line. Then, arguing as in the proof of Theorem 7.7.1, we have $S^{-1}\mathcal{N}(t) = [\tilde{\mathcal{N}}(t) \oplus 0 \oplus 0]$ for $t \in \Omega$ and for some subspace $\tilde{\mathcal{N}}(t) \subseteq \mathbb{C}^k$, and

$$S^{-1}\left(\operatorname{Im}\begin{bmatrix} I \\ X(t) \end{bmatrix}\right) = \left\{\begin{bmatrix} x \\ y \\ z \end{bmatrix} : x \in \tilde{\mathcal{N}}(t),\, y \in \tilde{\mathcal{Y}}_0,\, z \in (\tilde{\mathcal{N}}(t))^\perp\right\} \quad (7.8.4)$$

where $\tilde{\mathcal{Y}}_0$ is the unique $(n-k)$-dimensional B_0-invariant G_0-neutral subspace. By Lemma 7.8.1, $\tilde{\mathcal{N}}(t)$ is an analytic subspace valued function, and the formula (7.8.4) shows that $S^{-1}\left[\operatorname{Im}\begin{bmatrix} I \\ X(t) \end{bmatrix}\right]$ is also analytic. But then, again by Lemma 7.8.1, $\operatorname{Im}\begin{bmatrix} I \\ X(t) \end{bmatrix}$ is an analytic subspace valued function, so the orthogonal projector $P(t)$ on this subspace,

$$P(t) = \begin{bmatrix} Z(t) & Z(t)X(t)^* \\ X(t)Z(t) & X(t)Z(t)X(t)^* \end{bmatrix},\quad t \in \Omega \quad (7.8.5)$$

where $Z(t) = (I + X(t)^*X(t))^{-1/2}$, is an analytic matrix function. In particular, $X(t) = (X(t)Z(t)) \cdot Z(t)^{-1}$ is analytic as well. $\square$

As in Theorem 7.8.2, it can be proved that the maps Ric and Ric^{-1} are C^p (p times continuously differentiable), where p is a positive integer or $p = \infty$. Here, we say that a subspace valued function $\mathcal{L}(t) \subseteq \mathbb{C}^m$, $t \in \Omega$, where Ω is an open connected set in $\mathbb{R}^k$, belongs to the class C^p if every entry of the orthogonal projector $P(t)$ on $\mathcal{L}(t)$ is p times continuously differentiable. In other words, let $P(t) = [a_{ij}(t)]_{i,j=1}^m$; then all partial derivatives

$$\frac{\partial^{\alpha_1 + \cdots + \alpha_k} a_{ij}(t)}{\partial t_1^{\alpha_1} \ldots \partial t_m^{\alpha_m}},\quad \sum_{j=1}^k \alpha_j \le p$$

exist and are continuous on Ω. We omit the statement and proof of the C^p analogue of Theorem 7.8.2.

7.9 Relaxing the controllability condition

In Sections 7.2–7.8 we have studied the CARE

$$XDX + XA + A^*X - C = 0$$

under the general working hypotheses that $D \ge 0$, $C = C^*$, the pair (A, D) is controllable and, as usual, $A, C, D \in \mathbb{C}^{n \times n}$. The most notable exception is

RELAXING THE CONTROLLABILITY CONDITION

the existence statement of Theorem 7.3.7 in which controllability of (A, D) is replaced by the weaker *sign controllability*. An intermediate condition of greater practical significance is *stabilizability* of (A, D) (see Section 4.4, and recall that "controllable" $\Rightarrow$ "stabilizable" $\Rightarrow$ "sign controllable"). In this section results are obtained under the stabilizable hypothesis using methods developed earlier in this chapter. This weaker condition will also be used in Chapter 9 where a different line of attack is developed.

It will be more convenient to modify the Riccati equation slightly and consider

$$XDX - XA - A^*X - C = 0. \tag{7.9.1}$$

Clearly, replacing A by $-A$ does not affect controllability, but it does have some effect on stabilizability (see the remarks at the end of the Section 4.5).

To take advantage of theory already developed using the controllability condition it is natural to use the control normal form for the pair (A, D) in which, as we have seen in Proposition 4.5.1, a (possibly smaller) controllable pair appears. So first consider the controllable subspace of (A, D) (cf. equation (4.1.2)):

$$\mathcal{C} := \operatorname{Im}[D \ AD \ A^2D \ \ldots \ A^{n-1}D] = \sum_{j=0}^{\infty} \operatorname{Im}(A^j D).$$

By Proposition 4.1.2, $\mathcal{C}$ is A-invariant and contains $\operatorname{Im} D$. Write $\mathbb{C}^n = \mathcal{C} \oplus \mathcal{C}^\perp$ and considering representations of A, D and C with respect to this orthogonal decomposition we may write

$$A = \begin{bmatrix} A_1 & A_{12} \\ 0 & A_2 \end{bmatrix}, \quad D = \begin{bmatrix} D_1 & 0 \\ 0 & 0 \end{bmatrix}, \quad C = \begin{bmatrix} C_1 & C_{12} \\ C_{12}^* & C_2 \end{bmatrix}, \tag{7.9.2}$$

where $D_1 \geq 0$, $C_1 = C_1^*$, $C_2 = C_2^*$ and (see Proposition 4.5.1) the pair (A_1, D_1) is controllable. Furthermore (see Proposition 4.5.2) A_2 is c-stable.

We write any hermitian $n \times n$ matrix X in a similar way:

$$X = \begin{bmatrix} X_1 & X_{12} \\ X_{12}^* & X_2 \end{bmatrix} \tag{7.9.3}$$

where $X_1 = X_1^*$, $X_2 = X_2^*$. Using these representations in (7.9.1) three equivalent equations are obtained:

$$X_1 D_1 X_1 - X_1 A_1 - A_1^* X_1 - C_1 = 0 \tag{7.9.4}$$

$$(A_1^* - X_1 D_1) X_{12} + X_{12} A_2 = -(C_{12} + X_1 A_{12}) \tag{7.9.5}$$

$$A_2^* X_2 + X_2 A_2 = X_{12}^* D_1 X_{12} - X_{12}^* A_{12} - A_{12}^* X_{12} - C_2 \tag{7.9.6}$$

and observe that the first of these is a Riccati equation with the familiar controllability condition on (A_1, D_1). Theorem 7.6.1 therefore applies to (7.9.4) and shows that existence of a hermitian solution X_1 is equivalent to any one of the three following conditions:

(a) There exists an m-dimensional, H_0-neutral, M_1-invariant subspace of $\mathcal{C} \oplus \mathcal{C}$ where $m = \dim \mathcal{C}$,

$$H_0 = i \begin{bmatrix} 0 & I \\ -I & 0 \end{bmatrix}, \quad M_1 = \begin{bmatrix} -A_1 & D_1 \\ C_1 & A_1^* \end{bmatrix} \tag{7.9.7}$$

and I is the identity map on $\mathcal{C}$.

(b) The partial multiplicities of real eigenvalues (if any) of iM_1 are all even.

(c) The partial multiplicities of real eigenvalues (if any) of iM_1 are all even and the sign characteristic of (iM_1, H_0) consists of $+1$'s only.

It turns out that these properties also determine the existence of hermitian solutions of our original equation (7.9.1).

Theorem 7.9.1. *If $D \geq 0$, $C^* = C$ and (A, D) is stabilizable, then equation (7.9.1) has a hermitian solution if and only if there is an m-dimensional, H_0-neutral invariant subspace of M_1 or, equivalently, one of the conditions* (b) *or* (c) *holds.*

Proof. Assume that (7.9.1) has a hermitian solution. By Theorem 7.5.1 there is a maximal solution X_1 for (7.9.4) and $A_1 - D_1 X_1$ has all eigenvalues in the closed left half-plane. Since A_2 is stable it follows (from Theorem 5.2.2) that (7.9.5) has a unique solution X_{12}. Then A_2 stable also yields a unique solution X_2 of (7.9.6). Thus, a hermitian solution X for (7.9.1) is obtained.

Conversely, if (7.9.1) is satisfied by a hermitian matrix X then, as we have seen, (7.9.4) follows. The existence of this solution X_1 and Theorem 7.6.1 give our conclusions. □

The extremal solutions of Riccati equations, when they exist, are particularly important and so are their "stabilizing" properties. For example, it has been seen in Theorem 7.5.1 that, under a controllability condition on (A, D), there exist both maximal and minimal solutions, X_+, and X_-, and X_+ is almost stabilizing in the sense that $\sigma(A - DX_+)$ is in the closed left half-plane. It is important to know when $A - DX_+$ is actually stable, i.e. when $\sigma(A - DX_+)$ is in the open left half-plane. When this is the case, the hermitian solution X_+ is said to be stabilizing (or c-stabilizing).

In this connection observe that, under no essential condition other than $D^* = D$ we have the simple uniqueness property:

Proposition 7.9.2. *If the CARE (7.9.1) has a stabilizing solution then it is unique in the class of almost stabilizing solutions.*

Proof. If X_1, X_2 are hermitian and

$$X_j D X_j - X_j A - A^* X_j - C = 0$$

for $j = 1$ and 2 then, after subtraction, it is found that

$$(X_1 - X_2)(A - DX_1) + (A - DX_2)^*(X_1 - X_2) = 0.$$

Now this is a Lyapunov equation and if $A - DX_1$ and $A - DX_2$ have all their eigenvalues in the closed left half-plane and at least one of them is c-stable, then $A - DX_1$ and $-(A - DX_2)^*$ have no eigenvalues in common. Thus, it follows from Theorem 5.2.2 that $X_1 - X_2 = 0$. □

In the following theorem it is shown that when (A, D) is merely stabilizable one can assert the existence of a maximal solution which is also almost stabilizing. Then it is shown by example that a minimal solution need not exist.

Theorem 7.9.3. *If $D \geq 0$, $C^* = C$, (A, D) is stabilizable, and there exists a hermitian solution of (7.9.1), then there is a maximal hermitian solution, X_+. Moreover, X_+ coincides with the unique solution of (7.9.1) for which $\sigma(A - DX_+)$ lies in the closed left half-plane.*

Proof. Using the representation (7.9.4)–(7.9.6) for the equation (7.9.1), for any solution $X = X^*$ of (7.9.1) there is the representation

$$A - DX = \begin{bmatrix} A_1 - D_1 X_1 & A_{12} - D_1 X_{12} \\ 0 & A_2 \end{bmatrix}. \qquad (7.9.8)$$

As A_2 is necessarily stable, $\sigma(A - DX)$ lies in the closed left half-plane if and only if $\sigma(A_1 - D_1 X_1)$ has this property. By Theorem 7.5.1 $\sigma(A_1 - D_1 X_1)$ lies in the closed left half-plane only if X_1 is the maximal solution of (7.9.4). Thus, existence and uniqueness of the almost stabilizing solution of (7.9.1) follows from the proof of Theorem 7.9.1 (indeed, we have seen there that, when X_1 is the maximal solution of (7.9.4), the equations (7.9.5) and (7.9.6) are uniquely solvable for X_{12} and X_2).

That the almost stabilizing solution X_+ is maximal will follow as a by-product of the constructive comparison theorems that will be proved in Chapter 9. □

In contrast, under the stabilizable condition on (A, D), the minimal hermitian solution of (7.9.1) need not exist:

Example 7.9.1. Let

$$D = \begin{bmatrix} 1 & 0 \\ 0 & 0 \end{bmatrix}, \quad A = \begin{bmatrix} 0 & 0 \\ 0 & -1 \end{bmatrix}, \quad C = \begin{bmatrix} 1 & 0 \\ 0 & 0 \end{bmatrix}.$$

A calculation shows that the hermitian solutions of (7.9.1) are

$$\begin{bmatrix} 1 & 0 \\ 0 & 0 \end{bmatrix}, \begin{bmatrix} -1 & b \\ \bar{b} & -|b|^2/2 \end{bmatrix}, \quad b \in \mathcal{C}.$$

Clearly, $\begin{bmatrix} 1 & 0 \\ 0 & 0 \end{bmatrix}$ is the maximal solution, but there is no minimal solution. □

The next result clarifies those situations in which $A - DX_+$ (of Theorem 7.9.3) is stable, i.e. X_+ is stabilizing in the strict sense.

Theorem 7.9.4. *Suppose that $D \geq 0$, $C^* = C$, (A, D) is stabilizable, and there is a hermitian solution of (7.9.1). Then for the maximal hermitian solution X_+ of (7.9.1), $A - DX_+$ is stable if and only if the matrix*

$$M := \begin{bmatrix} -A & D \\ C & A^* \end{bmatrix}$$

has no eigenvalues on the imaginary axis.

Observe that, from Theorem 7.2.8 (and being wary of changes in the meanings of A and M), the matrix M has no eigenvalues on the imaginary axis if, in addition, we have $C \geq 0$ and $(C, -A)$ is detectable.

Proof. Note first of all that X_+ exists by the preceding theorem. Suppose that M has no eigenvalues on the imaginary axis. Using the representations (7.9.2) it is easily seen that

$$\sigma(M) = \sigma(M_1) \cup \sigma(-A_2) \cup \sigma(A_2^*). \tag{7.9.9}$$

Now any solution X_1 of (7.9.4) has the property that $\sigma(A - DX_1) \subseteq \sigma(M_1)$. In particular, because (A_1, D_1) is controllable, (7.9.4) has a maximal solution X_{1+} and $\sigma(A_1 - D_1 X_{1+})$ must be stable. Let X_+ be the corresponding solution of (7.9.1) (obtained from equations (7.9.4)–(7.9.6)). It then follows from the triangular representation (7.9.8) that $A - DX_+$ is also stable.

Conversely, given that $A - DX_+$ is stable where X_+ is the maximal solution of (7.9.1), define X_{1+} as in (7.9.3). Since A_2 is stable it follows from (7.9.8) that $A_1 - D_1 X_{1+}$ is stable and also that X_{1+} is a maximal solution of (7.9.4). Hence M_1 has no eigenvalues on the imaginary axis and, from (7.9.9), the same is true of M. □

RELAXING THE CONTROLLABILITY CONDITION

Now, instead of focussing on extremal solutions of (7.9.1), we consider general hermitian solutions and develop characterizations of these solutions in geometric terms. It will be convenient to retain the notation introduced in (7.9.2), (7.9.3) and (7.9.7). Also, let P be the orthogonal projector onto the controllable subspace (considered as a map from $\mathbb{C}^n$ to $\mathcal{C}$).

The equation (7.9.1) will be called *regular* if for every hermitian solution X the matrices $-A|\mathcal{C} + PDPXP$ and A_2^* have no common eigenvalues.

A sufficient condition (which does not involve solutions) for regularity is that

$$\sigma(M_1) \cap \overline{\sigma(A_2)} = \emptyset. \tag{7.9.10}$$

Indeed, if (7.9.10) holds, and if $X = \begin{bmatrix} X_1 & X_{12} \\ X_{12}^* & X_2 \end{bmatrix}$ is a hermitian solution of (7.9.1), then

$$-A|\mathcal{C} + PDPXP = -A_1 + D_1 X_1$$

is similar to the restriction of M_1 to its invariant subspace $\operatorname{Im} \begin{bmatrix} I \\ X_1 \end{bmatrix}$, and therefore

$$\sigma(-A|\mathcal{C} + PDPXP) \cap \overline{\sigma(A_2)} = \emptyset. \tag{7.9.11}$$

It turns out that in the regular case there is a one-to-one correspondence between the set of hermitian solutions X of (7.9.1) and the set of M_1-invariant subspaces $\mathcal{L}$ such that $\mathcal{L} \subseteq \mathcal{L}_+$, where $\mathcal{L}_+$ is the spectral M_1-invariant subspace corresponding to its eigenvalues in the open right half-plane. Moreover, this correspondence is well-behaved with respect to topology, partial order and analyticity:

Theorem 7.9.5. *Given the hypotheses of Theorem 7.9.4, assume also that the CARE (7.9.1) is regular.*

(i) *For every M_1-invariant subspace $\mathcal{L} \subseteq \mathcal{L}_+$ there exists a unique hermitian solution $X = \varphi(\mathcal{L})$ of (7.9.1) satisfying*

$$\operatorname{Im} \begin{bmatrix} I \\ PXP \end{bmatrix} \cap \mathcal{L}_+ = \mathcal{L}, \tag{7.9.12}$$

where P is the orthogonal projector on $\mathcal{C}$. Conversely, for every hermitian solution X of (7.9.1) there is a unique M_1-invariant subspace $\mathcal{L} \subseteq \mathcal{L}_+$ such that (7.9.12) holds, i.e. $X = \varphi(\mathcal{L})$.

(ii) *The correspondence φ introduced in (i) is a homeomorphism between the set of hermitian solutions of (7.9.1) and the set of all M_1-invariant subspaces which are contained in $\mathcal{L}_+$, i.e. φ and φ^{-1} are continuous functions, when the set of M_1-invariant subspaces contained in $\mathcal{L}_+$ is considered as a metric space with the metric induced by the gap.*

(iii) Let $\mathcal{L}_1$ and $\mathcal{L}_2$ be M_1-invariant subspaces contained in $\mathcal{L}_+$, and let $X_1 = \varphi(\mathcal{L}_1)$, $X_2 = \varphi(\mathcal{L}_2)$ be the corresponding hermitian solutions of (7.9.1). Then $\mathcal{L}_1 \supseteq \mathcal{L}_2$ if and only if $X_1 \geq X_2$.

(iv) Let $\mathcal{L}(t)$, $t \in \Omega$ be an analytic subspace valued function defined on an open connected set $\Omega \subseteq \mathbb{R}^k$ such that for every $t \in \Omega$ the subspace $\mathcal{L}(t)$ is M_1-invariant and contained in $\mathcal{L}_+$. Then $X(t) = \varphi(\mathcal{L}(t))$ is an analytic $n \times n$ matrix function whose values are hermitian solutions of (7.9.1). Conversely, if $X(t)$, $t \in \Omega$, is an analytic matrix function whose values are hermitian solutions of (7.9.1), then the subspace valued function $\varphi^{-1}(X(t))$ is analytic on $t \in \Omega$ as well.

As in Section 7.8, analyticity of a subspace valued function is understood in the sense of analyticity of the orthogonal projector on that subspace. We also remark that a result similar to part (iv) of Theorem 7.9.5 holds for C^p (rather than analytic) dependence on t, where p is a positive integer or $p = \infty$.

Proof. (i) Write (7.9.1) in the equivalent form (7.9.4)–(7.9.6). By Theorem 7.6.2 there is a one-to-one correspondence between the set of hermitian solutions X_1 of (7.9.2), and the set of all M_1-invariant subspaces $\mathcal{L}$ which are contained in $\mathcal{L}_+$ given by the formula

$$\operatorname{Im} \begin{bmatrix} I \\ X_1 \end{bmatrix} \cap \mathcal{L}_+ = \mathcal{L}. \tag{7.9.13}$$

Further note that the restriction of M_1 to $\operatorname{Im} \begin{bmatrix} I \\ X_1 \end{bmatrix}$ is similar to $-(A_1 - D_1 X_1)$. Therefore the regularity condition (7.9.11) ensures that equation (7.9.5) can be uniquely solved for X_{12} for any hermitian solution X_1 of (7.9.4). Also (7.9.6) can be uniquely solved for X_2 (where X_2 is necessarily hermitian) in view of the fact that $\sigma(A_2)$ is contained in the open left half-plane (see Section 5.3).

(ii) Use Theorem 7.7.1 by which the correspondence between X_1 and $\mathcal{L}$ is continuous both ways. Furthermore, note that the solutions X_{12} and X_2 of (7.9.5) and (7.9.6) are also continuous provided X_1 is so (see Theorem 5.4.1).

(iv) The proof is similar to that of (ii) using the analyticity result of Theorem 7.8.2 and using Corollary 5.4.2 instead of Theorem 5.4.1.

(iii) By Theorem 7.6.4, the correspondence between hermitian solutions $X_1 = \psi(\mathcal{L})$ of (7.9.4) and the subspaces

$$\mathcal{L} = \operatorname{Im} \begin{bmatrix} I \\ X_1 \end{bmatrix} \cap \mathcal{L}_+$$

is partial order preserving in the sense that $\mathcal{L}_1 \supseteq \mathcal{L}_2$ holds if and only if $\psi(\mathcal{L}_1) - \psi(\mathcal{L}_2)$ is positive semidefinite. In view of this property, the proof of statement (iii) will follow from the following result which concerns the possibility of extending (or dilating) solutions of (7.9.4) to solutions of (7.9.1) in such a way as to retain their partial ordering. □

Lemma 7.9.6. *Let D_1, A_1, C_1, A_2, C_2, A_{12}, C_{12} be complex matrices of sizes $m \times m$, $m \times m$, $m \times m$, $p \times p$, $p \times p$, $m \times p$, $m \times p$, respectively, where D_1, C_1 and C_2 are hermitian. Assume that $\sigma(-A_2) \cap \overline{\sigma(A_2)} = \emptyset$, and that*

$$\sigma(-(A_1 - D_1 X_1)) \cap \overline{\sigma(A_2)} = \emptyset$$

for every hermitian solution X_1 of the equation

$$X_1 D_1 X_1 - X_1 A_1 - A_1^* X_1 - C_1 = 0. \qquad (7.9.14)$$

Then for every hermitian solution X_1 of (7.9.14) there are unique matrices X_{12} and X_2 such that

$$X = \begin{bmatrix} X_1 & X_{12} \\ X_{12}^* & X_2 \end{bmatrix}$$

satisfies the equation

$$X \begin{bmatrix} D_1 & 0 \\ 0 & 0 \end{bmatrix} X - X \begin{bmatrix} A_1 & A_{12} \\ 0 & A_2 \end{bmatrix} - \begin{bmatrix} A_1^* & 0 \\ A_{12}^* & A_2^* \end{bmatrix} X - \begin{bmatrix} C_1 & C_{12} \\ C_{12}^* & C_2 \end{bmatrix} = 0,$$
$$(7.9.15)$$

and if $X_1 \geq Y_1$ are hermitian solutions of (7.9.14) then the corresponding solutions X and Y of (7.9.15) satisfy $X \geq Y$.

We will not present the proof of Lemma 7.9.6. A complete proof is found in Gohberg et al. (1986b).

We conclude this section with some remarks concerning the minimal solution of (7.9.1). As we have seen in Example 7.9.1, the minimal solution does not always exist. It exists, however, if the equation is regular (indeed, by Theorem 7.9.5 (iii) the minimal solution is the one that corresponds to the zero subspace). More generally, we have the following result:

Theorem 7.9.7. *The following statements are equivalent for the CARE (7.9.1) under the hypotheses of Theorem 7.9.4 (we do not assume regularity):*

(i) *the set of hermitian solutions is compact;*

(ii) *there exists a minimal (hermitian) solution;*

(iii) *the equation is regular.*

Proof. We already observed that (iii) $\Rightarrow$ (ii). Assume now that (ii) holds, and let $X_- = [a_{ij}]_{i,j=1}^n$ and $X_+ = [b_{ij}]_{i,j=1}^n$ be the minimal and maximal solutions of (7.9.1), respectively. Then for any hermitian solution $X = [x_{ij}]_{i,j=1}^n$ we have $X_- \leq X \leq X_+$, which implies $a_{ii} \leq x_{ii} \leq b_{ii}$ for $i = 1, \ldots, n$, and

$$|x_{ij} - a_{ij}|^2 \leq (x_{ii} - a_{ii})(x_{jj} - a_{jj}) \leq (b_{ii} - a_{ii})(b_{jj} - a_{jj})$$

for $i \neq j$. These inequalities clearly show that the set of hermitian solutions is bounded. Since the closedness of this set is evident (if $\{X_m\}_{m=1}^\infty$ are hermitian solutions of (7.9.1) and $X_m \to X$ as $m \to \infty$, then X is a hermitian solution as well), we obtain (i).

It remains to prove that (i) $\Rightarrow$ (iii). Suppose (iii) is false, and let $X_0 = \begin{bmatrix} X_{01} & X_{012} \\ X_{012}^* & X_{02} \end{bmatrix}$ be a hermitian solution of (7.9.1) such that

$$\sigma(-A|\mathcal{C} + PDPX_0P) \cap \overline{\sigma(A_2)} \neq \emptyset,$$

which can be rewritten in the form $\sigma(-A_1 + D_1 X_{01}) \cap \sigma(A_2^*) \neq \emptyset$. By Theorem 5.2.2, the linear equation

$$(A_1^* - X_{01} D_1)Y + Y A_2 = -(C_{12} + X_{01} A_{12}) \qquad (7.9.16)$$

has more than one solution Y (the equation is consistent because X_{012} is a solution). Thus, the set of all solutions of (7.9.16) is of the form

$$X_{012} + \{\text{subspace of dimension} \geq 1\},$$

which is obviously noncompact. As each of the solutions Y of (7.9.16), together with the solution Z of

$$A_2^* Z + Z A_2 = Y^* D_1 Y - Y^* A_{12} - A_{12}^* Y - C_2$$

produces a hermitian solution $\begin{bmatrix} X_{01} & Y \\ Y^* & Z \end{bmatrix}$ of (7.9.1), we obtain a contradiction with (i). □

7.10 The LQR form and matrix pencils

The continuous algebraic Riccati equation takes a special form in the linear-quadratic regulator (LQR) problem (see Chapter 16). Let us write it here as

$$XDX - X\hat{A} - \hat{A}^* X - \hat{Q} = 0 \qquad (7.10.1)$$

where

$$\hat{A} = A - BR^{-1}C, \quad \hat{Q} = Q - C^* R^{-1} C, \quad D = BR^{-1}B^* \qquad (7.10.2)$$

A, Q are $n \times n$, B, C^* are $n \times m$, and R is $m \times m$, hermitian and nonsingular. Of course, $n \times n$ solutions X are sought for equation (7.10.1).

The Hamiltonian matrix M of Section 7.2 takes the form

$$M = i \begin{bmatrix} -\hat{A} & D \\ \hat{Q} & \hat{A}^* \end{bmatrix} \qquad (7.10.3)$$

and keeping the definitions (7.10.2) in mind we may write

$$(-i)M = \begin{bmatrix} -A & 0 \\ Q & A^* \end{bmatrix} - \begin{bmatrix} -B \\ C^* \end{bmatrix} R^{-1} [C \ B^*]$$

and the expression on the right suggests a Schur complement (see p. 46 of Lancaster and Tismenetsky (1985), for example). Indeed, this clue leads one to the dilation:

$$\begin{bmatrix} -\hat{A} & D & 0 \\ \hat{Q} & \hat{A}^* & 0 \\ 0 & 0 & R \end{bmatrix}$$
$$= \begin{bmatrix} 0 & -I & BR^{-1} \\ I & 0 & -C^*R^{-1} \\ 0 & 0 & I \end{bmatrix} \begin{bmatrix} Q & A^* & C^* \\ A & 0 & B \\ C & B^* & R \end{bmatrix} \begin{bmatrix} I & 0 & 0 \\ 0 & I & 0 \\ -R^{-1}C & -R^{-1}B^* & I \end{bmatrix}, \qquad (7.10.4)$$

(an idea that originates with Van Dooren (1981)).

If we define hermitian matrices

$$G = \begin{bmatrix} 0 & -iI & 0 \\ iI & 0 & 0 \\ 0 & 0 & 0 \end{bmatrix}, \quad H = \begin{bmatrix} Q & A^* & C^* \\ A & 0 & B \\ C & B^* & R \end{bmatrix}, \qquad (7.10.5)$$

equation (7.10.4) leads to a strict equivalence relation:

$$\begin{bmatrix} 0 & -I & BR^{-1} \\ I & 0 & -C^*R^{-1} \\ 0 & 0 & I \end{bmatrix} (\lambda G - H) \begin{bmatrix} I & 0 & 0 \\ 0 & I & 0 \\ -R^{-1}C & -R^{-1}B^* & I \end{bmatrix}$$
$$= (-i) \begin{bmatrix} \lambda I - M & 0 \\ 0 & -iR \end{bmatrix}. \qquad (7.10.6)$$

It follows immediately that, since R is nonsingular, $\lambda G - H$ is a *regular* matrix pencil (see Section 1.6). Furthermore, the finite spectrum of $\lambda G - H$ (including all partial multiplicities) simply reproduces the corresponding spectral properties of the Hamiltonian M (see the Appendix of Gohberg et al. (1986a), for example).

The objective is now to relate solutions of (7.10.1) to deflating subspaces of the pair (G, H), as described in Section 1.6. This description of solutions of the CARE avoids inversion of R, which provides a considerable numerical advantage. Also, it suggests possible generalizations for problems in which R is singular.

Our first result concerns solutions X of (7.10.1) which are not necessarily hermitian and imposes only very weak hypotheses on the coefficients. Note that it is not even assumed that Q is hermitian. This theorem is the basis for useful numerical algorithms for the solution of the CARE.

It will be convenient to extend the notion of a graph subspace as follows: a subspace $\mathcal{S}$ of $\mathbb{C}^{2n+m}$ will be called an *extended graph subspace* if it can be written in the form

$$\mathcal{S} = \operatorname{Im} \begin{bmatrix} I_n \\ X \\ Z \end{bmatrix}$$

for some $n \times n$ matrix X and an $m \times n$ matrix Z. Notice that an extended graph subspace is necessarily n-dimensional and could also be written in the form

$$\mathcal{S} = \operatorname{Im} \begin{bmatrix} X_1 \\ X_2 \\ X_3 \end{bmatrix} \qquad (7.10.7)$$

where X_1 is an $n \times n$ invertible matrix, X_2 is $n \times n$, and X_3 is $m \times m$.

Theorem 7.10.1. *Let R be hermitian and invertible. Then equation (7.10.1) has a solution if and only if there is a deflating subspace $\mathcal{S}_0$ for (G, H) of the form (7.10.7) which is an extended graph subspace. When this is the case, the matrix $X = X_2 X_1^{-1}$ is a solution of (7.10.1).*

Proof. If equation (7.10.1) has a solution X put $X_1 = I$, $X_2 = X$, $X_3 = -R^{-1}(B^*X + C)$ and form the extended graph subspace $\mathcal{S}_0$ as in (7.10.7). Then

$$H \begin{bmatrix} I \\ X \\ X_3 \end{bmatrix} = \begin{bmatrix} Q + A^*X - C^*R^{-1}(B^*X + C) \\ A - BR^{-1}(B^*X + C) \\ 0 \end{bmatrix} = \begin{bmatrix} \hat{Q} + \hat{A}^*X \\ \hat{A} - DX \\ 0 \end{bmatrix}$$

$$= \begin{bmatrix} -X \\ I \\ 0 \end{bmatrix} (\hat{A} - DX) = G \begin{bmatrix} I \\ X \\ X_3 \end{bmatrix} (-i)(\hat{A} - DX). \qquad (7.10.8)$$

It has already been observed that $\lambda G - H$ is regular, so it follows from Lemma 1.6.1 that $\mathcal{S}_0$ is deflating for (G, H).

Conversely, if $\mathcal{S}_0$ has the form (7.10.7) with X_1 invertible then define $X = X_2 X_1^{-1}$, $Z = X_3 X_1^{-1}$ and we may write

$$\mathcal{S}_0 = \operatorname{Im} \begin{bmatrix} I \\ X \\ Z \end{bmatrix}.$$

Clearly, $(\operatorname{Ker} G) \cap \mathcal{S}_0 = \{0\}$ and, when $\mathcal{S}_0$ is deflating for (G, H), it follows from Corollary 1.6.3 that there is an $n \times n$ matrix K such that

THE LQR FORM AND MATRIX PENCILS

$$G \begin{bmatrix} I \\ X \\ Z \end{bmatrix} K = H \begin{bmatrix} I \\ X \\ Z \end{bmatrix}.$$

Thus, as above, it is found that

$$\begin{bmatrix} -X \\ I \\ 0 \end{bmatrix} iK = \begin{bmatrix} \hat{Q} + \hat{A}^*X \\ \hat{A} - DX \\ 0 \end{bmatrix}$$

whence $iK = \hat{A} - DX$ and $-X\hat{A} + XDX = \hat{Q} + \hat{A}^*X$, which is equation (7.10.1). □

Since

$$\begin{bmatrix} I \\ X \\ Z \end{bmatrix}^* G \begin{bmatrix} I \\ X \\ Z \end{bmatrix} = i(X^* - X)$$

we immediately obtain:

Proposition 7.10.2. *The deflating subspace S_0 of Theorem 7.10.1 is G-neutral if and only if X is hermitian.*

A deflating subspace S_0 of the form (7.10.7) obviously satisfies (Ker G)∩S_0 = {0} and, since $\lambda G - H$ is regular, it follows from Corollary 1.6.3 that $HS_0 \subseteq GS_0$. Thus, if S_0 is G-neutral, as in the last proposition, then it is necessarily H-neutral as well.

Recall the notation

$$\hat{H} = \begin{bmatrix} 0 & iI \\ -iI & 0 \end{bmatrix}$$

used in Section 7.2 and let us make a connection with the geometrical ideas of Propositions 7.1.1 and 7.2.1.

Theorem 7.10.3. *The Hamiltonian matrix M of (7.10.3) has an n-dimensional, invariant, and $\hat{H}$-neutral subspace of the form $S = \text{Im} \begin{bmatrix} I \\ X \end{bmatrix}$, where X is hermitian, if and only if the pair (G, H) (of (7.10.5)) has a deflating subspace of the form*

$$S_e = \text{Im} \begin{bmatrix} I \\ X \\ -R^{-1}(C + B^*X) \end{bmatrix}$$

which is G-neutral.

Proof. Given the M-invariant, $\hat{H}$-neutral subspace $\mathcal{S}$ let K be the $n \times n$ matrix for which

$$i \begin{bmatrix} -\hat{A} & D \\ \hat{Q} & \hat{A}^* \end{bmatrix} \begin{bmatrix} I \\ X \end{bmatrix} = \begin{bmatrix} I \\ X \end{bmatrix} K. \tag{7.10.9}$$

Also, let $Z = -R^{-1}(C + B^*X)$. It follows from (7.10.4) that

$$H = \begin{bmatrix} 0 & I & C^*R^{-1} \\ -I & 0 & BR^{-1} \\ 0 & 0 & I \end{bmatrix} \begin{bmatrix} -\hat{A} & D & 0 \\ \hat{Q} & \hat{A}^* & 0 \\ C & B^* & R \end{bmatrix}$$

and, hence,

$$H \begin{bmatrix} I \\ X \\ Z \end{bmatrix} = \begin{bmatrix} 0 & I & C^*R^{-1} \\ -I & 0 & BR^{-1} \\ 0 & 0 & I \end{bmatrix} \begin{bmatrix} K \\ XK \\ 0 \end{bmatrix} (-i)$$

$$= \begin{bmatrix} 0 & -iI & 0 \\ iI & 0 & 0 \\ 0 & 0 & 0 \end{bmatrix} \begin{bmatrix} I \\ X \\ Z \end{bmatrix} K = G \begin{bmatrix} I \\ X \\ Z \end{bmatrix} K,$$

so that $\mathcal{S}_e$ is deflating for (G, H). By Proposition 7.10.2, $\mathcal{S}_e$ is also G-neutral.

Conversely, if it is given that

$$\begin{bmatrix} Q & A^* & C^* \\ A & 0 & B \\ C & B^* & R \end{bmatrix} \begin{bmatrix} I \\ X \\ Z \end{bmatrix} = \begin{bmatrix} 0 & -iI & 0 \\ iI & 0 & 0 \\ 0 & 0 & 0 \end{bmatrix} \begin{bmatrix} I \\ X \\ Z \end{bmatrix} K$$

for some K, it follows from the last row that $Z = -R^{-1}(C + B^*X)$. Then it is easily deduced from the first two rows that (7.10.9) follows so that $\mathcal{S}$ is M-invariant. The hypothesis that $\mathcal{S}_e$ is G-neutral trivially implies that $\mathcal{S}$ is $\hat{H}$-neutral and (by Proposition 7.10.2) that X is hermitian. $\square$

We can now write down an analogue of (part of) the fundamental existence Theorem 7.3.7 in terms of the pencil $\lambda G - H$.

Corollary 7.10.4. *If $R > 0$, $Q^* = Q$ and (A, B) is sign controllable then equation (7.10.1) has a hermitian solution if and only if there is a deflating n-dimensional $\mathcal{S}$ subspace for (G, H) which is G-neutral and for which $(\operatorname{Ker} G) \cap \mathcal{S} = \{0\}$.*

Proof. Clearly $R > 0$ implies $D = BR^{-1}B^* \geq 0$, and since $\operatorname{Im} B = \operatorname{Im}(BR^{-1}B) = \operatorname{Im} D$ the sign controllability of (A, D) follows from that of (A, B). Now it follows from the equivalence of statements (i) and (ii) of Theorem 7.3.7 that there is a hermitian solution X for (7.10.1) if and only if there is an n-dimensional M-invariant and $\hat{H}$-neutral subspace and, furthermore, such a subspace can always be assumed to be a graph subspace of the form $\operatorname{Im} \begin{bmatrix} I \\ X \end{bmatrix}$ (see Lemma 7.2.7). Thus, the Corollary becomes an immediate consequence of Theorem 7.10.3. $\square$

This Corollary provides a generalization of results of Hammarling and Singer (1984) if we recall the implications:

(A, B) controllable $\Rightarrow$ (A, B) c-stabilizable $\Rightarrow$ (A, B) sign controllable.

Concerning statement (iii) of Theorem 7.3.7, it follows from the strict equivalence of equation (7.10.6) that, when (7.10.1) has a hermitian solution, all (finite) real eigenvalues of $\lambda G - H$ (if any) have only even partial multiplicities. We omit translation of the sign-characteristic property to the context of matrix pencils.

7.A Appendix: the metric space of subspaces

In this appendix we recall the basic facts about the topological properties of the set of subspaces in $\mathbb{C}^n$. All the results (except for Theorem 7.A.9) and their proofs hold for the set of subspaces in $\mathbb{R}^n$ as well. Some results will be presented without proof in such cases a complete proof is found in Chapter 13 of Gohberg et al. (1986a).

We consider $\mathbb{C}^n$ endowed with the standard scalar product. If $x = (x_1, \ldots, x_n)^T$, $y = (y_1, \ldots, y_n)^T \in \mathbb{C}^n$, then $\langle x, y \rangle = \sum_{i=1}^n x_i \bar{y}_i$, and the corresponding norm is

$$\|x\| = \left(\sum_{i=1}^n |x_i|^2 \right)^{1/2}.$$

The norm of an $n \times n$ matrix A (or a transformation $A : \mathbb{C}^n \to \mathbb{C}^n$) is defined accordingly:

$$\|A\| = \max_{x \in \mathbb{C}^n \setminus \{0\}} \|Ax\|/\|x\|.$$

Now we introduce a concept that serves as a measure of distance between subspaces. The *gap* between subspaces $\mathcal{L}$ and $\mathcal{M}$ (in $\mathbb{C}^n$) is defined as

$$\theta(\mathcal{L}, \mathcal{M}) = \|P_{\mathcal{M}} - P_{\mathcal{L}}\| \qquad (7.A.1)$$

where $P_{\mathcal{L}}$ and $P_{\mathcal{M}}$ are the orthogonal projectors on $\mathcal{L}$ and $\mathcal{M}$, respectively. It is clear from the definition that $\theta(\mathcal{L}, \mathcal{M})$ is a *metric* in the set of all subspaces in $\mathbb{C}^n$; that is, $\theta(\mathcal{L}, \mathcal{M})$ enjoys the following properties: (a) $\theta(\mathcal{L}, \mathcal{M}) > 0$ if $\mathcal{L} \neq \mathcal{M}$, $\theta(\mathcal{L}, \mathcal{L}) = 0$; (b) $\theta(\mathcal{L}, \mathcal{M}) = \theta(\mathcal{M}, \mathcal{L})$; (c) $\theta(\mathcal{L}, \mathcal{M}) \leq \theta(\mathcal{L}, \mathcal{N}) + \theta(\mathcal{N}, \mathcal{M})$ (the triangle inequality).

It follows from (7.A.1) that

$$\theta(\mathcal{L}, \mathcal{M}) = \theta(\mathcal{L}^\perp, \mathcal{M}^\perp)$$

where $\mathcal{L}^\perp$ and $\mathcal{M}^\perp$ denote orthogonal complements. Indeed, $P_{\mathcal{L}^\perp} = I - P_{\mathcal{L}}$, so $\|P_{\mathcal{M}} - P_{\mathcal{L}}\| = \|P_{\mathcal{M}^\perp} - P_{\mathcal{L}^\perp}\|$.

In the following paragraphs denote by $S_\mathcal{L}$ the unit sphere in a nonzero subspace $\mathcal{L} \subseteq \mathbb{C}^n$, that is, $S_\mathcal{L} = \{x \in \mathcal{L} \mid \|x\| = 1\}$. We also need the concept of the distance of $d(x, Z)$ from $x \in \mathbb{C}^n$ to a set $Z \subseteq \mathbb{C}^n$. This is defined by $d(x, Z) = \inf_{t \in Z} \|x - t\|$.

Theorem 7.A.1. *Let $\mathcal{M}, \mathcal{L}$ be nonzero subspaces in $\mathbb{C}^n$. Then*

$$\theta(\mathcal{L}, \mathcal{M}) = \max\{\sup_{x \in S_\mathcal{M}} d(x, \mathcal{L}), \sup_{x \in S_\mathcal{L}} d(x, \mathcal{M})\}.$$

As an immediate consequence of this theorem we find that $\theta(\mathcal{L}, \mathcal{M}) \leq 1$ for any two subspaces $\mathcal{L}$ and $\mathcal{M}$ (this inequality is also trivially satisfied in the case when at least one of $\mathcal{L}$ and $\mathcal{M}$ is the zero subspace).

Theorem 7.A.2. *If $\mathcal{M}, \mathcal{L}$ are subspaces in $\mathbb{C}^n$, and P_1 and P_2 are projectors with $\operatorname{Im} P_2 = \mathcal{L}$ and $\operatorname{Im} P_2 = \mathcal{M}$, not necessarily orthogonal, then*

$$\theta(\mathcal{L}, \mathcal{M}) \leq \|P_1 - P_2\|.$$

Theorem 7.A.3. *If $\theta(\mathcal{L}, \mathcal{M}) < 1$, then $\dim \mathcal{L} = \dim \mathcal{M}$.*

We have already seen that the set $\mathbf{G}(\mathbb{C}^n)$ of all subspaces in $\mathbb{C}^n$ is a metric space with respect to the gap $\theta(\mathcal{L}, \mathcal{M})$. Now we investigate some topological properties of $\mathbf{G}(\mathbb{C}^n)$, that is, those properties that depend on convergence (or divergence) in the sense of the gap metric.

Theorem 7.A.4. *The metric space $\mathbf{G}(\mathbb{C}^n)$ is compact and, therefore, complete (as a metric space).*

Recall that compactness of $\mathbf{G}(\mathbb{C}^n)$ means that for every sequence $\mathcal{L}_1, \mathcal{L}_2, \ldots$, of subspaces in $\mathbf{G}(\mathbb{C}^n)$ there exists a converging subsequence $\mathcal{L}_{i_1}, \mathcal{L}_{i_2}, \ldots$, that is, such that

$$\lim_{k \to \infty} \theta(\mathcal{L}_{i_k}, \mathcal{L}_0) = 0$$

for some $\mathcal{L}_0 \in \mathbf{G}(\mathbb{C}^n)$. Completeness of $\mathbf{G}(\mathbb{C}^n)$ means that every sequence of subspaces $\mathcal{L}_i$, $i = 1, 2, \ldots$, for which $\lim_{i, j \to \infty} \theta(\mathcal{L}_i, \mathcal{L}_j) = 0$ is convergent to some subspace.

The following characterization of limits in $\mathbf{G}(\mathbb{C}^n)$ is very useful.

Theorem 7.A.5. *Let $\mathcal{M}_1, \mathcal{M}_2, \ldots$ be a sequence of subspaces in $\mathbf{G}(\mathbb{C}^n)$, such that $\theta(\mathcal{M}_p, \mathcal{M}) \to 0$ as $p \to \infty$ for some subspace $\mathcal{M} \subseteq \mathbb{C}^n$. Then $\mathcal{M}$ consists of exactly those vectors $x \in \mathbb{C}^n$ for which there exists a sequence of vectors $x_p \in \mathbb{C}^n$, $p = 1, 2, \ldots$ such that $x_p \in \mathcal{M}_p$, $p = 1, 2, \ldots$ and $x = \lim_{p \to \infty} x_p$.*

We are particularly interested in invariant and semidefinite subspaces.

APPENDIX: THE METRIC SPACE OF SUBSPACES

Corollary 7.A.6. *Let $\{M_m\}_{m=1}^{\infty}$ be a sequence of $n \times n$ matrices having the limit $M = \lim_{m \to \infty} M_m$. For $m = 1, 2, \ldots$ let $\mathcal{M}_m$ be an M_m-invariant subspace, and assume that there is a limit $\mathcal{M} = \lim_{m \to \infty} \mathcal{M}_m$ (i.e. $\theta(\mathcal{M}_m, \mathcal{M}) \to 0$ as $m \to \infty$). Then $\mathcal{M}$ is M-invariant, and*

$$\dim \mathcal{M} = \dim \mathcal{M}_m \qquad (7.A.2)$$

(at least for m large enough).

Proof. The equality (7.A.2) follows immediately from Theorem 7.A.3. To prove the M-invariance of $\mathcal{M}$, let $x \in \mathcal{M}$. Using Theorem 7.A.5, form a sequence $\{x_m\}_{m=1}^{\infty}$, $x_m \in \mathcal{M}_m$ such that $\lim_{m \to \infty} x_m = x$. Now

$$\|M_m x_m - Mx\| \leq \|M_m x_m - M_m x\| + \|M_m x - Mx\|$$
$$\leq \left(\max_{m \geq 1} \|M_m\|\right) \|x_m - x\| + \|M - M_m\| \, \|x\|,$$

and therefore $\lim_{m \to \infty} M_m x_m = Mx$. As $\mathcal{M}_m$ is M_m-invariant, we have $M_m x_m \in \mathcal{M}_m$, and again by Theorem 7.A.5, $Mx \in \mathcal{M}$. □

Corollary 7.A.7. *Let H be an invertible hermitian $n \times n$ matrix, and let $\{H_m\}_{m=1}^{\infty}$ be a sequence of hermitian matrices converging to H. For every $m = 1, 2, \ldots$ let $\mathcal{M}_m$ be a maximal H_m-nonnegative subspace. If $\mathcal{M} = \lim_{m \to \infty} \mathcal{M}_m$, then $\mathcal{M}$ is a maximal H-nonnegative subspace.*

Proof. For m large enough, H_m is invertible and has the same number of positive eigenvalues (counted with multiplicities) as H. Therefore, the maximal dimension of an H_m-nonnegative subspace is the same as that for H (if m is large enough). So we only have to prove that $\mathcal{M}$ is H-nonnegative. Let $x \in \mathcal{M}$, and (using Theorem 7.A.5) find a sequence $\{x_m\}_{m=1}^{\infty}$, $x_m \in \mathcal{M}_m$ such that $x = \lim_{m \to \infty} x_m$. Now

$$\langle Hx, x \rangle = x^* H x = \lim_{m \to \infty} (x_m^* H_m x_m) = \lim_{m \to \infty} \langle H_m x_m, x_m \rangle,$$

and the limit is nonnegative because $\langle H_m x_m, x_m \rangle \geq 0$ for all m. □

Results analogous to Corollary 7.A.7 also hold for the following classes of subspaces: H-nonnegative, H-neutral, H-nonpositive, maximal H-nonpositive. The proof is essentially the same as that of Corollary 7.A.7.

We restate Corollary 7.A.7 for real skew-symmetric matrices and corresponding neutral subspaces:

Corollary 7.A.8. *Let $\{H_m\}_{m=1}^{\infty}$ be a sequence of real $n \times n$ skew-symmetric matrices such that $H = \lim_{m \to \infty} H_m$ exists and is invertible (H is necessarily skew-symmetric). If $\mathcal{M}_m \subseteq \mathbb{R}^n$ is an H_m-neutral subspace for $m = 1, 2, \ldots$ and if $\mathcal{M} = \lim_{m \to \infty} \mathcal{M}_m$, then $\mathcal{M}$ is H-neutral.*

Again, the proof is essentially the same as that of Corollary 7.A.7.

Our next result compares the gap between two invariant subspaces and the gap between their intersections with spectral subspaces.

Theorem 7.A.9. *Let M be an $n \times n$ (complex) matrix. Partition the set of eigenvalues of M into a union of disjoint sets:*

$$\sigma(M) = \Omega_1 \cup \ldots \cup \Omega_r, \quad \Omega_i \cap \Omega_j = \emptyset \ \text{if} \ i \neq j \qquad (7.A.3)$$

and let $\mathcal{R}_i \subseteq \mathbf{C}^n$ be the spectral M-invariant subspace corresponding to the eigenvalues in Ω_i ($i = 1, \ldots, r$). Then there exist positive constants C_1 and C_2 (depending on M only) such that for any two M-invariant subspaces $\mathcal{M}$ and $\mathcal{N}$ the inequalities

$$C_1 \sum_{i=1}^{r} \theta(\mathcal{M} \cap \mathcal{R}_i, \mathcal{N} \cap \mathcal{R}_i) \leq \theta(\mathcal{M}, \mathcal{N}) \leq C_2 \sum_{i=1}^{r} \theta(\mathcal{M} \cap \mathcal{R}_i, \mathcal{N} \cap \mathcal{R}_i)$$

hold.

Proof. Observe that it is sufficient to prove the theorem for a matrix M_0 similar to M rather than for M itself. Indeed, let $M_0 = S^{-1}MS$ for an invertible S; then $\mathcal{M}$ is an M-invariant subspace if and only if $S^{-1}\mathcal{M}$ is M_0-invariant, and $S^{-1}\mathcal{R}_i$ is the spectral M_0-invariant subspace corresponding to the eigenvalues of M_0 lying in Ω_i. Further, let $P_\mathcal{M}$ (resp. $P_\mathcal{N}$) be the orthogonal projector on $\mathcal{M}$ (resp. $\mathcal{N}$). Then by Theorem 7.A.2

$$\theta(S^{-1}\mathcal{M}, S^{-1}\mathcal{N}) \leq \|S^{-1}P_\mathcal{M}S - S^{-1}P_\mathcal{N}S\|$$
$$\leq \|S\| \cdot \|S^{-1}\| \cdot \|P_\mathcal{M} - P_\mathcal{N}\| = \|S\| \cdot \|S^{-1}\|\theta(\mathcal{M}, \mathcal{N}), \qquad (7.A.4)$$

and similarly

$$\theta(\mathcal{M}, \mathcal{N}) \leq \|S\| \cdot \|S^{-1}\|\theta(S^{-1}\mathcal{M}, S^{-1}\mathcal{N}). \qquad (7.A.5)$$

In view of (7.A.4) and (7.A.5) we can consider $M_0 = S^{-1}MS$ in place of M, for any choice of the invertible matrix S.

Observe that

$$\mathbf{C}^n = \mathcal{R}_1 \dotplus \cdots \dotplus \mathcal{R}_r$$

is a direct sum decomposition. Therefore, we can choose S in such a way that the subspaces $\mathcal{R}_{0i} = S^{-1}\mathcal{R}_i$ ($i = 1, \ldots, r$) are orthogonal to each other. Let $M_0 = S^{-1}MS$. For any M_0-invariant subspace $\mathcal{M}_0$ we have

$$\mathcal{M}_0 = (\mathcal{M}_0 \cap \mathcal{R}_{01}) \oplus \cdots \oplus (\mathcal{M}_0 \cap \mathcal{R}_{0r})$$

where the sum is orthogonal. Thus,

$$P_{\mathcal{M}_0} = \sum_{i=1}^{r} P_{\mathcal{M}_0 \cap \mathcal{R}_{0i}},$$

where P_χ stands for the orthogonal projector on the subspace χ. Consequently, for M_0-invariant subspaces $\mathcal{M}_0$ and $\mathcal{N}_0$,

$$\theta(\mathcal{M}_0, \mathcal{N}_0) = \|P_{\mathcal{M}_0} - P_{\mathcal{N}_0}\| = \|\sum_{i=1}^{r}(P_{\mathcal{M}_0 \cap \mathcal{R}_{0i}} - P_{\mathcal{N}_0 \cap \mathcal{R}_{0i}})\|$$

$$\leq \sum_{i=1}^{r} \|P_{\mathcal{M}_0 \cap \mathcal{R}_{0i}} - P_{\mathcal{N}_0 \cap \mathcal{R}_{0i}}\| = \sum_{i=1}^{r} \theta(\mathcal{M}_0 \cap \mathcal{R}_{0i}, \mathcal{N}_0 \cap \mathcal{R}_{0i}). \quad (7.\text{A}.6)$$

Conversely, let j be such that

$$\|P_{\mathcal{M}_0 \cap \mathcal{R}_{0j}} - P_{\mathcal{N}_0 \cap \mathcal{R}_{0j}}\| = \max_{1 \leq i \leq r} \|P_{\mathcal{M}_0 \cap \mathcal{R}_{0i}} - P_{\mathcal{N}_0 \cap \mathcal{R}_{0i}}\|,$$

and let $x \in \mathcal{M}_0 \cap \mathcal{R}_{0j}$, $\|x\| = 1$, be such that

$$\|P_{\mathcal{M}_0 \cap \mathcal{R}_{0j}} - P_{\mathcal{N}_0 \cap \mathcal{R}_{0j}}\| = \|(P_{\mathcal{M}_0 \cap \mathcal{R}_{0j}} - P_{\mathcal{N}_0 \cap \mathcal{R}_{0j}})x\|.$$

Now

$$\sum_{i=1}^{r} \|P_{\mathcal{M}_0 \cap \mathcal{R}_{0i}} - P_{\mathcal{N}_0 \cap \mathcal{R}_{0i}}\| \leq r\|(P_{\mathcal{M}_0 \cap \mathcal{R}_{0j}} - P_{\mathcal{N}_0 \cap \mathcal{R}_{0j}})x\|$$

$$\leq r\|\sum_{i=1}^{r}(P_{\mathcal{M}_0 \cap \mathcal{R}_{0i}} - P_{\mathcal{N}_0 \cap \mathcal{R}_{0i}})x\| \leq r\|\sum_{i=1}^{r}(P_{\mathcal{M}_0 \cap \mathcal{R}_{0i}} - P_{\mathcal{N}_0 \cap \mathcal{R}_{0i}})\|,$$

so, as in (7.A.6), we obtain

$$r^{-1}\sum_{i=1}^{r} \theta(\mathcal{M}_0 \cap \mathcal{R}_{0i}, \mathcal{N}_0 \cap \mathcal{R}_{0i}) \leq \theta(\mathcal{M}_0, \mathcal{N}_0).$$

This proves Theorem 7.A.9 for $M_0 = S^{-1}MS$. □

A useful particular case of Theorem 7.A.9 deserves a separate statement:

Corollary 7.A.10. *Let M be an $n \times n$ matrix, and let $\mathcal{R}$ be a spectral M-invariant subspace corresponding to a subset of $\sigma(M)$. Then there is a $C > 0$ (depending on M only) such that for any two M-invariant subspaces $\mathcal{M}$ and $\mathcal{N}$ the inequality*

$$\theta(\mathcal{M} \cap \mathcal{R}, \mathcal{N} \cap \mathcal{R}) \leq C\theta(\mathcal{M}, \mathcal{N})$$

holds.

If M is an $n \times n$ real matrix, then the result analogous to Theorem 7.A.9 holds for real M-invariant subspaces, provided the partition (7.A.3) is such that Ω_j are symmetric relative to the real axis: $\lambda_0 \in \Omega_j$ implies $\bar{\lambda}_0 \in \Omega_j$. The statement of the real analogue of Theorem 7.A.9 is left to the reader; its proof is the same.

7.11 Exercises

7.11.1. Prove that the set of solutions of the Riccati equation (7.1.1) is either finite (possibly empty) or a continuum, and has a finite number of connected components.

7.11.2. Prove that every invertible matrix A has a square root, i.e. there is a matrix X such that $X^2 = A$.

7.11.3. Find the Jordan forms of 4×4 nilpotent matrices that have a square root. (A necessary and sufficient condition for existence of a square root of a nilpotent matrix is given by Cross and Lancaster (1974)).

7.11.4. Construct the matrix M (of Section 7.2) for a scalar equation $dx^2 + (a+\bar{a})x - c = 0$, where $d > 0$ and c are real numbers, and verify directly the result of Theorem 7.3.7 for this case.

7.11.5. State and prove the C^p analogue of Theorem 7.8.2.

7.11.6. Consider the CARE (7.2.1) with

$$D = \begin{bmatrix} 1 & 1 \\ 1 & 1 \end{bmatrix}, \quad A = \begin{bmatrix} 0 & -1 \\ 1 & 0 \end{bmatrix}, \quad C = \begin{bmatrix} 1 & 1 \\ 1 & 1 \end{bmatrix}.$$

(a) Find all hermitian solutions X of this equation.

(b) Identify the maximal and the minimal hermitian solutions.

(c) Find all solutions X of this CARE satisfying

$$(X^* - X)(A + DX) \leq 0.$$

(d) Find all solutions X satisfying $(X^* - X)(A + DX) \geq 0$.

(In this example, the matrix M of Section 7.2 has no real eigenvalues but is not diagonable.)

7.11.7. Repeat Exercise 7.11.6, but with

$$D = \begin{bmatrix} 1 & 0 \\ 0 & 1 \end{bmatrix}, \quad A = \begin{bmatrix} 0 & 0 \\ 0 & -1 \end{bmatrix}, \quad C = \begin{bmatrix} 1 & 0 \\ 0 & 3 \end{bmatrix}.$$

In this example, the matrix M has no real eigenvalues and is diagonable.

7.11.8. Repeat Exercise 7.11.6, but with

$$D = \begin{bmatrix} 1 & 1 \\ 1 & 1 \end{bmatrix}, \quad A = \begin{bmatrix} 0 & 0 \\ 0 & 0 \end{bmatrix}, \quad C = \begin{bmatrix} 1 & 0 \\ 0 & 0 \end{bmatrix}.$$

In this example, some of the eigenvalues of M are real. Note that the pair (A, D) is not sign-controllable.

7.11.9. Find all hermitian solutions of the CARE (7.2.1) with
$$D = \begin{bmatrix} 1 & 0 \\ 0 & 0 \end{bmatrix}, \quad A = C = 0.$$
Note that maximal and minimal hermitian solutions do not exist.

7.11.10. Find maximal and minimal hermitian solutions X_+ and X_-, respectively, for the CARE (7.2.1), with
$$D = \begin{bmatrix} 1 & 0 \\ 0 & 1 \end{bmatrix}, \quad A = \begin{bmatrix} -1 & 0 \\ 0 & 1 \end{bmatrix}, \quad C = 0.$$
In this example, $X_+ \geq 0 \geq X_-$, both X_+ and X_- are singular; however, the difference $X_+ - X_-$ is positive definite.

7.11.11. Prove the following generalization of a part of Lemma 7.3.3 (with the sign $-$). (This was originally proved by Wimmer (1982).) For an $n \times n$ rational matrix function $H(z) = [h_{ik}(z)]_{i,k=1}^n$, denote $\tilde{H}(z) = [\bar{h}_{ki}(-z)]_{i,k=1}^n$. Let $G(z)$ and $D(z)$ be $n \times n$ rational matrix functions which have no pole at α, $\alpha \in i\mathbb{R}$, and satisfy the following conditions: $\det G(z) \not\equiv 0$; $D = \tilde{D}$; $D(\alpha) \geq 0$; $\text{rank}\,[G(\alpha) \; D(\alpha)] = n$. Then the partial multiplicities of
$$M(z) = \begin{bmatrix} G(z) & D(z) \\ 0 & -\tilde{G}(z) \end{bmatrix}$$
at α are twice the partial multiplicities of $G(z)$ at α.

Hints for the proof: Use the local Smith form for $G(z)$ (Theorem 6.2.1) to reduce the proof to the case when $G(z)$ is diagonal:
$$G(z) = \text{diag}\,[(z-\alpha)^{k_1}, \ldots, (z-\alpha)^{k_r}]; \quad 0 < k_1 \leq \ldots \leq k_r.$$
Further, show that $D(\alpha) > 0$. Because of
$$\begin{bmatrix} D^{-1} & 0 \\ \tilde{G}D^{-1} & I \end{bmatrix} \begin{bmatrix} G & D \\ 0 & -\tilde{G} \end{bmatrix} \begin{bmatrix} I & 0 \\ -D^{-1}G & I \end{bmatrix} = \begin{bmatrix} 0 & I \\ \tilde{G}D^{-1}G & 0 \end{bmatrix},$$
we can consider $W(z) := \tilde{G}(z)D(z)^{-1}G(z)$ in place of $M(z)$. Now $D(\alpha)^{-1} > 0$ implies that the principal minors of $W(z)$ are not identically zero, and therefore $r = 2k_1 + \cdots + 2k_m$ is the largest exponent with the property that $(z-\alpha)^r$ divides all $m \times m$ minors of $W(z)$.

7.11.12. State and prove a result analogous to Exercise 7.11.11 concerning rational matrix functions of the form $\begin{bmatrix} G(z) & D(z) \\ 0 & (G(\bar{z}))^* \end{bmatrix}$ (a generalization of the "+ part" of Lemma 7.3.3).

7.11.13. We denote by $\mathcal{X}_X$ the characteristic polynomial of a matrix X. Let $N = \begin{bmatrix} A & D \\ C & -A^* \end{bmatrix}$, and assume that

$$\mathcal{X}_N(z) = (-1)^n q(z)\overline{q(-\bar{z})}, \tag{7.11.1}$$

where the polynomial $q(z)$ is such that both α and $-\bar{\alpha}$ are roots of $q(z)$ only if $\alpha \in i\mathbb{R}$. Prove that, under the hypotheses that $D \geq 0$, $C = C^*$ and (A, D) is controllable, the CARE (7.2.1) has a unique hermitian solution X such that $\mathcal{X}_{A+DX} = q$ if and only if all partial multiplicities of N corresponding to its pure imaginary (or zero) eigenvalues are even.

Hints: The "only if" part follows from the implication (i) $\Rightarrow$ (iii) of Theorem 7.3.7. Conversely, assume all pure imaginary (or zero) eigenvalues of N have only even partial multiplicities. By Theorem 2.5.1 there exists an n-dimensional N-invariant $\hat{H}$-neutral subspace, and by Theorem 7.2.4 the CARE has a hermitian solution. Finally, existence of a unique hermitian solution X with $\mathcal{X}_{A+DX} = q$ follows from Theorem 7.4.1.

7.11.14. [We use the notation of Execise 7.11.13.] Let $q(z)$ be a polynomial as in Exercise 7.11.13. Prove that, under the hypotheses that $D \geq 0$ and $C = C^*$, the CARE has a unique hermitian solution X such that $\mathcal{X}_{A+DX} = q$ if and only if (A, D) is sign controllable and all partial multiplicities of N corresponding to its pure imaginary (or zero) eigenvalues are even.

Hint: Reduce to the case of controllable (A, D) (Exercise 7.11.13) by using (7.9.4) and (7.9.5) (replacing there $-A$ by A).

7.11.15. Let $\mathcal{R}(X) = XDX - XA - A^*X - C$ with $D^* = D$, $C \geq 0$ and (C, A) observable. Show that all hermitian solutions of $\mathcal{R}(X) = 0$ (if any) are nonsingular.

7.11.16. Consider the CARE $XDX + XA + A^*X - C = 0$ in which $D^* = D$ and is invertible, $DA^* = AD$ (A is D-self-adjoint), $C^* = C$ and $C + A^*D^{-1}A = \beta^2 D^{-1}$, $\beta > 0$. With M defined as in Section 7.2 show that $\sigma(M) = \{i\beta, -i\beta\}$ and that (if $n > 1$) the CARE has a continuum of solutions which includes the two hermitian matrices $X_{\pm} = -D^{-1}A \pm \beta D^{-1}$. Show also that, if $D > 0$ then X_+, X_- are extremal solutions.

7.12 Notes

The geometric theory developed in this chapter can be traced back to Potter (1966), and Theorem 7.1.2 contains his basic ideas. Important papers developing this idea are by Wonham (1968), Martensson (1971), Willems (1971), Kučera

(1972a), and Coppel (1974). The necessary and sufficient conditions for existence of a solution of the CARE of Theorem 7.2.4 evolved from this approach and appeared in papers of Curilov (1978 and 1979) and Lancaster and Rodman (1980). The latter paper includes results like Theorem 7.3.7 but with the stronger controllability hypothesis.

The sign characteristic of the pair (M, H) (equivalently, of the pair $(M, \hat{H})$; see Theorems 7.3.5, 7.3.7, 7.6.1) was first computed in Rodman (1983), under the controllability hypothesis. The notion of "sign controllability" was introduced and exploited by Faibusovich (1986). A related notion of "nabla stabilizability" was introduced earlier by Curilov (1978). Here, the proof of the key Lemma 7.3.2 is taken from Gohberg et al. (1979) (see also Gohberg et al. (1986b)).

The exposition of Theorem 7.5.1 and its proof follow Section II.4.6 of Gohberg et al. (1983). Theorems 7.5.3 and 7.5.4 are due to Coppel (1974) and generalize earlier results of Willems (1971). Our exposition uses Coppel's arguments. Theorem 7.6.2 was proved in Rodman (1981). The proof of Theorem 7.6.4 follows the exposition of Ran and Rodman (1984a). The material on continuity and analyticity of solutions (Sections 7.7 and 7.8) is based on ideas and works of Shayman (1983) and Ran and Rodman (1984a,b).

The main results of Section 7.9 are taken from Gohberg et al. (1986b). The introduction of matrix pencils, as described in Section 7.10, in the analysis of the CARE was initiated by Van Dooren (1981). The introduction of the sign-controllability condition in this context, as in Corollary 7.10.4, seems to be new. See Mehrmann (1991) for further literature and numerical methods.

The Riccati equations of Exercises 7.11.6, 7.11.7, 7.11.10 are taken from the paper of Rodriguez–Canabal (1973). A more detailed analysis of Exercises 7.11.6, 7.11.7, as well as Exercises 7.11.8 and 7.11.9 is found in Kučera (1991). Full proofs of the statements of Exercises 7.11.13 and 7.11.14 (originally due to Curilov (1978) and Wimmer (1984)) are also given by Kučera (1991). The CARE of Exercise 7.11.16 arises in a paper of Elsner et al. (1994).

Review articles containing material of this chapter include Kučera (1973 and 1991), Singer and Hammarling (1983), Chapter 4 of Gohberg et al. (1983), Ran and Rodman (1984a), Ando (1988), and Lancaster and Rodman (1991). Note also the monograph of Mehrmann (1991).

8
GEOMETRIC THEORY: THE REAL CASE

Using the geometric methods developed in Chapter 7 we now study symmetric Riccati equations of the form (7.2.1) for which the coefficient matrices A, C, D are assumed to be real and, accordingly, only real solutions X are sought. The theory will parallel that of Chapter 7 and so, when the real case does not differ significantly from the complex case already investigated, proofs may be omitted or merely outlined. However, for some purposes it will be necessary to call on the analysis of real matrix pairs carried out in Section 2.7 and Chapter 3.

8.1 Solutions and invariant subspaces

Let $A, C, D \in \mathbb{R}^{n \times n}$ and $C^T = C$, $D^T = D$. Consider the CARE

$$XDX + XA + A^T X - C = 0 \tag{8.1.1}$$

and define the $2n \times 2n$ real matrices

$$M_r = \begin{bmatrix} A & D \\ C & -A^T \end{bmatrix}, \quad \hat{H}_r = \begin{bmatrix} 0 & I \\ -I & 0 \end{bmatrix}, \quad H_r = \begin{bmatrix} -C & A^T \\ A & D \end{bmatrix}, \tag{8.1.2}$$

where the subscript r denotes "real". Then $\hat{H}_r M_r = -M_r^T \hat{H}_r$, $H_r M_r = -M_r^T H_r$, and $H_r = -\hat{H}_r M_r$.

Since $\hat{H}_r^T = -\hat{H}_r$ and $\hat{H}_r M_r = -M_r^T \hat{H}_r$, M_r is $\hat{H}_r$-skew-symmetric in the sense of Section 3.2. Also, $H_r^T = H_r$ and $H_r M_r = -M_r^T H_r$ implies that M_r is H_r-skew-symmetric in the sense of Section 2.7, provided that H_r is nonsingular. These observations open up two different, but closely connected, lines of investigation which take advantage of the real canonical forms of Theorems 3.2.4 and 2.7.1, respectively.

The notion of "graph subspace" introduced in Section 7.1 continues to play a fundamental role and, as we are concerned with real solutions X of (8.1.1), they will now be subspaces of $\mathbb{R}^{2n}$. In particular, it is clear that X is a (real) solution of (8.1.1) if and only if

$$G(X) := \operatorname{Im} \begin{bmatrix} I \\ X \end{bmatrix} \subseteq \mathbb{R}^{2n}$$

is M_r-invariant (cf. Proposition 7.1.1).

For the distinction between symmetric and nonsymmetric solutions the following proposition is vital (cf. Proposition 7.2.1). Note the different roles of H_r and $\hat{H}_r$.

Proposition 8.1.1. *Let X be a real solution of* (8.1.1). *Then*

(α) *X is symmetric if and only if $G(X)$ is $\hat{H}_r$-neutral*

(β) *The subspace $G(X)$ is H_r-nonpositive (resp. H_r-nonnegative) if and only if*

$$(X^T - X)(A + DX) \leq 0 \qquad (8.1.3)$$

(resp. $(X^T - X)(A + DX) \geq 0$).

The inequality (8.1.3) means that the real matrix $(X^T - X)(A + DX)$ is symmetric and negative semidefinite. The proof of the proposition is straightforward.

When applying statement (β) the theory diverges from the complex case because it can no longer be assumed without loss of generality that H_r (or M_r) is invertible. We get away with this in the complex case by using a transformation $A \to A + i\alpha I$ where $\alpha \in \mathbb{R}$. This transformation is not available for the real theory.

To develop a theory parallel to that of Chapter 7 using the H_r-scalar product we need to *assume* that H_r is invertible and we also impose a condition to ensure that H_r has n positive eigenvalues and n negative eigenvalues. This will appear as Proposition 8.1.4, followed by the main theorem of this section (Theorem 8.1.5), which is the real analogue of Theorem 7.2.3. But we also need some technical preparations.

First recall Theorem 2.1.3 which implies that for any nonsingular real symmetric matrix H the maximal dimension of an H-neutral subspace is $\min(k, \ell)$ where H has k and ℓ positive and negative eigenvalues, respectively.

Lemma 8.1.2. *Among the following statements, (ii) and (iii) are equivalent, and each of them implies (i):*

(i) *there exists an n-dimensional M_r-invariant H_r-neutral subspace;*

(ii) *there exists an n-dimensional M_r-invariant $\hat{H}_r$-neutral subspace;*

(iii) *for every nonzero pure imaginary eigenvalue ib of M_r (if any) we have $\nu([z_i^T H_r z_j]_{i,j=1}^p) = \frac{p}{2}$, where $z_1, \ldots, z_p$ is a basis in the even dimensional subspace $\mathcal{R}_{\pm ib}(M_r)$.*

If, in addition, M_r is invertible, then all three statements (i) – (iii) are equivalent (in fact, an M_r-invariant subspace is H_r-neutral if and only if it is $\hat{H}_r$-neutral).

Proof. First consider statement (iii) and apply Theorem 3.3.1 to the pair $\hat{H}_r, M_r$. It is clear that statement (iii) holds if and only if the parameter δ_k associated with nonzero pure imaginary eigenvalues ib_k is zero (see Theorem 3.2.4 IV). Then the equivalence of (ii) and (iii) follows from Theorem 3.3.1.

It follows easily from the relation $H_r = -\hat{H}_r M_r$ that (ii) implies (i).

Finally, if it is assumed that M_r is invertible then the relation $\hat{H}_r = -H_r M_r^{-1}$ can be used to show that (i) implies (ii). □

The dimension of a maximal M_r-invariant H_r-nonpositive subspace will also be required. This is obtained immediately from Theorem 2.7.3 in the case that M_r (and hence H_r) is invertible. For convenience, define the parameter

$$\alpha(M_r, H_r) = \sum \epsilon_j \tag{8.1.4}$$

where the ϵ_j are those members of the sign characteristic of (M_r, H_r) associated with the odd partial multiplicities (and corresponding to the pure imaginary eigenvalues of M_r, if any); see Theorems 2.7.1 and 2.7.3.

Lemma 8.1.3. *If M_r is invertible then the maximal dimension of an M_r-invariant and H_r-nonpositive subspace is*

$$n - \alpha(M_r, H_r). \tag{8.1.5}$$

The case when H_r has n positive and n negative eigenvalues is of particular interest. In view of Theorem 2.1.2 and the preceding lemma, this happpens if and only if $\alpha(M_r, H_r) = 0$, or, if and only if there exists an n-dimensional M_r-invariant H_r-nonpositive subspace.

Another, possibly more constructive, condition ensuring this property follows from the definition of M_r (see equation (8.1.2)), namely, when A is invertible and suitably large compared to C and D. To make this statement precise we use the following simple fact. In the statement of this proposition the matrix norm must be "power submultiplicative", i.e. for any $n \times n$ matrix W, and any positive integer m, $\|W^m\| \leq \|W\|^m$.

Proposition 8.1.4. *Let*

$$Y = \begin{bmatrix} -C & A^T \\ A & D \end{bmatrix}$$

be a $2n \times 2n$ real symmetric matrix partitioned into $n \times n$ blocks. Assume that A is invertible and

$$\min(\|CA^{-1}D(A^T)^{-1}\|, \|(A^T)^{-1}CA^{-1}D\|) < 1. \tag{8.1.6}$$

Then Y is invertible and has n positive eigenvalues and n negative eigenvalues.

Proof. Suppose that
$$\|CA^{-1}D(A^T)^{-1}\| < 1. \tag{8.1.7}$$

We have the congruence
$$\begin{bmatrix} I & 0 \\ 0 & A^{-1} \end{bmatrix} \begin{bmatrix} -C & A^T \\ A & D \end{bmatrix} \begin{bmatrix} I & 0 \\ 0 & (A^T)^{-1} \end{bmatrix} = \begin{bmatrix} -C & I \\ I & A^{-1}D(A^T)^{-1} \end{bmatrix}$$

and also
$$\begin{bmatrix} I & C \\ 0 & I \end{bmatrix} \begin{bmatrix} -C & I \\ I & A^{-1}D(A^T)^{-1} \end{bmatrix} = \begin{bmatrix} 0 & I + CA^{-1}D(A^T)^{-1} \\ I & A^{-1}D(A^T)^{-1} \end{bmatrix}.$$

So in view of (8.1.7) Y is invertible. Let
$$Y(\alpha) = \begin{bmatrix} -\alpha C & A^T \\ A & \alpha D \end{bmatrix}, \quad 0 \le \alpha \le 1.$$

Then (again by (8.1.7)) $Y(\alpha)$ is invertible for all $\alpha \in [0,1]$, and therefore by the continuity of its eigenvalues as functions of α, the number of positive (resp. negative) eigenvalues of $Y(\alpha)$ does not depend on α. But $Y(0)$ clearly has n positive and n negative eigenvalues, therefore the same is true for $Y(1) = Y$. □

We now state and prove one of the main results of this section (cf. Theorem 7.2.3).

Theorem 8.1.5. *Assume that $D \ge 0$, the pair (A, D) is controllable, and the matrix M_r is invertible. Then the equation (8.1.1) admits a solution X such that*

$$(X^T - X)(A + DX) \le 0 \tag{8.1.8}$$

if and only if $\alpha(M_r, H_r) = 0$. If this condition is satisfied, then the formula

$$\mathcal{M} = \mathrm{Im} \begin{bmatrix} I \\ X \end{bmatrix} \tag{8.1.9}$$

establishes a one-to-one correspondence between the set of solutions X with the property (8.1.8) and the set of M_r-invariant maximal H_r-nonpositive subspaces $\mathcal{M}$.

The proof of this theorem is obtained by simply combining the results of Lemma 7.2.2 (which holds verbatim for the real case), Proposition 8.1.1 and Lemma 8.1.3.

Corollary 8.1.6. *Assume the hypotheses of Theorem 8.1.5 and assume, in addition, that A is invertible and $\min(\|CA^{-1}D(A^T)^{-1}\|, \|(A^T)^{-1}CA^{-1}D\|) < 1$ for some power submultiplicative norm $\|\cdot\|$; then (8.1.1) admits solutions X with the property (8.1.8), and the set of all such X is described by the formula (8.1.9), where $\mathcal{M}$ is any n-dimensional M_r-invariant H_r-nonpositive subspace.*

This corollary is obtained using Proposition 8.1.4.

Theorem 8.1.7. *Assume that $D \geq 0$, and the pair (A, D) is controllable. Then equation (8.1.1) admits a real symmetric solution if and only if, for every pair of pure imaginary nonzero eigenvalues $\pm ib$ of M_r, the equality*

$$\nu([z_i^T(\hat{H}_r M_r)z_j]_{i,j=1}^{2p}) = p$$

holds, where $z_1, \ldots, z_{2p}$ is a basis in $\mathcal{R}_{\pm ib}(M_r)$. In this case the same formula, (8.1.9), establishes a one-to-one correspondence between the set of real symmetric solutions X of (8.1.1) and the set of M_r-invariant maximal $\hat{H}_r$-neutral subspaces $\mathcal{M}$.

Again, the proof is obtained by combining Lemma 7.2.2, Proposition 8.1.1 and Lemma 8.1.2.

If we are interested only in the existence of symmetric solutions, or of solutions satisfying (8.1.8), then the controllability hypothesis in Theorems 8.1.5 and 8.1.7 can be relaxed to sign controllability (in the spirit of Theorem 7.2.5).

In contrast to the definition of (7.2.12) a pair of *real* matrices (A, B), where A is $n \times n$ and B is $n \times m$, is called *sign controllable* if for every complex number $\alpha + i\beta$ either

$$\mathcal{R}_{\alpha \pm i\beta}(A) \subseteq \mathcal{C}_{A,B} \qquad (8.1.10)$$

or

$$\mathcal{R}_{-\alpha \pm i\beta}(A) \subseteq \mathcal{C}_{A,B}. \qquad (8.1.11)$$

(As usual, we interpret $\mathcal{R}_{\alpha \pm i\beta}(A)$ as the zero subspace if $\alpha \pm i\beta$ are not eigenvalues of A.)

Theorem 8.1.8. *Assume that $D \geq 0$, the pair (A, D) is sign controllable, and M_r is invertible. Then the conditions (i)–(iii) below are equivalent.*

(i) *The equation (8.1.1) admits a real solution X such that $(X^T - X)(A + DX) \leq 0$.*

(ii) *There exists an n-dimensional M_r-invariant H_r-nonpositive subspace.*

(iii) *The equality $\alpha(M_r, H_r) = 0$ holds.*

If, in addition, A is invertible and condition (8.1.6) is satisfied, then the conditions (i)–(iii) above hold.

Proof. The equivalence of (ii) and (iii) follows from the remark after Lemma 8.1.3. The implication (i) $\Rightarrow$ (ii) was proved in Proposition 8.1.1. The implication (ii) $\Rightarrow$ (i) is ensured by Lemma 8.1.9 below. Finally, the last statement of Theorem 8.1.8 follows from Proposition 8.1.4. $\square$

Lemma 8.1.9. *Assume the hypotheses of Theorem 8.1.8. If there exists an n-dimensional M_r-invariant H_r-nonpositive subspace, then there exists such a subspace which is also a graph subspace.*

Proof. The line of argument used to prove Lemma 7.2.6 is followed. Let $\mathcal{L}$ be an n-dimensional M_r-invariant H_r-nonpositive subspace such that the eigenvalues of $-M_r|\mathcal{L}$ with nonzero real parts are in $\bar{S} = \{\bar{\lambda}|\lambda \in S\}$ where S is a maximal subset of the complex plane with the properties that $\lambda \in S \Rightarrow \bar{\lambda} \in S$, $\mathcal{R}_{\alpha \pm i\beta}(A) \subseteq \mathcal{C}_{A,D}$ for every $\alpha \pm i\beta \in S$, and if $\alpha + i\beta \in S$, $\alpha, \beta \in \mathcal{R}$, $\alpha \neq 0$, then $-\alpha + i\beta \notin S$. It turns out that such a subspace $\mathcal{L}$ is a graph subspace. $\square$

The final result of this section concerns existence of real symmetric solutions:

Theorem 8.1.10. *Assume that $D \geq 0$ and the pair (A, D) is sign controllable. Then the conditions (i)–(iii) below are equivalent.*

(i) *The CARE (8.1.1) admits a real symmetric solution.*

(ii) *There exists an n-dimensional M_r-invariant $\hat{H}_r$-neutral subspace.*

(iii) *For every pair $\pm ib$ of nonzero pure imaginary eigenvalues $\pm ib$ of M_r the equality*
$$\nu([z_i^T(\hat{H}_r M_r)z_j]_{i,j=1}^{2p}) = p$$
holds, where $z_1, \ldots, z_{2p}$ is a basis in $\mathcal{R}_{\pm ib}(M_r)$.

Proof. The equivalence (ii) $\Leftrightarrow$ (iii) is part of Lemma 8.1.2, and the implication (i) $\Rightarrow$ (ii) follows from Proposition 8.1.1.

It remains to prove that (ii) implies (i). This can be accomplished by exhibiting an n-dimensional M_r-invariant $\hat{H}_r$-neutral subspace which is also a graph subspace, and this can be done by following the argument used in the proof of Lemma 7.2.6. $\square$

Corollary 8.1.11. *Assume $D \geq 0$, $C \geq 0$, (A, D) is stabilizable and (C, A) detectable (in the sense of c-stability of Section 4.4). Then the CARE (8.1.1) admits a real symmetric solution. Moreover, there exists a unique real symmetric solution X_0 of (8.1.1) which stabilizes, i.e. such that $A + DX_0$ is c-stable.*

Proof. By Theorem 7.2.8 the matrix M_r has no pure imaginary (or zero) eigenvalues. Therefore by Theorem 8.1.10 the CARE (8.1.1) has symmetric solutions. Let $\mathcal{L}$ be the n-dimensional M_r-invariant H_r-nonpositive subspace constructed as in the proof of Lemma 8.1.9, taking

$$S = \bar{S} = \{\lambda \in \mathbb{C} : \operatorname{Re}\lambda \geq 0\}.$$

The proof of Lemma 8.1.9 ensures that $\mathcal{L}$ is a graph subspace:

$$\mathcal{L} = \operatorname{Im} \begin{bmatrix} I \\ X_0 \end{bmatrix}.$$

Here X_0 is a symmetric solution of (8.1.1) such that $-M_r|\mathcal{L}$, which is similar to $-(A+DX_0)$ has all its eigenvalues in S. We conclude that $\sigma(A+DX_0)$ is in the closed left half-plane and, since M_r has no eigenvalues on the imaginary axis, in fact $A+DX_0$ is c-stable.

It remains to prove uniqueness of the stabilizing symmetric solution of (8.1.1). Let X_0 be such a solution. Then $\mathcal{L} := \operatorname{Im} \begin{bmatrix} I \\ X_0 \end{bmatrix}$ is an n-dimensional M_r-invariant $\hat{H}_r$-neutral subspace such that $M_r|\mathcal{L}$ has all eigenvalues in the open left half-plane. Since $\sigma(M_r) \cap i\mathbb{R} = \emptyset$, such a subspace is unique (it coincides with the spectral subspace for M_r corresponding to the eigenvalues in the open left half-plane). □

8.2 Existence of symmetric solutions

In this section it will be shown that, when the CARE (8.1.1) admits a real symmetric solution, the matrix M_r of equation (8.1.2) has certain special spectral characteristics. As in the previous section, A, $C = C^T$ and $D = D^T$ are real $n \times n$ matrices, and only real $n \times n$ matrices X are admitted as solutions of the CARE. Recall the notation from the previous section:

$$M_r = \begin{bmatrix} A & D \\ C & -A^T \end{bmatrix}, \quad \hat{H}_r = \begin{bmatrix} 0 & I \\ -I & 0 \end{bmatrix}.$$

As $\hat{H}_r = -\hat{H}_r^T$ and $\hat{H}_r M_r = -M_r^T \hat{H}_r$, the canonical form of the pair $(\hat{H}_r, M_r)$ (of Theorem 3.2.4) determines their "sign characteristic". Namely a sign ± 1 for every real Jordan block in the real Jordan form of M_r corresponding to a pair of nonzero pure imaginary eigenvalues $\pm ib$, and also for every real Jordan block of even size with eigenvalue zero.

The main result of this section is:

Theorem 8.2.1. *Assume that $D \geq 0$. If the CARE (8.1.1) has a real symmetric solution X, and*

$$\mathcal{R}_{\pm i\beta}(A + DX) \subseteq \mathcal{C}_{A,D} \tag{8.2.1}$$

for every pair of pure imaginary or zero eigenvalues $\pm i\beta$ of $A + DX$, then all the partial multiplicities corresponding to pure imaginary and zero eigenvalues of M_r (if any) are even. Moreover, the sign characteristic of the pair $(M_r, \hat{H}_r)$ consists of -1's for pure imaginary nonzero eigenvalues, and the sign attached to any even partial multiplicity corresponding to the zero eigenvalue of M_r is $(-1)^{m-1}$, where $2m$ is the partial multiplicity.

Proof. The first statement follows from Theorem 7.3.1. The second statement (concerning the sign characteristic) follows from Theorem 3.4.1 and Theorem 7.3.5. The latter theorem asserts that the sign characteristic of $(iM_r, i\hat{H}_r)$, considered as complex matrices, consists of $+1$'s only. □

Using Theorem 8.2.1, we can augment the characterizations of existence of symmetric solutions given in Theorem 8.1.10.

Theorem 8.2.2. *Assume that $D \geq 0$ and the pair (A, D) is sign controllable. Then the CARE (8.1.1) admits a symmetric solution if and only if all multiplicities corresponding to the pure imaginary or zero eigenvalues of M_r are even and the sign characteristic of $(M_r, \hat{H}_r)$ consists of -1's with the exception of signs corresponding to multiplicities of the zero eigenvalues which are not divisible by 4, and these signs are $+1$.*

Proof. The part "if" follows from the implication (iii) $\Rightarrow$ (i) in Theorem 8.1.10. Conversely, assume that the CARE admits a real symmetric solution X. As in the proof of Theorem 7.3.5 we show that $\mathcal{R}_{\pm ib}(A + DX) \subseteq \mathcal{C}_{A,D}$ for every pair of pure imaginary or zero eigenvalues $\pm ib$ of $A + DX$. Thus, Theorem 8.2.1 can be applied. □

8.3 Special and extremal symmetric solutions

It will be assumed in this section that $C = C^T$, $D \geq 0$ and the pair (A, D) is controllable (recall that throughout this chapter the matrices A, D and C are assumed to be $n \times n$ and real, and the solutions X are required to be real as well). As in Section 7.4, we say that a solution X of the CARE (8.1.1) is called *special* if

$$(X^T - X)(A + DX) \leq 0.$$

According to Theorem 8.1.5, under certain conditions, special solutions are in one-to-one correspondence with M_r-invariant maximal H_r-nonpositive subspaces. Based on this correspondence, we have the following analogue of Theorem 7.4.1.

Theorem 8.3.1. *Assume that M_r is invertible, and that*

$$\alpha(M_r, H_r) := \sum_{j \in J} \kappa_j = 0,$$

where J is the collection of all Jordan blocks J_{2m} ($\pm ib$) ($b > 0$) in the Jordan form of M_r with odd m, and κ_j are the corresponding signs in the sign characteristic of (M_r, H_r) (see equations (2.7.5)). Let S be a set of eigenvalues of M_r with nonzero real parts such that $\lambda \in S$ implies $\bar{\lambda} \in S$, $-\bar{\lambda} \notin S$, and which is maximal (subject to these properties). Then there exists a special solution X of the CARE such that S is exactly the set of eigenvalues of $A + DX$ having nonzero real parts. If, in addition, the pair (M_r, H_r) satisfies the sign condition (as defined before Theorem 2.4.2), then the special solution X is unique.

This result follows from Theorems 8.1.5 and 2.4.2.

The counterpart of Theorem 8.3.1 concerning symmetric solutions reads as follows.

Theorem 8.3.2. *Assume that the CARE has real symmetric solutions. Then for every set S as in Theorem 8.3.1 there exists a unique real symmetric solution X such that S is exactly the set of eigenvalues of $A + DX$ having nonzero real parts.*

This result follows by combining Theorems 3.4.1, 8.2.1 and 8.1.7.

We have seen in Section 7.5 (Theorem 7.5.1) that, assuming the CARE admits symmetric solutions, it has maximal and minimal hermitian solutions X_+ and X_-, respectively. It turns out that both X_+ and X_- are in fact real. Indeed, since $\bar{X}_+$ is also a hermitian solution of the CARE, we find that $\bar{X}_+ \leq X_+$. So $X_+ - \bar{X}_+$ is positive semidefinite. But the diagonal entries of $X_+ - \bar{X}_+$ are zeros; so we must conclude that $X_+ - \bar{X}_+ = 0$, i.e. X_+ is real. Similarly, it is found that X_- is real.

Recall from Theorem 7.5.1 that X_+ (resp. X_-) is characterized by the property that $\sigma(A + DX_+)$ (resp. $\sigma(A + DX_-)$) lies in the closed right (resp. left) half-plane, and is obtained in Theorem 8.3.2 by choosing

$$S = \{\lambda \in \sigma(M_r)| \operatorname{Re} \lambda > 0\}$$

(resp. $S = \{\lambda \in \sigma(M_r)| \operatorname{Re} \lambda < 0\}$). Now the results of Theorems 7.5.3 and 7.5.4 are valid in the real case as well, simply by restricting these theorems to the real symmetric solutions of the CARE.

8.4 Partial order, continuity, and analyticity

Throughout this section it will be assumed that the CARE (8.1.1) admits real symmetric solutions, and we retain the hypotheses ($C = C^T$, $D \geq 0$, (A, D)

controllable) of the previous section. It follows from Theorem 7.6.2 that there is a one-to-one correspondence between the real symmetric solutions X of the CARE and real M_r-invariant subspaces $\mathcal{N}$ such that $\mathcal{N} \subseteq \mathcal{N}_+$. Here $\mathcal{N}_+$ is the spectral M_r-invariant subspace corresponding to its eigenvalues in the open right half-plane. This one-to-one correspondence is expressed by the formula

$$\text{Im} \begin{bmatrix} I \\ X \end{bmatrix} \cap \mathcal{N}_+ = \mathcal{N}. \tag{8.4.1}$$

(cf. Theorem 8.1.7 and formula (8.1.9)). As in Section 7.6, we express the correspondence (8.4.1) by writing

$$X = \text{Ric}(\mathcal{N}) = \text{Ric}(\mathcal{N}; A, D, C).$$

Real analogues of several results of Sections 7.6, 7.7 and 7.8 (for example, Theorems 7.6.4, 7.7.1 and 7.8.2) can be obtained by simply restricting the "complex" results to the real symmetric solutions of the CARE (8.1.1). Thus:

Theorem 8.4.1. (a) *For every two M_r-invariant subspaces $\mathcal{N}, \mathcal{L} \subseteq \mathcal{N}_+$ we have $\text{Ric}(\mathcal{N}) \leq \text{Ric}(\mathcal{L})$ if and only if $\mathcal{N} \subseteq \mathcal{L}$.*
(b) *The functions Ric and Ric^{-1} are continuous. Here the set of real M_r-invariant subspaces contained in $\mathcal{N}_+$ is considered in the topology induced by the gap topology (see Appendix 7.A), and the set of all symmetric solutions is considered in the topology induced by the usual topology on the $n \times n$ real matrices.*
(c) *The functions Ric and Ric^{-1} are smooth in the following sense: Let Ω be an open connected set in $\mathbb{R}^k$, and let $\mathcal{N}(t)$, $t \in \Omega$, be a family of M_r-invariant subspaces contained in $\mathcal{N}_+$ such that the orthogonal projector onto $\mathcal{N}(t)$ is p times continuously differentiable (or analytic) as a function of $t \in \Omega$. Then $X(t) = \text{Ric}(\mathcal{N}(t))$ is also p times continuously differentiable (or analytic) on Ω. Conversely, if $X(t)$ is p times continuously differentiable (or analytic) as a function on Ω, then so is the orthogonal projector onto $\mathcal{N}(t) = \text{Im} \begin{bmatrix} I \\ X(t) \end{bmatrix} \cap \mathcal{N}_+$.*

The description of isolated real symmetric solutions follows the pattern established in Theorem 7.7.2:

Theorem 8.4.2. *The following statements are equivalent for a (real) symmetric solution X_0 of the CARE (8.1.1):*

(i) *X_0 is isolated, i.e. there is an open neighborhood of X_0 in the set of $n \times n$ real matrices containing no other symmetric solutions;*

(ii) *X_0 is complex isolated, i.e. there is an open neighborhood of X_0 in the set of $n \times n$ complex matrices containing no other hermitian solutions;*

(iii) *each common eigenvalues λ with nonzero real part of $A + DX_0$ and of $-(A^T + X_0 D)$ is an eigenvalue of M_r of geometric multiplicity one;*

(iv) *for every nonzero real eigenvalue λ of M_0 with $\dim \mathrm{Ker}\,(\lambda I - M_r) > 1$ we have $\mathcal{R}_\lambda(M_r) \subseteq \mathrm{Im}\,\begin{bmatrix} I \\ X_0 \end{bmatrix}$ or $\mathcal{R}_\lambda(M_r) \cap \mathrm{Im}\,\begin{bmatrix} I \\ X_0 \end{bmatrix} = \{0\}$, and for every pair $\alpha \pm i\beta$ ($\alpha \neq 0$) of non-real complex conjugate eigenvalues of M_r with*

$$\dim \mathrm{Ker}\,((\alpha^2 + \beta^2)I - 2\alpha M_r + M_r^2) > 2 \qquad (8.4.2)$$

we have either

$$\mathcal{R}_{\alpha \pm i\beta}(M_r) \subseteq \mathrm{Im}\,\begin{bmatrix} I \\ X_0 \end{bmatrix} \quad \text{or} \quad \mathcal{R}_{\alpha \pm i\beta}(M_r) \cap \mathrm{Im}\,\begin{bmatrix} I \\ X_0 \end{bmatrix} = \{0\}. \qquad (8.4.3)$$

Proof. Clearly, (ii) implies (i). Since $\mathcal{R}_{\alpha+i\beta}(M_r)$ and $\mathcal{R}_{\alpha-i\beta}(M_r)$ (considered as subspaces in $\mathbb{C}^{2n}$) are related by

$$\mathcal{R}_{\alpha+i\beta}(M_r) = \overline{\mathcal{R}_{\alpha-i\beta}(M_r)},$$

we find that condition (iii) of Theorem 7.7.2 is actually equivalent to (iv), and therefore all three statements (ii), (iii) and (iv) are equivalent.

Finally, to prove (i) $\Rightarrow$ (iv) assume that (iv) is not valid. Say,

$$\{0\} \neq \mathcal{R}_\lambda(M_r) \cap \mathrm{Im}\,\begin{bmatrix} I \\ X_0 \end{bmatrix} \neq \mathcal{R}_\lambda(M_r)$$

for some $\lambda \in \sigma(M_r)$, where $\lambda > 0$ and

$$\dim \mathrm{Ker}\,(\lambda_0 I - M_r) > 1.$$

By Lemma 7.7.3 the subspace

$$\mathrm{Im}\,\begin{bmatrix} I \\ X_0 \end{bmatrix} \cap \mathcal{N}_+$$

is not isolated in the set of all M_r-invariant subspaces contained in $\mathcal{N}_+$. (More exactly, the real version of this lemma is used here; see Theorem 14.6.5 of Gohberg et al. (1986a)). Since the function Ric is continuous (Theorem 8.4.1(b)), it follows that X_0 is not isolated in the set of all symmetric solutions, i.e. (i) does not hold. If (8.4.2) is not valid for some pair $\alpha \pm i\beta \in \sigma(M)$ with $\alpha > 0$ and satisfying (8.4.2), then (using Theorem 14.6.5 of Gohberg et al. (1986a) once more) an analogous argument applies. □

We conclude this section by counting the symmetric solutions:

Theorem 8.4.3. *The number of symmetric solutions of the CARE (8.1.1) is finite if and only if*
$$\dim \operatorname{Ker}(M_r - \lambda_0 I) = 1 \qquad (8.4.4)$$
for every real positive eigenvalue λ_0 of M_r, and
$$\dim \operatorname{Ker}((\alpha^2 + \beta^2)I - 2\alpha M_r + M_r^2) = 2 \qquad (8.4.5)$$
for every pair of complex conjugate eigenvalues $\alpha \pm i\beta$ of M_r with $\alpha > 0$. In this case, the number of symmetric solutions is given by
$$\left(\prod_{i=1}^{p}(\dim \mathcal{R}_{\lambda_i}(M_r) + 1)\right) \cdot \left(\prod_{j=1}^{q}(\frac{1}{2}\dim \mathcal{R}_{\alpha_j \pm i\beta_j}(M_r) + 1)\right),$$
where $\lambda_1, \ldots, \lambda_p$ are all the distinct real positive eigenvalues of M_r, and $\alpha_1 \pm i\beta_1, \ldots, \alpha_q \pm i\beta_q$ are all the distinct pairs of complex conjugate eigenvalues of M_r having positive real parts.

If at least one of the conditions (8.4.4) and (8.4.5) does not hold, then the CARE (8.1.1) has a continuum of symmetric solutions.

The proof of Theorem 8.4.3 is analogous to that of Corollary 7.6.3 and therefore is omitted.

8.5 Relaxing the controllability assumption

Consider the equation
$$XDX - XA - A^T X - C = 0 \qquad (8.5.1)$$
where the controllability assumption employed in the previous section is relaxed. Thus, we assume that $D \geq 0$, $C = C^T$ and (A, D) is c-stabilizable. In this framework, all the results of Section 7.9 are valid, with proofs obtained by specializing those results to real solutions X. We will state here just one result — the real analogue of Theorem 7.9.3.

Theorem 8.5.1. *Assume that $D \geq 0$, $C = C^T$, (A, D) is c-stabilizable, and that (8.5.1) has a real symmetric solution. Then there exists a maximal real symmetric solution X_+, which can be characterized also as the unique stabilizing solution.*

Recall that a solution X_0 of (8.5.1) is called *stabilizing* if $X_0 = X_0^T$ and $\sigma(A - DX_0)$ lies in the closed left half-plane.

8.6 The pencil approach

The CARE with real coefficients in the linear-quadratic regulator form can be studied using the pencil approach, in much the same way as the CARE with complex coefficients, studied in Section 7.10. However, some changes in the presentation will be neceessary. Thus, we consider the equation

$$XDX - X\hat{A} - \hat{A}^T X - \hat{Q} = 0, \qquad (8.6.1)$$

where

$$\hat{A} = A - BR^{-1}C, \quad \hat{Q} = Q - C^T R^{-1} C, \quad D = BR^{-1}B^T,$$

and where A, Q are $n \times n$, B, C^T are $n \times m$, and $R = R^T$ is $m \times m$. All matrices A, B, C, D, Q and R are assumed to be real. Real $n \times n$ solutions X are sought for the equation (8.6.1).

The strict equivalence relation (7.10.6) now takes the form

$$\begin{bmatrix} 0 & -I & BR^{-1} \\ I & 0 & -C^T R^{-1} \\ 0 & 0 & I \end{bmatrix} (\lambda \hat{G} - H) \begin{bmatrix} I & 0 & 0 \\ 0 & I & 0 \\ -R^{-1}C & -R^{-1}B^T & I \end{bmatrix} = \begin{bmatrix} \lambda I - \hat{M} & 0 \\ 0 & -R \end{bmatrix}, \qquad (8.6.2)$$

where

$$\hat{G} = \begin{bmatrix} 0 & I & 0 \\ -I & 0 & 0 \\ 0 & 0 & 0 \end{bmatrix}, \quad \hat{M} = \begin{bmatrix} -\hat{A} & D \\ \hat{Q} & \hat{A}^T \end{bmatrix}, \quad H = \begin{bmatrix} Q & A^T & C^T \\ A & 0 & B \\ C & B^T & R \end{bmatrix}$$

Using the relation (8.6.2), the following real analogue of Theorem 7.10.1 can be proved using the same ideas.

Theorem 8.6.1. *Let $R = R^T$ be nonsingular. Then the CARE (8.6.1) has a real solution if and only if there is an n-dimensional deflating subspace S_0 for $(\hat{G}, H)$ of the form*

$$S_0 = \mathrm{Im} \begin{bmatrix} X_1 \\ X_2 \\ X_3 \end{bmatrix}$$

where X_1 and X_2 are $n \times n$ real matrices with X_1 invertible. When this is the case the matrix $X = X_2 X_1^{-1}$ is a real solution of (8.6.1).

The other results of Section 7.10 can be adapted in a similar way for the real case. We leave this interpretation to the interested reader.

8.7 Exercises

8.7.1. Assume $D \geq 0$, (A, D) controllable, and the CARE (8.1.1) has real symmetric solutions. Prove that every hermitian solution of this equation is real if and only if all eigenvalues of M_r with nonzero real part are in fact real.

8.7.2. Supply the details for the proof of Lemma 8.1.9.

8.7.3. Supply the details for the proof of Theorem 8.1.10.

8.7.4. (cf. Exercise 7.11.6.) Consider the CARE (8.1.1) with

$$D = \begin{bmatrix} 1 & 1 \\ 1 & 1 \end{bmatrix}, \quad A = \begin{bmatrix} 0 & -1 \\ 1 & 0 \end{bmatrix}, \quad C = \begin{bmatrix} 1 & 1 \\ 1 & 1 \end{bmatrix}.$$

(a) Find all real symmetric solutions.

(b) Identify the maximal and minimal real symmetric solutions.

(c) Find all real solutions X satisfying $(X^T - X)(A + DX) \leq 0$.

(c) Find all real solutions X satisfying $(X^T - X)(A + DX) \geq 0$.

8.7.5. (cf. Exercise 7.11.7, 7.11.8.)

(a) The same as Exercise 8.7.4 but with

$$D = \begin{bmatrix} 1 & 0 \\ 0 & 1 \end{bmatrix}, \quad A = \begin{bmatrix} 0 & 0 \\ 0 & -1 \end{bmatrix}, \quad C = \begin{bmatrix} 1 & 0 \\ 0 & 3 \end{bmatrix}.$$

(b) The same as Exercise 8.7.4 but with

$$D = \begin{bmatrix} 1 & 1 \\ 1 & 1 \end{bmatrix}, \quad A = \begin{bmatrix} 0 & 0 \\ 0 & 0 \end{bmatrix}, \quad C = \begin{bmatrix} 1 & 0 \\ 0 & 0 \end{bmatrix}.$$

8.7.6. State and prove the real analogue of the result of Exercise 7.11.13.

8.7.7. State and prove the real analogue of the result of Exercise 7.11.14.

8.7.8. Find real symmetric solutions of the CARE with

$$D = \begin{bmatrix} 1 & 0 \\ 0 & 0 \end{bmatrix}, \quad A = \begin{bmatrix} 0 & 0 \\ 1 & 0 \end{bmatrix}, \quad C = \begin{bmatrix} -2 & 0 \\ 0 & 1 \end{bmatrix}.$$

Prove that this equation has a unique (complex) hermitian solution which turns out to be real.

8.8 Notes

In physical problems involving Riccati equations those of real type arise most frequently. Consequently, the literature (including most of the sources quoted in Chapter 7) is dominated numerically by papers devoted to the real case. Our analysis shows that the extra generality included in the treatment of complex equations is attained at little or no extra expense of effort. Furthermore, the contrasts between these two cases are illuminating. Exercise 8.7.8 is taken from Lancaster and Rodman (1980).

9
CONSTRUCTIVE EXISTENCE AND COMPARISON THEOREMS

Consider the symmetric algebraic Riccati equation

$$XDX - XA - A^*X - C = 0.$$

The main theme of this chapter concerns comparison results of the following nature: suppose the coefficients of the equation are changed, resulting in the equation

$$XD'X - XA' - A'^*X - C' = 0.$$

If the first equation has hermitian solutions, when can we say that the second also has solutions? Is there some relationship between maximal solutions of the two equations?

The methods of this chapter are based on a constructive iterative procedure, in contrast with Chapter 7 where the methods use geometrical properties of invariant subspaces.

9.1 Comparison theorems

In this section we study the continuous algebraic Riccati equation

$$\mathcal{R}(X) := XDX - XA - A^*X - C = 0 \qquad (9.1.1)$$

under the hypotheses that $D \geq 0$, $C = C^*$, the pair (A, D) is stabilizable, and the size of matrices A, D and C is $n \times n$. Recall that, under these conditions, the *existence* of hermitian solutions X of $\mathcal{R}(X) = 0$ can be characterized using spectral properties of the matrix $\begin{bmatrix} -A & D \\ C & A^* \end{bmatrix}$ as described in Theorem 7.9.1.

Together with the equation (9.1.1) consider the inequality

$$\mathcal{R}(X) = XDX - XA - A^*X - C \leq 0. \qquad (9.1.2)$$

Naturally, an $n \times n$ matrix X is called a *solution* of $\mathcal{R}(X) \leq 0$ if the left-hand side of (9.1.2) is a negative semidefinite matrix. Obviously, the solution set of $\mathcal{R}(X) = 0$ is always contained in that of $\mathcal{R}(X) \leq 0$, but not conversely.

232 CONSTRUCTIVE EXISTENCE AND COMPARISON THEOREMS

Theorem 9.1.1. *Assume that $D \geq 0$, $C^* = C$, (A, D) is stabilizable, and there exists a hermitian solution of the inequality $\mathcal{R}(X) \leq 0$. Then there exists a hermitian solution X_+ of $\mathcal{R}(X) = 0$ such that $X_+ \geq X$ for every hermitian solution X of $\mathcal{R}(X) \leq 0$. In particular, X_+ is the maximal hermitian solution of $\mathcal{R}(X) = 0$. Moreover, all the eigenvalues of $A - DX_+$ are in the closed left half-plane.*

Proof. By the hypotheses, there exists an $X = X^*$ such that

$$XDX - XA - A^*X - C' = 0, \tag{9.1.3}$$

where C' is a hermitian matrix such that $C' \leq C$.

As the pair (A, D) is stabilizable, it follows from Lemma 4.5.4 that there exists an $X_0 \geq 0$ such that $A - DX_0$ is stable, i.e. has all its eigenvalues in the open left half-plane.

Starting with X_0, we shall define a nonincreasing sequence of hermitian matrices $\{X_\nu\}_{\nu=0}^\infty$ satisfying $X_\nu \geq X$ as well as the equalities

$$X_{\nu+1}(A - DX_\nu) + (A - DX_\nu)^* X_{\nu+1} = -X_\nu DX_\nu - C, \quad \nu = 0, 1, \cdots. \tag{9.1.4}$$

The reason for choosing this equation will be clarified in the next section. The sequence $\{X_\nu\}_{\nu=0}^\infty$ will also have the property that $A - DX_\nu$ is stable for all ν. We know that $A - DX_0$ is stable and, assuming inductively that we have already defined $X_\nu = X_\nu^*$ with $A - DX_\nu$ stable, it follows from Theorem 5.2.2 that (9.1.4) has a unique solution $X_{\nu+1}$ which is necessarily hermitian.

We are now to show that $A - DX_{\nu+1}$ is stable. To this end note the following identity which holds for any hermitian matrices Y and $\hat{Y}$:

$$Y(A - DY) + (A - DY)^* Y + YDY$$
$$= Y(A - D\hat{Y}) + (A - D\hat{Y})^* Y + \hat{Y}D\hat{Y} - (Y - \hat{Y})D(Y - \hat{Y}). \tag{9.1.5}$$

By assumption, there exists a hermitian solution X of (9.1.3). Letting $Y = X$, and $\hat{Y} = X_\nu$ in (9.1.5), we get

$$X(A - DX_\nu) + (A - DX_\nu)^* X + X_\nu DX_\nu - (X - X_\nu)D(X - X_\nu) = -C'. \tag{9.1.6}$$

Subtract (9.1.6) from (9.1.4):

$$(X_{\nu+1} - X)(A - DX_\nu) + (A - DX_\nu)^* (X_{\nu+1} - X) = -(X - X_\nu)D(X - X_\nu) - (C - C').$$

As $A - DX_\nu$ is stable it follows from Theorem 5.3.1(a), that $X_{\nu+1} \geq X$.

Next, use (9.1.5) again with $Y = X_{\nu+1}$, $\hat{Y} = X_\nu$ and apply (9.1.4) to get

$$X_{\nu+1}(A - DX_{\nu+1}) + (A - DX_{\nu+1})^* X_{\nu+1} + X_{\nu+1} DX_{\nu+1}$$
$$= -C - (X_{\nu+1} - X_\nu)D(X_{\nu+1} - X_\nu). \qquad (9.1.7)$$

Subtracting (9.1.6) with X_ν replaced by $X_{\nu+1}$, we obtain

$$(X_{\nu+1} - X)(A - DX_{\nu+1}) + (A - DX_{\nu+1})^*(X_{\nu+1} - X)$$
$$= -(X_{\nu+1} - X_\nu)D(X_{\nu+1} - X_\nu) - (X_{\nu+1} - X)D(X_{\nu+1} - X) - C + C'. (9.1.8)$$

(Note that if the right-hand side here is strictly negative definite then Theorem 5.3.2(a) would show that $A - DX_{\nu+1}$ is stable; but it is not necessary to make this assumption.)

Assume that $(A - DX_{\nu+1})x = \lambda x$ for some λ with $\operatorname{Re}\lambda \geq 0$ and some $x \neq 0$. Then

$$(\overline{\lambda} + \lambda)x^*(X_{\nu+1} - X)x = x^* W x, \qquad (9.1.9)$$

where $W \leq 0$ is the right-hand side of (9.1.8). As $X_{\nu+1} - X \geq 0$, equation (9.1.9) implies $x^* W x = 0$ which, using the definition of W, in turn implies

$$x^*(X_{\nu+1} - X_\nu)D(X_{\nu+1} - X_\nu)x = 0.$$

But $D \geq 0$, so $D(X_{\nu+1} - X_\nu)x = 0$. Now

$$(A - DX_\nu)x = (A - DX_{\nu+1})x = \lambda x,$$

a contradiction with the stability of $A - DX_\nu$. Hence $A - DX_{\nu+1}$ is stable as well.

Next it will be shown that the sequence $\{X_\nu\}_{\nu=0}^\infty$ is nonincreasing. Consider the equality (9.1.7) with ν replaced by $\nu - 1$, and subtract from it equation (9.1.4) to get

$$(X_\nu - X_{\nu+1})(A - DX_\nu) + (A - DX_\nu)^*(X_\nu - X_{\nu+1}) = -(X_\nu - X_{\nu-1})D(X_\nu - X_{\nu-1}).$$

As $A - DX_\nu$ is stable,

$$X_\nu - X_{\nu+1} = \int_0^\infty e^{(A-DX_\nu)t}(X_\nu - X_{\nu-1})D(X_\nu - X_{\nu-1})e^{(A-DX_\nu)^* t}\, dt \geq 0.$$

So $\{X_\nu\}_{\nu=0}^\infty$ is a nonincreasing sequence of hermitian matrices bounded below by X. Hence the limit $X'_+ = \lim_{\nu \to \infty} X_\nu$ exists. Passing to the limit in (9.1.4) when $\nu \to \infty$ shows that X'_+ is a hermitian solution of $\mathcal{R}(X) = 0$. Since $A - DX_\nu$ is stable for all $\nu = 0, 1, \ldots$, the matrix $A - DX'_+$ has all its eigenvalues in the closed left half-plane. Also $X'_+ \geq X$ for every hermitian solution of (9.1.2). □

We now make several important deductions from Theorem 9.1.1. First of all, we complete the proof of Theorem 7.9.3.

Proof of Theorem 7.9.3 (completion). Theorem 9.1.1 shows that the maximal solution X_+ of (7.9.1) exists, and $\sigma(A-DX_+)$ is in the closed left half-plane. Since we have seen already in the proof of Theorem 7.9.3 that the almost stabilizing solution is unique, this theorem is now proved completely. □

In many applications equation (9.1.1) has the property $C \geq 0$, or $C > 0$, in addition to the hypotheses already made on D, C and the pair (A, D). Then inequality (9.1.2) is trivially satisfied by the hermitian matrix $X = 0$ and we have access to the results of Theorem 9.1.1. These important cases are examined in more detail in the next two theorems.

Theorem 9.1.2. *If $D \geq 0$, $C \geq 0$ and the pair (A, D) is stabilizable then there exist hermitian solutions of $\mathcal{R}(X) = 0$. Moreover, the maximal hermitian solution X_+ (which exists by Theorem 7.9.1) also satisfies $X_+ \geq 0$. If, in addition, (C, A) is detectable then $A - DX_+$ is stable.*

Proof. As noted above, $C \geq 0$ implies $X = 0$ is a solution of (9.1.2) and we obtain the conclusions of Theorem 7.9.1. Furthermore, in the proof of that theorem we obtain $X_\nu \geq 0$ for $\nu = 0, 1, 2, \ldots$, and hence $X_+ = \lim_{\nu \to \infty} X_\nu \geq 0$.

If (C, A) is detectable then it follows from Theorem 7.2.8 that the matrix

$$M := \begin{bmatrix} -A & D \\ C & A^* \end{bmatrix}$$

has no eigenvalues on the imaginary axis (note a change in the sign of A and the definition of M from Sections 7.2–7.9 and this chapter). Then Theorem 7.9.4 implies that $A - DX_+$ is stable. □

When $C > 0$ the pair (C, A) is certainly detectable and both statements of the last theorem apply. However, a little more information can be squeezed out as a corollary of the following theorem.

Theorem 9.1.3. *Assume that $D \geq 0$, $C^* = C$, and the pair (A, D) is stabilizable. If there is a hermitian matrix X such that $\mathcal{R}(X) < 0$, then the maximal hermitian solution X_+ of $\mathcal{R}(X) = 0$ exists and $A - DX_+$ is stable.*

Conversely, if there is a hermitian solution X_+ of $\mathcal{R}(X) = 0$ for which $A - DX_+$ is stable, then there is a hermitian matrix X for which $\mathcal{R}(X) < 0$.

Proof. It follows readily from the basic recursion relation (9.1.4) that

$$(X_{\nu+1} - X)(A - DX_\nu) + (A - DX_\nu)^*(X_{\nu+1} - X) = \mathcal{R}(X) - (X - X_\nu)D(X - X_\nu).$$

for $\nu = 0, 1, 2, \ldots$. In the limit as $\nu \to \infty$ we obtain

$$(X_+ - X)(A - DX_+) + (A - DX_+)^*(X_+ - X) \leq \mathcal{R}(X).$$

Since $\mathcal{R}(X) < 0$ it follows from Theorem 5.3.2 (a) that $X_+ > X$ and $A - DX_+$ is stable.

COMPARISON THEOREMS

Conversely, if $A - DX_+$ is stable it follows from Theorem 7.9.3 that the matrix

$$M := \begin{bmatrix} -A & D \\ C & A^* \end{bmatrix}$$

has no eigenvalues on the imaginary axis. Then, for $\epsilon > 0$ sufficiently small the same is true of

$$M_\epsilon := \begin{bmatrix} -A & D \\ C - \epsilon I & A^* \end{bmatrix}$$

and, by the same theorem, there must be a hermitian solution X_ϵ of the perturbed equation

$$XDX - XA - A^*X - C + \epsilon I = 0.$$

But this is just $\mathcal{R}(X) = -\epsilon I < 0$. □

Lemma 9.1.4. *Let $A \in \mathbb{C}^{n \times n}$, $B \in \mathbb{C}^{n \times m}$ and (A, B) be controllable. Let $B_1 \in \mathbb{C}^{n \times \ell}$ and define B_0 so that $BB^* + B_1 B_1^* = B_0 B_0^*$. Then $(A + B_1 G, B_0)$ is controllable for any $G \in \mathbb{C}^{\ell \times n}$.*

Proof. It is easily seen that $\text{Ker}(B_0 B_0^*) \subseteq \text{Ker}(BB^*) \cap \text{Ker}(B_1 B_1^*)$ and, taking orthogonal complements, it follows that $\text{Im } B + \text{Im } B_1 \subseteq \text{Im } B_0$. Using definition (4.1.2) of a controllable subspace it follows that, for any $G \in \mathbb{C}^{\ell \times n}$, $\text{Im}(B_1 G) \subseteq \text{Im } B_0$ and $\mathcal{C}_{A+B_1 G, B_1} = \mathcal{C}_{A, B_0} \supseteq \mathcal{C}_{A,B}$. Hence the result. □

Theorem 9.1.5. *If $D \geq 0$, $C \geq 0$, (A, D) is stabilizable, and (C, A) is observable, then the maximal hermitian solution X_+ of $\mathcal{R}(X) = 0$ exists, and we have $X_+ > 0$ and $A - DX_+$ stable.*

Proof. Since observability implies detectability, it follows from Theorem 9.1.2 that $A - DX_+$ is stable, X_+ exists and $X_+ \geq 0$. Also, we have

$$X_+(-A + DX_+) + (-A + DX_+)^* X_+ = C + X_+ DX_+.$$

Now (C, A) observable implies $(-A^*, C^{1/2})$ controllable and an application of Lemma 9.1.4 yields the controllability of $(-A^* + X_+ D^{1/2} K, (C + X_+ DX_+)^{1/2})$. Taking $G = D^{1/2}$ we find that $((-A + DX_+)^*, (C + X_+ DX_+)^{1/2}$ is controllable. Now apply the *coup de grâce* with Theorem 5.3.2 to obtain $X_+ > 0$. □

Now we formulate an apparently more comprehensive comparison result than Theorem 9.1.1 although it is, in fact, an easy consequence of the latter theorem. Consider the equation

$$XD'X - XA' - A'^*X - C' = 0. \tag{9.1.10}$$

where D' and C' are hermitian, along with equation (9.1.1). Again A', D', C' are $n \times n$ complex matrices. Define matrices

$$K = \begin{bmatrix} C & A^* \\ A & -D \end{bmatrix}, \quad K' = \begin{bmatrix} C' & A'^* \\ A' & -D' \end{bmatrix} \tag{9.1.11}$$

corresponding to equations (9.1.1) and (9.1.10), respectively, and observe that K and K' are $2n \times 2n$ hermitian matrices.

Corollary 9.1.6. *Assume $D \geq 0$, $C^* = C$, (A, D) is stabilizable. Further, assume that $K \geq K'$ and that (9.1.10) has a hermitian solution X. Then the CARE (9.1.1) has a maximal solution X_+ and $X_+ \geq X$.*

If, in addition, (A', D') is stabilizable, then (9.1.10) has a maximal solution X'_+ and $X_+ \geq X'_+$.

Proof. Let X be a hermitian solution of (9.1.10), then

$$[I \ \ X^*]K' \begin{bmatrix} I \\ X \end{bmatrix} = 0.$$

Since $K \geq K'$ we have

$$[I \ \ X^*]K \begin{bmatrix} I \\ X \end{bmatrix} = [I \ \ X^*](K - K') \begin{bmatrix} I \\ X \end{bmatrix} \geq 0.$$

Thus, X is a solution of the inequality (9.1.2). Now apply Theorem 9.1.1 to establish the first statement.

Observe now that $K \geq K'$ implies $D' \geq D \geq 0$. If, in addition, (A', D') is stabilizable, then by Theorem 7.9.3 the equation (9.1.10) has the maximal solution X'_+. The first part of this corollary implies $X'_+ \leq X_+$. □

Further applications of Theorem 9.1.2 will be given in Section 9.3 for the special form of the algebraic Riccati equation that appears in the LQR problem (see Chapter 16).

9.2 The rate of convergence

As in (9.1.1), let $\mathcal{R}(X) = XDX - XA - A^*X - C$, the Riccati function which, if $D^* = D$, $C^* = C$, maps hermitian matrices to hermitian matrices. The set of all hermitian matrices of size n forms a linear vector space $\mathcal{H}$ over $\mathbb{R}$, and it is possible to formulate the Frechét derivatives of the function $\mathcal{R}$. The first Frechét derivatives at a matrix X is a linear map $\mathcal{R}'_X : \mathcal{H} \to \mathcal{H}$ and is easily found to be (see Ostrowski (1973), Ortega and Rheinboldt (1970), Hille and Phillips (1957), for example):

$$\mathcal{R}'_X(H) = -\{H(A - DX) + (A - DX)^*H\}. \tag{9.2.1}$$

Also the second derivative at X, $\mathcal{R}''_X : (\mathcal{H} \times \mathcal{H}) \to \mathcal{H}$ is given by

$$\mathcal{R}''_X(H_1, H_2) = H_1DH_2 + H_2DH_1.$$

The Newton–Kantorovich procedure for the solution of $\mathcal{R}(X) = 0$ is now

$$X_{\nu+1} = X_\nu - (\mathcal{R}'_{X_\nu})^{-1}\mathcal{R}(X_\nu), \quad \nu = 0, 1, 2, \ldots \tag{9.2.2}$$

or, $\mathcal{R}'_{X_\nu}(X_{\nu+1} - X_\nu) = -\mathcal{R}(X_\nu)$. Using the representation (9.2.1) it is easily seen that this recurrence (when it is well-defined) is simply equation (9.1.4), the basis of the constructive methods of this chapter.

Once the recurrence (9.1.4) is understood in this way there is some expectation of the high rate of local convergence generally associated with Newton's method, given the existence of $(\mathcal{R}'_X)^{-1}$ in a neighborhood of the solution of $\mathcal{R}(X)$ obtained from (9.2.2) in the limit as $\nu \to \infty$. In this context, the surprising feature of the convergence proof obtained in the preceding section is that it is *not* a local argument. Let us show that, using the techniques established in the proof of Theorem 9.1.1, quadratic convergence is ensured for the whole sequence $\{X_\nu\}_{\nu=0}^\infty$ in the following sense:

Theorem 9.2.1. *Assume $D \geq 0$, $C^* = C$, (A, D) is stabilizable, and there exists a maximal hermitian solution X_+ of $\mathcal{R}(X) = 0$ for which $A - DX_+$ is stable. Then, for the sequence $\{X_\nu\}_{\nu=0}^\infty$ defined by (9.1.4) and a hermitian matrix X_0 for which $A - DX_0$ is stable, there is a constant $\kappa > 0$ such that, for $\nu = 0, 1, 2, \ldots$,*

$$\|X_{\nu+1} - X_+\| \leq \kappa \|X_\nu - X_+\|^2, \qquad (9.2.3)$$

and $\|\ \|$ is the spectral norm.

Proof. First rewrite equation (9.1.7) in the form

$$X_{\nu+1}(A - DX_{\nu+1}) + (A - DX_{\nu+1})^* X_{\nu+1} =$$
$$-C - X_{\nu+1} D X_{\nu+1} - (X_{\nu+1} - X_\nu) D (X_{\nu+1} - X_\nu).$$

Then set $Y = X_+$ and $\hat{Y} = X_{\nu+1}$ in the identity (9.1.5) to obtain

$$X_+(A - DX_{\nu+1}) + (A - DX_{\nu+1})^* X_+ =$$
$$-C - X_{\nu+1} D X_{\nu+1} + (X_{\nu+1} - X_+) D (X_{\nu+1} - X_+).$$

Subtract these equations to get

$$(X_+ - X_{\nu+1})(A - DX_{\nu+1}) + (A - DX_{\nu+1})^* (X_+ - X_{\nu+1})$$
$$= (X_{\nu+1} - X_+) D (X_{\nu+1} - X_+) + (X_{\nu+1} - X_\nu) D (X_{\nu+1} - X_\nu).$$

Then put $A - DX_{\nu+1} = (A - DX_+) - D(X_{\nu+1} - X_+)$ on the left to obtain

$$(X_+ - X_{\nu+1})(A - DX_+) + (A - DX_+)^* (X_+ - X_{\nu+1})$$
$$= -(X_{\nu+1} - X_+) D (X_{\nu+1} - X_+) + (X_{\nu+1} - X_\nu) D (X_{\nu+1} - X_\nu).$$

Since $(A - DX_+)$ is stable and we have already proved $X_{\nu+1} \geq X_+$, equation (5.3.3) gives

$$0 \leq X_{\nu+1} - X_+ \leq \int_0^\infty e^{(A-DX_+)t} (X_{\nu+1} - X_\nu) D (X_{\nu+1} - X_\nu) e^{(A-DX_+)^* t} \, dt.$$

As $A - DX_+$ is stable there exist $\alpha > 0$ and $\kappa_0 > 0$ such that $\|e^{A-DX_+}\| \leq \kappa_0 e^{-\alpha t}$ for all $t \geq 0$. Consequently, there is a $\kappa > 0$ such that

$$\|X_{\nu+1} - X_+\| \leq \kappa \|X_{\nu+1} - X_\nu\|^2.$$

Finally, we know that $X_\nu \downarrow X_+$ so that $\|X_\nu - X_{\nu+1}\| \leq \|X_\nu - X_+\|$, and (9.2.3) is obtained. □

238 CONSTRUCTIVE EXISTENCE AND COMPARISON THEOREMS

An interesting question remains concerning the rate of convergence when $A - DX_+$ has eigenvalues on the imaginary axis, as well as the left half-plane. For example, there is a theorem of Ostrowski (Theorem 40.1 of Ostrowski (1973)) showing that under certain hypotheses (which are not readily verified for our problem), and when $\mathcal{R}'_{X_+}$ fails to be invertible, the convergence may be linear (in the limit as $\nu \to \infty$).

Note that Theorems 9.1.2, 9.1.3 and Theorem 9.1.5 give useful conditions under which quadratic convergence can be guaranteed. The following example shows that linear convergence can also occur:

Example 9.2.1. Let

$$D = \begin{bmatrix} 1 & 0 \\ 0 & 0 \end{bmatrix}, \quad A = \begin{bmatrix} 0 & 0 \\ 0 & -1 \end{bmatrix}, \quad C = \begin{bmatrix} 0 & 1 \\ 1 & 2 \end{bmatrix}.$$

It is easily verified that there is a unique hermitian solution of $\mathcal{R}(X) = 0$, namely,

$$X_+ = \begin{bmatrix} 0 & 1 \\ 1 & 1/2 \end{bmatrix}.$$

Thus, (A, D) is stabilizable, and $A - DX_+ = \begin{bmatrix} 0 & -1 \\ 0 & -1 \end{bmatrix}$ and is not stable. Note also that, if

$$M = i \begin{bmatrix} -A & D \\ C & A^T \end{bmatrix}$$

then $\sigma(M) = \{i, -i, 0, 0\}$.

Start Newton iterations (i.e. apply the recursion (9.1.4)) with $X_0 = \begin{bmatrix} 1 & 0 \\ 0 & 0 \end{bmatrix}$, and it can be proved by induction that, for $n = 1, 2, 3, \ldots$,

$$X_n = \begin{bmatrix} 2^{-n} & 1 - 2^{-n} \\ 1 - 2^{-n} & \frac{1}{2} + 2^{-n} \end{bmatrix}.$$

Consequently, for $n = 1, 2, 3, \ldots$, it is found that

$$\frac{\|X_{n+1} - X_+\|}{\|X_n - X_+\|} = \frac{1}{2}. \quad \square$$

9.3 Stabilizing and almost stabilizing solutions

Recall that a hermitian solution X of the CARE (9.1.1) is called stabilizing (resp. almost stabilizing) if all the eigenvalues of $A - DX$ have negative (resp. nonpositive) real parts. Such solutions play important roles in applications, therefore

STABILIZING AND ALMOST STABILIZING SOLUTIONS

conditions that guarantee existence and/or uniqueness of stabilizing and almost stabilizing solutions are of considerable interest. Several results of this kind have already appeared in Chapters 7–9 (for example, Theorem 9.1.5, Proposition 7.9.2). Here we present further results concerning stabilizing and almost stabilizing solutions. In the proofs we will rely heavily on the geometric theory developed in Chapter 7 and on the comparison theorems of Section 9.1. Standing assumptions in this section are $D \geq 0$ and $C = C^*$; additional hypotheses will be assumed when necessary.

First of all, recall that by Proposition 7.9.2 the stabilizing solution, if it exists, is unique. On the other hand, an almost stabilizing solution need not be unique:

Example 9.3.1. If

$$D = A = \begin{bmatrix} 1 & 0 \\ 0 & 0 \end{bmatrix}; \quad C = \begin{bmatrix} -1 & 0 \\ 0 & 0 \end{bmatrix}$$

then all hermitian solutions of (9.1.1) are given by $X = \begin{bmatrix} 1 & 0 \\ 0 & \alpha \end{bmatrix}$, $\alpha \in \mathbb{R}$, and each one of them is almost stabilizing. □

A criterion for existence and uniqueness of an almost stabilizing solution is given in the following theorem.

Theorem 9.3.1. *There exists a unique almost stabilizing solution of the CARE (9.1.1) if and only if (A, D) is stabilizable and the partial multiplicities of pure imaginary (or zero) eigenvalues of the matrix $M := \begin{bmatrix} -A & D \\ C & A^* \end{bmatrix}$ are all even. In this case, the almost stabilizing solution is the maximal hermitian solution.*

It is convenient to state separately a lemma which will be used in the proof of Theorem 9.3.1.

Lemma 9.3.2. *Let*

$$X = \begin{bmatrix} X_{11} & X_{12} \\ X_{21} & X_{22} \end{bmatrix}$$

be an $n \times n$ matrix partitioned into matrix blocks, as shown. Assume that at least one of X_{12} and X_{21} is the zero matrix. If $\lambda_0 \in \sigma(X)$ but $\lambda_0 \notin \sigma(X_{22})$, then the partial multiplicities of X corresponding to λ_0 coincide with those of X_{11} corresponding to the same λ_0.

We leave the proof of this lemma as an exercise.

Proof of Theorem 9.3.1. Assume that (A, D) is stabilizable and the condition on the eigenvalues of M is satisfied. Let $\mathcal{C}$ be the controllable subspace of (A, D) and, with respect to the decomposition $\mathbb{C}^n = \mathcal{C} \oplus \mathcal{C}^\perp$, write (as in (7.9.2)):

$$A = \begin{bmatrix} A_1 & A_{12} \\ 0 & A_2 \end{bmatrix}, \quad D = \begin{bmatrix} D_1 & 0 \\ 0 & 0 \end{bmatrix}, \quad C = \begin{bmatrix} C_1 & C_{12} \\ C_{12}^* & C_2 \end{bmatrix}. \quad (9.3.1)$$

Observe that A_2 is c-stable. The structure of A, D and C displayed by equations (9.3.1) implies that the partial multiplicities of pure imaginary (or zero) eigenvalues of M coincide with those for the matrix $M := \begin{bmatrix} -A_1 & D_1 \\ C_1 & A_1^* \end{bmatrix}$. Indeed, we have

$$\begin{bmatrix} I & 0 & 0 & 0 \\ 0 & 0 & I & 0 \\ 0 & I & 0 & 0 \\ 0 & 0 & 0 & I \end{bmatrix} M \begin{bmatrix} I & 0 & 0 & 0 \\ 0 & 0 & I & 0 \\ 0 & I & 0 & 0 \\ 0 & 0 & 0 & I \end{bmatrix} = \begin{bmatrix} -A_1 & D_1 & -A_2 & 0 \\ C_1 & A_1^* & C_{12} & 0 \\ 0 & 0 & -A_2 & 0 \\ C_{12}^* & A_{12}^* & C_2 & A_2^* \end{bmatrix}, \quad (9.3.2)$$

and it remains to apply Lemma 9.3.2 twice to the right-hand side of (9.3.2). By Theorem 7.9.1 the CARE (9.1.1) has a hermitian solution and an application of Theorem 7.9.3 shows existence and uniqueness of an almost stabilizing solution which coincides with the maximal hermitian solution.

Conversely, let X be the unique almost stabilizing solution of (9.1.1). Partition

$$X = \begin{bmatrix} X_1 & X_{12} \\ X_{12}^* & X_2 \end{bmatrix}$$

with respect to $\mathbb{C}^n = \mathcal{C} \oplus \mathcal{C}^\perp$; then equations (7.9.4)–(7.9.6) are satisfied, where the partitions (9.3.1) of A, D and C are used. Since

$$A - DX = \begin{bmatrix} A_1 - D_1 X_1 & A_{12} - D_1 X_{12} \\ 0 & A_2 \end{bmatrix},$$

it follows that $\sigma(A_2)$ lies in the closed left half-plane. Now the *uniqueness* of X implies that the hermitian solution X_2 of (7.9.6) is unique. It follows from Theorem 5.2.2 that

$$\sigma(A_2) \cap \sigma(-A_2^*) = \emptyset. \quad (9.3.3)$$

Indeed, if (9.3.3) does not hold, there will exist two different solutions Z_1 and Z_2 of the equation

$$ZA_2 + A_2^* Z = Q, \quad (9.3.4)$$

where Q is the right-hand side of (7.9.6). Then (9.3.4) has two different *hermitian* solutions $\frac{1}{2}(Z_1 + Z_1^*)$ and $\frac{1}{2}(Z_2 + Z_2^*)$, or the corresponding homogeneous equation

$$ZA_2 + A_2^* Z = 0$$

has two different *hermitian* solutions $\frac{1}{2i}(Z_1 - Z_1^*)$ and $\frac{1}{2i}(Z_2 - Z_2^*)$. In either case a contradiction is obtained with our assumption that X is unique. Thus,

A_2 must be stable, and therefore the pair (A, D) is stabilizable. Now Theorem 7.9.1 implies that the partial multiplicities corresponding to the pure imaginary (or zero) eigenvalues of M_1 are all even. In view of equation (9.3.2) and, making use of Lemma 9.3.2 again, the same is true for M. □

Of particular interest are positive semidefinite almost stabilizing and stabilizing solutions of the CARE (9.1.1). We have:

Theorem 9.3.3. *Assume $C \geq 0$. The following statements are equivalent:*

(i) *there is a stabilizing positive semidefinite solution;*

(ii) (A, D) *is stabilizable and* $M = \begin{bmatrix} -A & D \\ C & A^* \end{bmatrix}$ *has no pure imaginary (or zero) eigenvalues;*

(iii) (A, D) *is stabilizable and the pure imaginary (or zero) eigenvalues of A, if any, are C-observable, i.e.*

$$Au = i\alpha u, \quad Cu = 0, \quad \alpha \in \mathbb{R} \Longrightarrow u = 0.$$

Proof. Assume (ii) holds. Then by Theorem 9.1.2, the maximal hermitian solution X_+ is positive semidefinite and is almost stabilizing by Theorem 9.3.1. The formula

$$\begin{bmatrix} -A & D \\ C & A^* \end{bmatrix} \begin{bmatrix} I & 0 \\ X_+ & I \end{bmatrix} = \begin{bmatrix} I & 0 \\ X_+ & I \end{bmatrix} \begin{bmatrix} -A + DX_+ & D \\ 0 & A^* - X_+ D \end{bmatrix} \quad (9.3.5)$$

shows that in fact $A - DX_+$ is stable. So (i) holds.

Conversely, assume (i) holds. By Proposition 7.9.2 there is a unique almost stabilizing solution, so Theorem 9.3.1 is applicable. In particular, the stabilizing solution is maximal, and therefore is positive semidefinite by Theorem 9.1.2. Now equality (9.3.5) guarantees that M has no eigenvalues on the imaginary axis.

It remains to relate the conditions (i) and (ii) to (iii). This is left as an exercise. □

9.4 Comparison theorems: the LQR form

In this section we study comparison theorems for Riccati equations of the form

$$XBR^{-1}B^*X - X(A - BR^{-1}C) - (A - BR^{-1}C)X - (Q - C^*R^{-1}C) = 0, \quad (9.4.1)$$

where the matrices A, B, C, Q, and R are of sizes $n \times n$, $n \times m$, $m \times n$, $n \times n$ and $m \times m$, respectively, R is positive definite, and Q is hermitian. Again, X is

the $n \times n$ matrix to be found. The connection with equations of the form (9.1.1) is clear:
$$X\tilde{D}X - X\tilde{A} - \tilde{A}^*X - \tilde{C} = 0, \qquad (9.4.2)$$
where $\tilde{A} = A - BR^{-1}C$, $\tilde{D} = BR^{-1}B^*$, $\tilde{C} = Q - C^*R^{-1}C$. The stabilizability condition translates well:

Lemma 9.4.1. *The pair (A, B) is stabilizable if and only if the pair $(\tilde{A}, \tilde{D})$ is stabilizable.*

Proof. This follows easily from Lemma 4.5.3. □

We now state the comparison theorem. Together with (9.4.1), consider the equation
$$XB\tilde{R}^{-1}B^*X - X(A - B\tilde{R}^{-1}\tilde{C}) - (A - B\tilde{R}^{-1}\tilde{C})^*X - (\tilde{Q} - \tilde{C}^*\tilde{R}^{-1}\tilde{C}) = 0, \quad (9.4.3)$$
with the same A and B, but generally different $\tilde{R}$, $\tilde{C}$ and $\tilde{Q}$.

Theorem 9.4.2. *Assume that $R > 0$, $Q = Q^*$ and (A, B) is stabilizable. Further assume that $\tilde{R} > 0$ and $\tilde{Q} = \tilde{Q}^*$. If*
$$\begin{bmatrix} \tilde{Q} & \tilde{C}^* \\ \tilde{C} & \tilde{R} \end{bmatrix} \le \begin{bmatrix} Q & C^* \\ C & R \end{bmatrix} \qquad (9.4.4)$$
and (9.4.3) has a hermitian solution, then both (9.4.1) and (9.4.3) have maximal hermitian solutions X_+ and $\tilde{X}_+$, respectively, and $X_+ \ge \tilde{X}_+$.

For the proof of Theorem 9.4.2 it will be convenient to introduce the notation
$$\tilde{T} = \begin{bmatrix} \tilde{Q} & \tilde{C}^* \\ \tilde{C} & \tilde{R} \end{bmatrix}, \quad T = \begin{bmatrix} Q & C^* \\ C & R \end{bmatrix}.$$

First we prove another lemma.

Lemma 9.4.3. *Let X be a hermitian solution of (9.4.3). Then*
$$-XBR^{-1}B^*X + X(A - BR^{-1}C) + (A - BR^{-1}C)^*X$$
$$+ (Q - C^*R^{-1}C) = \begin{bmatrix} I \\ -L \end{bmatrix}^* (T - \tilde{T}) \begin{bmatrix} I \\ -L \end{bmatrix} + G^*\tilde{R}G, \quad (9.4.5)$$
where
$$L := R^{-1}(C + B^*X), \quad G := L - \tilde{R}^{-1}(\tilde{C} + B^*X).$$

Proof. We denote $\tilde{R}^{-1}(\tilde{C} + B^*X)$ by $\tilde{L}$. Also, denote the left-hand side of (9.4.5) by $\mathcal{R}(X)$. Rewrite $\mathcal{R}(X)$ and the equation (9.4.3) as follows:

$$\mathcal{R}(X) = XA + A^*X - L^*RL + Q,$$

and

$$XA + A^*X - \tilde{L}^*\tilde{R}\tilde{L} + \tilde{Q} = 0.$$

Now a computation completes the proof of this lemma.

$$\mathcal{R}(X) - \begin{bmatrix} I \\ -L \end{bmatrix}^* \{T - \tilde{T}\} \begin{bmatrix} I \\ -L \end{bmatrix} - G^*\tilde{R}G$$
$$= Q - L^*RL - \tilde{Q} + \tilde{L}^*\tilde{R}\tilde{L} - Q + \tilde{Q} + L^*C - L^*\tilde{C} + C^*L - \tilde{C}^*L$$
$$\quad - L^*RL + L^*\tilde{R}\tilde{L} - L^*\tilde{R}L - \tilde{L}^*\tilde{R}\tilde{L} + L^*\tilde{R}\tilde{L} + \tilde{L}^*\tilde{R}L$$
$$= -L^*(RL - C - \tilde{R}\tilde{L} + \tilde{C}) - (RL - C - \tilde{R}\tilde{L} + \tilde{C})^*L = 0. \quad \square$$

Proof of Theorem 9.4.2. Since (9.4.4) holds, it follows from Lemma 9.4.3 that for any hermitian solution X of (9.4.3) the inequality

$$XBR^{-1}B^*X - X(A - BR^{-1}C) - (A - BR^{-1}C)^*X - (Q - C^*R^{-1}C) \leq 0 \quad (9.4.6)$$

holds. By Theorem 9.1.1 (which is applicable in view of Lemma 9.4.1) there is a maximal solution X_+ of (9.4.1), and $X_+ \geq X$ for any hermitian solution X of (9.4.3). Finally, the existence of the maximal solution of (9.4.3) follows from Theorem 7.9.3. $\square$

We indicate one important case when the existence of a positive semidefinite maximal solution is ensured.

Theorem 9.4.4. *Assume that $R > 0$, $Q = Q^*$, (A, B) is stabilizable and*

$$\begin{bmatrix} Q & C^* \\ C & R \end{bmatrix} \geq 0.$$

Then the equation (9.4.1) has a maximal solution X_+, and $X_+ \geq 0$.

Proof. The Schur decomposition

$$\begin{bmatrix} I & -C^*R^{-1} \\ 0 & I \end{bmatrix} \begin{bmatrix} Q & C^* \\ C & R \end{bmatrix} \begin{bmatrix} I & 0 \\ -R^{-1}C & I \end{bmatrix} = \begin{bmatrix} Q - C^*R^{-1}C & 0 \\ 0 & R \end{bmatrix}$$

shows that $Q - C^*R^{-1}C \geq 0$. So the inequality (9.4.6) has the solution $X = 0$. Now argue as in the proof of Theorem 9.4.2. $\square$

As in Theorem 9.1.1, the maximal solution X_+ of (9.4.1) has the property that all eigenvalues of $A - BR^{-1}(C + B^*X_+)$ lie in the closed left half-plane. Also, Theorems 9.1.2, 9.1.3 and 9.1.5 can be applied to the equation (9.4.1) to yield results concerning the c-stability of $A - BR^{-1}(C + B^*X_+)$. We omit the straightforward statements of these results.

9.5 The real case

All the results in this chapter apply to the CARE

$$XDX - XA - A^T X - C = 0 \tag{9.5.1}$$

with real matrix coefficients A, D and C (where D is symmetric positive semidefinite, $C = C^T$, and (A, D) is stabilizable). These results are obtained by simply applying the corresponding results for the complex equation (9.1.1) and, by noticing that the maximal solution of (9.5.1) in the set of all hermitian solutions (when it exists) must be real. We leave the statements of these results to the interested reader.

9.6 Exercises

9.6.1. Prove that under the hypotheses of Corollary 9.1.6, the stabilizability of (A', D'') is ensured if $A' = A$.

9.6.2. Verify that the Newton–Kantorovich scheme of equation (9.2.2) does, indeed, determine the recurrence relation of equation (9.1.4).

9.6.3. Prove the equivalence of (i) and (ii) to (iii) in Theorem 9.3.3.

Hint: Use equality (9.3.5), which implies the similarity of $\begin{bmatrix} -A & D \\ C & A^* \end{bmatrix}$ and $\begin{bmatrix} -A + DX_+ & D \\ 0 & A^* - X_+D \end{bmatrix}$.

9.6.4. Prove Lemma 9.3.2.

9.6.5. The results of Section 9.3 have counterparts concerning antistabilizing solutions of (9.1.1), i.e. hermitian solutions X for which $A - DX$ has all its eigenvalues in the open right half-plane. Prove the counterpart of Theorem 9.3.3: Assume $C \geq 0$. Then the following statements are equivalent:

(i) there is an antistabilizing negative semidefinite solution of (9.1.1);

(ii) $(-A, D)$ is stabilizable and M has no pure imaginary (or zero) eigenvalues;

(iii) $(-A, D)$ is stabilizable and the pure imaginary (or zero) eigenvalues of A are C-observable.

Hint: Apply Theorem 9.3.3 to the equation

$$XDX + XA + A^*X - C = 0. \qquad (9.6.1)$$

The equality

$$\begin{bmatrix} I & 0 \\ 0 & -I \end{bmatrix} \begin{bmatrix} -A & D \\ C & A^* \end{bmatrix} \begin{bmatrix} -I & 0 \\ 0 & I \end{bmatrix} = \begin{bmatrix} A & D \\ C & -A^* \end{bmatrix}$$

shows that

$$\sigma(M) = -\sigma \begin{bmatrix} A & D \\ C & -A^* \end{bmatrix}.$$

9.7 Notes

The methods of this chapter originate with a paper of Kleinman (1968) concerning controllable systems. The theory has gone through a sequence of refinements admitting stabilizable systems and weakening the hypothesis of existence of hermitian solutions of the Riccati *equation* to the Riccati *inequality* (as in Theorem 9.1.3). In this sequence we mention works of Coppel (1974), Wimmer (1976 and 1985), Gohberg et al. (1986b), and Ran and Vreugdenhil (1988).

It is clear from the treatment given in Section 9.1 that there are close connections between the study of Riccati equations and inequalities. There is extensive literature on the inequalities of which we mention only a few which are most closely connected with the equations: Willems (1971), Faibusovich (1987), Scherer (1991), Trentelman and Willems (1991), and Lindquist et al. (1994). In particular, the paper of Scherer contains extensive analysis of the Riccati inequality under the mild hypothesis of sign controllability.

The proof of Theorem 9.1.1 follows that given by Gohberg et al.(1986b), and is based on the second proof of Theorem 2.1 of Coppel (1974). Theorem 9.1.3 is due to Kaashoek and Ran (1991). Lemma 9.1.4 is due to Wonham (1968). The proof of Theorem 9.2.1 is based on the exposition of Mehrmann (1991) who attributes the argument to L. Elsner. Theorem 9.3.1 is well-known. It can be viewed as a particular case of the more general result described in Exercise 7.11.14. Theorem 9.3.3 was originally proved by Kučera (1972a). The proof of Theorem 9.4.2 is taken from the paper of Ran and Vreugdenhil (1988).

10
HERMITIAN SOLUTIONS AND FACTORIZATIONS OF RATIONAL MATRIX FUNCTIONS

Chapters 7 and 8 focus on a geometric theory connecting hermitian solutions of the symmetric algebraic Riccati equation (7.2.1) with invariant subspaces of the associated Hamiltonian matrix $M = i \begin{bmatrix} A & D \\ C & -A^* \end{bmatrix}$. In Chapter 9 a quite different approach is developed using an iterative process. In the present chapter a third, apparently quite different, line of attack is developed. It begins with the easy observation that, if (7.2.1) (or (10.1.1) below) has a hermitian solution X then the rational matrix function

$$Z(\lambda) := I + D^{1/2}(\lambda I + iA^*)^{-1}C(\lambda I - iA)^{-1}D^{1/2}$$

is positive semidefinite on the real axis, i.e. $Z(\lambda) \geq 0$ for all $\lambda \in \mathbb{R}$ at which $Z(\lambda)$ is defined (see the first step in the proof of Theorem 10.1.1). Theorem 10.1.1 goes on to present necessary and sufficient conditions for the existence of hermitian solutions of the CARE in these terms.

Now it is well-known (and is the subject of Theorem 6.7.2) that certain rational matrix functions which are positive semidefinite on the real axis admit symmetric factorization (in the form of equation (6.7.4)). Developing this idea in our context leads to Theorem 10.1.2 and its Corollary 10.1.3.

Adaptation of the theory developed in Section 10.1 to the study of real symmetric solutions of the CARE with real coefficients is the subject of Section 10.2.

10.1 Nonnegative rational matrix functions

Our first objective in this section is to connect the hermitian solutions of the CARE

$$XDX + XA + A^*X - C = 0 \tag{10.1.1}$$

with certain rational matrix functions and their minimal factorizations. It is assumed throughout this chapter that A, C, D are $n \times n$, with $D \geq 0$, $C^* = C$ and (A, D) is a controllable pair.

The first result concerns the existence of hermitian solutions. We let D_0 be any $n \times n$ matrix for which $D = D_0 D_0^*$. For example, D_0 could be the unique positive semidefinite square root of D, as described in Section 1.5.

Theorem 10.1.1. *The following statements are equivalent:*

(i) *the CARE* (10.1.1) *has hermitian solutions;*

(ii) *the rational matrix function*

$$Z(\lambda) := I + D_0^*(\lambda I + iA^*)^{-1} C(\lambda I - iA)^{-1} D_0 \qquad (10.1.2)$$

is nonnegative on the real axis, i.e. $x^* Z(\lambda) x \geq 0$ for every $x \in \mathbb{C}^n$ and every real λ which is not a pole of $Z(\lambda)$;

(iii) *the rational matrix function*

$$W(\lambda) = I + (\lambda I - iA)^{-1} D_0 Z(\lambda)^{-1} D_0^* (\lambda I + iA^*)^{-1} \qquad (10.1.3)$$

is nonnegative on the real axis.

As the proof will show, the implications (i) $\Rightarrow$ (ii) $\Rightarrow$ (iii) are valid without the assumption that (A, D) is controllable. This assumption is essential only in the implication (iii) $\Rightarrow$ (i).

Proof. Let X be a hermitian solution of the CARE (10.1.1). Then the equality

$$X(i\lambda I + A) + (-i\lambda I + A^*)X + XDX = C$$

holds for a complex λ. Premultiplying by $D_0^*(-i\lambda I + A^*)^{-1}$, postmultiplying by $(i\lambda I + A)^{-1} D_0$, and adding I to both parts, we have

$$I + D_0^*(-i\lambda I + A^*)^{-1} X D_0$$
$$+ D_0^* X (i\lambda I + A)^{-1} D_0 + D_0^*(-i\lambda I + A^*)^{-1} X D_0 D_0^* X (i\lambda I + A)^{-1} D_0$$
$$= I + D_0^*(-i\lambda I + A^*)^{-1} C (i\lambda I + A)^{-1} D_0 \qquad (10.1.4)$$

for every λ which is not an eigenvalue of iA or of $-iA^*$. The right-hand side of (10.1.4) is just $Z(\lambda)$, while the left-hand side is equal to

$$(I + D_0^*(-i\lambda I + A^*)^{-1} X D_0)(I + D_0^* X (i\lambda I + A)^{-1} D_0),$$

which is positive semidefinite for real $\lambda \notin \sigma(iA) \cup \sigma(-iA^*)$ in view of the equality $X = X^*$. Thus, $Z(\lambda) \geq 0$ for every real $\lambda \notin \sigma(iA) \cup \sigma(-iA^*)$. If some real λ_0 is not a pole of $Z(\lambda)$ but is an eigenvalue of at least one of iA and $(-iA^*)$, then $Z(\lambda_0) \geq 0$ still holds since $Z(\lambda) \geq 0$ for all real $\lambda \neq \lambda_0$ sufficiently close to λ_0, and since the set of positive semidefinite matrices is closed. Thus, (i) $\Rightarrow$ (ii). The implication (ii) $\Rightarrow$ (iii) is evident.

NONNEGATIVE RATIONAL MATRIX FUNCTIONS

Now assume that $W(\lambda)$ is nonnegative on the real axis. Rewrite $W(\lambda)$ in the form

$$\begin{aligned} W(\lambda) &= I + (\lambda I - iA)^{-1}\{I + D(\lambda I + iA^*)^{-1}C(\lambda I - iA)^{-1}\}^{-1}D(\lambda I + iA^*)^{-1} \\ &= I + \{\lambda I - iA + D(\lambda I + iA^*)^{-1}C\}^{-1}D(\lambda I + iA^*)^{-1} \\ &= I - iV(\lambda), \end{aligned}$$

where

$$V(\lambda) = -\{\lambda I - iA + D(\lambda I + iA^*)^{-1}C\}^{-1}(-iD)(\lambda I + iA^*)^{-1}.$$

Now observe that $V(\lambda)$ is just the $n \times n$ upper right quarter in the $2n \times 2n$ matrix

$$\begin{bmatrix} \lambda I - iA & -iD \\ -iC & \lambda I + iA^* \end{bmatrix}^{-1} = (\lambda I - M)^{-1},$$

where $M = i\begin{bmatrix} A & D \\ C & -A^* \end{bmatrix}$, as in Section 7.2. Indeed, this follows from the general fact that the inverse of an $(n+m) \times (n+m)$ matrix

$$S = \begin{bmatrix} S_{11} & S_{12} \\ S_{21} & S_{22} \end{bmatrix},$$

where the size of S_{11} (resp. S_{22}) is $n \times n$ (resp. $m \times m$), is given by the formula

$$S^{-1} = \begin{bmatrix} T^{-1} & -T^{-1}S_{12}S_{22}^{-1} \\ -S_{22}^{-1}S_{21}T^{-1} & S_{22}^{-1}S_{21}T^{-1}S_{12}S_{22}^{-1} + S_{22}^{-1} \end{bmatrix}, \quad (10.1.5)$$

where $T = S_{11} - S_{12}S_{22}^{-1}S_{21}$, provided the matrices S_{22} and T are invertible. So

$$W(\lambda) = I - i[I\ 0](\lambda I - M)^{-1}\begin{bmatrix} 0 \\ I \end{bmatrix} = I + R^*\hat{H}(\lambda I - M)^{-1}R, \quad (10.1.6)$$

where $R = \begin{bmatrix} 0 \\ I \end{bmatrix}$ and $\hat{H} = i\begin{bmatrix} 0 & I \\ -I & 0 \end{bmatrix}$.

Our next step is to prove that realization (10.1.6) is minimal (see Section 6.1 for a review of the notion of a minimal realization). Since M is $\hat{H}$-self-adjoint, i.e. $\hat{H}M = M^*\hat{H}$, it follows from Theorem 6.1.5 that we have only to check that $\dim \mathcal{C}_{M,R} = \operatorname{rank}[R, MR, \ldots, M^m R] = 2n$ for m large enough. As we know (see Theorem 7.2.3) equation (10.1.1) always has a (not necessarily hermitian) solution X. Then observe that

$$\begin{bmatrix} I & 0 \\ -X & I \end{bmatrix} M \begin{bmatrix} I & 0 \\ X & I \end{bmatrix} = i \begin{bmatrix} A+DX & D \\ 0 & -XD-A^* \end{bmatrix}, \quad \begin{bmatrix} I & 0 \\ -X & I \end{bmatrix} R = R$$

and, clearly, it will suffice to prove that

$$\operatorname{rank}[R, \tilde{M}R, \ldots, \tilde{M}^m R] = 2n \tag{10.1.7}$$

for m large enough, where

$$\tilde{M} = \begin{bmatrix} A+DX & D \\ 0 & -XD-A^* \end{bmatrix}.$$

A simple induction argument shows that, for $k = 1, 2, \ldots,$

$$\tilde{M}^k R = \begin{bmatrix} \sum_{j=0}^{k-1}(A+DX)^j D(-XD-A^*)^{k-1-j} \\ * \end{bmatrix}. \tag{10.1.8}$$

Denoting $Y_1 = A + DX$, $Y_2 = -XD - A^*$, we have

$$[D, Y_1 D + DY_2, \ldots, \sum_{j=0}^{k-1} Y_1^j DY_2^{k-1-j}] \begin{bmatrix} I & -Y_2 & \ldots & 0 \\ 0 & I & & \vdots \\ \vdots & \vdots & \ddots & \\ & & & -Y_2 \\ 0 & 0 & \ldots & I \end{bmatrix}$$

$$= [D, Y_1 D, \ldots, Y_1^{k-1} D];$$

so

$$\operatorname{rank}[D, Y_1 D + DY_2, \ldots, \sum_{j=0}^{k-1} Y_1^j DY_2^{k-1-j}] = \operatorname{rank}[D, Y_1 D, \ldots, Y_1^{k-1} D] \tag{10.1.9}$$

for $k = 1, 2, \ldots$. But, as (A, D) is controllable, so is (Y_1, D) (see Lemma 4.4.1), and so we have

$$\operatorname{rank}[D, Y_1 D, \ldots, Y_1^{k-1} D] = n$$

for sufficiently large k. Now (10.1.7) follows from (10.1.8) and (10.1.9), and the realization (10.1.6) is indeed minimal.

Now we can quickly finish the proof of Theorem 10.1.1. It follows from Lemma 6.5.1 that all partial multiplicities of $W(\lambda)$ corresponding to real points are even. Since (10.1.6) is a minimal realization it now follows from Theorem 6.2.2 that the partial multiplicities corresponding to the real eigenvalues (if any) of M are all even. Now Theorem 7.6.1 implies that the CARE (10.1.1) admits a hermitian solution. □

Let us analyse the equivalence of statements (i) and (iii) in Theorem 10.1.1 more closely. As we have seen in the proof, the function $W(\lambda)$ (given by (10.1.3)) can also be written in the form

$$W(\lambda) = I + R^* \hat{H}(\lambda I - M)^{-1} R, \qquad (10.1.10)$$

where

$$R = \begin{bmatrix} 0 \\ I \end{bmatrix}, \quad \hat{H} = i \begin{bmatrix} 0 & I \\ -I & 0 \end{bmatrix}, \quad M = i \begin{bmatrix} A & D \\ C & -A^* \end{bmatrix}, \qquad (10.1.11)$$

and the realization (10.1.10) is minimal. If (10.1.1) has hermitian solutions, then $W(\lambda)$ is nonnegative on the real axis, and therefore $W(\lambda)^{-1}$ is also nonnegative on the real axis. A minimal realization of $W(\lambda)^{-1}$ is found using Proposition 6.4.1:

$$W(\lambda)^{-1} = I - R^* \hat{H}(\lambda I - M^\times)^{-1} R,$$

where $M^\times := M - RR^* \hat{H} = i \begin{bmatrix} A & D \\ C+I & -A^* \end{bmatrix}$. By Lemma 6.5.1 and Theorem 6.2.2 the partial multiplicities of $M^\times$ corresponding to real eigenvalues (if any) are all even. Now by Theorem 7.6.1 the equation

$$XDX + A^*X + XA - (C+I) = 0 \qquad (10.1.12)$$

admits hermitian solutions as well. To sum up, if (10.1.1) has hermitian solutions, then so does (10.1.12). It turns out that this fact is a particular case of the comparison theorem (Theorem 9.1.1); but here we have established it by an easy manipulation of rational matrix functions.

Note that if $X = X^*$ is a solution of (10.1.1) and $\tilde{X} = \tilde{X}^*$ is a solution of (10.1.12), then the difference $X - \tilde{X}$ is invertible. This fact will be used below in the statement and proof of Theorem 10.1.2. Indeed, let $x \in \mathbb{C}^n$ be such that $Xx = \tilde{X}x$. Then

$$x^* XDXx + x^* XAx + x^* A^* Xx = x^* \tilde{X}D\tilde{X}x + x^* \tilde{X}Ax + x^* A^* \tilde{X}x.$$

But the left-hand side of this equality is x^*Cx, while the right-hand side is $x^*(C+I)x$. So $x = 0$.

There is a one-to-one correspondence between minimal factorizations of $W(\lambda)$ of the type $W(\lambda) = (L(\bar{\lambda}))^* L(\lambda)$ where $L(\lambda)$ is an $n \times n$ rational matrix function with $L(\infty) = I$ and pairs of matrices $(X, \tilde{X})$, where X is a hermitian solution of (10.1.1) and $\tilde{X}$ is a hermitian solution of (10.1.12). (See Section 6.4 for basic facts concerning minimal factorizations of rational matrix functions.) This correspondence is described in the following theorem.

Theorem 10.1.2. *Assume that the rational matrix function $W(\lambda)$ given by (10.1.3) is nonnegative on the real line. Let X and $\tilde{X}$ be hermitian solutions* of (10.1.1) and (10.1.12), respectively. Then $W(\lambda)$ admits a minimal factorization

$$W(\lambda) = (L(\bar{\lambda}))^* L(\lambda) \qquad (10.1.13)$$

with

$$L(\lambda) = I_n + R^* \hat{H} \begin{bmatrix} ZX & -Z \\ \tilde{X}ZX & -\tilde{X}Z \end{bmatrix} (\lambda I_{2n} - M)^{-1} R, \qquad (10.1.14)$$

where $Z = (X - \tilde{X})^{-1}$ and R, $\hat{H}$, M are defined in (10.1.11).

Conversely, any rational matrix function $L(\lambda)$ with $L(\infty) = I$ and for which (10.1.13) is a minimal factorization of $W(\lambda)$, has the form (10.1.14) for some pair X, $\tilde{X}$ of hermitian solutions of (10.1.1) and (10.1.12), respectively.

Moreover, the zeros of $L(\lambda)$ are the eigenvalues of $i(A + D\tilde{X})$, and the poles of $L(\lambda)$ are the eigenvalues of $-i(A + DX)^*$, counting multiplicities in both cases. More exactly, the partial zero (resp. pole) multiplicities at its zero (resp. pole) λ_0 are precisely the multiplicities of $i(A + D\tilde{X})$ (resp. $-i(A + DX)^*$) corresponding to λ_0.

Proof. By Theorem 6.5.3, all minimal factorizations of $W(\lambda)$ of the type $W(\lambda) = (L(\bar{\lambda}))^* L(\lambda)$ with a rational matrix function $L(\lambda)$ such that $L(\infty) = I$ are given by the formula

$$W(\lambda) = \{I + R^* \hat{H} (\lambda I - M)^{-1}(I - \pi) R\}\{I + R^* \hat{H} \pi (\lambda I - M)^{-1} R\}, \qquad (10.1.15)$$

where $\pi : \mathbb{C}^{2n} \to \mathbb{C}^{2n}$ is any projection with the properties that $\operatorname{Ker} \pi$ is M-invariant, $\operatorname{Im} \pi$ is $M^\times$-invariant and both subspaces $\operatorname{Ker} \pi$ and $\operatorname{Im} \pi$ are n-dimensional and $\hat{H}$-neutral. Here

$$M^\times = M - RR^* \hat{H} = i \begin{bmatrix} A & D \\ C + I & -A^* \end{bmatrix}.$$

Now let (10.1.13) be a minimal factorization, where $L(\infty) = I$. Then the subspace $\operatorname{Ker} \pi$ is M-invariant $\hat{H}$-neutral of dimension n. According to Theorem 7.2.4 there is a hermitian solution X of (10.1.1) such that $\operatorname{Ker} \pi = \operatorname{Im} \begin{bmatrix} I \\ X \end{bmatrix}$. Similarly one shows that $\operatorname{Im} \pi = \operatorname{Im} \begin{bmatrix} I \\ \tilde{X} \end{bmatrix}$ for a hermitian solution $\tilde{X}$ of (10.1.12). Note that the matrix

$$\begin{bmatrix} ZX & -Z \\ \tilde{X}ZX & -\tilde{X}Z \end{bmatrix}, \quad Z = (X - \tilde{X})^{-1} \qquad (10.1.16)$$

is a projection with image equal to $\operatorname{Im} \begin{bmatrix} I \\ \tilde{X} \end{bmatrix}$ and kernel equal to $\operatorname{Im} \begin{bmatrix} I \\ X \end{bmatrix}$, and therefore this projection is π. Now compare (10.1.15) with (10.1.16) to obtain the converse statement of the theorem.

THE REAL CASE

Assume now that X and $\tilde{X}$ are hermitian solutions of (10.1.1) and (10.1.12), respectively. Then the subspace $\text{Im} \begin{bmatrix} I \\ X \end{bmatrix}$ is M-invariant $\hat{H}$-neutral of dimension n, and the subspace $\text{Im} \begin{bmatrix} I \\ \tilde{X} \end{bmatrix}$ is $M^\times$-invariant $\hat{H}$-neutral of dimension n (see Proposition 7.2.1). We claim that

$$\mathbb{C}^{2n} = \text{Im} \begin{bmatrix} I \\ X \end{bmatrix} \dotplus \text{Im} \begin{bmatrix} I \\ \tilde{X} \end{bmatrix}.$$

Indeed, assume

$$\begin{bmatrix} x \\ Xx \end{bmatrix} = \begin{bmatrix} y \\ \tilde{X}y \end{bmatrix}$$

for some $x, y \in \mathbb{C}^n$. Then $x = y$ and $Xx = \tilde{X}x$. As $X - \tilde{X}$ is invertible, $x = y = 0$. Let π be the projection onto $\text{Im} \begin{bmatrix} I \\ \tilde{X} \end{bmatrix}$ along $\text{Im} \begin{bmatrix} I \\ X \end{bmatrix}$, then π is given by (10.1.16). According to the remark stated in the beginning of this proof, the factorization (10.1.15) is of the desired form.

It remains to prove the statements about the poles and zeros of $L(\lambda)$. It follows from Theorem 6.4.3 that the zeros of $L(\lambda)$ are the eigenvalues of $M^\times|_{\text{Im }\pi}$, multiplicities counted. Similarly, the poles of $L(\lambda)$ coincide (including multiplicities) with the complex conjugates of the eigenvalues of $M|_{\text{Ker }\pi}$. It remains to recall that $M|_{\text{Ker }\pi}$ is equal to the restriction of M to $\text{Im} \begin{bmatrix} I \\ X \end{bmatrix}$, and is similar to $i(A + DX)$. Similarly, $M^\times|_{\text{Im }\pi}$ is similar to $i(A + D\tilde{X})$. □

Choosing a fixed solution $\tilde{X}$ of (10.1.12) one obtains from Theorem 10.1.2 a one-to-one correspondence between the set of hermitian solutions of (10.1.1) and a set of certain minimal factorizations of type (10.1.13). For instance:

Corollary 10.1.3. *There is a one-one correspondence between hermitian solutions X of (10.1.1) and minimal factorizations of $W(\lambda)$ of the form $W(\lambda) = (L(\bar{\lambda}))^* L(\lambda)$ such that $L(\infty) = I_n$ and $L(\lambda)$ has all its zeros in the closed upper half-plane. This correspondence is given by formula (10.1.14), where $\tilde{X}$ is the maximal hermitian solution of (10.1.12).*

10.2 The real case

Here, we consider the continuous algebraic Riccati equation (10.1.1) with real $n \times n$ coefficients A, D, and C, and with the hypotheses of Section 10.1 (i.e. D is symmetric positive semidefinite, $C = C^T$, and (A, D) is a controllable pair). Let D_0 be any real $n \times n$ matrix such that $D = D_0 D_0^T$. It will be convenient to modify the definitions of the functions $Z(\lambda)$ and $W(\lambda)$ used in the preceding section. Define

$$\tilde{Z}(\lambda) = I - D_0^T(\lambda I + A^T)^{-1}C(\lambda I - A)^{-1}D_0,$$

$$\tilde{W}(\lambda) = I - (\lambda I - A)^{-1}D_0[\tilde{Z}(\lambda)^{-1}]D_0^T(\lambda I + A^T)^{-1}.$$

Observe that $\tilde{Z}(\lambda)$ and $\tilde{W}(\lambda)$ are real for real λ (not poles of $\tilde{Z}(\lambda)$ or $\tilde{W}(\lambda)$). The connection with $Z(\lambda)$ and $W(\lambda)$ defined by (10.1.2) and (10.1.3) is given simply by $\tilde{Z}(\lambda) = Z(i\lambda)$, $\tilde{W}(\lambda) = W(i\lambda)$. Theorem 10.1.1 is easily rewritten for the real case:

Theorem 10.2.1. *The following statements are equivalent:*

(i) *the real CARE admits real symmetric solutions;*

(ii) $\tilde{Z}(\lambda)$ *is nonnegative on the imaginary axis: i.e.* $x^*\tilde{Z}(\lambda_0)x \geq 0$ *for every* $x \in \mathbb{C}^n$ *and for every pure imaginary* λ_0 *which is not a pole of* $\tilde{Z}(\lambda)$;

(iii) $\tilde{W}(\lambda)$ *is nonnegative on the imaginary axis.*

The proof follows from Theorem 10.1.1 upon recalling that the CARE admits real symmetric solutions if and only if it admits (complex) hermitian solutions (see Theorem 8.1.7).

Observe that $\tilde{Z}(\lambda_0)$ is real for real λ_0 (not poles of $\tilde{Z}(\lambda)$). Therefore,

$$Z(-i\alpha) = \overline{Z(i\alpha)}, \quad \alpha \in \mathbb{R}. \tag{10.2.1}$$

Using this identity and the fact that a (complex) matrix Y is positive semidefinite if and only if $\bar{Y}$ is so, we see that the condition (ii) in Theorem 10.2.1 can be replaced by

(ii′) $x^*\tilde{Z}(i\alpha)x \geq 0$ *for every* $x \in \mathbb{C}^n$ *and every real positive* α *such that* $i\alpha$ *is not a pole of* $\tilde{Z}(\lambda)$.

Clearly, "positive" in (ii′) can be replaced by "negative". Similarly, (iii) can be replaced by

(iii′) $x^*\tilde{W}(i\alpha)x \geq 0$ *for every* $x \in \mathbb{C}^n$ *and every real positive* α *such that* $i\alpha$ *is not a pole of* $\tilde{W}(\lambda)$.

We consider now the real analogue of the factorization result of Theorem 10.1.2.

THE REAL CASE

Theorem 10.2.2. *Assume that $\tilde{W}(\lambda)$ is nonnegative on the imaginary axis. Let X and $\tilde{X}$ be real symmetric solutions of the real CARE (10.1.1) and of*

$$XDX + XA + A^T X - (C + I) = 0, \qquad (10.2.2)$$

respectively. Then $\tilde{W}(\lambda)$ admits a minimal factorization

$$\tilde{W}(\lambda) = (\tilde{L}(-\bar{\lambda}))^* \tilde{L}(\lambda), \qquad (10.2.3)$$

with

$$\tilde{L}(\lambda) = I_n + \begin{bmatrix} 0 & I \end{bmatrix} \begin{bmatrix} 0 & I \\ -I & 0 \end{bmatrix} \begin{bmatrix} ZX & -Z \\ \tilde{X}ZX & -\tilde{X}Z \end{bmatrix} \left(\lambda I_{2n} - \begin{bmatrix} A & D \\ C & -A^T \end{bmatrix} \right)^{-1} \begin{bmatrix} 0 \\ I \end{bmatrix} \qquad (10.2.4)$$

where $Z = (X - \tilde{X})^{-1}$. Note that $\tilde{L}(\lambda)$ is a rational matrix function with value I at infinity and which is real on the real line (i.e. $\tilde{L}(\lambda_0)$ is real for $\lambda_0 \in \mathbb{R}$ such that λ_0 is not a pole of $\tilde{L}(\lambda)$).

Conversely, any rational matrix function $\tilde{L}(\lambda)$ which is real on the real line and $\tilde{L}(\infty) = I$, and for which (10.2.3) is a minimal factorization has the form (10.2.4) for some pair X, $\tilde{X}$ of real symmetric solutions of (10.1.1) and of (10.2.2) respectively. Moreover, the zeros of $\tilde{L}(\lambda)$ are the eigenvalues of $A + D\tilde{X}$, and the poles of $\tilde{L}(\lambda)$ are the eigenvalues of $-(A + DX)^T$, counting multiplicities in both cases.

Note that, as explained in Section 10.1, the difference $X - \tilde{X}$ is indeed invertible for any pair of solutions X, $\tilde{X}$ as in Theorem 10.2.2. The phrase "counting multiplicities" is understood as in the statement of Theorem 10.1.2.

Proof. For the direct statement of the theorem apply Theorem 10.1.2 with $W(\lambda)$ and $L(\lambda)$ given by (10.1.12) and (10.1.13), respectively, and observe that $\tilde{W}(\lambda) = W(i\lambda)$, $\tilde{L}(\lambda) = L(i\lambda)$.

Conversely, let $\tilde{L}(\lambda)$ be as in the statement of the theorem. By Theorem 10.1.2 there exist (complex) hermitian solutions X and $\tilde{X}$ of (10.1.1) and (10.2.2), respectively, such that $\tilde{L}(\lambda)$ is given by the formula (10.2.4). It remains only to show that X and $\tilde{X}$ are actually real. To this end rewrite $\tilde{L}(\lambda)$ in the form

$$\tilde{L}(\lambda) = I + \begin{bmatrix} -ZX & Z \end{bmatrix} \left(\lambda I - \begin{bmatrix} A & D \\ C & -A^T \end{bmatrix} \right)^{-1} \begin{bmatrix} 0 \\ I \end{bmatrix}. \qquad (10.2.5)$$

By Theorem 6.4.4 the realization (10.2.5) is minimal. By taking complex conjugates of every entry of the matrices involved, and by using the hypothesis that $\tilde{L}(\lambda)$ is real on the real axis, we obtain

$$\tilde{L}(\lambda) = I + \begin{bmatrix} -\bar{Z}\bar{X} & \bar{Z} \end{bmatrix} \left(\lambda I - \begin{bmatrix} A & D \\ C & -A^T \end{bmatrix} \right)^{-1} \begin{bmatrix} 0 \\ I \end{bmatrix}. \qquad (10.2.6)$$

By Theorem 6.1.5 the realization (10.2.6) is also minimal, and therefore (Theorem 6.1.4) there is a unique invertible matrix S such that

$$[-ZX \quad Z] = [-\bar{Z}\bar{X} \quad \bar{Z}]S,$$

$$\begin{bmatrix} A & D \\ C & -A^T \end{bmatrix} = S^{-1} \begin{bmatrix} A & D \\ C & -A^T \end{bmatrix} S, \quad \begin{bmatrix} 0 \\ I \end{bmatrix} = S^{-1} \begin{bmatrix} 0 \\ I \end{bmatrix}. \quad (10.2.7)$$

Formula (6.1.10) shows that, in fact, $S = I$. Now (10.2.7) yields $Z = \bar{Z}$, $ZX = \bar{Z}\bar{X}$, i.e. both Z and X are real. But then $\tilde{X}$ is real as well. □

As in the complex case, by choosing a fixed real symmetric solution $\tilde{X}$ of (10.2.2) (the maximal one, for instance) a one-to-one correspondence can be established between certain minimal factorizations of the form (10.2.3) and real symmetric solutions of the real CARE. Thus, a real analogue of Corollary 10.1.3 is obtained. We leave the statement of this analogue for the reader.

10.3 Notes

The proof of the harder part of Theorem 10.1.1 (that statement (iii) implies statement (i)) is taken from Chapter II.4 of Gohberg et al. (1983). The discussion following Theorem 10.1.1, as well as Theorem 10.1.2 and its proof originate with Ran and Rodman (1984a). In the special case when M has no real eigenvalues a correspondence between hermitian solutions of (10.1.1) and factorizations of a nonnegative rational matrix function was treated by Finesso and Picci (1982). However, such correspondences have a longer history and were also discussed by Willems (1971) and Molinari (1973a, b).

11
PERTURBATION THEORY

In this chapter we present the basic results concerning behaviour of hermitian solutions of algebraic Riccati equations under perturbations, or changes, of the coefficient matrices of the CARE (10.1.1). In the first section we consider the problem of existence of hermitian solutions, while in the later sections the focus is on the behaviour of extremal solutions or, more generally, on unmixed solutions.

11.1 Existence of hermitian solutions

Our basic result here is the following:

Theorem 11.1.1. *Let a sequence of algebraic Riccati equations*

$$XD_mX + XA_m + A_m^*X - C_m = 0, \quad m = 1, 2, \ldots \quad (11.1.1)$$

be given, where $D_m \geq 0$, $C_m = C_m^*$ (for all m), and all matrices here are $n \times n$. Assume further that the limits

$$D = \lim_{m \to \infty} D_m, \quad C = \lim_{m \to \infty} C_m, \quad A = \lim_{m \to \infty} C_m$$

exist and that the pair (A, D) is sign controllable. Then, if each equation (11.1.1) admits a hermitian solution, then the equation

$$XDX + XA + A^*X - C = 0 \quad (11.1.2)$$

admits a hermitian solution as well.

Recall that sign controllability of (A, D) means that either $\mathcal{R}_\lambda(A) \subseteq \mathcal{C}_{A,D}$ or $\mathcal{R}_{-\bar\lambda} \subseteq \mathcal{C}_{A,D}$ for every $\lambda \in \mathbb{C}$. First let us show that the property of sign controllability is persistent under small perturbations:

Lemma 11.1.2. *If the pair (A, B) is sign controllable, then there is an $\epsilon > 0$ such that every pair (A', B') with*

$$\|A - A'\| + \|B - B'\| < \epsilon \quad (11.1.3)$$

is sign controllable as well.

Proof. Let $\mathcal{C} = \text{Im}\,[B, AB, \ldots, A^{n-1}B]$. Denote by $\lambda_1, \ldots, \lambda_p$ the distinct eigenvalues of A on the imaginary axis, and by $\mu_1, -\bar{\mu}_1, \ldots, \mu_r, -\bar{\mu}_r$ the distinct pairs of eigenvalues of A which are symmetric relative to the imaginary axis (we assume that none of the μ_j's is on the imaginary axis). By assumption, we have $\mathcal{R}_{\lambda_j}(A) \subseteq \mathcal{C}$ and for $j = 1, \ldots, r$ at least one of $\mathcal{R}_{\mu_j}(A) \subseteq \mathcal{C}$ and $\mathcal{R}_{-\bar{\mu}_j}(A) \subseteq \mathcal{C}$ holds. For simplicity of notation assume

$$\mathcal{R}_{\mu_j}(A) \subseteq \mathcal{C} \quad (j = 1, \ldots, r). \tag{11.1.4}$$

Let $\delta > 0$ be a fixed number, chosen sufficiently small so that any closed disc $\{\lambda \in \mathbf{C} : |\lambda - \lambda_0| \leq \delta\}$, where λ_0 is an eigenvalue of A, does not contain any other eigenvalues of A. Let

$$\Gamma := \bigcup_{j=1}^{p} \{\lambda \in \mathbf{C} : |\lambda - \lambda_j| \leq \delta\} \cup \bigcup_{j=1}^{r} \{\lambda \in \mathbf{C} : |\lambda - \mu_j| \leq \delta\}.$$

and denote by $\mathcal{R}_\Gamma(A)$ the spectral subspace of A corresponding to the eigenvalues in Γ. Then we have $\mathcal{R}_\Gamma(A) \subseteq \mathcal{C}$.

Letting $\mathcal{C}(A', B')$ be the controllable subspace of (A', B'), we claim that when $\epsilon > 0$ is sufficiently small and (11.1.3) holds, then

$$\mathcal{R}_\Gamma(A') \subseteq \mathcal{C}(A', B'). \tag{11.1.5}$$

We argue by contradiction, and assume that the claim is false. Then there exists a sequence of pairs of matrices $\{A'_m\}$, $\{B'_m\}$ with the following properties:

(i) $\lim_{m \to \infty} A'_m = A$; $\lim_{m \to \infty} B'_m = B$.

(ii) there is $\alpha_m \in \Gamma$ and $x_m \perp \mathcal{C}(A'_m, B'_m)$, $\|x_m\| = 1$ such that

$$(I - P_m)A'_m x_m = \alpha_m x_m, \tag{11.1.6}$$

where P_m is the orthogonal projector on $\mathcal{C}(A'_m, B'_m)$.

Since Γ consists of finitely many connected components, by choosing a subsequence, if necessary, we can assume that either $\alpha_m \to \lambda_j$ for some j or $\alpha_m \to \mu_j$ (for some j) as $m \to \infty$. Say, $\alpha_m \to \lambda_1$. Using the compactness of the unit sphere in $\mathbf{C}^n$, and passing to a subsequence again, we can assume that $x_m \to x$ as $m \to \infty$. Further, using the compactness of the set of subspaces in $\mathbf{C}^n$ (see Section 7.A), and passing again to a subsequence, we can assume that $\mathcal{C}(A'_m, B'_m) \to \mathcal{L}$ for some subspace $\mathcal{L} \subseteq \mathbf{C}^n$ (and convergence is in the gap metric). It is easy to see that $x \perp \mathcal{L}$. As each $\mathcal{C}(A'_m, B'_m)$ is A'_m-invariant and $\lim_{m \to \infty} A'_m = A$, the limit $\mathcal{L}$ is A-invariant (Corollary 7.A.6). Also, $\mathcal{L} \supseteq \mathcal{C}$ by

Theorem 7.A.5. Thus, denoting by $P_{\mathcal{L}}$ the orthogonal projector on $\mathcal{L}$, we have $I - P_m \to I - P_{\mathcal{L}}$ as $m \to \infty$. So, passing to the limit in (11.1.6) we obtain

$$(I - P_{\mathcal{L}})Ax = \lambda_1 x, \quad \|x\| = 1. \qquad (11.1.7)$$

With respect to the orthogonal decomposition $\mathbb{C}^n = \mathcal{C} \oplus (\mathcal{C}^\perp \cap \mathcal{L}) \oplus \mathcal{L}^\perp$ we have

$$A = \begin{bmatrix} A_1 & A_{12} & A_{13} \\ 0 & A_2 & A_{23} \\ 0 & 0 & A_{33} \end{bmatrix}$$

and (11.1.7) shows that $\lambda_1 \in \sigma(A_{33})$. This contradicts the sign controllability of (A, B). If $\alpha_m \to \mu_j$ as $m \to \infty$ then, similarly, we obtain a contradiction with (11.1.4).

We have proved that (11.1.5) holds for all (A', B') sufficiently close to (A, B). As the eigenvalues of A' are close to those of A (if A' is sufficiently close to A), it follows from the construction of Γ that $\sigma(A') \backslash \Gamma$ does not contain points on the imaginary axis or pairs of points symmetric with respect to the imaginary axis (if A' is sufficiently close to A). This proves Lemma 11.1.2. $\square$

Proof of Theorem 11.1.1. Define

$$M = i \begin{bmatrix} A & D \\ C & -A^* \end{bmatrix}, \quad H = \begin{bmatrix} -C & A^* \\ A & D \end{bmatrix}$$

and, for $m = 1, 2, \ldots$

$$M_m = i \begin{bmatrix} A_m & D_m \\ C_m & -A_m^* \end{bmatrix}, \quad H_m = \begin{bmatrix} -C_m & A_m^* \\ A_m^* & D_m \end{bmatrix}$$

We can assume without loss of generality that H is invertible and has n positive and n negative eigenvalues (counted with multiplicities); otherwise replace A by $A + i\alpha I$ and A_m by $A_m + i\alpha I$, where α is a sufficiently large real number. Since $H_m \to H$ as $m \to \infty$, the matrices H_m have the same properties (at least for large m). Further, by Lemma 11.1.2 the sign controllability of (A, D) implies that of (A_m, D_m) for sufficiently large m. By Theorem 7.2.5, for sufficiently large m there exists an n-dimensional M_m-invariant H_m-neutral subspace; call it $\mathcal{L}_m$. Since the set of subspaces is compact (see Theorem 7.A.4), one can choose a converging sequence $\mathcal{L}_{m_k} \to \mathcal{L}$, when $k \to \infty$. By Corollaries 7.A.6 and 7.A.7, $\mathcal{L}$ is n-dimensional M-invariant and H-neutral. Now apply Theorem 7.2.5 to the equation (11.1.2). $\square$

11.2 Extremal solutions: continuous dependence

We assume in this section that the coefficients of the equation

$$XDX - X\dot A - A^*X - C = 0 \tag{11.2.1}$$

are such that $D \geq 0$, $C = C^*$, and (A, D) stabilizable. We know from Theorem 7.9.3 that, if equation (11.2.1) has hermitian solutions, then it has a maximal hermitian solution X_+ which is characterized by the property that $\sigma(A - DX_+)$ lies in the closed left half-plane. The behaviour of X_+ as a function of A, C and D will be studied here so we write $X_+ = X_+(A, C, D)$.

Denote by Q the set of all ordered triples (A, C, D) of $n \times n$ (complex) matrices such that $D \geq 0$, $C = C^*$, (A, D) is stabilizable, and the equation (11.2.1) admits hermitian solutions. Since Q is a subset of $\mathbb{C}^{3n^2} = \mathbb{C}^{n^2} \times \mathbb{C}^{n^2} \times \mathbb{C}^{n^2}$, the natural topology on Q is induced by the norm $\|.\|$ in the set $\mathbb{C}^{n^2}$ of all $n \times n$ matrices. When we can speak about continuous functions defined on Q, the use of this topology will be implied.

Theorem 11.2.1. *The maximal hermitian solution $X_+(A, C, D)$ of (11.2.1) is a continuous function of $(A, C, D) \in Q$.*

The proof of Theorem 11.2.1 depends on the corresponding result for invariant neutral subspaces. Let Z be the set of all pairs of $2n \times 2n$ matrices (B, H), where $H = H^*$ is invertible, B is H-self-adjoint, and B has only even partial multiplicities corresponding to real eigenvalues (if any), with all signs $+1$'s in the sign characteristic of (B, H). It follows from Theorem 2.5.2 (by taking $\mathcal{C}$ to be the set of all eigenvalues of B with positive imaginary parts, and by taking $\mathcal{N}_+ = \mathcal{M}_+$), that there is a unique n-dimensional B-invariant H-neutral subspace $\mathcal{N}_u \subseteq \mathbb{C}^{2n}$ such that $\sigma(B|\mathcal{N}_u)$ is in the closed upper half-plane. Similarly, there exists a unique n-dimensional B-invariant H-neutral subspace $\mathcal{N}_l$ such that $\sigma(B/\mathcal{N}_l)$ is in the closed lower half-plane. We write $\mathcal{N}_u = \mathcal{N}_u(B, H)$ and $\mathcal{N}_l = \mathcal{N}_l(B, H)$ to emphasize the dependence of $\mathcal{N}_u$ and $\mathcal{N}_l$ on $(B, H) \in Z$.

Lemma 11.2.2. *The subspaces $\mathcal{N}_u(B, H)$ and $\mathcal{N}_l(B, H)$ are continuous functions of $(B, H) \in Z$.*

(Here, continuity is understood in the sense of the gap metric (see Section 7.A) and the topology on Q mentioned above.)

Proof. If $\mathcal{N}_u(B, H)$ is not continuous, then there exist $(B_0, H_0) \in Z$, $\epsilon > 0$ and a sequence $(B_m, H_m) \in Z$, $m = 1, 2, \ldots$ such that $\lim_{m \to \infty} B_m = B_0$, $\lim_{m \to \infty} H_m = H_0$, and

$$\theta(\mathcal{N}_u(B_m, H_m), \quad \mathcal{N}_u(B, H)) \geq \epsilon. \tag{11.2.2}$$

Since the set of all n-dimensional subspaces in $\mathbf{C}^{2n}$ is compact, there is a convergent subsequence of $\{\mathcal{N}_u(B_m, H_m)\}$. Without loss of generality we can assume that $\mathcal{N}_u(B_m, H_m)$ itself converges, say

$$\lim_{m \to \infty} \mathcal{N}_u(B_m, H_m) = \mathcal{N}_0.$$

By Corollaries 7.A.6 and 7.A.7 the subspace $\mathcal{N}_0$ is n-dimensional B_0-invariant and H_0-neutral.

In view of the continuity of eigenvalues (as functions of the matrix) all eigenvalues of $B_0|\mathcal{N}_0$ lie in the closed upper half-plane. Let us provide a more detailed proof of this statement. Pick a basis $\varphi_1, \ldots, \varphi_n$ in $\mathcal{N}_0$. Then (using Theorem 7.A.5) there exists a basis $\varphi_{1,m}, \ldots, \varphi_{n,m}$ in $\mathcal{N}_u(B_m, H_m)$ such that

$$\varphi_{j,m} \to \varphi_j \quad \text{as } m \to \infty \quad (j = 1, \ldots, n). \tag{11.2.3}$$

Write $B_m|\mathcal{N}_u(B_m, H_m)$ as an $n \times n$ matrix $\tilde{B}_m$ in the basis $\{\varphi_{1,m}, \ldots, \varphi_{n,m}\}$, and write $B_0|\mathcal{N}_0$ as an $n \times n$ matrix $\tilde{B}_0$ in the basis $\{\varphi_1, \ldots, \varphi_n\}$. Because of (11.2.3) we have

$$\lim_{m \to \infty} \tilde{B}_m = \tilde{B}_0. \tag{11.2.4}$$

Now pick $\lambda_0 \in \mathbf{C}$ in the open lower half-plane. We claim that $\lambda_0 I - \tilde{B}_0$ is invertible. Suppose not, and let Γ be a small circle around λ_0 which does not intersect the real line, and such that $\det(\lambda I - \tilde{B}_0) \neq 0$ for $\lambda \in \Gamma$. Then the polynomial $\det(\lambda I - \tilde{B}_0)$ has a zero inside Γ (namely, $\lambda = \lambda_0$). Hence the same is true for any polynomial of degree n whose coefficients are sufficiently close to those of $\det(\lambda I - \tilde{B}_0)$ (this follows from Rouché's theorem, for example). Hence, by (11.2.4) $\det(\lambda I - \tilde{B}_m)$ has a zero inside Γ for m large enough, but this contradicts the assumption that all eigenvalues of $B_m|\mathcal{N}_u(B_m, H_m)$ are in the closed upper half-plane. So $\lambda_0 I - \tilde{B}_0$ is invertible and, consequently, all eigenvalues of $B_0|\mathcal{N}_0$ are in the closed upper half-plane.

Now, because of the uniqueness of $\mathcal{N}_u(B, H)$ we must have $\mathcal{N}_0 = \mathcal{N}_u(B, H)$. But this contradicts (11.2.2), and Lemma 11.2.2 is proved. □

Proof of Theorem 11.2.1. As in Chapter 7, define

$$M = i \begin{bmatrix} -A & D \\ C & A^* \end{bmatrix}, \quad \hat{H} = i \begin{bmatrix} 0 & I \\ -I & 0 \end{bmatrix}.$$

By Theorem 7.3.7 M has only even partial multiplicities corresponding to the real eigenvalues, and the sign characteristic of $(M, \hat{H})$ consists of $+1$'s only. Also, the subspace

$$\mathcal{L}_+ = \text{Im} \begin{bmatrix} I \\ X_+ \end{bmatrix}$$

is n-dimensional M-invariant and $\hat{H}$-neutral, and $M|\mathcal{L}_+$ is similar to $i(-A + DX_+)$. By Theorem 7.9.3 $\sigma(i(-A + DX_+))$ lies in the closed upper half-plane and therefore the same is true for $\sigma(M|\mathcal{L}_+)$. By Lemma 11.2.2 $\mathcal{L}_+$ is a continuous function of (A, C, D). The orthogonal projector on $\mathcal{L}_+$ is given by the formula

$$P_{\mathcal{L}_+} = \begin{bmatrix} I & X_+^* \\ X_+ & X_+ X_+^* \end{bmatrix} \begin{bmatrix} (I + X_+^* X_+)^{-1} & 0 \\ 0 & (I + X_+ X_+^*)^{-1} \end{bmatrix},$$

(cf. equation (7.8.5)) from which the continuity of X_+ as a function of $(A, C, D) \in Q$ follows immediately.

It is natural to ask whether one can say more about the nature of the dependence of X_+ on $(A, C, D) \in Q$ than mere continuity. The following simple example shows that in general X_+ is not differentiable:

Example 11.2.1. Let

$$A(t) = \begin{bmatrix} 0 & 1 \\ 0 & 0 \end{bmatrix}, \quad D(t) = \begin{bmatrix} 0 & 0 \\ 0 & 1 \end{bmatrix}, \text{ and } C(t) = \begin{bmatrix} t^2 & 0 \\ 0 & 0 \end{bmatrix}, \quad t \in \mathbb{R}.$$

Then the associated matrix $M = M(t)$ is

$$M(t) = i \begin{bmatrix} 0 & -1 & 0 & 0 \\ 0 & 0 & 0 & 1 \\ t^2 & 0 & 0 & 0 \\ 0 & 0 & 1 & 0 \end{bmatrix}.$$

It is not hard to check that the eigenvalues of $M(t)$ are $\epsilon_1, \epsilon_2, \epsilon_3, \epsilon_4$, where $\epsilon_k = |t|^{1/2} \exp(\frac{1}{4}\pi i + \frac{1}{2} k \pi i)$, $k = 1, 2, 3, 4$. The corresponding eigenvectors are

$$x_k(t) = [1 \ \ i(\epsilon_k - \epsilon_k^3) \ \ \epsilon_k^2]^T.$$

For $i \neq j$ we have

$$\text{span}\{x_i(t), x_j(t)\} = \text{Im} \begin{bmatrix} I \\ x_{ij}(t) \end{bmatrix},$$

where

$$x_{ij}(t) = \begin{bmatrix} i\epsilon_i \epsilon_j (\epsilon_i + \epsilon_j) & -(\epsilon_i^2 + \epsilon_i \epsilon_j + \epsilon_j^2) \\ -(\epsilon_i \epsilon_j) & -i(\epsilon_j + \epsilon_i) \end{bmatrix}.$$

To obtain X_+ we must take ϵ_i and ϵ_j in the upper half-plane, and then $\epsilon_j + \epsilon_i = |t|^{1/2} c$ for some $c \neq 0$ independent of t. So $x_{ij}(t)$ cannot be differentiable at $t = 0$. Observe that in this example $(A(t), D(t))$ is controllable for all real values of t, the matrix $M(t)$ has no real eigenvalues for $t \neq 0$ and has the zero eigenvalue with algebraic multiplicity 4, and geometric multiplicity 1, when $t = 0$. Thus, by Theorem 7.2.4, for every real t there is a hermitian solution of

$$XD(t)X - XA(t) - A(t)^* X - C(t) = 0.$$

Observe that in this example $x_{ij}(t)$ behaves like $|t|^{1/2}$ in a neighborhood of $t_0 = 0$. □

This example motivates the following conjecture:

Conjecture 11.2.3. *The behaviour of $X_+(A, C, D)$ as a function of $(A, C, D) \in Q$ is Hölder-like with exponent $\frac{1}{2}$. In other words, for every $(A_0, C_0, D_0) \in Q$ there exist positive constants K and ϵ such that the inequality*

$$\|X_+(A, C, D) - X_+(A_0, C_0, C_0)\| \leq K(\|A - A_0\| + \|C - C_0\| + \|D - D_0\|)^{1/2}$$

holds for every $(A, C, D) \in Q$ satisfying

$$\|A - A_0\| + \|C - C_0\| + \|D - D_0\| < \epsilon.$$

We conclude this section with a remark concerning minimal solutions. As Example 7.9.1 shows, the minimal solutions of (11.2.1) under the stabilizability assumption need not exist. If, however, the pair (A, D) controllable, then the minimal solution $X_- = X_-(A, C, D)$ exists and an analogue of Theorem 11.2.1 holds:

Theorem 11.2.4. *The minimal solution $X_-(A, C, D)$ of (11.2.1) is a continuous function of the coefficients A, C and D, on the set of all ordered triples (A, C, D) of $n \times n$ matrices such that $D \geq 0$, $C = C^*$ and the pair (A, D) is controllable.*

11.3 Extremal solutions: analytic dependence

Suppose now that the coefficients $A = A(t)$, $D = D(t)$ and $C = C(t)$ of equation (11.2.1) are real analytic functions of the real parameter t.

As Example 11.2.1 shows, the maximal solution $X_+ = X_+(t)$ need not be an analytic function of t, even under the controllability condition. It turns out that the extra assumption needed to ensure the analyticity of maximal (or minimal) hermitian solutions is the invariance of the number of real eigenvalues of M (counting with multiplicities):

Theorem 11.3.1. *Let $A(t)$, $D(t)$, and $C(t)$ be analytic $n \times n$ matrix functions of t on a real interval (α, β), with $D(t)$ positive semidefinite hermitian, $C(t)$*

hermitian, and $(A(t), D(t))$ stabilizable for every $t \in (\alpha, \beta)$. Assume that for all $t \in (\alpha, \beta)$, the Riccati equation

$$X(t)D(t)X(t) - X(t)A(t) - A(t)^*X(t) - C(t) = 0 \qquad (11.3.1)$$

has a hermitian solution. Further assume that the number of pure imaginary or zero eigenvalues (counting multiplicities) of

$$M(t) := \begin{bmatrix} -A(t) & D(t) \\ C(t) & A(t)^* \end{bmatrix}$$

is constant. Then the maximal solution $X_+(t)$ of (11.3.1) is an analytic function of $t \in (\alpha, \beta)$. Conversely, if $X_+(t)$ is an analytic function of $t \in (\alpha, \beta)$, then the number of pure imaginary or zero eigenvalues of $M(t)$ is constant.

Proof. The first part of this theorem was proved by Ran and Rodman (1988a) (and previously by Rodman (1981) under the controllability assumption). Its proof in full generality requires methods which are beyond the scope of this book and therefore will not be given here. We prove the first part of the theorem only in the relatively simple case when $M(t)$ has no pure imaginary or zero eigenvalues. In this case, by Theorem 7.9.4, $\sigma(-A(t) + D(t)X_+(t))$ is in the open right half-plane; in other words, $\operatorname{Im} \begin{bmatrix} I \\ X_+(t) \end{bmatrix}$ is the spectral subspace of $M(t)$ corresponding to its eigenvalues in the open right half-plane. Proposition 1.3.3 shows that the Riesz projection on $\operatorname{Im} \begin{bmatrix} I \\ X_+(t) \end{bmatrix}$ is an analytic function of $t \in (\alpha, \beta)$. As in the proof of Theorem 7.8.2 we deduce that the orthogonal projection $P_+(t)$ on $\operatorname{Im} \begin{bmatrix} I \\ X_+(t) \end{bmatrix}$ is an analytic function of $t \in (\alpha, \beta)$. This projection is given by the formula (7.8.5) with $X(t)$ replaced by $X_+(t)$, and the analyticity of $X_+(t)$ for $t \in (\alpha, \beta)$ follows.

For the converse part of Theorem 11.3.1, assume that $X_+(t)$ is analytic. As the number of real eigenvalues of $M(t)$ is twice the number of pure imaginary (or zero) eigenvalues of $-A(t) + D(t)X_+(t)$, it is enough to prove that the number of pure imaginary eigenvalues of $-A(t) + D(t)X_+(t)$ is constant. By Theorem 7.9.3, $\sigma(-A(t) + D(t)X_+(t))$ is in the closed right half-plane. Let λ_0 be a pure imaginary eigenvalue of $-A(t_0) + D(t_0)X_+(t_0)$ for some $t_0 \in (\alpha, \beta)$, i.e. λ_0 is a pure imaginary root of the characteristic polynomial $f(\lambda, t)$ of $-A(t) + D(t)X_+(t)$ when $t = t_0$. The polynomial $f(\lambda, t)$ is monic (i.e. has leading coefficient 1) and its coefficients are analytic functions of $t \in (\alpha, \beta)$. It is a basic fact in the theory of algebraic functions (see, e.g., Chapter 8 of Markushevich (1965), or Baumgärtel (1985)) that the zeros of $f(\lambda, t)$, i.e. eigenvalues of $-A(t) + D(t)X_+(t)$, which are close to λ_0 when t is sufficiently close to t_0, admit developments into fractional power series (Puiseux series):

$$\lambda(t) = \lambda_0 + \sum_{j=1}^{\infty} \alpha_j (t-t_0)^{j/p}$$

for t in a neighborhood of t_0. It is easy to see that the only way $\lambda(t)$ can be in the closed right half-plane for all choices of the branch of $(t-t_0)^{1/p}$ is when $\lambda(t)$ stays pure imaginary. This proves the second part of Theorem 11.3.1. $\square$

If the pair $(A(t), D(t))$ is controllable for all $t \in (\alpha, \beta)$, so that the existence of the minimal solution $X_-(t)$ is guaranteed (under the hypotheses of Theorem 11.3.1), then $X_-(t)$ is an analytic function of t on (α, β) if and only if the number of eigenvalues of $M(t)$ on the imaginary axis is constant. The proof is similar to that of Theorem 11.3.1.

Consider now a more general situation when $A(t)$, $D(t)$ and $C(t)$ are analytic functions of several real variables for $t = (t_1, \ldots, t_d) \in \Omega$, where Ω is an open connected set in $\mathbb{R}^d$.

Conjecture 11.3.2. *Assuming that $A(t)$, $D(t)$ and $C(t)$ depend analytically on several real variables, and under the hypotheses of Theorem 11.3.1, the maximal solution $X_+(t)$ is analytic on Ω if and only if the number of eigenvalues of $M(t)$ on the imaginary axis, for $t \in \Omega$, does not depend on t.*

The conjecture is valid if $M(t)$ has no eigenvalues on the imaginary axis and can be proved in the same way as Theorem 11.3.1. This result originates with Delchamps (1980) and Rodman (1980).

Our final remark in this section concerns unmixed solutions. A hermitian solution X of (11.2.1) is called *unmixed* if $\sigma(A - DX)$ does not contain pairs of complex numbers $\lambda, -\bar{\lambda}$ with $\operatorname{Re} \lambda \neq 0$. By Theorem 7.9.3, the maximal solution is unmixed (under the hypotheses of that theorem). Being unmixed is a generic property; for if all nonpure imaginary eigenvalues of M are simple (algebraic multiplicity equals 1), then every hermitian solution is unmixed. The analytic behaviour of unmixed solutions is analogous to that of the maximal solution. We state (without proof) one result to this effect.

Theorem 11.3.3. *Assume the hypotheses of Theorem 11.3.1 and, in addition, that (in the sense of Section 7.9) equation (11.3.1) is regular for all $t \in (\alpha, \beta)$. Assume also that the number of pure imaginary or zero eigenvalues of $M(t)$ is independent of t. Then for every unmixed solution X_0 of equation (11.3.1) at $t = t_0 \in (\alpha, \beta)$ there exists a unique hermitian valued $n \times n$ matrix function $X(t)$, $t \in (\alpha, \beta)$ with the following properties:*

(i) *$X(t)$ is a solution of (11.3.1) for every $t \in (\alpha, \beta)$;*

(ii) *$X(t_0) = X_0$;*

(iii) *$X(t)$ is an unmixed solution of (11.3.1) for all t sufficiently close to t_0;*

(iv) $X(t)$ is analytic in a real neighborhood of t_0.

A full proof of this theorem can be found in a paper of Ran and Rodman (1988a).

11.4 The real case

The main results of Sections 11.2 and 11.3 (Theorems 11.2.1, 11.2.4, 11.3.1 and 11.3.3 in particular) are applicable to the equation

$$XDX - XA - A^T X - C = 0 \qquad (11.4.1)$$

where the matrices A, D and C are real. Since the extremal solutions of (11.4.1) (if they exist) are necessarily real, we obtain the real analogues of the theorems mentioned above.

Concerning the result of Theorem 11.1.1 observe that, in view of Theorems 7.3.7 and 8.1.10, under the assumptions that $D = D^T \geq 0$, $C = C^T$ and the pair (A, D) is sign controllable, the equation (11.4.1) has real symmetric solutions if and only if it has (complex) hermitian solutions. Thus, Theorem 11.1.1 is valid in the real case as well.

11.5 Exercises

11.5.1. Prove that the controllability condition in Theorem 11.2.4 can be replaced by a weaker condition that $(-A, D)$ is stabilizable.

Hint: Use the method of Exercise 9.6.4.

11.5.2. Prove that the triple of matrices

$$A = \begin{bmatrix} 0 & 1 \\ 0 & 0 \end{bmatrix}, \quad C = \begin{bmatrix} 0 & 0 \\ 0 & 0 \end{bmatrix}, \quad D = \begin{bmatrix} 0 & 0 \\ 0 & 1 \end{bmatrix}$$

(cf. Example 11.2.1) belongs to the boundary of Q.

Hint: Consider the following perturbation of C: $\begin{bmatrix} -\varepsilon & 0 \\ 0 & 0 \end{bmatrix}$, where $\varepsilon > 0$ is small.

11.5.3. Find an example of the CARE (11.1.2) (where $D \geq 0$, $C = C^*$ and (A, D) controllable) having hermitian solutions, and of a particular sequence (11.1.1) of CARE's (such that $D_m \geq 0$, $C_m = C_m^*$, $D_m \to D$, $C_m \to C$, $A_m \to A$) having no hermitian solutions.

11.5.4. The CARE (11.1.2) is called *robustly solvable* if there exists an $\varepsilon > 0$ such that every CARE

$$XD'X + XA' + A'^*X - C' = 0$$

with $D' \geq 0$, $C' = C'^*$ and

$$\|D - D'\| + \|A - A'\| + \|C - C'\| < \varepsilon$$

admits a hermitian solution. Prove that if $D \geq 0$, $C = C^*$, (A, D) is sign controllable, and $M = i \begin{bmatrix} A & D \\ C & -A^* \end{bmatrix}$ has no real eigenvalues then (11.1.2) is robustly solvable. [It is proved in Ran and Rodman (1992a) that M having no real eigenvalues is necessary for the robust solvability, under the additional hypothesis that (A, D) is controllable.]

11.6 Notes

The results of this chapter are based on robustness properties of invariant subspaces corresponding to hermitian solutions of the CARE, as developed by Ran and Rodman (1988b).

Example 11.2.1 is from Ran and Rodman (1989). Theorem 11.3.1 is taken from Lancaster and Rodman (1991). Conjecture 11.3.2 was originally stated in Ran and Rodman (1988a).

The study of parametrized linear control systems is an integral part of modern control theory. See, for example, the works of Tugnait (1985), Polderman (1986), Kamen and Khargonekar (1984) in this area.

The stability of hermitian solutions of the CARE in the sense of robustness (i.e. persistence under small perturbations of the coefficients of the CARE) has also been studied by Ran and Rodman (1984b, and 1992a). The concept of "unmixed solutions" defined in Section 11.3 was introduced and studied by Shayman (1983).

This chapter presents one theoretical approach to perturbation theory. Related ideas arising in the numerical analysis of Riccati equations concern the sensitivity of solutions to perturbations of the coefficients. Papers on this topic include those by Byers (1984), Gahinet and Laub (1990), and Kenney and Hewer (1990).

Part III

DISCRETE ALGEBRAIC RICCATI EQUATIONS

In this part we study matrix equations of the form

$$X = A^*XA + Q - (C + B^*XA)^*(R + B^*XB)^{-1}(C + B^*XA), \qquad (1)$$

where A, B, C, Q and R are given matrices of sizes $n \times n$, $n \times m$, $m \times n$, $n \times n$ and $m \times m$, respectively, such that R and Q are hermitian, and X is an $n \times n$ matrix to be found. By definition, a solution X is a matrix for which $R + B^*XB$ is invertible. We will be interested mostly in hermitian solutions X.

The developments in this part are parallel to those in Part 2 and, first of all, are based on the description of the solutions of (1) in terms of invariant subspaces of a certain matrix. In contrast to Part 2, this matrix turns out to be unitary in an indefinite scalar product, rather than self-adjoint. This geometric theory is the content of Chapter 12.

In Chapter 13 the ideas developed in Chapter 9 for the continuous Riccati equation (9.1.1) are applied in the context of equation (1). Thus, related Riccati *inequalities* make an appearance, along with analysis of a constructive iterative procedure for the solution of (1), and comparison theorems.

The exposition in this part is less detailed, and contains less material, than in Part II, due mainly to our more limited knowledge of the theory of matrix equations of the form (1).

Part III

DISCRETE ALGEBRAIC RICCATI EQUATIONS

12
GEOMETRIC THEORY FOR THE DISCRETE ALGEBRAIC RICCATI EQUATION

It is to be shown in Chapter 16 that, to solve the "linear-quadratic regulator" (or LQR) problem for a time-invariant differential system, it is necessary to solve an algebraic Riccati equation of the kind discussed in Part II of this book, equation (7.2.1) for example. The discrete analogue of the LQR problem requires the solution of a rather different algebraic equation for X:

$$X = A^*XA + Q - (C + B^*XA)^*(R + B^*XB)^{-1}(C + B^*XA).$$

It is therefore tempting to describe this equation as a discrete algebraic Riccati equation (or DARE) and, when convenient, we shall do so.

The geometric theory for the DARE turns out to be considerably more complicated than for the symmetric CARE. The identification of a main operator (whose invariant subspaces determine solutions of a DARE) is already a significant task when compared to analysis of CARE (see Proposition 12.2.2). Having done this, it is possible to obtain several conclusions for the DARE by a suitable transformation between equations of CARE and DARE types (see Lemma 12.3.2). In contrast to analysis of the CARE it is convenient to make use of techniques involving rational matrix functions at an early stage of the analysis.

12.1 Preliminaries

We start with some elementary algebraic transformations of the discrete algebraic Riccati equation under which the set of hermitian solutions is invariant. It is convenient to get these computations out of the way so that they do not clutter subsequent arguments.

Proposition 12.1.1. *Let A, B, C, Q, R be given complex matrices of sizes $n \times n$, $n \times m$, $m \times n$, $n \times n$ and $m \times m$, respectively, with $Q^* = Q$, $R^* = R$ and R invertible. Then the three following equations, in which*

$$A_1 := A - BR^{-1}C, \quad Q_1 := Q - C^*R^{-1}C,$$

have the same set of solutions X:

$$X = A^*XA + Q - (C + B^*X^*A)^*(R + B^*XB)^{-1}(C + B^*XA), \quad (12.1.1)$$
$$X = A_1^*XA_1 + Q_1 - A_1^*XB(R + B^*XB)^{-1}B^*XA_1, \quad (12.1.2)$$
$$X = A_1^*X(I + BR^{-1}B^*X)^{-1}A_1 + Q_1. \quad (12.1.3)$$

Proof. From (12.1.1) we have

$$\begin{aligned} X = A^*XA + Q_1 &- A^*XB(R + B^*XB)^{-1}B^*XA \\ &- C^*(R + B^*XB)^{-1}B^*XA - C^*(R + B^*XB)^{-1}C \\ &- A^*XB(R + B^*XB)^{-1}C + C^*R^{-1}C \end{aligned} \quad (12.1.4)$$

Now

$$(R + B^*XB)^{-1} = R^{-1}(I + B^*XBR^{-1})^{-1} = R^{-1}(I - B^*XB(R + B^*XB)^{-1}).$$

Also

$$-R(R + B^*XB)^{-1}R + R = B^*XB - B^*XB(R + B^*XB)^{-1}B^*XB,$$

so

$$\begin{aligned} &-C^*(R + B^*XB)^{-1}C + C^*R^{-1}C \\ &= C^*R^{-1}\{B^*XB - B^*XB(R + B^*XB)^{-1}B^*XB\}R^{-1}C. \end{aligned}$$

Note also the equality

$$(R + B^*XB)^{-1} = (I - (R + B^*XB)^{-1}B^*XB)R^{-1}.$$

These facts allow us to rewrite equation (12.1.4) in the form

$$\begin{aligned} X = A^*XA + Q_1 &- A^*XB(R + B^*XB)^{-1}B^*XA \\ &- C^*R^{-1}\{I - B^*XB(R + B^*XB)^{-1}\}B^*XA \\ &- A^*XB(I - (R + B^*XB)^{-1}B^*XB)R^{-1}C \\ &+ C^*R^{-1}\{B^*XB - B^*XB(R + B^*XB)^{-1}B^*XB\}R^{-1}C, \end{aligned}$$

which, after simple algebra, is easily seen to be equal to the right-hand side of (12.1.2).

Finally, the equivalence of (12.1.2) and (12.1.3) is easily verified because of the equalities

$$I - B(R + B^*XB)^{-1}B^*X = (I + BR^{-1}B^*X)^{-1},$$
$$X(I + BR^{-1}B^*X)^{-1} = (I + XBR^{-1}B^*)^{-1}X. \quad (12.1.5)$$

$\square$

In view of Proposition 12.1.1 it will often be assumed in the sequel that $C = 0$ in equation (12.1.1).

12.2 Hermitian solutions and invariant Lagrangian subspaces

We give here a description of the hermitian solutions X of the discrete algebraic Riccati equation

$$X = A^*XA + Q - A^*XB(R + B^*XB)^{-1}B^*XA \qquad (12.2.1)$$

in terms of certain subspaces. The matrices A, Q and X here are $n \times n$, the matrix R is $m \times m$, and it is assumed that Q and R are hermitian. Some properties of the term $R + B^*XB$ are developed first of all. Of course, if X is a solution of the equation then $R + B^*XB$ is necessarily invertible. The question of when $R + B^*XB > 0$ is of particular interest, and an answer will be given in Corollary 12.2.4.

It is useful here, and in the sequel, to associate with equation (12.2.1) the $n \times n$ rational matrix valued function

$$\Psi(z) = R + B^*(z^{-1}I - A^*)^{-1}Q(zI - A)^{-1}B. \qquad (12.2.2)$$

One checks easily that

$$(\Psi(z))^* = \Psi(\bar{z}^{-1})$$

and, hence, that $\Psi(z)$ takes hermitian values on the unit circle.

The role of the function $\Psi(z)$ in the study of equation (12.2.1) is underscored by the interesting formula

$$\Psi(z) = (\Delta_X(\bar{z}^{-1}))^*(R + B^*XB)\Delta_X(z), \qquad (12.2.3)$$

where X is any hermitian solution of (12.2.1), and

$$\Delta_X(z) = I + (R + B^*XB)^{-1}B^*XA(zI - A)^{-1}B. \qquad (12.2.4)$$

The verification of formula (12.2.3) is straightforward. On multiplying out the terms on the right it is found that $\Psi(z)$ is, indeed, independent of X. In particular, formula (12.2.3) implies the following result. Denote by $n_+(Z)$, $n_-(Z)$ and $n_0(Z)$ the number of positive, negative, and zero eigenvalues, respectively, of a hermitian matrix Z (the eigenvalues being counted with algebraic multiplicities). Also, let **T** denote the unit circle in the complex plane.

Theorem 12.2.1. *Assume that (12.2.1) has a hermitian solution. Then the numbers $n_+ := n_+(\Psi(z))$, $n_- := n_-(\Psi(z))$, $n_0 := n_0(\Psi(z))$ are constant on* **T**, *with the possible exception of a finite number of points. Moreover, for every hermitian solution X of (12.2.1) we have*

$$n_\eta(R + B^*XB) = n_\eta; \quad \eta = +, -, 0.$$

Proof. Observe that $\det(\Delta_X(z)) \not\equiv 0$, where $\Delta_X(z)$ is given by (12.2.4) (indeed, for $|z|$ large $\Delta_X(z)$ is close to I and therefore invertible). Thus, $\Delta_X(z)$ is analytic and invertible on $\mathbf{T}$, with the possible exception of a finite set Ω of values of z. Formula (12.2.3) shows that, for $\eta = +, -,$ or 0,

$$n_\eta(\Psi(z)) = n_\eta(R + B^*XB)$$

for every $z \in \mathbf{T}\setminus\Omega$. □

The subspaces that will describe the hermitian solution of the DARE are first of all Lagrangian with respect to the indefinite scalar product in $\mathbf{C}^{2n}$ induced by the hermitian matrix $J = \begin{bmatrix} 0 & -iI \\ iI & 0 \end{bmatrix}$. A subspace $\mathcal{M} \subseteq \mathbf{C}^{2n}$ is called *Lagrangian with respect to J* (or *J-Lagrangian*) if $\dim \mathcal{M} = n$ and $\langle Jx, y \rangle = 0$ for all $x, y \in \mathcal{M}$, i.e., $\mathcal{M}$ is a *maximal J-neutral subspace*.

We need an additional invariance property of the subspaces and, at this point, we make the assumption that the matrices A and R are invertible. (Other approaches to the DARE will be considered in Chapters 13 and 15 in which this assumption will be removed). Let

$$T = \begin{bmatrix} A + BR^{-1}B^*A^{*-1}Q & -BR^{-1}B^*A^{*-1} \\ -A^{*-1}Q & A^{*-1} \end{bmatrix}. \qquad (12.2.5)$$

A straightforward calculation shows that

$$T^*JT = J.$$

In other words, T is J-unitary. In particular, T is invertible.

The assumption of invertibility of A and R allows us to work with realizations of the function $\Psi(z)$ of equation (12.2.2) and its inverse as follows:

$$\Psi(z) = R + [-B^*A^{*-1}Q, B^*A^{*-1}] \left(zI - \begin{bmatrix} A & 0 \\ -A^{*-1}Q & A^{*-1} \end{bmatrix} \right)^{-1} \begin{bmatrix} B \\ 0 \end{bmatrix}, \qquad (12.2.6)$$

$$\Psi(z)^{-1} = R^{-1} - [-R^{-1}B^*A^{*-1}Q, R^{-1}B^*A^{*-1}](zI - T)^{-1} \begin{bmatrix} BR^{-1} \\ 0 \end{bmatrix}. \qquad (12.2.7)$$

In fact, (12.2.7) is obtained from (12.2.6) by using the formula (6.3.14). The realizations (12.2.6) and (12.2.7) need not be minimal.

The following proposition is the basis of the geometric theory for the DARE and is the discrete analogue of Proposition 7.2.1(β). As in Section 7.1, if $X \in \mathbf{C}^{n \times n}$, the subspace $\operatorname{Im} \begin{bmatrix} I \\ X \end{bmatrix}$ is called the *graph* of X.

Proposition 12.2.2. *Let X be a hermitian solution of (12.2.1). Then the graph of X is J-Lagrangian and T-invariant. Conversely, if the graph of X is J-Lagrangian and T-invariant, then X is a hermitian solution of (12.2.1).*

Proof. Assume X is a hermitian solution of (12.2.1). The J-Lagrangian property of $\mathcal{M}$ is a simple consequence of the hermitian property of X. Consider

$$T\begin{bmatrix} I \\ X \end{bmatrix} = \begin{bmatrix} A + BR^{-1}B^*A^{*-1}(Q - X) \\ -A^{*-1}(Q - X) \end{bmatrix}.$$

So, in order to show that $\operatorname{Im} \begin{bmatrix} I \\ X \end{bmatrix}$ is T-invariant, we need to show that

$$X(A + BR^{-1}B^*A^{*-1}(Q - X)) = -A^{*-1}(Q - X), \qquad (12.2.8)$$

or equivalently

$$Q - X + A^*XA + A^*XBR^{-1}B^*A^{*-1}(Q - X) = 0, \qquad (12.2.9)$$

which we rewrite as

$$A^*XA + A^*(I + XBR^{-1}B^*)A^{*-1}(Q - X) = 0. \qquad (12.2.10)$$

Now rewrite (12.2.1) using Proposition 12.1.1 and equation (12.1.5):

$$X = Q + A^*(I + XBR^{-1}B^*)^{-1}XA. \qquad (12.2.11)$$

Since X solves (12.2.1) we can use (12.2.11) in (12.2.10) to replace $(Q - X)$ by $-A^*(I + XBR^{-1}B^*)^{-1}XA$. This shows that, indeed, (12.2.10) holds.

Conversely, assume that $\operatorname{Im} \begin{bmatrix} I \\ X \end{bmatrix}$ is J-Lagrangian and T-invariant. Then X is clearly hermitian and (12.2.8) or, equivalently, (12.2.9) holds. Rewrite (12.2.8) in the form

$$XA + (I + XBR^{-1}B^*)A^{*-1}(Q - X) = 0. \qquad (12.2.12)$$

Next, we show that $I + XBR^{-1}B^*$ is invertible. Indeed, let the row vector y be such that $y(I + XBR^{-1}B^*) = 0$. Premultiplying (12.2.12) by y, we see that $yX = 0$. But then $y = -yXBR^{-1}B^* = 0$. So $I + XBR^{-1}B^*$ is invertible. Therefore $I + B^*XBR^{-1}$ and $R + B^*XB$ are invertible as well. Thus, (12.2.12) implies

$$A^{*-1}(Q - X) = -(I + XBR^{-1}B^*)^{-1}XA.$$

Using this in (12.2.9) we obtain

$$\begin{aligned} 0 &= Q - X + A^*XA - A^*XBR^{-1}B^*(I + XBR^{-1}B^*)^{-1}XA \\ &= Q - X + A^*(I - XBR^{-1}B^*(I + XBR^{-1}B^*)^{-1})XA \\ &= Q - X + A^*(I + XBR^{-1}B^*)^{-1}XA, \end{aligned}$$

which is (12.2.11). By Proposition 12.1.1 and equation (12.1.5), X is a solution of (12.2.1). $\square$

We would like to make a stronger statement than Proposition 12.2.2; namely, that every J-langrangian T-invariant subspace has the form $\mathcal{M} = \text{Im} \begin{bmatrix} I \\ X \end{bmatrix}$ for some hermitian solution X of the DARE. However, in general this statement is not true. We need additional hypotheses to ensure that every J-Lagrangian T-invariant subspace is a graph subspace. By analogy with the continuous case (see Theorem 7.6.1), a controllability hypothesis is expected. In addition to this, an important role is played by the matrix valued function $\Psi(z)$ of (12.2.2).

We now state and prove the main result of this section connecting hermitian solutions of the DARE with properties of the J-unitary matrix T of (12.2.5):

Theorem 12.2.3. *Assume that (A, B) is controllable, A and R are invertible and suppose there exists an $\eta \in \mathbf{T}$ such that $\Psi(\eta) > 0$, where $\Psi(z)$ is given by (12.2.2). Then the following statements are equivalent:*

(i) *the DARE (12.2.1) has a hermitian solution;*

(ii) *there exists a T-invariant J-Lagrangian subspace;*

(iii) *the partial multiplicities of T (i.e., the sizes of Jordan blocks in the Jordan form of T) corresponding to its unimodular eigenvalues, if any, are all even.*

When the statements (i)–(iii) hold, there is a one-to-one correspondence between the set of hermitian solutions of the DARE (12.2.1) and the set of T-invariant J-Lagrangian subspaces as follows: every T-invariant J-Lagrangian subspace $\mathcal{M}$ is of the form $\mathcal{M} = \text{Im} \begin{bmatrix} I \\ X \end{bmatrix}$ for a hermitian solution X of the DARE and conversely, if X is a hermitian solution of the DARE then $\text{Im} \begin{bmatrix} I \\ X \end{bmatrix}$ is T-invariant and J-Lagrangian.

For future reference note that, under the hypotheses of this theorem, $T|\mathcal{M}$ is similar to $A + BR^{-1}B^*(A^*)^{-1}(Q - X)$, where $\mathcal{M}$ is a T-invariant J-Lagrangian subspace and is the graph subspace of X.

Proof. In Proposition 12.2.2 we have already seen that (i) implies (ii).

For the converse, let $\mathcal{M} = \text{Im} \begin{bmatrix} X_1 \\ X_2 \end{bmatrix}$ be a T-invariant J-Lagrangian subspace. Here X_1 and X_2 are $n \times n$ matrices. Since $\dim \mathcal{M} = n$, we have $\text{Ker}\, X_1 \cap \text{Ker}\, X_2 = \{0\}$. We shall show that $\text{Ker}\, X_1 = \{0\}$. In that case we can take $X = X_2 X_1^{-1}$ and, by Proposition 12.2.2, X is a solution of the DARE (12.2.1).

Choose a point η on the unit circle such that $\Psi(\eta) > 0$ and the matrices $T - \eta I$ and $A^{*-1} - \eta I$ are invertible. Let $\varphi(\lambda) = (\lambda + \eta)(\lambda - \eta)^{-1}$. Since $\mathcal{M}$ is T-invariant, it is also $\varphi(T)$-invariant.

We shall compute the second block column of $\varphi(T)$, in other words, $\varphi(T)\begin{bmatrix} 0 \\ y \end{bmatrix}$, where $y \in \mathbb{C}^n$ is arbitrary (to appear as equation (12.2.15)). Let

$$(T-\eta)^{-1}\begin{bmatrix} 0 \\ y \end{bmatrix} = \begin{bmatrix} z_1 \\ z_2 \end{bmatrix}, \quad \begin{bmatrix} 0 \\ y \end{bmatrix} = (T-\eta)\begin{bmatrix} z_1 \\ z_2 \end{bmatrix}.$$

(Here, and elsewhere, we write η for the scalar matrix ηI.) Then

$$(A-\eta)z+1+BR^{-1}B^*A^{*-1}(Qz_1-z_2)=0, \tag{12.2.13}$$

$$-A^{*-1}Qz_1+(A^{*-1}-\eta)z_2=y. \tag{12.2.14}$$

Solving the last equation for z_2 we have $z_2 = (A^{*-1}-\eta)^{-1}(A^{*-1}Qz_1+y)$ and so

$$A^{*-1}(Qz_1-z_2) = A^{*-1}(I-(A^{*-1}-\eta)^{-1}A^{*-1})Qz_1 - A^{*-1}(A^{*-1}-\eta)^{-1}y$$
$$= -(\eta^{-1}-A^*)^{-1}Qz_1 - A^{*-1}(A^*-\eta)^{-1}y.$$

Substitute this in (12.2.13) to obtain

$$(A-\eta-BR^{-1}B^*(\eta^{-1}-A^*)^{-1}Q)z_1 = BR^{-1}B^*A^{*-1}(A^{*-1}-\eta)^{-1}y,$$

i.e.,

$$z_1 = (A-\eta-BR^{-1}B^*(\eta^{-1}-A^*)^{-1}Q)^{-1}BR^{-1}B^*A^{*-1}(A^{*-1}-\eta)^{-1}y,$$
$$z_2 = (A^{*-1}-\eta)^{-1}A^{*-1}Q(A-\eta-BR^{-1}B^*(\eta^{-1}-A^*)^{-1}Q)^{-1}$$
$$\times BR^{-1}B^*A^{*-1}(A^{*-1}-\eta)^{-1}y + (A^{*-1}-\eta)^{-1}y.$$

Now we have to find

$$\varphi(T)\begin{bmatrix} 0 \\ y \end{bmatrix}$$

$$= (T+\eta)(T-\eta)^{-1}\begin{bmatrix} 0 \\ y \end{bmatrix} = (T+\eta)\begin{bmatrix} z_1 \\ z_2 \end{bmatrix}$$

$$= (T-\eta)\begin{bmatrix} z_1 \\ z_2 \end{bmatrix} + \begin{bmatrix} 2\eta z_1 \\ 2\eta z_2 \end{bmatrix} = \begin{bmatrix} 2\eta z_1 \\ 2\eta z_2 + y \end{bmatrix}$$

$$= \begin{bmatrix} 2(A-\eta-BR^{-1}B^*(\eta^{-1}-A^*)^{-1}Q)^{-1}BR^{-1}B^*(\eta^{-1}-A^*)^{-1} \\ 2(A^{*-1}-\eta)^{-1}A^{*-1}Q(A-\eta-BR^{-1}B^*(\eta^{-1}-A^*)^{-1}Q)^{-1} \\ \cdot BR^{-1}B^*(\eta^{-1}-A^*)^{-1} + (A^{*-1}+\eta)(A^{*-1}-\eta)^{-1} \end{bmatrix} y.$$

We shall rewrite this using realization (12.2.7) for $\Psi(z)^{-1}$ which, after some simple algebra (taking advantage of equation (10.1.5), for example), can be rewritten in the form

$$\Psi(z)^{-1} = R^{-1} - R^{-1}B^*(z^{-1}-A^*)^{-1}Q(z-A+BR^{-1}B^*(z^{-1}-A^*)^{-1}Q)^{-1}BR^{-1}.$$

It follows that

$$B\Psi(\eta)^{-1}B^* = (\eta-A)(\eta-A+BR^{-1}B^*(\eta^{-1}-A^*)^{-1}Q)^{-1}BR^{-1}B^*.$$

Hence we can rewrite (using $\bar\eta = \eta^{-1}$)

$$\varphi(T)\begin{bmatrix} 0 \\ y \end{bmatrix}$$
$$= \begin{bmatrix} -2(\eta-A)^{-1}B\Psi(\eta)^{-1}B^*(\bar{\eta}-A^*)^{-1} \\ (A^{*-1}+\eta)(A^{*-1}-\eta)^{-1}+2(\eta-A^{*-1})^{-1}A^{*-1}Q(\eta-A)^{-1} \\ \cdot B\Psi(\eta)^{-1}B^*(\bar{\eta}-A^*)^{-1} \end{bmatrix} y. \quad (12.2.15)$$

We now return to the main line of the proof. We have $\mathcal{M} = \operatorname{Im} \begin{bmatrix} X_1 \\ X_2 \end{bmatrix}$ and if it is assumed that $X_1 x = 0$, then

$$\begin{bmatrix} 0 \\ X_2 x \end{bmatrix} \in \mathcal{M} \quad \text{and} \quad \varphi(T) \begin{bmatrix} 0 \\ X_2 x \end{bmatrix} \in \mathcal{M},$$

and by the J-Lagrangian property of $\mathcal{M}$,

$$0 = \left\langle J \begin{bmatrix} 0 \\ X_2 x \end{bmatrix}, \varphi(T) \begin{bmatrix} 0 \\ X_2 x \end{bmatrix} \right\rangle$$
$$= 2i \langle X_2 x, (\eta - A)^{-1} B \Psi(\eta)^{-1} B^* (\bar{\eta} - A^*)^{-1} X_2 x \rangle.$$

Since $\Psi(\eta) > 0$ by assumption, it follows that $B^*(\bar{\eta} - A^*)^{-1} X_2 x = 0$. So, for y with $\begin{bmatrix} 0 \\ y \end{bmatrix} \in \mathcal{M}$ we have

$$\varphi(T) \begin{bmatrix} 0 \\ y \end{bmatrix} = \begin{bmatrix} 0 \\ (A^{*-1}+\eta)(A^{*-1}-\eta)^{-1} y \end{bmatrix} \in \mathcal{M}$$

and $B^*(\bar{\eta} - A^*)^{-1} y = 0$. It then follows by induction on j that

$$B^*(A^{*-1}+\eta)^j (A^{*-1}-\eta)^{-j} (\bar{\eta}-A^*)^{-1} y = 0, \quad j = 0, 1, 2, \ldots$$

In particular, $B^*(A^{*-1}+\eta)^j (A^{*-1}-\eta)^{-j} (\bar{\eta}-A^*)^{-1} X_2 x = 0$. Now any power of A^{*-1} can be expressed as a linear combination of powers of $(A^{*-1}+\eta)(A^{*-1}-\eta)^{-1}$. So $(\bar{\eta}-A^*)^{-1} X_2 x \in \bigcap_{j=0}^{n-1} \operatorname{Ker} B^*(A^{*-1})^j = \{0\}$, because (A, B) is controllable. Hence $X_2 x = 0$, and $x \in \operatorname{Ker} X_1 \cap \operatorname{Ker} X_2 = \{0\}$. It follows that X_1 is invertible as desired. Hence (ii) implies (i).

We now show that (i) and (iii) are equivalent. First note that, by Theorem 1.7.5, the partial multiplicities of T at its unimodular eigenvalues are those of $i\varphi(T)$ at its real eigenvalues. Further, the above proof that (ii) implies (i) shows, in particular, that the second block column of $\varphi(T)$ has the form $\begin{bmatrix} D \\ C^* \end{bmatrix}$, where $D \leq 0$ and the pair (C, D) is controllable. Indeed, we have

$$D = -2(\eta - A)^{-1} B \Psi(\eta)^{-1} B^* (\bar{\eta} - A^*)^{-1}, \quad (12.2.16)$$

$$C = (A^{-1} - \bar{\eta})^{-1}(A^{-1} + \bar{\eta}) - DQA^{-1}(\bar{\eta} - A^{-1})^{-1}, \quad (12.2.17)$$

and in view of Proposition 4.1.2 the controllability of (C, D) is reduced to that of $((A^{-1} - \bar{\eta})^{-1}(A^{-1} + \bar{\eta}), (\eta - A)^{-1} B)$. The controllability of the latter pair follows easily from that of (A, B), using the fact that the matrix $(A^{-1} - \bar{\eta})^{-1}(A^{-1} + \bar{\eta})$ is a linear combination of powers of A (cf. the preceding paragraph).

Now suppose that (iii) holds. As before, let $\varphi(\lambda) = (\lambda + \eta)(\lambda - \eta)^{-1}$. Since $i\varphi(T)$ is J-self-adjoint (i.e., $J(i\varphi(T)) = (i\varphi(T))^*J$), and all partial multiplicities of $i\varphi(T)$ at its real eigenvalues are even, we can use Theorem 2.5.2 to obtain the existence of a J-Lagrangian subspace $\mathcal{M}$ which is $i\varphi(T)$-invariant. But then $\mathcal{M}$ is also T-invariant. So (ii) follows.

Conversely, suppose (ii) holds. Then also (i) holds, as we have shown above. Let X be a solution of (12.2.1), then $\mathcal{M} = \text{Im} \begin{bmatrix} I \\ X \end{bmatrix}$ is $\varphi(T)$-invariant. Let us denote

$$\varphi(T) = \begin{bmatrix} -C & D \\ E & C^* \end{bmatrix}, \qquad (12.2.18)$$

where $E = E^*$, $D = D^*$ (this form of $\varphi(T)$ follows from the J- self-adjointness of $i\varphi(T)$). As we have observed earlier, actually $D \leq 0$ and the pair (C, D) is controllable. Then the $\varphi(T)$- invariance of $\mathcal{M}$ shows that X solves the continuous algebraic Riccati equation

$$XDX - XC - C^*X - E = 0. \qquad (12.2.19)$$

We can then apply Theorem 7.3.7 to see that $\varphi(T)$ has only even partial multiplicities at its pure imaginary eigenvalues. But this implies that T has only even partial multiplicities at its unimodular eigenvalues. So (iii) holds. □

Observe also that, if X is a hermitian solution of (12.2.1), then equation (12.2.3) holds. As $\Psi(\eta) > 0$ it follows that $R + B^*XB > 0$, and hence $\Psi(z) \geq 0$ for all z on the unit circle which are not poles of $\Psi(z)$.

Thus, an application of Theorem 12.2.3 yields:

Corollary 12.2.4. *Given the hypotheses of Theorem 12.2.3, and also one of the conditions* (i), (ii) *and* (iii) *of that theorem, then* $\Psi(z) > 0$ *for all z on the unit circle with the possible exception of a finite number of points. Furthermore, $R + B^*XB > 0$ for every hermitian solution X of* (12.2.1).

An important question is whether, conversely, the condition $\Psi(z) \geq 0$ for all z on the unit circle which are not poles of $\Psi(z)$ implies the existence of hermitian solutions of the DARE (12.2.1). We return to this question in Section 12.4.

Under the hypotheses of Theorem 12.2.3 the matrix T has additional properties involving its sign characteristic with respect to J.

Theorem 12.2.5. *Assume the hypotheses of Theorem 12.2.3, and assume that the DARE* (12.2.1) *has hermitian solutions. Then the sign characteristic of the pair* (T, J) *consists of -1's only.*

Proof. If X is a hermitian solution of the DARE, then X solves the CARE (12.2.19).

By Theorem 7.6.1, the sign characteristic of the pair $\left(i \begin{bmatrix} C & -D \\ -E & -C^* \end{bmatrix}, -J \right)$ consists of $+1$'s only. In view of Theorem 2.3.4 the sign characteristic of (T, J) is equal to that of

$$(-i\varphi(T), J) = \left(i \begin{bmatrix} C & -D \\ -E & -C^* \end{bmatrix}, J \right).$$

It remains to observe that, in general, for a J-self-adjoint matrix Z, the signs in the sign characteristic of (Z, J) corresponding to the real eigenvalue λ_0 of Z, are opposite to the signs in the sign characteristic of $(Z, -J)$ corresponding to λ_0. □

12.3 Description of solutions in terms of invariant subspaces

In Theorem 7.6.2 the hermitian solutions of the continuous algebraic Riccati equation are described in terms of the matrix $\begin{bmatrix} A & D \\ C & -A^* \end{bmatrix}$ which is self-adjoint in the indefinite scalar product defined by $\begin{bmatrix} 0 & -iI \\ iI & 0 \end{bmatrix} = J$. Here, an analogous theory is developed for hermitian solutions of the DARE (12.2.1) in terms of invariant subspaces of the matrix T of (12.2.5) which is J-unitary. We continue to use the notation introduced in the preceding section.

In the next theorem, $\mathbb{D}$ stands for the open unit disk.

Theorem 12.3.1. *Given the hypotheses of Theorem 12.2.3 assume, in addition, that equation (12.2.1) has a hermitian solution. Then for every T-invariant subspace $\mathcal{N}$ such that $\sigma(T|\mathcal{N}) \subset \mathbb{D}$ there is a unique hermitian solution X of (12.2.1) with $\operatorname{Im} \begin{bmatrix} I \\ X \end{bmatrix} \cap \mathcal{R}(T, \mathbb{D}) = \mathcal{N}$. Here $\mathcal{R}(T, \mathbb{D})$ denotes the sum of root subspaces of T with respect to its eigenvalues in $\mathbb{D}$:*

$$\mathcal{R}(T, \mathbb{D}) = \sum_{\|\lambda_0\| < 1} \operatorname{Ker} (T - \lambda_0 I)^{2n}.$$

In other words, under the hypotheses of Theorem 12.3.1, there exists a one-to-one correspondence between hermitian solutions X of (12.2.1) and T-invariant subspaces $\mathcal{N} \subseteq \mathcal{R}(T, \mathbb{D})$, given by the formula

$$\operatorname{Im} \begin{bmatrix} I \\ X \end{bmatrix} \cap \mathcal{R}(T, \mathbb{D}) = \mathcal{N}. \qquad (12.3.1)$$

The proof of Theorem 12.3.1, as well as that of several subsequent results, will proceed by reduction to the corresponding results for the continuous algebraic

DESCRIPTION OF SOLUTIONS   281

Riccati equation. (This line of attack requires the additional assumption that
A is invertible.) To make the reduction more explicit, we state here a lemma
that describes the correspondence between the discrete and continuous Riccati
equations.

Lemma 12.3.2. *Consider the DARE* (12.2.1) *and assume that A and R are
invertible. Then a hermitian matrix X is a solution of* (12.2.1) *if and only if X
is a solution of the CARE*

$$XDX - XC - C^*X - E = 0, \qquad (12.3.2)$$

where

$$D = -2(\eta - A)^{-1}B\Psi(\eta)^{-1}B^*(\bar{\eta} - A^*)^{-1} \qquad (12.3.3)$$

$$C = (A^{-1} - \bar{\eta})^{-1}(A^{-1} + \bar{\eta}) + 2(\eta - A)^{-1}B\Psi(\eta)^{-1}B^*(\bar{\eta} - A^*)^{-1}QA^{-1}(\bar{\eta} - A^{-1})^{-1}, \qquad (12.3.4)$$

*and E is a suitable hermitian matrix (the exact form of E is not important here).
The function $\Psi(z)$ is defined in equation* (12.2.2), *and η is a unimodular number
such that the matrices $\Psi(\eta)$, $\eta - A$, $\bar{\eta} - A^{-1}$ and $\eta - T$ are invertible, where T
is given by equation* (12.2.5) *(and the existence of such an η is guaranteed).*

Proof. Let $\varphi(\lambda) = (\lambda + \eta)(\lambda - \eta)^{-1}$. Then $\varphi(T)$ is well-defined. If $X = X^*$ is a solution of (12.2.1), then $\mathcal{M} = \text{Im} \begin{bmatrix} I \\ X \end{bmatrix}$ is T-invariant by Proposition 12.2.2 and, therefore, is also $\varphi(T)$-invariant. As in the proof of Theorem 12.2.3 we verify that $\varphi(T) = \begin{bmatrix} -C & D \\ E & C^* \end{bmatrix}$, where D and C are given by (12.3.3) and (12.3.4), respectively. The $\varphi(T)$-invariance of $\mathcal{M}$ means that X is a solution of (12.3.2).

Conversely, assume X is a solution of (12.3.2). It follows that $\text{Im} \begin{bmatrix} I \\ X \end{bmatrix}$ is $\varphi(T)$-invariant. But then $\text{Im} \begin{bmatrix} I \\ X \end{bmatrix}$ is T-invariant as well. Indeed, since the function $\varphi(\lambda)$ is such that $\varphi'(\lambda_0) \neq 0$ for every $\lambda_0 \neq \eta$ and $\lambda_1 \neq \lambda_2 \Rightarrow \varphi(\lambda_1) \neq \varphi(\lambda_2)$, it follows that T and $\varphi(T)$ have exactly the same set of invariant subspaces (see Theorem 1.7.5). By Proposition 12.2.2, X is a solution of (12.2.1). □

Proof of Theorem 12.3.1. By Theorem 12.2.3, for every hermitian solution X of (12.2.1) the subspace (12.3.1) is indeed T-invariant and, obviously, is contained in $\mathcal{R}(T, \mathbb{D})$.

Conversely, let $\mathcal{N}$ be a T-invariant subspace contained in $\mathcal{R}(T, \mathbb{D})$. By Lemma 12.3.2 every solution $X = X^*$ of (12.2.1) is also a solution of (12.3.2), which we rewrite in the form

$$X(-D)X + XC + C^*X - (-E) = 0. \qquad (12.3.5)$$

Observe (as in the proof of Theorem 12.2.3) that $-D \geq 0$ and the pair $(C, -D)$ is controllable. Let

$$M_0 = \begin{bmatrix} C & -D \\ -E & -C^* \end{bmatrix}.$$

By Theorem 7.6.2, the formula

$$\operatorname{Im} \begin{bmatrix} I \\ X \end{bmatrix} \cap \mathcal{M}_+ = \mathcal{M}$$

gives a one-to-one correspondence between the hermitian solutions X of (12.3.5) and M_0-invariant subspaces $\mathcal{M}$ such that $\mathcal{M} \subseteq \mathcal{M}_+$. Here $\mathcal{M}_+$ is the spectral M_0-invariant subspace corresponding to the eigenvalues of M_0 in the open right half-plane. It remains to observe that $M_0 = -\varphi(T)$, and therefore M_0 and T have the same invariant subspaces and, by the spectral mapping theorem (Proposition 1.7.3),

$$\mathcal{M}_+ = \mathcal{R}(T; \mathbb{D}). \qquad \square$$

Several remarks concerning Theorem 12.3.1 and its proof are in order. Firstly, by analogy with Theorem 7.6.2, one can generalize Theorem 12.3.1 by replacing $\mathcal{R}(T, \mathbb{D})$ with the spectral T-invariant subspace corresponding to a set of eigenvalues $\mathcal{S}$ of T with the following properties:

(i) $\lambda \in \mathcal{S} \Rightarrow \bar{\lambda}^{-1} \notin \mathcal{S}$ (in particular, $\mathcal{S}$ does not contain unimodular eigenvalues of T, if any).

(ii) $\mathcal{S}$ is a maximal set of eigenvalues of T subject to (i), i.e., if $\mathcal{S} \subseteq \mathcal{S}_1 \subseteq \sigma(T)$, and $\mathcal{S}_1$ has the property (i), then in fact $\mathcal{S}_1 = \mathcal{S}$.

Secondly, the correspondence between the solutions of the DARE (12.2.1) and of the CARE (12.3.2) established in Lemma 12.3.2, allows us to transfer many results obtained in Chapter 7 for the equation (12.3.2) to the solutions of (12.2.1), just as we did in the proof of Theorem 12.3.1. Let us denote by $\operatorname{Ric}(A, B, Q, R)$ the set of hermitian solutions of (12.2.1), and by $\operatorname{Inv}(T \mid \mathcal{R}(T, \mathbb{D}))$ the set of T-invariant subspaces $\mathcal{N}$ such that $\sigma(T \mid \mathcal{N}) \subset \mathbb{D}$. The correspondence between $\operatorname{Ric}(A, B, Q, R)$ and $\operatorname{Inv}(T \mid \mathcal{R}(T, \mathbb{D}))$ given by (12.3.1) can be expressed in terms of the function $\varphi : \operatorname{Inv}(T \mid \mathcal{R}(T, \mathbb{D})) \to \operatorname{Ric}(A, B, Q, R)$ (so that $X = \varphi(\mathcal{N})$) and the inverse function φ^{-1} exists. We have:

Theorem 12.3.3. *Given the hypotheses of Theorem 12.2.3 assume, in addition, that the DARE (12.2.1) has a hermitian solution. Then the function φ and its inverse have the following properties:*

(i) *φ and φ^{-1} are order preserving: $X_1 \leq X_2$ for two hermitian solutions X_1 and X_2 of (12.2.1) if and only if $\varphi^{-1}(X_1) \subseteq \varphi^{-1}(X_2)$.*

(ii) φ and φ^{-1} are Lipschitz continuous; i.e., there exist positive constants K_1 and K_2 depending only on the coefficients of the equation (12.2.1) such that

$$\|X_1 - X_2\| \leq K_1 \theta(\varphi^{-1}(X_1), \varphi^{-1}(X_2))$$

for all $X_1, X_2 \in \mathrm{Ric}\,(A, B, Q, R)$ and

$$\theta(\mathcal{N}_1, \mathcal{N}_2) \leq K_2 \|\varphi(\mathcal{N}_1) - \varphi(\mathcal{N}_2)\|$$

for all $\mathcal{N}_1, \mathcal{N}_2 \in \mathrm{Inv}\,(T\,|\,\mathcal{R}(T, \mathbb{D}))$.

(iii) φ and φ^{-1} are real analytic: if $X(t)$ is a real analytic function of the real variables $t \in \Omega$, where Ω is an open connected set in $\mathbb{R}^k$, and if $X(t) \in \mathrm{Ric}\,(A, B, Q, R)$ for all $t \in \Omega$, then the subspace valued function $\varphi^{-1}(X(t))$ is real analytic on $t \in \Omega$ as well. Conversely, if $\mathcal{N}(t)$, $t \in \Omega$ is a real analytic subspace valued function, and if $\mathcal{N}(t) \in \mathrm{Inv}\,(T\,|\,\mathcal{R}(T, \mathbb{D}))$ for all $t \in \Omega$, then $\varphi(\mathcal{N}(t))$, $t \in \Omega$, is real analytic as well.

In particular, there exist maximal and minimal solution X_+ and X_-, respectively, of (12.2.1). In fact, $X_+ = \varphi(\mathcal{R}(T, \mathbb{D}))$, $X_- = \varphi(\{0\})$. Further, the maximal and minimal solutions of (12.2.1) are characterized by the property that all eigenvalues of $A + BR^{-1}B^*A^{*-1}(Q - X_+)$ are in the closed unit disk, and all eigenvalues λ of $A + BR^{-1}B^*A^{*-1}(Q - X_-)$ satisfy $|\lambda| \geq 1$.

Note that using equation (12.2.11), which is equivalent to the DARE (12.2.1), we also have

$$A + BR^{-1}B^*A^{*-1}(Q - X_\pm) = A - B(R + B^*X_\pm B)^{-1}B^*X_\pm A. \qquad (12.3.6)$$

In Section 12.5 it will be convenient to use the right-hand side of this algebraic equality.

12.4 Positive definiteness of Ψ and existence of hermitian solutions

We return now to the question asked in Section 12.2: whether the condition $\Psi(z) \geq 0$ for all $|z| = 1$ which are not poles of $\Psi(z)$ implies that the equation (12.2.1) has hermitian solutions. It turns out that the answer is affirmative provided $R > 0$.

Theorem 12.4.1. *Given the hypotheses of Theorem 12.2.3 assume in addition that $R > 0$. Then the conditions (i)–(iii) of Theorem 12.2.3 are equivalent to*

(iv) $\Psi(z) > 0$ *on* $\mathbf{T}$ *(with the possible exception of a finite number of points).*

In fact, we know by Corollary 12.2.4 that (i)–(iii) of Theorem 12.2.3 implies (iv). So the new content in Theorem 12.4.1 is that the converse also holds provided $R > 0$.

In the proof of Theorem 12.4.1 the following result concerning positive semidefinite rational matrix functions will be used. In anticipation of future use, we present this result under less restrictive hypotheses than those of Theorem 12.4.1.

Theorem 12.4.2. *Let $W(z)$ be an $n \times n$ rational matrix function such that $W(z) \geq 0$ for every z on the unit circle (which is not a pole of $W(z)$). Further, assume that $W(z)$ is analytic and invertible at infinity, let*

$$W(z) = D + C(zI - A)^{-1}B \qquad (12.4.1)$$

be a realization of $W(z)$, where the matrix A of size $m \times m$ is invertible and the equalities

$$SA = A^{*-1}S, \quad A^{*-1}C^* = SB, \quad -B^*A^{*-1}S = C \qquad (12.4.2)$$

hold for some invertible skew-hermitian matrix S and let $A^\times = A - BD^{-1}C$.

Assume that there exists a projection $\pi : \mathbb{C}^m \to \mathbb{C}^m$ such that $\operatorname{Ker} \pi$ is A-invariant, $\operatorname{Im} \pi$ is $A^\times$-invariant, the equality

$$S\pi = (I - \pi^*)S \qquad (12.4.3)$$

holds, and the pair $(\pi A \pi, \pi B)$, where $\pi A \pi : \operatorname{Im} \pi \to \operatorname{Im} \pi$ and $\pi B : \mathbb{C}^n \to \operatorname{Im} \pi$, is controllable. Then the partial multiplicities of the matrices A and $A^\times$ corresponding to their eigenvalues on the unit circle (if any) are all even.

Note that the realization (12.4.1) need not be minimal. On the other hand, if the realization (12.4.1) is minimal, then A is automatically invertible and the equalities (12.4.2) hold for a unique invertible matrix S. Indeed, the function $W(z)$ takes hermitian values on the unit circle, and therefore

$$W(z) = (W(\bar{z}^{-1}))^* \qquad (12.4.4)$$

for all $z \in \mathbb{C} \cup \{\infty\}$ (which are not poles of $W(z)$). By hypothesis, infinity is not a pole of $W(z)$, and by virtue of (12.4.4), zero is also not a pole of $W(z)$. But in a minimal realization (12.4.1) the eigenvalues of A coincide with the poles of $W(z)$ (see Theorem 6.2.2). So A is invertible. Further,

$$D - CA^{-1}B = W(0) = (W(\infty))^* = D^*, \qquad (12.4.5)$$

and using (12.4.4) and (12.4.5), we obtain

$$\begin{aligned} W(z) &= D^* + B^*(z^{-1} - A^*)^{-1}C^* = D^* - B^*A^{*-1}(z - A^{*-1})^{-1}zC^* \\ &= D^* - B^*A^{*-1}C^* - B^*A^{*-1}(z - A^{*-1})^{-1}A^{*-1}C^* \\ &= D - B^*A^{*-1}(z - A^{*-1})^{-1}A^{*-1}C^*. \end{aligned} \qquad (12.4.6)$$

If (12.4.1) is a minimal realization, then (12.4.6) is minimal as well and, by Theorem 6.1.4, there exists a unique invertible matrix S satisfying (12.4.2). Moreover,

POSITIVE DEFINITENESS OF Ψ

taking adjoints in (12.4.2) we conclude by the uniqueness of S that $S = -S^*$. Furthermore, if (12.4.1) is minimal, then the existence of the projector π with the properties required in Theorem 12.4.2 is guaranteed by Corollary 6.6.2.

The proof of Theorem 12.4.2 will proceed by reduction to a problem in matrix theory (not involving rational matrix functions). The matrix theory problem and its solution are given by the following lemma:

Lemma 12.4.3. *Let A_1, B_1, C_1 and D_1 be matrices of sizes $m \times m$, $m \times n$, $n \times m$ and $n \times n$, respectively, such that $D_1 > 0$ and*

$$HA_1 = A_1^* H, \quad HB_1 = C_1^* \tag{12.4.7}$$

for some invertible matrix H. Further, let $\pi : \mathbb{C}^m \to \mathbb{C}^m$ (where $m \times m$ is the size of A_1) be a projection with the properties that $\operatorname{Ker} \pi$ is A_1-invariant, $\operatorname{Im} \pi$ is $(A_1 - B_1 D_1^{-1} C_1)$-invariant and

$$H\pi = (I - \pi^*)H, \quad (I - \pi)H^{*-1}H = I - \pi. \tag{12.4.8}$$

If the pair $(\pi A_1 \pi, \pi B_1)$ (understood as a pair of linear transformations into $\operatorname{Im} \pi$) is controllable, i.e. $\sum_{j=0}^{\infty} \operatorname{Im} ((\pi A_1 \pi)^j \pi B_1) = \operatorname{Im} \pi$, then the partial multiplicities of $A_1 - B_1 D_1^{-1} C_1$ corresponding to the real eigenvalues (if any) are all even.

Note that the second equality of (12.4.8) is always satisfied as long as H is hermitian.

Proof. Denote by $\operatorname{Im} \pi \oplus \operatorname{Im} \pi^*$ the exterior sum of the subspaces $\operatorname{Im} \pi$ and $\operatorname{Im} \pi^*$, i.e., the linear space of all column vectors $\begin{bmatrix} x \\ y \end{bmatrix}$ with $x \in \operatorname{Im} \pi$ and $y \in \operatorname{Im} \pi^*$. Let

$$S = \begin{bmatrix} \pi A_1^\times \pi & \pi B_1 D_1^{-1} B_1^* \pi^* \\ 0 & \pi^* (A_1^\times)^* \pi^* \end{bmatrix} : \operatorname{Im} \pi \oplus \operatorname{Im} \pi^* \to \operatorname{Im} \pi \oplus \operatorname{Im} \pi^* \tag{12.4.9}$$

where $A_1^\times = A_1 - B_1 D_1^{-1} C_1$. From the properties of π it follows that $\pi A_1(I - \pi) = 0$ and $(I - \pi) A_1^\times \pi = 0$. Since (12.4.8) gives $(I - \pi) H^{*-1} = H^{*-1} \pi^*$, it follows from (12.4.7) that

$$S = \begin{bmatrix} \pi A_1^\times \pi & -\pi A_1^\times (I - \pi) H^{*-1} \\ -H^*(I - \pi) A_1^\times \pi & H^*(I - \pi) A_1^\times (I - \pi) H^{*-1} \end{bmatrix}.$$

For example:

$$\begin{aligned}
\pi^*(A_1^\times)^*\pi^* &= \pi^* H^*(H^{*-1}A_1^*H^* - H^{*-1}C_1^*D_1^{-1}B_1^*H^*)H^{*-1}\pi^* \\
&= H^*(I-\pi)(A_1 - H^{*-1}HB_1 D_1^{-1}C_1)(I-\pi)H^{*-1} \\
&= H^*(I-\pi)(A_1 - B_1 D_1^{-1}C_1)(I-\pi)H^{*-1} \\
&= H^*(I-\pi)A_1^\times(I-\pi)H^{*-1}.
\end{aligned}$$

Note that the linear transformation

$$U = \begin{bmatrix} \pi \\ -H^*(I-\pi) \end{bmatrix} : \text{Im } \pi \dotplus \text{Ker } \pi \to \text{Im } \pi \oplus \text{Im } \pi^*$$

is invertible with the inverse $U^{-1} = [\pi, -(I-\pi)H^{*-1}]$. As $S = UA_1^\times U^{-1}$, it is sufficient to prove that the partial multiplicities of S corresponding to the real eigenvalues are all even.

Let $J : \text{Im } \pi \to \mathbf{C}^k$ be an invertible transformation (here $k = \dim \text{Im } \pi$), and let $J^* : \mathbf{C}^k \to \text{Im } \pi^*$ be the adjoint transformation to J with respect to the sesquilinear form $[x, y] = y^*x$, where $x \in \text{Im } \pi$, $y \in \text{Im } \pi^*$. In other words, J^* is defined by the property that $z^*(Jx) = [x, J^*z]$ for every $x \in \text{Im } \pi$ and every $z \in \mathbf{C}^k$. This definition of J^* is correct because the form $[\cdot,\cdot]$ is nondegenerate (if for some $y \in \text{Im } \pi^*$ we have $y^*x = 0$ for all $x \in \text{Im } \pi$, then $y \in (\text{Im } \pi)^\perp = \text{Ker } \pi^*$ and hence $y = 0$). Put

$$V = \begin{bmatrix} J & 0 \\ 0 & J^{*-1} \end{bmatrix} : \text{Im } \pi \oplus \text{Im } \pi^* \to \mathbf{C}^k \oplus \mathbf{C}^k$$

then

$$VSV^{-1} = \begin{bmatrix} J\pi A_1^\times \pi J^{-1} & J\pi B_1 D_1^{-1}B_1^*\pi^* J^* \\ 0 & J^{*-1}\pi^*(A_1^\times)^*\pi^* J^* \end{bmatrix} : \mathbf{C}^k \oplus \mathbf{C}^k \to \mathbf{C}^k \oplus \mathbf{C}^k.$$

Denoting $Q = J\pi A_1^\times \pi J^{-1}$ and $R = J\pi B_1$, it follows from the definitions that $Q^* = J^{*-1}\pi^*(A_1^\times)^*\pi^* J^*$ and $R^* = B_1^*\pi^* J^*$. The controllability of $(\pi A_1^\times \pi, \pi B_1)$ follows from Lemma 4.4.1 and, consequently, (Q, R) is also controllable. Now

$$VSV^{-1} = \begin{bmatrix} Q & RD_1^{-1}R^* \\ 0 & Q^* \end{bmatrix},$$

and the conclusion of the lemma follows from Lemma 7.3.3. $\square$

Proof of Theorem 12.4.2. Let η be a point on the unit circle which is not an eigenvalue of A or $A^\times$ and such that $W(\eta) > 0$. Introduce the Mobius transformation $\varphi(z) = \eta(z-i)(z+i)^{-1}$, $z \in \mathbf{C}\setminus\{i\}$, and let $W_1(z) = W(\varphi(z))$. Obviously, $W_1(z)$ is a rational matrix function. Using some simple algebraic manipulations (or see Theorem 1.9 of Bart et al. (1979)), it is not difficult to check that $W_1(z)$ admits a realization

$$W_1(z) = W(\eta) + C_1(z - A_1)^{-1}B_1 \qquad (12.4.10)$$

where

$$C_1 = \sqrt{2}\,i\eta C(\eta - A)^{-1}; \quad B_1 = \sqrt{2}\,(\eta - A)^{-1}B; \quad A_1 = -i(A + \eta)(A - \eta)^{-1}.$$

Using the equalities (12.4.2) one checks without difficulty that

$$(iS)A_1 = A_1^*(iS), \quad (iS)B_1 = C_1^*. \qquad (12.4.11)$$

(It is convenient here to use the equation

$$S(z - A)^{-1} = (z - A^{*-1})^{-1}S$$

which holds for all $z \in \mathbb{C}$ for which both sides of this equation are defined. To verify this equation note that $S(z - A) = (z - A^{*-1})S$ for all complex z.)

Further, if π is a projection as in Theorem 12.4.2, then obviously Ker π is A_1-invariant. Also, Im π is $A_1^\times$-invariant, where $A_1^\times = A_1 - B_1 W(\eta)^{-1} C_1$. This follows from the easily verified assertion that

$$A_1^\times = -i(A^\times + \eta)(A^\times - \eta)^{-1}. \qquad (12.4.12)$$

(This also follows from Theorem 1.9 in Bart et al. (1979) applied to the realization

$$W_1(z)^{-1} = W(\eta)^{-1} - W(\eta)^{-1} C_1 (z - A_1^\times) B_1 W(\eta)^{-1}$$

(cf. (6.3.14))).

Finally, let us check that the pair $(\pi A_1 \pi, \pi B_1)$ is controllable. Observe that, with respect to the direct sum decomposition $\mathbb{C} = \text{Im } \pi \dotplus \text{Ker } \pi$, the matrix A has the form $A = \begin{bmatrix} \pi A \pi & 0 \\ * & * \end{bmatrix}$. Since $\eta - A$ is invertible, it follows that $\pi(\eta - A)\pi$ is invertible (as a linear transformation of Im π into itself), and hence the controllability of $(\pi A_1 \pi, \pi B_1)$ is equivalent to the controllability of

$$\left(\pi(\eta - A)\pi \cdot \pi A_1 \pi \cdot (\pi(\eta - A)\pi)^{-1}, \; \frac{1}{\sqrt{2}} \pi(\eta - A)\pi \cdot \pi B_1 \right) = (\pi A_1 \pi, \pi B).$$

However, the controllability of $(\pi A_1 \pi, \pi B)$ follows easily from that of $(\pi A\pi, \pi B)$. Indeed, $\pi A_1 \pi = f(\pi A \pi)$ where $f(\lambda) = -i(\lambda + \eta)(\lambda - \eta)^{-1}$. Hence $\pi A_1 \pi$ and $\pi A \pi$ have the same set of invariant subspaces (Theorem 1.7.5) and, by Proposition 4.1.2, the pairs $(\pi A_1 \pi, \pi B)$ and $(\pi A\pi, \pi B)$ are simultaneously controllable.

Now we can apply Lemma 12.4.3 (with $D_1 = W(\eta)$ and $H = iS$) and deduce that the partial multiplicities of $A_1^\times$ corresponding to the real eigenvalues are all even. In view of Theorem 1.7.5, it follows from (12.4.12) that this is true also for the partial multiplicities of $A^\times$ corresponding to the eigenvalues on the unit circle.

It remains to prove the part of Theorem 12.4.2 concerning A. The realization (12.4.1) leads to the following realization for $W(z)^{-1}$ (see equation (6.3.14)):

$$W(z)^{-1} = D^{-1} - D^{-1}C(zI - A^\times)^{-1}BD^{-1}. \qquad (12.4.13)$$

We shall verify that the hypotheses of Theorem 12.4.2 are satisfied with respect to (12.4.13). Indeed, let

$$\tilde{\pi} = I - \pi, \quad \tilde{S} = S^*, \quad \tilde{D} = D^{-1}, \quad \tilde{C} = -D^{-1}C, \quad \tilde{B} = BD^{-1}.$$

Then clearly $A = A^\times - \tilde{B}\tilde{D}^{-1}\tilde{C}$. The equality $\tilde{S}\tilde{\pi} = (I - \tilde{\pi}^*)\tilde{S}$ (a counterpart of (12.4.3)) is also clear. To verify the equality

$$(A^\times)^{*-1}\tilde{C}^* = \tilde{S}\tilde{B}, \qquad (12.4.14)$$

take the conjugate transpose and rewrite it in the form $-D^{-1}C(A^\times)^{-1} = D^{*-1}B^*S$, or

$$-D^*D^{-1}C = B^*SA^\times. \qquad (12.4.15)$$

The right-hand-side of (12.4.15) is equal to

$$B^*SA - B^*SBD^{-1}C = B^*A^{*-1}S - B^*A^{*-1}C^*D^{-1}C$$
$$= -C + CS^{-1}C^*D^{-1}C = -C + CA^{-1}BD^{-1}C, \qquad (12.4.16)$$

where we have used equalities (12.4.2), and in the last step the equality $S^{-1}C^* = A^{-1}B$ has also been used. (This follows from $C^* = A^*SB$ and $A^*S = SA^{-1}$). In view of (12.4.5), $-C + CA^{-1}BD^{-1}C = -D^*D^{-1}C$, and (12.4.15) is proved.

The equality $-\tilde{B}^*\tilde{S}A^\times = \tilde{C}$ is proved in a similar way, first by rewriting it in the form

$$B^*S^*(A - BD^{-1}C) = D^*D^{-1}C,$$

and then using the equalities (12.4.5) and $B^*S^*A = C$ (this latter follows from $C^* = A^*SB$ by taking the conjugate transpose). Finally, we check that

$$\tilde{S}A^\times = (A^\times)^{*-1}\tilde{S}. \qquad (12.4.17)$$

For this purpose, compute

$$A^{\times *}\tilde{S}A^{\times} - \tilde{S} = (A^* - C^*D^{*-1}B^*)S^*(A - BD^{-1}C) - S^*$$
$$= A^*S^*A - C^*D^{*-1}B^*S^*A - A^*S^*BD^{-1}C$$
$$+ C^*D^{*-1}B^*S^*BD^{-1}C - S^*$$
$$= -C^*D^{*-1}B^*S^*A - A^*S^*BD^{-1}C + C^*D^{*-1}B^*S^*BD^{-1}C, \quad (12.4.18)$$

and we have used $A^*S^*A = S^*$, which follows from (12.4.2). Now, using (12.4.2) again,

$$A^*S^*B = S^*A^{-1}B = -C^*, \quad B^*S^*A = CA^{-1}A = C, \quad (12.4.19)$$

and substituting these equalities in (12.4.18), we obtain

$$A^{\times *}\tilde{S}A^{\times} - \tilde{S} = C^*(-D^{*-1} + D^{-1} + D^{*-1}B^*S^*BD^{-1})C$$
$$= C^*D^{*-1}(-D + D^* + B^*S^*B)D^{-1}C \quad \text{(by (12.4.5))}$$
$$= C^*D^{*-1}(-CA^{-1} + B^*S^*)BD^{-1}C = 0$$

by (12.4.19). Thus, (12.4.17) is verified. □

The proof of Theorem 12.4.1 requires some analysis of another rational matrix function closely related to $\Psi(z)$, namely,

$$W(z) = I + (z - A)^{-1}B\Psi(z)^{-1}B^*(z^{-1} - A^*)^{-1}. \quad (12.4.20)$$

Recall the standing hypotheses that A and R are invertible matrices.

Lemma 12.4.4. *The matrix function $W(z)$ admits a realization*

$$W(z) = I + \frac{1}{2}[I \ 0](zI - T)^{-1}T\begin{bmatrix} 0 \\ I \end{bmatrix}, \quad (12.4.21)$$

and T is defined in (12.2.5).

If, moreover, the pair (A, B) is controllable and the matrix R is positive definite, then the realization (12.4.21) is minimal.

Proof. Using formula (12.2.15) (with η replaced by $z \in \mathbb{C}$) we obtain

$$[I \ 0](T + z)(T - z)^{-1}\begin{bmatrix} 0 \\ I \end{bmatrix} = -2(z - A)^{-1}B\Psi(z)^{-1}B^*(z^{-1} - A^*)^{-1}.$$

This formula was first proved for all $z \in \mathbf{T}$ but, since both sides are rational matrix functions, it extends to all $z \in \mathbb{C}$ for which both sides make sense. It follows that

$$\frac{1}{2}[I\ 0](T+z)(z-T)^{-1}\begin{bmatrix}0\\I\end{bmatrix} = (z-A)^{-1}B\Psi(z)^{-1}B^*(z^{-1}-A^*)^{-1},$$

and, consequently, $W(z) = I + \frac{1}{2}[I\ 0](T+z)(z-T)^{-1}\begin{bmatrix}0\\I\end{bmatrix}$. In turn,

$$\frac{1}{2}[I\ 0](T+z)(z-T)^{-1}\begin{bmatrix}0\\I\end{bmatrix} = \frac{1}{2}[I\ 0][(z-T+2T)(z-T)^{-1}]\begin{bmatrix}0\\I\end{bmatrix}$$

$$= \frac{1}{2}[I\ 0][I+2(z-T)^{-1}T]\begin{bmatrix}0\\I\end{bmatrix}$$

$$= [I\ 0](z-T)^{-1}T\begin{bmatrix}0\\I\end{bmatrix},$$

and (12.4.21) follows.

Assume now in addition that $R > 0$ and (A, B) is a controllable pair. It will be verified that

$$\bigcap_{j=0}^{\infty} \operatorname{Ker}([I\ 0]T^j) = \{0\} \qquad (12.4.22)$$

and

$$\bigcap_{j=0}^{\infty} \operatorname{Ker}([0\ I]T^{*j+1}) = \{0\}. \qquad (12.4.23)$$

In fact, because T^* is invertible and the subspace on the left of (12.4.23) is T^*-invariant (and hence T^{*-1}-invariant), (12.4.23) is equivalent to

$$\bigcap_{j=0}^{\infty} \operatorname{Ker}([0\ I]T^{*j}) = \{0\}. \qquad (12.4.24)$$

In view of Theorem 6.1.5 this will show that the realization (12.4.21) is indeed minimal. So let $[I\ 0]\begin{bmatrix}x\\y\end{bmatrix} = 0, T\begin{bmatrix}x\\y\end{bmatrix} = \lambda\begin{bmatrix}x\\y\end{bmatrix}$. Then $x = 0$ and $\lambda \neq 0$, as T is invertible. So $BR^{-1}B^*A^{*-1}y = 0$ and $A^{*-1}y = \lambda y$. We obtain $BR^{-1}B^*y = 0$, which gives $B^*y = 0$ (because $R > 0$). But then, since (A, B) is controllable, $y = 0$. Hence (12.4.22) follows. To prove (12.4.24), assume $T^*\begin{bmatrix}x\\y\end{bmatrix} = \lambda\begin{bmatrix}x\\y\end{bmatrix}$ and $[0\ I]\begin{bmatrix}x\\y\end{bmatrix} = 0$. Then $\lambda \neq 0$ and $y = 0$, from which we get $A^*x = \lambda x$, $BR^{-1}B^*x = 0$. As before, this yields $x = 0$ because of the controllability of (A, B). □

The proof of Theorem 12.4.1 is now easily accomplished: Under the hypotheses of Theorem 12.4.1, assume that (iv) holds. Formula (12.4.20) shows that also $W(z) > 0$ on **T** (with the possible exception of a finite number of points). By Lemma 12.4.4 the realization (12.4.21) is minimal, and therefore Theorem 12.4.2 is applicable (see the remarks after the statement of Theorem 12.4.2). We conclude that the partial multiplicities of T corresponding to its unimodular eigenvalues are all even. □

12.5 Relaxing the controllability condition

In the main results of this chapter so far (Theorems 12.2.3, 12.3.1, 12.3.3, 12.4.1) the assumptions of controllability of (A, B) played a significant role. We show here that this assumption can be relaxed (at the expense of weaker conclusions). Similarly, many results which are parallels of those obtained in Chapter 7 can be obtained for the discrete Riccati equation. Once again, the key is Lemma 12.3.2 which allows us to reduce discrete Riccati equations to continuous ones.

We start with sign controllable pairs, and adapt the definition to suit the DARE. A pair of matrices (A, B), where A is $n \times n$ and B is $n \times m$, will be called *d-sign controllable* if $\mathcal{R}_0(A) \subseteq \mathcal{C}(A, B)$ and for any nonzero $\lambda \in \mathbb{C}$ at least one of the subspaces $\mathcal{R}_\lambda(A)$ and $\mathcal{R}_{\bar{\lambda}^{-1}}(A)$ is contained in $\mathcal{C}(A, B)$ (as defined in equation (4.1.2)). As usual, in this book, $\mathcal{R}_\lambda(A)$ is understood to be the zero subspace if λ is not an eigenvalue of A.

As in the previous sections, we consider the DARE

$$X = A^*XA + Q - A^*XB(R + B^*XB)^{-1}B^*XA \qquad (12.5.1)$$

where A, B, Q, R are given matrices of sizes $n \times n$, $n \times m$, $n \times n$, and $m \times m$, respectively, with Q and R hermitian. It will be assumed throughout this section that A and R are invertible. As in equation (12.2.2) the rational matrix function

$$\Psi(z) = R + B^*(z^{-1}I - A^*)^{-1}Q(zI - A)^{-1}B \qquad (12.5.2)$$

is associated with equation (12.5.1). The matrices T of equation (12.2.5)) and J are defined as in Section 12.2.

Theorem 12.5.1. *Assume that the pair (A, B) is d-sign controllable and $\Psi(\eta) > 0$ for some η on the unit circle. Then the following statements are equivalent:*

(i) *the DARE (12.5.1) has a hermitian solution X;*

(ii) *there exists a T-invariant J-Lagrangian subspace;*

(iii) *T has all even multiplicities corresponding to the unimodular eigenvalues (if any), and the sign characteristic of (T, J) consists of -1's only.*

Proof. (iii) $\Rightarrow$ (ii) follows from Theorem 2.5.1 (adapted to H-unitary matrices rather than H-self-adjoint ones).

Assume that (i) holds and let $\varphi(\lambda) = (\lambda + \eta)(\lambda - \eta)^{-1}$, where η is as in Lemma 12.3.2. We have

$$i \begin{bmatrix} C & -D \\ -E & -C^* \end{bmatrix} = -i\varphi(T)$$

where D and C are given by (12.3.3) and (12.3.4), respectively (see the proof of Theorem 12.2.3). Lemma 12.3.2 shows that the CARE (12.3.2) has hermitian solutions. Using formulas (12.2.16) and (12.2.17), one verifies that the pair $(C, -D)$ is sign controllable (the following fact, whose proof is analogous to that of Lemma 4.5.3, is useful here: a pair (Y, Z) is sign controllable if and only if $(Y + ZK, ZL)$ is sign controllable for any matrices K and L of suitable dimensions such that Im (BL) = Im B). Thus, Theorem 7.3.7 is applicable. We find that $-i\varphi(T)$ has only even multiplicities corresponding to its real eigenvalues (if any), and the sign characteristic of $(-i\varphi(T), J)$ consists of -1's only (observe that $\hat{H} = -J$, where $\hat{H}$ is taken from Theorem 7.3.7). By Theorem 2.3.4 the sign characteristic of (T, J) consists of -1's only, and Theorem 1.7.5 ensures that T has only even multiplicities corresponding to its unimodular eigenvalues. In other words, (iii) holds.

Applying Theorem 7.3.7 for

$$\varphi(T) = \begin{bmatrix} -C & D \\ E & C^* \end{bmatrix}$$

we find that (ii) implies the existence of a hermitian solution X of (12.3.2). By Lemma 12.3.2 X is also a solution of (12.5.1). $\square$

The situation in which the pair (A, B) is d-stabilizable is important in some applications and is, at once, weaker than the controllability of (A, B), and more restrictive than d-sign controllability (see Section 4.4). Recall that a pair of matrices (A, B) (where A is $n \times n$, B is $n \times m$) is called d-stabilizable if there exists a matrix K such that $A + BK$ is d-stable, or equivalently (Proposition 4.5.2) the matrix A_2 in the control normal form

$$\begin{bmatrix} A_1 & A_{12} \\ 0 & A_2 \end{bmatrix}, \begin{bmatrix} B_1 \\ 0 \end{bmatrix} \qquad (12.5.3)$$

of (A, B) is d-stable.

To extend the geometric theory to the d-stabilizable case, we employ the method used in Section 7.9 for the continuous Riccati equations, and assume (without loss of generality) that A and B have the form (12.5.3) with respect to the orthogonal decomposition $\mathbb{C}^n = \mathcal{C} \oplus \mathcal{C}^\perp$, where $\mathcal{C}$ is the controllable subspace of (A, B).

RELAXING THE CONTROLLABILITY CONDITION 293

Partition X and Q conformally with (12.5.3):

$$X = \begin{bmatrix} X_1 & X_2 \\ X_2^* & X_3 \end{bmatrix}, \quad Q = \begin{bmatrix} Q_1 & Q_2 \\ Q_2^* & Q_3 \end{bmatrix} \tag{12.5.4}$$

(we assume that X is hermitian, while $Q = Q^*$ is a standing assumption). Then equation (12.5.1) is equivalent to the following system of equations:

$$X_1 = A_1^* X_1 A_1 + Q_1 - A_1^* X_1 B_1 (R + B_1^* X_1 B_1)^{-1} B_1^* X_1 A_1 \tag{12.5.5}$$

$$X_2 = A_1^* X_1 A_{12} + Q_2 - A_1^* X_1 B_1 (R + B_1^* X_1 B_1)^{-1} B_1^* X_1 A_{12}$$
$$\quad + A_1^* [I - X_1 B_1 (R + B_1^* X_1 B_1)^{-1} B_1^*] X_2 A_2 \tag{12.5.6}$$

$$X_3 = (A_{12}^* X_1 + A_2^* X_2^*) A_{12} + (A_{12}^* X_2 + A_2^* X_3) A_2$$
$$\quad + Q_3 + (A_{12}^* X_1 B_1 + A_2^* X_2^* B_1)(R + B_1^* X_1 B_1)^{-1}(B_1^* X_2 A_2 + B_1^* X_1 A_{12}). \tag{12.5.7}$$

The equation (12.5.5) is another DARE but with the controllable pair (A_1, B_1). If X_1 is a hermitian solution of (12.5.5) then equation (12.5.6) is just a linear equation in X_2 with the form of equation (5.2.4). Since A_2 is d-stable it follows from Theorem 5.2.3 that equation (12.5.6) is uniquely solvable for X_2 provided all eigenvalues of

$$A_1^* - A_1^* X_1 B_1 (R + B_1^* X_1 B_1)^{-1} B_1^*$$

lie in the closed unit disk; or, equivalently, this property holds for

$$A_1 - B_1(R + B_1^* X_1 B_1)^{-1} B_1^* X_1 A_1. \tag{12.5.8}$$

Finally, (12.5.7) is a symmetric Stein equation with the unknown X_3 (of the form (5.3.5)) which, in view of the d-stability of A_2, is uniquely solvable by the same Theorem 5.2.3 and the solution X_3 is hermitian.

Note also that, because of the special form of A and B in (12.5.3), the function $\Psi(z)$ of (12.2.2) remains invariant when A, B, Q are replaced by A_1, B_1, Q_1. We can therefore apply Theorem 12.2.3 to the equation (12.5.5). If there exists a hermitian solution X_1 of (12.5.5), then by Theorem 12.3.3, the maximal hermitian solution X_+ exists, and (see equation (12.3.6)) for this solution the matrix (12.5.8) does, indeed, have all its eigenvalues in the closed unit disk. Incorporating Theorem 12.2.5 as well, the following result is obtained:

Theorem 12.5.2. *Assume that the pair (A, B) is d-stabilizable and $\Psi(\eta) > 0$ for some $\eta \in \mathbf{T}$, and write A and B as in (12.5.3) with respect to the decomposition $\mathbf{C}^n = \mathcal{C} \oplus \mathcal{C}^\perp$, where $\mathcal{C}$ is the controllable subspace of (A, B). Assume also that A_1 and R are invertible. Then the following statements are equivalent:*

(i) *the DARE (12.5.1) has a hermitian solution;*

(ii) there exists a T_1-invariant J_1-Lagrangian subspace, where

$$T_1 = \begin{bmatrix} A_1 + B_1 R^{-1} B_1^* A_1^{*-1} Q_1 & -B_1 R^{-1} B_1^* A_1^{*-1} \\ -A_1^{*-1} Q_1 & A_1^{*-1} \end{bmatrix}$$

and $J_1 = \begin{bmatrix} 0 & -iI \\ iI & 0 \end{bmatrix}$, where I is the identity map on $\mathcal{C}$;

(iii) T_1 has all even partial multiplicities corresponding to its unimodular eigenvalues (if any);

(iv) T_1 has all even partial multiplicities corresponding to the unimodular eigenvalues, and the sign characteristic of (T_1, J_1) consists of -1's only.

Using the representation of the DARE (12.5.1) in the form (12.5.5)–(12.5.7), results completely analogous to Theorems 7.9.5 and 7.9.7 can be proved. We will state only the analogue of Theorem 7.9.5(i).

The equation (12.5.1) will be called *regular* if for every hermitian solution X of (12.5.1) (partitioned as in (12.5.4)) the matrices $A_1^* - A_1^* X_1 B_1 (R + B_1^* X_1 B_1)^{-1} B_1^*$ and A_2 have no eigenvalues λ and μ, respectively, such that $\lambda\mu = 1$. This condition guarantees (by Theorem 5.2.3) that the equation (12.5.6) is always uniquely solvable for X_2. A sufficient condition for regularity (which does not involve the solutions X) is that $\sigma(T_1)$ and $\overline{\sigma(A_2)}$ do not contain numbers λ and μ, respectively, such that $\lambda\mu = 1$. Indeed, the restriction

$$T_1 | \operatorname{Im} \begin{bmatrix} I \\ X_1 \end{bmatrix}$$

is similar to $A_1 - B_1(R + B_1^* X_1 B_1)^{-1} B_1^* X_1 A_1$ (cf. formula (12.3.6)), and therefore

$$\sigma(A_1 - B_1(R + B_1^* X_1 B_1)^{-1} B_1^* X_1 A_1) \subseteq \sigma(T_1).$$

Theorem 12.5.3. *Assume that the DARE (12.5.1) has hermitian solutions and is regular. Denote by $P : \mathbb{C}^n \to \mathcal{C}$ the orthogonal projector on the controllable subspace $\mathcal{C}$, and let $\mathcal{L}_+$ be the spectral T_1-invariant subspace corresponding to the eigenvalues of T_1 in the open unit disk. Then for every hermitian solution X of (12.5.1) the subspace*

$$\mathcal{L} = \operatorname{Im} \begin{bmatrix} I \\ PXP \end{bmatrix} \cap \mathcal{L}_+ \qquad (12.5.9)$$

is T_1-invariant. Conversely, for every T_1-invariant subspace $\mathcal{L} \subseteq \mathcal{L}_+$ there exists a unique hermitian solution X of (12.5.1) such that (12.5.9) holds.

We omit the details of the proof.

12.6 A more general DARE

In Sections 12.2–12.5 we have developed the geometric theory of the DARE (12.2.1) under the basic assumptions of controllability of (A, B) (relaxed to some extent in Section 12.5) and invertibility of A and R. This theory can be extended to the DARE

$$X = A^*XA + Q - (C + B^*XA)^*(R + B^*XB)^{-1}(C + B^*XA) \qquad (12.6.1)$$

((12.2.1) is obtained by putting $C = 0$ in (12.6.1)). We indicate in this section how such an extension works for (12.6.1). Many details will be omitted. Our standing assumptions in this section are $Q = Q^*$ and $R = R^*$; the size of A, Q and X is $n \times n$ while the size of R is $m \times m$.

The role of the function $\Psi(z)$ of (12.2.2) is now played by the $n \times n$ rational matrix function

$$\hat{\Psi}(z) = [B^*(z^{-1} - A^*)^{-1}, \; I] \begin{bmatrix} Q & C^* \\ C & R \end{bmatrix} \begin{bmatrix} (z-A)^{-1}B \\ I \end{bmatrix} \qquad (12.6.2)$$

$$= R + C(z-A)^{-1}B + B^*(z^{-1} - A^*)^{-1}C^*$$
$$+ B^*(z^{-1} - A^*)^{-1}Q(z-A)^{-1}B.$$

The function $\hat{\Psi}(z)$ is hermitian on the unit circle. Every hermitian solution X of (12.6.1) leads to a factorization

$$\hat{\Psi}(z) = (\Phi_X(\bar{z}^{-1}))^*(R + B^*XB)\Phi_X(z), \qquad (12.6.3)$$

where

$$\Phi_X(z) = I + (R + B^*XB)^{-1}(C + B^*XA)(z-A)^{-1}B.$$

The verification of (12.6.3) is straightforward. As a consequence, the full analogue of Theorem 12.2.1 is valid for (12.6.1). Observe that when $C = 0$ in (12.6.1), the functions $\hat{\Psi}(z)$ and $\Phi_X(z)$ become $\Psi(z)$ and $\Delta_X(z)$, respectively, of Section 12.2.

We now make the assumption that the matrices A and $D = R - CA^{-1}B$ are invertible. Define the matrix

$$U = \begin{bmatrix} A - BD^{*-1}(C - B^*A^{*-1}Q) & -BD^{*-1}B^*A^{*-1} \\ A^{*-1}(-Q + C^*D^{*-1}(C - B^*A^{*-1}Q)) & A^{*-1}(I + C^*D^{*-1}B^*A^{*-1}) \end{bmatrix}. \qquad (12.6.4)$$

The matrix U plays the role of T of Section 12.2 and, indeed, $U = T$ if $C = 0$. Introduce also the matrix

$$V = \begin{bmatrix} A & 0 \\ -A^{*-1}Q & A^{*-1} \end{bmatrix}.$$

Then we have the following (not necessarily minimal) realizations for $\hat{\Psi}(z)$ and $\hat{\Psi}(z)^{-1}$:

$$\hat{\Psi}(z) = D^* + [C - B^* A^{*-1}Q, B^* A^{*-1}](zI - V)^{-1} \begin{bmatrix} B \\ -A^{*-1}C^* \end{bmatrix}, \quad (12.6.5)$$

$$\hat{\Psi}(z)^{-1} = D^{*-1} - [D^{*-1}(C - B^* A^{*-1}Q), D^{*-1}B^* A^{*-1}] \cdot$$
$$\cdot (zI - U)^{-1} \begin{bmatrix} BD^{*-1} \\ -A^{*-1}C^* D^{*-1} \end{bmatrix}. \quad (12.6.6)$$

Equation (12.6.5) can be easily verified using the definition (12.6.2) of $\hat{\Psi}(z)$, while (12.6.6) follows from (12.6.5) using the general formula (6.3.14) for the inverse of a rational matrix function in a realized form. Note also that

$$U = V - \begin{bmatrix} B \\ -A^{*-1}C^* \end{bmatrix} D^{*-1}[C - B^* A^{*-1}Q, B^* A^{*-1}]. \quad (12.6.7)$$

As in Section 12.2, let $J = \begin{bmatrix} 0 & -iI \\ iI & 0 \end{bmatrix}$.

Proposition 12.6.1. *Let X be a hermitian solution of (12.6.1). Then the graph of X, i.e. the subspace* $\operatorname{Im} \begin{bmatrix} I \\ X \end{bmatrix}$, *is J-Lagrangian and U-invariant.*

Proof. The graph of X is U-invariant if and only if the equality

$$X(A - BD^{*-1}C + BD^{*-1}B^* A^{*-1}(Q - X))$$
$$= A^{*-1}(-Q + C^* D^{*-1}(C - B^* A^{*-1}Q)) + X + C^* D^{*-1} B^* A^{*-1} X$$

holds. Multiplying both sides by A^* on the left and regrouping terms it follows that this equality is equivalent to

$$X = A^* X A + Q - (C^* + A^* X B) D^{*-1}(C - B^* A^{*-1}(Q - X)). \quad (12.6.8)$$

On the other hand, if X is a hermitian solution of (12.6.1), then

$$Q - X = (C + B^* X A)^* (R + B^* X B)^{-1} (C + B^* X A) - A^* X A,$$

consequently,

$$C - B^* A^{*-1}(Q - X) = [I - B^* A^{*-1}(C + B^* X A)^* (R + B^* X B)^{-1}](C + B^* X A)$$
$$= D^*(R + B^* X B)^{-1}(C + B^* X A), \quad (12.6.9)$$

and substituting into (12.6.8) we see that (12.6.8) is satisfied, and hence the graph of X is U-invariant. □

A MORE GENERAL DARE

It is not clear whether the converse statement holds, i.e. whether for every J-Lagrangian U-invariant graph subspace $\mathrm{Im}\begin{bmatrix} I \\ X \end{bmatrix}$ the hermitian matrix X is a solution of (12.6.1). Proposition 12.2.2 shows that the answer is affirmative if $C = 0$.

Note that both U and V are J-unitary:

$$U^*JU = J; \quad V^*JV = J. \tag{12.6.10}$$

The verification of these equalities is straightforward albeit tedious (especially for $U^*JU = J$ where it is convenient to use (12.6.7) together with $V^*JV = J$ to verify $U^*JU = J$). In particular, U is invertible.

We now state the main result of this section — a full analogue of Theorems 12.2.3 and 12.2.5 and Corollary 12.2.4.

Theorem 12.6.2. *Assume that (A, B) is controllable, the matrices A and $D = R - CA^{-1}B$ are invertible, and that there exists an $\eta \in \mathbf{T}$ such that $\hat{\Psi}(\eta) > 0$. Then the following statements (i)–(iii) are equivalent:*

(i) *the DARE (12.6.1) has a hermitian solution;*

(ii) *there exists a U-invariant J-Lagrangian subspace;*

(iii) *the partial multiplicities of U corresponding to its unimodular eigenvalues (if any) are all even.*

When the statements (i)–(iii) hold, there is a one-to-one correspondence between the set of hermitian solutions X of (12.6.1) and the set of U-invariant J-Lagrangian subspaces $\mathcal{M}$ given by $\mathcal{M} = \mathrm{Im}\begin{bmatrix} I \\ X \end{bmatrix}$. Moreover, if (i)–(iii) hold, then:

(iv) $\hat{\Psi}(z) > 0$ *for all $z \in \mathbf{T}$ with the possible exception of a finite number of points;*

(v) $R + B^*XB > 0$ *for every hermitian solution X of (12.6.1);*

(vi) *the sign characteristic of the pair (U, J) consists of -1's only.*

Proof. The statements (iv) and (v) follow from (12.6.3) in view of our hypotheses that $\hat{\Psi}(\eta) > 0$.

(i) $\Rightarrow$ (ii) is proved in Proposition 12.6.1.

For the proof of (ii) $\Rightarrow$ (i) we follow the approach used in the proof of Theorem 12.2.3. Let $\mathcal{M} = \mathrm{Im}\begin{bmatrix} X_1 \\ X_2 \end{bmatrix}$ be a U-invariant J-Lagrangian subspace. We have to prove that $\mathrm{Ker}\, X_1 = \{0\}$. Let $\eta \in \mathbf{T}$ be such that $\Psi(\eta) > 0$, $\eta \notin \sigma(U)$ and

$\eta \notin \sigma(A^{*-1})$ (consequently, also $\eta \notin \sigma(A)$). Let $\varphi(\lambda) = (\lambda + \eta)(\lambda - \eta)^{-1}$. A computation shows that

$$\varphi(U)\begin{bmatrix} 0 \\ y \end{bmatrix} = \begin{bmatrix} -2(\eta-A)^{-1}B\hat{\Psi}(\eta)^{-1}B^*(\bar{\eta}-A^*)^{-1} \\ (A^{*-1}+\eta)(A^{*-1}-\eta)^{-1} \\ +2(\eta-A^{*-1})^{-1}A^{*-1}Z\hat{\Psi}(\eta)^{-1}B^*(\bar{\eta}-A^*)^{-1} \end{bmatrix} y \quad (12.6.11)$$

where $Z = Q(z-A)^{-1}B + C^*$. We omit the computation (the formula (12.6.11) is an extension of (12.2.15) and can be proved in a similar way; see also the proof of Theorem 4.4 in Lancaster et al. (1986)). Using (12.6.11), the proof of (ii) $\Rightarrow$ (i) is completed as in Theorem 12.2.3.

The remaining parts of Theorem 12.6.2 are proved again in the same way as in Theorems 12.2.3 and 12.2.5, by writing $\varphi(U)$ in the form:

$$\varphi(U) = \begin{bmatrix} -\tilde{C} & \tilde{D} \\ \tilde{E} & \tilde{C}^* \end{bmatrix},$$

where $\tilde{E} = \tilde{E}^*$, $\tilde{D} \leq 0$, and $(\tilde{C}, \tilde{D})$ is controllable. $\square$

Note that for the subspace $\mathcal{M} = \text{Im}\begin{bmatrix} I \\ X \end{bmatrix}$ of the statement of Theorem 12.6.2, $U|\mathcal{M}$ is similar to

$$A - BD^{*-1}C + BD^{*-1}B^*A^{*-1}(Q - X)$$
$$= A - B(R + B^*XB)^{-1}(C + B^*XA). \quad (12.6.12)$$

Indeed, $U|\mathcal{M}$ is similar to the left-hand side of (12.6.12) just by the form of U (see (12.6.4)). The equality in (12.6.12) follows in view of the formula

$$B^*A^{*-1}(Q - X) = C - D^*(R + B^*XB)^{-1}(C + B^*XA)$$

(cf. (12.6.9)).

We leave it to the reader to state and prove the results for equation (12.6.1) analogous to those obtained in Section 12.3.

As in Section 12.4, we introduce the function

$$W(z) = I + (z - A)^{-1}B\hat{\Psi}(z)^{-1}B^*(z^{-1} - A^*)^{-1}.$$

Using the formula (12.6.9), one obtains in complete analogy with Lemma 12.4.4 the realization

$$W(z) = I + \frac{1}{2}[I \ 0](z - U)^{-1}U\begin{bmatrix} 0 \\ I \end{bmatrix}. \quad (12.6.13)$$

The realization (12.6.13) is not necessarily minimal, but:

A MORE GENERAL DARE

Lemma 12.6.3. *If (A, B) is controllable, and the matrix D is positive definite, then (12.6.13) is a minimal realization.*

The proof of this lemma is analogous to that of Lemma 12.4.4.
We are now able to complement Theorem 12.6.2:

Theorem 12.6.4. *Given the hypotheses of Theorem 12.6.2, assume in addition that $D > 0$. Then the statements (i)–(iii) in Theorem 12.6.2 are equivalent to each of the statements (iv) and (v) in that theorem, as well as to the following statement:*

(vii) $W(z) > 0$ *for all $z \in \mathbf{T}$ with the possible exception of a finite number of points.*

The proof is similar to that of Theorem 12.4.1. Theorem 12.5.1 and other results of Section 12.5 can be extended to DARE (12.6.1) as well.

We conclude this section with an important transformation which can further extend applicability of the results of this section. Let the DARE (12.6.1) be given, and let K be an $m \times n$ matrix. Define

$$A_K = A - BK; \quad C_K = C - RK;$$
$$Q_K = Q - C^*K - C^*C + K^*RK, \quad (12.6.14)$$

and consider the transformed DARE

$$X = A_K^* X A_K + Q_K - (C_K + B^* X A_K)^*(R + B^* X B)^{-1}(C_K + B^* X A_K). \quad (12.6.15)$$

The transformation (12.6.14), (12.6.15) of the original equation (12.6.1) will be called the *feedback transformation*. The terminology stems from the LQR problem (see Section 16.5), where $A \to A - BK$ represents a closed loop feedback transformation of the discrete system

$$x_{k+1} = Ax_k + Bu_k \quad (k = 0, 1, \ldots).$$

Observe that the transformation described in Proposition 12.1.1 is a feedback transformation, with $K = R^{-1}C$. The quantities associated with (12.6.15) (such as $\hat{\Psi}_K(z)$) are denoted by appending the subscript "K" to the corresponding quantities associated with (12.6.1).

Lemma 12.6.5. (i) *A hermitian matrix X is a solution of (12.6.1) if and only if X is a solution of (12.6.15).*
(ii) *Let $\hat{\Psi}(z)$ be defined by (12.6.3); then*

$$\hat{\Psi}(z) = (Y(\bar{z})^{-1}))^* \hat{\Psi}_K(z) Y(z),$$

where $Y(z) = I + K(z - A)^{-1}B$.

(iii) *For every hermitian solution* X *of* (12.6.1) *(or of* (12.6.15)*),*

$$A - B(R + B^*XB)^{-1}(C + B^*XA)$$
$$= A_K - B(R + B^*XB)^{-1}(C_K + B^*XA_K).$$

Proof. Subtracting the right-hand side of (12.6.15) from the right-hand side of (12.6.1) and writing $Z = (R + B^*XB)^{-1}$ we obtain:

$$-(BK)^*XA - A^*XBK + (BK)^*XBK - C^*K - K^*C + K^*RK$$
$$-(-RK - B^*XBK)^*Z(C + B^*XA) - (C + B^*XA)^*Z(-RK - B^*XBK)$$
$$-(RK + B^*XBK)^*Z(RK + B^*XBK)$$
$$= -K^*B^*XA - A^*XBK + K^*B^*XBK - C^*K - K^*C + K^*RK$$
$$+K^*(C + B^*XA) + (C^* + A^*XB)K - (K^*R + K^*B^*XB)K = 0.$$

This proves (i). As for (ii), note first the equalities

$$\begin{bmatrix} Q_K & C_K^* \\ C_K & R \end{bmatrix} = \begin{bmatrix} I & -K^* \\ 0 & I \end{bmatrix} \begin{bmatrix} Q & C^* \\ C & R \end{bmatrix} \begin{bmatrix} I & 0 \\ -K & I \end{bmatrix} \quad (12.6.16)$$

and

$$\begin{bmatrix} I & 0 \\ -K & I \end{bmatrix} \begin{bmatrix} (zI - A_K)^{-1}B \\ I \end{bmatrix} = \begin{bmatrix} (zI - A_K)^{-1}B \\ I - K(zI - A_K)^{-1}B \end{bmatrix}. \quad (12.6.17)$$

Now

$$(zI - A_K)^{-1}B\{I - K(zI - A_K)^{-1}B\}^{-1} = (zI - A_K)^{-1}B\{I + K(zI - A)^{-1}B\}$$
$$= (zI - A_K)^{-1}B + (zI - A_K)^{-1}\{(zI - A_K) - (zI - A)\}(zI - A)^{-1}B$$
$$= (zI - A)^{-1}B. \quad (12.6.18)$$

So

$$(Y(\bar{z}^{-1}))^*\hat{\Psi}_K(z)Y(z)$$
$$= (I + B^*(z^{-1}I - A^*)K^*)[B^*(z^{-1}I - A_K^*)^{-1}, I]\begin{bmatrix} Q_K & C_K^* \\ C_K & R \end{bmatrix}$$
$$\cdot \begin{bmatrix} (zI - A_K)^{-1}B \\ I \end{bmatrix}(I + K(zI - A)^{-1}B) = \hat{\Psi}(z)$$

in view of (12.6.16)–(12.6.18). Finally, the verification of (iii) is straightforward. □

The feedback transformation allows us to extend some of the results of this section. For example:

Theorem 12.6.6. *Assume that (A, B) is controllable, and there exists an $\eta \in \mathbf{T}$ such that $\Psi(\eta) > 0$. Assume further that there exists an $m \times n$ matrix K such that $A - BK$ and $R - (C - RK)(A - BK)^{-1}B$ are invertible. If the DARE (12.6.1) has a hermitian solution, then $\Psi(z) > 0$ for all $z \in \mathbf{T}$ with the possible exception of a finite number of points, $R + B^*XB > 0$ for every hermitian solution X of (12.6.1), and there exist unique hermitian solutions X_+, X_- such that all eigenvalues λ of*
$$A - B(R + B^*X_\pm B)^{-1}(C + B^*X_\pm A)$$
satisfy $|\lambda| \leq 1$ for X_+ and $|\lambda| \geq 1$ for X_-. Moreover, X_+ (resp. X_-) is the maximal (resp. minimal) hermitian solution of (12.6.1).

The proof is obtained by applying Theorem 12.6.2 to the equation (12.6.15) and using Lemma 12.6.5.

12.7 The real case

In this section we briefly review some results on the DARE

$$X = A^T X A + Q - (C + B^T X A)^T (R + B^T X B)^{-1} (C + B^T X A), \quad (12.7.1)$$

with real coefficients which are analogous to those for the DARE (12.2.1) (if $C = 0$) or (12.6.1) with complex coefficients, and have already been obtained in Sections 12.1–12.6. Thus, A, B, C, Q and R are now real matrices of sizes $n \times n$, $n \times m$, $m \times n$, $n \times n$ and $m \times m$, respectively, with R and Q real and symmetric. The matrix X is an $n \times n$ real symmetric matrix to be found.

All the main results of this chapter remain valid in the real case as well, when properly reformulated. For simplicity, we focus on equation (12.2.1) only in this section. The $2n \times 2n$ matrix T of (12.2.5) now takes the form

$$T = \begin{bmatrix} A + BR^{-1}B^T(A^T)^{-1}Q & -BR^{-1}B^T(A^T)^{-1} \\ -(A^T)^{-1}Q & (A^T)^{-1} \end{bmatrix};$$

(under the assumption that the matrices A and R are invertible). Also, it is more convenient to use the real $2n \times 2n$ skew-symmetric matrix $\hat{J} = \begin{bmatrix} 0 & -I \\ I & 0 \end{bmatrix}$ rather than $J = \begin{bmatrix} 0 & -iI \\ iI & 0 \end{bmatrix}$. A real subspace $\mathcal{M} \subseteq \mathbb{R}^{2n}$ is called $\hat{J}$-Lagrangian if $\dim \mathcal{M} = n$ and $x^T \hat{J} y = 0$ for all vectors $x, y \in \mathcal{M}$.

An invariant subspace $\mathcal{M} \subseteq \mathbb{C}^m$ of a real $m \times m$ matrix X is called *real*, if there is a basis in $\mathcal{M}$ consisting of vectors with real coordinates. In this case, it is convenient to represent $\mathcal{M}$ as a real vector subspace of $\mathbb{R}^m$, spanned by a basis consisting of real vectors. We will not distinguish these two guises of a real invariant subspace by notation. They can be distinguished by writing "$\mathcal{M} \subseteq \mathbb{C}^m$" or "$\mathcal{M} \subseteq \mathbb{R}^m$", as the case may be.

The real analogue of Theorem 12.2.3 is

Theorem 12.7.1. *Assume that (A, B) is controllable, A and R are invertible, and $\Psi(\eta) > 0$ for some η, $|\eta| = 1$, where $\Psi(z)$ is the rational matrix function given by*
$$\Psi(z) = R + B^T(z^{-1} - A^T)^{-1}Q(z - A)^{-1}B.$$
Then the following statements are equivalent:

(i) *the DARE (12.7.1) has a real symmetric solution;*

(ii) *there exists a real T-invariant $\hat{J}$-Lagrangian subspace;*

(iii) *the partial multiplicities of T corresponding to its unimodular eigenvalues, if any, are all even.*

In case (i)–(iii) *hold, then there is a one-to-one correspondence between the set of real symmetric solutions X of (12.7.1) and the set of real T-invariant $\hat{J}$-Lagrangian subspaces $\mathcal{M}$, which is given by the formula*
$$\mathcal{M} = \mathrm{Im} \begin{bmatrix} I \\ X \end{bmatrix}. \tag{12.7.2}$$

Proof. The implications (i) $\Rightarrow$ (iii) and (ii) $\Rightarrow$ (iii) follow immediately from Theorem 12.2.3.

Assume now that (iii) holds. Then by Theorem 12.2.3 the equation (12.7.1) has a hermitian solution. By Theorem 12.3.3 this equation also has the maximal hermitian solution X_+. Taking complex conjugates entrywise in (12.7.1) (with X replaced by X_+) we see that $\bar{X}_+$ is also a hermitian solution of (12.7.1). By maximality of X_+, it follows that $\bar{X}_+ \leq X_+$. The positive semidefinite matrix $X_+ - \bar{X}_+$ has zeros on the main diagonal, and therefore it must be the zero matrix. Thus, $X_+ = \bar{X}_+$ is a real symmetric solution of (12.7.1), and (i) holds. So (iii) $\Rightarrow$ (i) is proved.

It remains to prove (i) $\Rightarrow$ (ii). If X is a real symmetric solution of (12.7.1), then by Theorem 12.3.3 the corresponding T-invariant J-Lagrangian subspace $\mathcal{M}$ given by (12.7.2) is real. This proves that (i) $\Rightarrow$ (ii). $\square$

To state the real analogue of Theorem 12.3.1, we denote by $\mathcal{R}(T, \mathbb{D})$ the sum of the root subspaces of the real matrix T with respect to the eigenvalues of T in the open unit disk $\mathbb{D}$. Observe that $\mathcal{R}(T, \mathbb{D})$ is a real T-invariant subspace; this follows easily from the real Jordan form of T (see Theorem 1.4.1).

Theorem 12.7.2. *Given the hypotheses of Theorem 12.7.1 assume, in addition, that the DARE (12.7.1) has a real symmetric solution. Then for every real T-invariant subspace $\mathcal{N}$ such that $\sigma(T(\mathcal{N})) \subset \mathbb{D}$ there is a unique real symmetric solution X of (12.7.1) with*
$$\mathrm{Im} \begin{bmatrix} I \\ X \end{bmatrix} \cap \mathcal{R}(T, \mathbb{D}) = \mathcal{N}. \tag{12.7.3}$$

Conversely, for every real symmetric solution X of (12.7.1) the subspace $\mathcal{N} := \mathrm{Im} \begin{bmatrix} I \\ X \end{bmatrix} \cap \mathcal{R}(T, \mathbb{D})$ is a real T-invariant subspace with $\sigma(T|\mathcal{N}) \subset \mathbb{D}$.

Proof. Given $\mathcal{N}$ as in Theorem 12.7.2, by Theorem 12.3.1 there exists a unique hermitian solution X of (12.7.1) such that (12.7.3) holds. But $\bar{X}$ is also a hermitian solution of (12.7.1). It is easy to see that for any basis $y_1, \ldots, y_p$ in $\operatorname{Im} \begin{bmatrix} I \\ X \end{bmatrix} \cap \mathcal{R}(T, \mathbb{D}) \subseteq \mathbb{C}^{2n}$ the set of complex conjugate vectors $\bar{y}_1, \ldots, \bar{y}_p$ is a basis in $\operatorname{Im} \begin{bmatrix} I \\ \bar{X} \end{bmatrix} \cap \mathcal{R}(T, \mathbb{D}) \subseteq \mathbb{C}^{2n}$. Indeed, it is clear that $\bar{y}_j \in \operatorname{Im} \begin{bmatrix} I \\ \bar{X} \end{bmatrix} \cap \mathcal{R}(T, \mathbb{D})$ and, moreover, $\bar{y}_1, \ldots, \bar{y}_p$ are linearly independent because $y_1, \ldots, y_p$ are. Furthermore, given any vector $z \in \operatorname{Im} \begin{bmatrix} I \\ \bar{X} \end{bmatrix} \cap \mathcal{R}(T, \mathbb{D})$, we know that $\bar{z}$ is a linear combination (with complex coefficients) of $y_1, \ldots, y_p$, and therefore z is a linear combination of $\bar{y}_1, \ldots, \bar{y}_p$.

In particular, by choosing $y_1, \ldots, y_p$ to be a basis in $\operatorname{Im} \begin{bmatrix} I \\ X \end{bmatrix} \cap \mathcal{R}(T, \mathbb{D})$ consisting of vectors with real components, we conclude that $y_1, \ldots, y_p$ is also a basis in $\operatorname{Im} \begin{bmatrix} I \\ \bar{X} \end{bmatrix} \cap \mathcal{R}(T, \mathbb{D})$, and therefore

$$\operatorname{Im} \begin{bmatrix} I \\ \bar{X} \end{bmatrix} \cap \mathcal{R}(T, \mathbb{D}) = \operatorname{Im} \begin{bmatrix} I \\ X \end{bmatrix} \cap \mathcal{R}(T, \mathbb{D}).$$

By the uniqueness part of Theorem 12.3.1 we conclude that $X = \bar{X}$, i.e. X is real symmetric.

Finally, the converse statement of Theorem 12.7.2 follows immediately from Proposition 12.2.2. □

12.8 Exercises

12.8.1. Find all hermitian solutions of the DARE (12.2.1) with

$$Q = \begin{bmatrix} 1 & 0 \\ 0 & 0 \end{bmatrix}, \quad A = \begin{bmatrix} 0 & 1 \\ -0.5 & 1.5 \end{bmatrix}, \quad B = \begin{bmatrix} 1 \\ 1 \end{bmatrix}, \quad R = 2.$$

12.8.2. [Connection between the CARE and the DARE]. Prove the following result: Consider the CARE

$$XF + F^*X - XGR^{-1}G^*X + S = 0, \qquad (12.8.1)$$

where $S = S^*$ and $R > 0$. Assume that there exists an $\alpha > 0$ such that matrices $\alpha I - F$ and

$$R + G^*(\alpha - F^*)^{-1}S(\alpha - F)^{-1}G$$

are invertible. Define

$$A = (\alpha - F)^{-1}(\alpha + F); \qquad B = \alpha\sqrt{2}(\alpha - F)^{-1}G,$$

$$Q = 2\alpha(\alpha - F^*)^{-1}S(\alpha - F)^{-1}; \qquad U = \alpha R + \frac{1}{2}G^*QG,$$

and assume, in addition, that α is such that the matrices U and

$$W := \sqrt{\alpha}R^{1/2} + \alpha^{-1/2}R^{-1/2}B^*(A^* + I)^{-1}X$$

are invertible for every hermitian solution X of (12.8.1) (in particular, such an α always exists if the set of hermitian solutions of (12.8.1) is bounded). Then every solution $X = X^*$ of (12.8.1) is a solution of

$$X = A_1^* X A_1 - A_1 X B(U + B^* X B)^{-1} B X A_1 + Q_1,$$

where

$$A_1 = A - BU^{-1}H^*; \qquad Q_1 = Q - HU^{-1}H^*,$$

and where $H = Q(A + I)^{-1}B = \sqrt{2}QG$.

12.8.3. [The converse result.] Consider the DARE

$$X = A^*XA + Q - (A^*XB + H)(U + B^*XB)^{-1}(B^*XA + H^*) \quad (12.8.2)$$

where $Q = Q^*$, the matrices $U = U^*$ and $A + I$ are invertible, and $H = 2Q(A + I)^{-1}B$. Assume, in addition, that the matrix

$$Z := U - B^*(A^* + I)^{-1}Q(A + I)^{-1}B$$

is positive definite, and the set of hermitian solutions of (12.8.2) is non-empty and bounded. Choose a positive number α large enough so that

$$I + \alpha^{-2}ZB^*(A^* + I)^{-1}XB$$

is invertible for every hermitian solution X of (12.8.2). Prove that every hermitian solution of (12.8.2) is also a solution of (12.8.1), with

$$F = \alpha(A - I)(A + I)^{-1}; \qquad S = 2\alpha(A^* + I)^{-1}Q(A + I)^{-1}$$

$$G = \sqrt{2}(A + I)^{-1}B; \qquad R = \alpha^{-1}(U - B^*(A^* + I)^{-1}Q(A + I)^{-1}B).$$

12.8.4. Prove that if (12.2.6) is a minimal realization, then the factorization (12.2.3) is necessarily minimal.

12.8.5. Find all hermitian solutions of the DARE (12.2.1) with

$$A = \begin{bmatrix} 1 & 0 \\ 0 & -1 \end{bmatrix}; \quad B = \begin{bmatrix} 0 \\ 1 \end{bmatrix}; \quad Q = \begin{bmatrix} 1 & 0 \\ 0 & 0 \end{bmatrix}; \quad R = \alpha,$$

where $\alpha \neq 0$ is a real parameter.

12.8.6. Supply the details for the proof of (12.6.11).

12.8.7. Prove Lemma 12.6.3.

12.8.8. Prove that, under the hypotheses of Theorem 12.6.4, the statement (vii) of that theorem is equivalent to (iv) of Theorem 12.6.2.

12.8.9. Prove that the condition $D > 0$ in Lemma 12.6.3 and Theorem 12.6.4 can be replaced by a weaker condition (and keeping the same conclusion), namely, that the numerical range of D does not contain 0.

[Hint: The numerical range, also called the field of values, of an $n \times n$ matrix Y is defined by $NR(Y) = \{x^*Yx \mid \|x\| = 1\}$; here the euclidean norm of $x \in \mathbb{C}^n$ is used. Note that if $0 \notin NR(Y)$, then Y is invertible and $0 \notin NR(Y^{-1})$. See Chapter 1 in Horn and Johnson (1991) for basic properties of the field of values.]

12.8.10. Provide details for the proof of Theorem 12.5.3.

12.8.11. State and prove the C^p analogue of Theorem 12.3.3.

12.8.12. Assume the hypotheses of Theorem 12.2.3. A hermitian solution X_0 of the DARE

$$X = A^*XA + Q - A^*XB(R + B^*XB)^{-1}B^*XA \qquad (12.8.3)$$

is called *isolated* if there exists an $\varepsilon > 0$ such that the equation (12.8.3) has no hermitian solutions Y different from X_0 and satisfying $\|Y - X_0\| < \varepsilon$. Identify the isolated hermitian solutions of (12.8.3) in terms of the corresponding subspaces $\mathcal{N}$, as in Theorem 12.3.1.

[Hint: Compare with the description of the isolated hermitian solutions for the CARE (Section 7.7)].

12.8.13. Identify the isolated real symmetric solutions of the DARE (12.8.3) with real coefficients.

12.9 Notes

The systematic study of the DARE began in the early 1970s, motivated primarily by linear optimal filtering problems (see the groundbreaking paper by Kalman (1960a) and the exposition by Silverman (1976)) and by optimal control of sample-data systems (see Dorato and Lewis (1971)). An early iterative scheme (of the type used in the next chapter) appeared in Hewer (1971), while seeds of the invariant subspaces analysis (of the type used in this chapter) are seen in a paper of Vaughan (1970). The invariant subspace approach was further developed in papers by Kučera (1972b), Pappas et al. (1980), Chan et al. (1984), and Lancaster et al. (1986). Analysis of the stabilizing solution of the DARE, based on the factorization (12.6.3) of $\hat{\Psi}(z)$, appeared in Molinari (1975). Results concerning existence of positive semidefinite solutions of the DARE in terms of spectral factorizations of the corresponding operator function were obtained by Helton (1978) (in the more general framework of Hilbert spaces). The paper by Dorato (1983) contains further information and a list of additional references (up to 1983) concerning the DARE.

The close connections between CARE and DARE have been studied several times in the literature. We quote just one source: Salgado et al. (1988), where the delta operator is used to study both CARE and DARE from a unified point of view.

Most of the exposition in this chapter is based on Lancaster et al. (1986), and Ran and Rodman (1992b). Theorem 12.6.2 appears to be new, although a similar result (with more restrictive hypotheses) was proved in Lancaster et al. (1986). Exercise 12.8.1 is taken from Kučera (1979). Connections between continuous and discrete equations like those in Exercises 12.8.2 and 12.8.3 can also be found in the literature (see Section 15.3 of Anderson and Moore (1971), for example).

Theorem 12.5.1 (with the exception of the statement concerning the sign characteristic) has been proved in Curilov (1986).

In a paper by Wimmer (1992a), the hermitian solutions of DARE are characterized in terms of maximal and minimal hermitian solutions (cf. Theorem 12.3.3) and projectors on suitable invariant subspaces, in analogy with Theorem 7.5.4. We also mention the paper Wimmer (1994a), where the existence of an unmixed solution of DARE having a prescribed unmixed characteristic polynomial is proved, provided certain observability and detectability conditions are satisfied.

13
CONSTRUCTIVE EXISTENCE AND COMPARISON THEOREMS

This chapter is an analogue of Chapter 9, in which existence and comparison theorems for the DARE (12.1.1) are developed, rather than for the CARE of (9.1.1). Throughout this chapter it is assumed that, in (12.1.1), $Q = Q^*$ and $R = R^*$. Additional hypotheses on A and B (such as d-stabilizability, as defined in Section 4.4) are assumed as required.

The most important results of this kind for the DARE with complex coefficient matrices are stated and proved. The corresponding results for real solutions of the DARE with real coefficients (i.e. equation (12.7.1)) can be obtained as simple special cases of the results for the case of complex coefficient matrices, and are not considered explicitly. In particular, observe that the maximal solution X_+ in the class of hermitian solutions of (12.7.1) (when it exists) is necessarily real. The reason is that, when X_+ is a maximal solution, so is $\bar{X}_+$, and the uniqueness of the maximal solution implies that $\bar{X}_+ = X_+$, i.e. that X_+ is real.

13.1 Existence of maximal hermitian solutions

Together with the basic DARE:

$$X = A^*XA + Q - (C + B^*XA)^*(R + B^*XB)^{-1}(C + B^*XA) \qquad (13.1.1)$$

we consider the discrete Riccati inequality

$$-X + A^*XA + Q - (C + B^*XA)^*(R + B^*XB)^{-1}(C + B^*XA) \geq 0, \quad (13.1.2)$$

(cf. equation (9.1.1) and inequality (9.1.2)). The fundamental existence theorem for a maximal solution of (13.1.1) is related to the much weaker condition of existence of any hermitian solution of the inequality (13.1.2) as follows:

Theorem 13.1.1. *Let (A, B) be a d-stabilizable pair, R be invertible, and assume that there is a hermitian solution $\tilde{X}$ of (13.1.2) for which $R + B^*\tilde{X}B > 0$. Then there exists a unique solution $X_+ = X_+^*$ of (13.1.1) such that $R + B^*X_+B > 0$ and $X_+ \geq X$ for all hermitian solutions of (13.1.2).*

Moreover, all the eigenvalues of

$$A - B(R + B^*X_+B)^{-1}(C + B^*X_+A)$$

lie in the closed unit disk.

Observe that under these hypotheses it follows from Theorem 12.2.1 (and Proposition 12.1.1) that $R + B^*XB > 0$ for every hermitian solution of (13.1.2).

Proof. The uniqueness of X_+ follows easily from the maximality property $X_+ \geq X$.

The bulk of the proof consists in showing the *existence* of X_+. We follow an approximation procedure (the discrete counterpart of the procedure used in Section 9.1), and obtain X_+ as the limit of a sequence of hermitian matrices. It will be convenient to write $\hat{Q} = Q - C^*R^{-1}C$, and let $\mathcal{R}(X)$ be the left-hand side of (13.1.2).

Since the pair (A, B) is d-stabilizable, there is an L_0 such that $A_0 := A - BL_0$ is d-stable (i.e., all eigenvalues of A_0 lie in the open unit disk). Let X_0 be the unique solution of the Stein equation

$$X_0 - A_0^*X_0A_0 = \hat{Q} + (L_0 - R^{-1}C)^*R(L_0 - R^{-1}C). \tag{13.1.3}$$

Then X_0 is hermitian (see equation (5.3.5)).

A straightforward computation serves to verify that

$$(X_0 - \tilde{X}) - A_0^*(X_0 - \tilde{X})A_0 = \mathcal{R}(\tilde{X}) + (SL_0 - E)^*S^{-1}(SL_0 - E), \tag{13.1.4}$$

where

$$S := R + B^*\tilde{X}B, \quad E := C + B^*\tilde{X}A. \tag{13.1.5}$$

Now $S > 0$ by hypothesis and if $\tilde{X}$ satisfies (13.1.2), (i.e., $\mathcal{R}(\tilde{X}) \geq 0$), then (13.1.4) implies

$$(X_0 - \tilde{X}) - A_0^*(X_0 - \tilde{X})A_0 \geq 0.$$

Since A_0 is d-stable, part (a) of Theorem 5.3.5 shows that $X_0 \geq \tilde{X}$. Consequently, $R + B^*X_0B \geq R + B^*\tilde{X}B = S > 0$.

Starting with X_0, A_0, L_0, induction is used to construct three sequences of matrices $\{X_i\}_{i=0}^\infty$, $\{A_i\}_{i=0}^\infty$, $\{L_i\}_{i=0}^\infty$, with certain properties. Thus, assume that for some $m \geq 1$ we have already determined matrices $\{X_i\}_{i=0}^{m-1}$, $\{A_i\}_{i=0}^{m-1}$, $\{L_i\}_{i=0}^{m-1}$, with $X_i = X_i^*$ for $(i = 0, \ldots, m-1)$,

$$X_0 \geq X_1 \geq \ldots \geq X_{m-1} \geq \tilde{X},$$

$$A_i = A - BL_i \quad (i = 0, \ldots, m-1),$$

where

$$L_i = (R + B^* X_{i-1} B)^{-1} (C + B^* X_{i-1} A) \quad (i = 1, \ldots, m-1), \qquad (13.1.6)$$

$$X_i - A_i^* X_i A_i = \hat{Q} + (L_i - R^{-1} C)^* R (L_i - R^{-1} C) \quad (i = 0, \ldots, m-1), \quad (13.1.7)$$

and the matrices $A_0, A_1, \ldots, A_{m-1}$ are d-stable. In particular, $X_{m-1} \geq \tilde{X}$ implies

$$R + B^* X_{m-1} B \geq R + B^* \tilde{X} B > 0,$$

and therefore $R + B^* X_{m-1} B$ is positive definite as well. Now define

$$L_m = (R + B^* X_{m-1} B)^{-1} (C + B^* X_{m-1} A), \qquad (13.1.8)$$

$$A_m = A - B L_m. \qquad (13.1.9)$$

It is to be shown that A_m is d-stable. The first step is a straightforward but tedious calculation to show that

$$\begin{aligned}(X_{m-1} - \tilde{X}) &- A_m^* (X_{m-1} - \tilde{X}) A_m \\ &= (L_m - L_{m-1})^* (R + B^* X_{m-1} B)^{-1} (L_m - L_{m-1}) \\ &\quad + \mathcal{R}(\tilde{X}) + (S L_m - E)^* S^{-1} (S L_m - E),\end{aligned} \qquad (13.1.10)$$

where S and E are defined in (13.1.5).

Now assume that A_m is not d-stable. Thus, $A_m x = \lambda x$ for some λ with $|\lambda| \geq 1$ and $x \neq 0$. But then, denoting the right-hand side of (13.1.10) by W, this equation implies

$$0 \geq (1 - |\lambda|^2) x^* (X_{m-1} - \tilde{X}) x^* = x^* W x.$$

But, in (13.1.10), $(R + B^* X_{m-1} B)^{-1}$, $\mathcal{R}(\tilde{X})$ and S^{-1} are all positive semidefinite. We conclude, therefore, that $W \geq 0$ and $W x = 0$. Hence (taking into account the form of W) $L_m x = L_{m-1} x$. But now

$$A_{m-1} x = (A - B L_{m-1}) x = (A - B L_m) x = A_m x = \lambda x,$$

which contradicts the d-stability of A_{m-1}. So A_m is d-stable as well.

We now define X_m as the unique solution (necessarily hermitian) of the Stein equation

$$X_m - A_m^* X_m A_m = \hat{Q} + (L_m - R^{-1} C)^* R (L_m - R^{-1} C). \qquad (13.1.11)$$

As in (13.1.4) it is found that

$$(X_m - \tilde{X}) - A_m^* (X_m - \tilde{X}) A_m = \mathcal{R}(\tilde{X}) + (S L_m - E)^* S^{-1} (S L_m - E) \geq 0. \quad (13.1.12)$$

Furthermore,

$$(X_{m-1} - X_m) - A_m^*(X_{m-1} - X_m)A_m$$
$$= (L_m - L_{m-1})^*(R + B^*X_{m-1}B)(L_m - L_{m-1}). \quad (13.1.13)$$

Indeed, using (13.1.11) and (13.1.7) successively,

$$(X_{m-1} - A_m^* X_{m-1} A_m) - (X_m - A_m^* X_m A_m)$$
$$= A_{m-1}^* X_{m-1} A_{m-1} - A_m^* X_{m-1} A_m$$
$$+ (L_{m-1} - R^{-1}C)^* R(L_{m-1} - R^{-1}C) - (L_m - R^{-1}C)^* R(L_m - R^{-1}C)$$
$$= (A - BL_{m-1})^* X_{m-1} (A - BL_{m-1}) - (A - BL_m)^* X_{m-1} (A - BL_m)$$
$$+ (L_{m-1} - R^{-1}C)^* R(L_{m-1} - R^{-1}C) - (L_m - R^{-1}C)^* R(L_m - R^{-1}C).$$

On expanding terms, cancellation, and recollecting some of the terms, this expression becomes

$$-L_{m-1}^*(C + B^*X_{m-1}A) - (C^* + A^*X_{m-1}B)L_{m-1}$$
$$+ L_m^*(C + B^*X_{m-1}A) + (C^* + A^*X_{m-1}B)L_m$$
$$+ L_{m-1}^*(R + B^*X_{m-1}B)L_{m-1} - L_m^*(R + B^*X_{m-1}B)L_m,$$

and using (13.1.8) it takes the form

$$(L_m^* - L_{m-1}^*)(R + B^*X_{m-1}B)L_m + L_m^*(R + B^*X_{m-1}B)(L_m - L_{m-1})$$
$$+ L_{m-1}^*(R + B^*X_{m-1}B)L_{m-1} - L_m^*(R + B^*X_{m-1}B)L_m$$
$$= (L_m^* - L_{m-1}^*)(R + B^*X_{m-1}B)(L_m - L_{m-1}).$$

This proves (13.1.13). Now, the equations (13.1.12) and (13.1.13), together with the d-stability of A_m imply (in view of Theorem 5.3.5) that $X_{m-1} \geq X_m \geq \tilde{X}$.

We have obtained a nonincreasing sequence $\{X_n\}_{n=0}^{\infty}$ of hermitian matrices bounded below by $\tilde{X}$. Hence

$$X_+ := \lim_{m \to \infty} X_m$$

exists and is a hermitian matrix, and $X_+ \geq \tilde{X}$. Since $R + B^*\tilde{X}B > 0$, the inequality $R + B^*X_+B > 0$ follows. Passing to the limit in (13.1.11) when $m \to \infty$, and, for brevity, writing $R_+ = R + B^*X_+B$, it is found that

$$X_+ - [A - BR_+^{-1}(C + B^*X_+A)]^*X_+[A - BR_+^{-1}(C + B^*X_+A)]$$
$$= Q - C^*R^{-1}C$$
$$+ [R_+^{-1}(C + B^*X_+A) - R^{-1}C]^*R[R_+^{-1}(C + B^*X_+A) - R^{-1}C].$$

This equality can be rewritten in the form

$$X_+ - A^*X_+A - Q = -(C+B^*X_+A)^* R_+^{-1} B^* X_+ A - A^* X_+ B R_+^{-1}(C+B^*X_+A)$$
$$-(C+B^*X_+A)^* R_+^{-1} C - C^* R_+^{-1}(C+B^*X_+A)$$
$$+(C+B^*X_+A)^* R_+^{-1} B^* X_+ B R_+^{-1}(C+B^*X_+A)$$
$$+(C+B^*X_+A)^* R_+^{-1} R R_+^{-1}(C+B^*X_+A)$$
$$= -(C+B^*XA)^* R_+^{-1}(C+B^*XA),$$

and so X_+ satisfies (13.1.1).

Finally, since A_m is d-stable for all $m \geq 0$, the eigenvalues of

$$\hat{A} := A - B(R + B^* X_+ B)^{-1}(C + B^* X_+ A) \qquad (13.1.14)$$

lie in the closed unit disk. □

In particular, the matrix X_+ obtained in Theorem 13.1.1 is the *maximal hermitian solution*. In applications, the situations when $X_+ \geq 0$ are important.

Corollary 13.1.2. *Assume that $R > 0$, (A, B) is a d-stabilizable pair, and*

$$\begin{bmatrix} Q & C^* \\ C & R \end{bmatrix} \geq 0. \qquad (13.1.15)$$

Then the maximal hermitian solution X_+ of (13.1.1) exists, is unique, and is positive semidefinite. Moreover, all the eigenvalues of the matrix (13.1.14) lie in the closed unit disk.

Proof. Because of the equality

$$\begin{bmatrix} I & -C^*R^{-1} \\ 0 & I \end{bmatrix} \begin{bmatrix} Q & C^* \\ C & R \end{bmatrix} \begin{bmatrix} I & 0 \\ -R^{-1}C & I \end{bmatrix} = \begin{bmatrix} Q - C^*R^{-1}C & 0 \\ 0 & R \end{bmatrix},$$

the condition (13.1.15) is equivalent to $Q - C^*R^{-1}C \geq 0$. Thus, $\tilde{X} = 0$ satisfies (13.1.2), and an application of Theorem 13.1.1 yields the corollary. □

Theorem 13.1.3. *Assume, in addition to the hypotheses of Corollary 13.1.2, that $(Q - C^*R^{-1}C, A - BR^{-1}C)$ is d-detectable. Then the maximal solution X_+ of the DARE (13.1.1) exists, and we have $X_+ \geq 0$ and the matrix $\hat{A}$ of (13.1.14) is d-stable. If, furthermore, $(Q - C^*R^{-1}C, A - BR^{-1}C)$ is observable, then we have $X_+ > 0$.*

Proof. We know from Corollary 13.1.2 that X_+ exists, $X_+ \geq 0$, and $\hat{A}$ is almost d-stable i.e. all eigenvalues of $\hat{A}$ are in the closed unit disk. Now we refer forward to Theorem 15.1.2. The hypotheses of this theorem are satisfied so the pencil

$$\lambda \begin{bmatrix} I & BR^{-1}B^* \\ 0 & (A - BR^{-1}C)^* \end{bmatrix} - \begin{bmatrix} A - BR^{-1}C & 0 \\ -Q + C^*R^{-1}C & I \end{bmatrix}$$

has no eigenvalues on the unit circle. Then it is easily verified (using Proposition 12.1.1) that

$$\begin{bmatrix} I & BR^{-1}B^* \\ 0 & (A - BR^{-1}C)^* \end{bmatrix} \begin{bmatrix} I \\ X_+ \end{bmatrix} K = \begin{bmatrix} A - BR^{-1}C & 0 \\ -Q + C^*R^{-1}C & I \end{bmatrix} \begin{bmatrix} I \\ X_+ \end{bmatrix}$$

where $K = (I + BR^{-1}B^*X_+)^{-1}(A - BR^{-1}C)$. But, using the first formula in (12.1.5) it can be seen that $K = \hat{A}$, of (13.1.4). Thus, $\hat{A}$ has no eigenvalues on the unit circle and is therefore d-stable.

For the last statement of the theorem introduce the abbreviations $\hat{Q} = Q - C^*R^{-1}C$ and $K = (R + B^*X_+B)^{-1}(C + B^*X_+A)$ for the feedback matrix defined by X_+ so that we have $\hat{A} = A - BK$ in (13.1.14). Now a (formidable) computation shows that equation (13.1.1) (with $X = X_+$) is equivalent to

$$X_+ - \hat{A}^*X_+\hat{A} = \hat{Q} + (C - RK)^*R^{-1}(C - RK), \tag{13.1.16}$$

which is a symmetric Stein equation and, by the first part of the theorem, $\hat{A}$ is d-stable.

The hypothesis that $(\hat{Q}, A - BR^{-1}C)$ is observable is equivalent to the controllability of $(A^* - C^*R^{-1}B^*, \hat{Q})$. Apply Lemma 9.1.4 to this pair letting $A^* - C^*R^{-1}B^* \to A$, $\hat{Q} \to B$, $(C - RK)^*R^{-1/2} \to B_1$, and

$$\{\hat{Q} + (C - RK)^*R^{-1}(C - RK)\}^{1/2} \to B_0.$$

In the conclusion set $G = R^{-1/2}B^*$ and it is found that the pair

$$(A^* - C^*R^{-1}B^* + (C^* - K^*R)R^{-1}B^*, \{\hat{Q} + (C - RK)^*R^{-1}(C - RK)\}^{1/2})$$
$$= ((A - BK)^*, \{\hat{Q} + (C - RK)^*R^{-1}(C - RK)\}^{1/2})$$
$$= (\hat{A}^*, \{\hat{Q} + (C - RK)^*R^{-1}(C - RK)\}^{1/2})$$

is controllable, and so $(\{\hat{Q} + (C - RK)^*R^{-1}(C - RK)\}^{1/2}, \hat{A})$ is observable. Finally, Theorem 5.3.5 can be applied to equation (13.1.16) and it follows that $X_+ > 0$. □

Using deeper methods, it has recently been shown by Wimmer that, in fact, $X_+ > 0$ under the weaker hypotheses of the first part of the theorem (see Wimmer (1994b)).

13.2 The rate of convergence

In Section 13.1 an existence theorem for the maximal hermitian solution X_+ of the DARE (13.1.1) is proved using an iterative procedure. Indeed, a monotone sequence $X_0 \geq X_1 \geq \ldots$ of hermitian matrices is constructed whose limit is X_+. In this section we analyze the rate of convergence of the sequence $\{X_m\}_{m=0}^\infty$ to X_+. Under the additional hypothesis that the matrix

$$A - B(R + B^*X_+B)^{-1}(C + B^*X_+A) \tag{13.2.1}$$

is d-stable, the rate of convergence is quadratic. This is a complete analogue with the continuous case of Theorem 9.2.1.

Theorem 13.2.1. *Assume that (A, B) is d-stabilizable and that equation (13.1.1) has a solution $\tilde{X} = \tilde{X}^*$ for which $R + B^*\tilde{X}B > 0$. Let L_0 be a matrix for which $A_0 := A - BL_0$ is d-stable and construct the sequence $\{X_m\}_{m=0}^\infty$ of hermitian matrices as follows:*

$$A_i = A - BL_i \quad (i = 0, 1, \ldots) \tag{13.2.2}$$

where

$$L_i = (R + B^*X_{i-1}B)^{-1}(C + B^*X_{i-1}A) \quad (i = 1, 2, \ldots) \tag{13.2.3}$$

and for $i = 0, 1, 2, \ldots$,

$$X_i - A_i^* X_i A_i = Q - C^* R^{-1} C + (L_i - R^{-1}C)^* R(L_i - R^{-1}C). \tag{13.2.4}$$

Assume further that the limit $X_+ = \lim_{m \to \infty} X_m$ (which exists by Theorem 13.1.1) has the property that the matrix (13.2.1) is d-stable. Then there exists a constant $\kappa > 0$ such that

$$\|X_{i+1} - X_+\| \leq \kappa \|X_i - X_+\|^2. \tag{13.2.5}$$

Proof. We start with some preliminary remarks. In the course of proof of Theorem 13.1.1 it has been proved that $X_i \geq \tilde{X}$ for all i. Therefore, by hypothesis,

$$R + B^*X_iB \geq R + B^*\tilde{X}B > 0$$

It follows that
$$\lim_{i \to \infty}(R + B^*X_iB)^{-1} = (R + B^*X_+B)^{-1}$$

and, therefore, the limit $L_+ = \lim_{i \to \infty} L_i$ exists and

$$L_+ = (R + B^*X_+B)^{-1}(C + B^*X_+A).$$

By hypothesis, the matrix $A_+ := A - BL_+$ is d-stable.

In complete analogy with equation (13.1.13), the following equality is established:

$$(X_i - X_+) - A_+^*(X_i - X_+)A_+ = (L_+ - L_i)^*(R + B^*X_iB)(L_+ - L_i).$$

Now

$$X_i - X_+ = \sum_{j=0}^{\infty} A_+^{*j}(L_+ - L_i)^*(R + B^*X_iB)(L_+ - L_i)A_+^j.$$

Because A_+ is d-stable, we may choose a norm $\|\cdot\|$ on matrices so that $\|A_+\| < 1$ (see Lemma 5.6.10 of Horn and Johnson (1985), for example). In this norm we have

$$\|X_i - X_+\| \le \kappa_1 \|L_+ - L_i\| \; \|L_+^* - L_-^*\| \; \|R + B^*X_iB\| \qquad (13.2.6)$$

for all i, where $\kappa_1 = \sum_{j=0}^{\infty} \|A_+^{*j}\| \, \|A_+^j\|$, and $\kappa_1 < \infty$. Since all norms on a finite dimensional space are equivalent we can then admit the spectral matrix norm in this inequality, at the expense of a different constant $\kappa_1 < \infty$.

We now analyze the right-hand side of (13.2.6). The sequence $\{X_i\}_{i=0}^{\infty}$ is bounded above by X_0, and since $R + B^*X_iB > 0$ for all i, we in fact have

$$\|R + B^*X_iB\| \le \|R + B^*X_0B\|.$$

Furthermore, using (13.2.3) we obtain

$$\begin{aligned}L_+ - L_i &= (R + B^*X_+B)^{-1}(C + B^*X_+A) - (R + B^*X_{i-1}B)^{-1}(C + B^*X_+A) \\ &\quad + (R + B^*X_{i-1}B)^{-1}B^*(X_+ - X_{i-1})A \\ &= (R + B^*X_+B)^{-1}B^*(X_{i-1} - X_+)B(R + B^*X_{i-1}B)^{-1}(C + B^*X_+A) \\ &\quad + (R + B^*X_{i-1}B)^{-1}B^*(X_+ - X_{i-1})A,\end{aligned}$$

and therefore

$$\|L_+ - L_i\| \le \kappa_2 \|X_+ - X_{i-1}\|$$

where

$$\kappa_2 = \|(R + B^*X_+B)^{-1}\|^2 \, \|B\|^2 \, \|C + B^*X_+A\| + \|(R + B^*X_+B)^{-1}\| \, \|B\| \, \|A\|. \quad \square$$

Again, in complete analogy with the continuous case of Section 9.2, the hypotheses of Theorem 13.1.1 do not imply that the convergence of $\{X_i\}_{i=0}^{\infty}$ is quadratic. The extra hypothesis in Theorem 13.2.1 that ensures quadratic convergence is the d-stability of the matrix (13.2.1). We give an example of a discrete Riccati equation satisfying all the hypotheses of Theorem 13.1.1, but where the convergence of $\{X_i\}_{i=0}^{\infty}$ is merely linear.

Example 13.2.1. Let $Q = 0$; $C = 0$; $A = B = R = I$. Then the iterations (13.2.2)–(13.2.4) have the form

$$L_i = (I + X_{i-1})^{-1} X_{i-1} \quad (i = 1, 2, \ldots) \tag{13.2.7}$$

$$X_i - (I - L_i)^* X_i (I - L_i) = L_i^* L_i \quad (i = 0, 1, \ldots), \tag{13.2.8}$$

where L_0 is any matrix such that $I - L_0$ is d-stable. Taking $L_0 = I$, the equations (13.2.7) and (13.2.8) are easily solved explicitly to yield

$$X_0 = I, \quad X_i = X_{i-1}(2I + X_{i-1})^{-1}, \quad i = 1, 2, \ldots,$$

which gives

$$X_i = (2^{i+1} - 1)^{-1} I, \quad i = 1, 2, \ldots$$

Thus, $X_+ = 0$, and

$$\|X_{i+1} - X_+\| \leq \frac{1}{2}\|X_i - X_+\|, \quad i = 0, 1, \ldots$$

So the convergence is linear but not quadratic. Note that the matrix (13.2.1) is I in this example which is not d-stable (as expected in view of Theorem 13.2.1). □

13.3 Comparison theorems

Theorem 13.1.1 leads easily to results concerning comparisons between two discrete Riccati equations. First, we keep the same matrices A and B in both equations. Thus, along with equation (13.1.1), consider another DARE

$$X = A^* X A + \tilde{Q} - (\tilde{C} + B^* X A)^* (\tilde{R} + B^* X B)^{-1} (\tilde{C} + B^* X A) \tag{13.3.1}$$

where $\tilde{Q} = \tilde{Q}^*$, $\tilde{R} = \tilde{R}^*$ and is invertible.

Theorem 13.3.1. *Assume that (A, B) is d-stabilizable, that R is invertible, and assume that the inequality*

$$\begin{bmatrix} Q & C^* \\ C & R \end{bmatrix} \geq \begin{bmatrix} \tilde{Q} & \tilde{C}^* \\ \tilde{C} & \tilde{R} \end{bmatrix} \tag{13.3.2}$$

holds. If the DARE (13.3.1) admits a hermitian solution $\tilde{X}$ such that $\tilde{R} + B^* \tilde{X} B > 0$, then both equations (13.1.1) and (13.3.1) admit maximal hermitian solutions X_+ and $\tilde{X}_+$, respectively, and

$$X_+ \geq \tilde{X}_+. \tag{13.3.3}$$

It will be convenient to establish first a formula for the left-hand side of (13.1.2) (which will be denoted $\mathcal{R}(X)$).

Lemma 13.3.2. *Let $X = X^*$ be a solution of (13.3.1) such that $R + B^*XB$ is invertible. Then*

$$\mathcal{R}(X) = [I \quad -L^*] \begin{bmatrix} Q - \tilde{Q} & C^* - \tilde{C}^* \\ C - \tilde{C} & R - \tilde{R} \end{bmatrix} \begin{bmatrix} I \\ -L \end{bmatrix} + G^*(\tilde{R} + B^*XB)G,$$

where

$$L := (R + B^*XB)^{-1}(C + B^*XA), \quad G := L - (\tilde{R} + B^*XB)^{-1}(\tilde{C} + B^*XA).$$

Proof. Define

$$S := R + B^*XB, \quad \tilde{S} := \tilde{R} + B^*XB, \quad E := C + B^*XA, \quad \tilde{E} := \tilde{C} + B^*XA.$$

Then by using the hypothesis that X solves (13.3.1), we obtain

$$\mathcal{R}(X) - [I \quad -L^*] \begin{bmatrix} Q - \tilde{Q} & C^* - \tilde{C}^* \\ C - \tilde{C} & R - \tilde{R} \end{bmatrix} \begin{bmatrix} I \\ -L \end{bmatrix} - G^*\tilde{S}G$$

$$= (-X + A^*XA + Q - E^*S^{-1}E) + (X - A^*XA - \tilde{Q} + \tilde{E}^*\tilde{S}^{-1}\tilde{E}) - (Q - \tilde{Q})$$
$$+ L^*(C - \tilde{C}) + (C^* - \tilde{C}^*)L - L^*(R - \tilde{R})L - (L - \tilde{S}^{-1}\tilde{E})^*\tilde{S}(L - \tilde{S}^{-1}\tilde{E})$$
$$= -E^*S^{-1}E + \tilde{E}^*\tilde{S}^{-1}\tilde{E} + L^*(C - \tilde{C}) + (C^* - \tilde{C}^*)L$$
$$- L^*(R - \tilde{R})L - L^*\tilde{S}L + \tilde{E}^*L + L^*\tilde{E} - \tilde{E}^*\tilde{S}^{-1}\tilde{E}.$$

Since $L = S^{-1}E$ the right-hand side can be rewritten in the form

$$-E^*S^{-1}E + E^*S^{-1}(C - \tilde{C}) + (C^* - \tilde{C}^*)S^{-1}E$$
$$- E^*S^{-1}(R - \tilde{R})S^{-1}E - E^*S^{-1}\tilde{S}S^{-1}E + \tilde{E}^*S^{-1}E + E^*S^{-1}\tilde{E}.$$

Now use $C - \tilde{C} = E - \tilde{E}$ and $R - \tilde{R} = S - \tilde{S}$ to further rewrite the above expression:

$$-E^*S^{-1}E + E^*S^{-1}(E - \tilde{E}) + (E^* - \tilde{E}^*)S^{-1}E$$
$$- E^*S^{-1}(S - \tilde{S})S^{-1}E - E^*S^{-1}\tilde{S}S^{-1}E + \tilde{E}^*S^{-1}E + E^*S^{-1}\tilde{E} = 0. \quad \square$$

Proof of Theorem 13.3.1. Let $\tilde{X} = \tilde{X}^*$ be a solution of (13.3.1) such that $\tilde{R} + B^*\tilde{X}B > 0$. By Lemma 13.3.2, $\mathcal{R}(\tilde{X}) \geq 0$. The inequality (13.3.2) implies, in particular, that $R \geq \tilde{R}$, and consequently

$$R + B^*\tilde{X}B \geq \tilde{R} + B^*\tilde{X}B > 0.$$

Thus, Theorem 13.1.1 is applicable, and so there exists the maximal hermitian solution X_+ of (13.1.1), and $X_+ \geq \tilde{X}$. By the same token, there exists the maximal hermitian solution $\tilde{X}_+$ of (13.3.1). As in the first part of the proof we also have $X_+ \geq \tilde{X}_+$. $\quad \square$

Next consider the new Riccati equation

$$X = \tilde{A}^*X\tilde{A} + \tilde{Q} - (\tilde{C} + \tilde{B}^*X\tilde{A})^*(\tilde{R} + \tilde{B}^*X\tilde{B})^{-1}(\tilde{C} + \tilde{B}^*X\tilde{A}) \qquad (13.3.4)$$

in which all matrices A, B, C, Q and R of (13.1.1) are subject to change resulting in new coefficients $\tilde{A}, \tilde{B}, \tilde{C}, \tilde{Q}$ and $\tilde{R}$. The basic assumptions of this chapter, $\tilde{Q} = \tilde{Q}^*$ and $\tilde{R} = \tilde{R}^*$ are retained. Associate the function

$$\tilde{\mathcal{R}}(X) = -X + \tilde{A}^*X\tilde{A} + \tilde{Q} - (\tilde{C} + \tilde{B}^*X\tilde{A})^*(\tilde{R} + \tilde{B}^*X\tilde{B})^{-1}(\tilde{C} + \tilde{B}^*X\tilde{A})$$

with equation (13.3.4), just as the function $\mathcal{R}(X)$ is associated with (13.1.1).

We start with a lemma concerning the comparison between $\tilde{\mathcal{R}}(X)$ and $\mathcal{R}(X)$. Note that in this lemma the d-stabilizability of (A, B) (or of $\tilde{A}, \tilde{B}$)) is not required.

Lemma 13.3.3. *Assume that X is a hermitian matrix such that $\tilde{R} + \tilde{B}^*X\tilde{B} > 0$. If both R and $\tilde{R}$ are positive definite and*

$$\begin{bmatrix} Q - C^*R^{-1}C & A^* - C^*R^{-1}B^* \\ A - BR^{-1}C & -BR^{-1}B^* \end{bmatrix} \geq \begin{bmatrix} \tilde{Q} - \tilde{C}^*\tilde{R}^{-1}\tilde{C}^* & \tilde{A}^* - \tilde{C}^*\tilde{R}^{-1}\tilde{B}^* \\ \tilde{A} - \tilde{B}\tilde{R}^{-1}\tilde{C} & -\tilde{B}\tilde{R}^{-1}\tilde{B}^* \end{bmatrix}, \qquad (13.3.5)$$

then also $R + B^*XB > 0$ and $\mathcal{R}(X) \geq \tilde{\mathcal{R}}(X)$.

We postpone the (rather long) proof of this lemma to the next section.

Lemma 13.3.4. *If (A, B) is d-stabilizable and (13.3.5) holds with $R > 0$, then $(\tilde{A}, \tilde{B})$ is also d-stabilizable.*

Proof. The inequality (13.3.5) implies

$$\operatorname{Ker} \begin{bmatrix} \tilde{A}^* - \tilde{C}^*\tilde{R}^{-1}\tilde{B}^* - \lambda I \\ \tilde{B}\tilde{R}^{-1}\tilde{B}^* \end{bmatrix} \subseteq \operatorname{Ker} \begin{bmatrix} A^* - C^*R^{-1}B^* - \lambda I \\ BR^{-1}B^* \end{bmatrix} \qquad (13.3.6)$$

for every $\lambda \in \mathbb{C}$. Indeed, let

$$(\tilde{A}^* - \tilde{C}^*\tilde{R}^{-1}\tilde{B}^*)x = \lambda x, \quad \tilde{B}^*\tilde{R}^{-1}\tilde{B}^*x = 0.$$

Because $0 \leq BR^{-1}B^* \leq \tilde{B}\tilde{R}^{-1}\tilde{B}^*$ in view of (13.3.5), we also have $BR^{-1}B^*x = 0$. Denote the difference between the left-hand side and the right-hand side of (13.3.5) by Δ; then

$$[0 \ x^*]\Delta \begin{bmatrix} 0 \\ x \end{bmatrix} = 0.$$

Since $\Delta \geq 0$, it follows that $\Delta \begin{bmatrix} 0 \\ x \end{bmatrix} = 0$, which implies

$$[(A^* - C^*R^{-1}B^*) - (\tilde{A}^* - \tilde{C}^*\tilde{R}^{-1}\tilde{B}^*)]x = 0.$$

So $(A^* - C^*R^{-1}B^*)x = \lambda x$, and (13.3.6) follows.

Now by Lemma 4.5.3 the pair $(A - BR^{-1}C, BR^{-1}B^*)$ is d-stabilizable. By part (b) of Theorem 4.5.6,

$$\text{Ker} \begin{bmatrix} A^* - C^*R^{-1}B^* - \lambda I \\ BR^{-1}B^* \end{bmatrix} = \{0\}$$

for all $\lambda \in \mathbb{C}$ with $|\lambda| \geq 1$. In view of (13.3.6) and the same Theorem 4.5.6 it follows that $(\tilde{A} - \tilde{B}\tilde{R}^{-1}\tilde{C}, \tilde{B}\tilde{R}^{-1}\tilde{B}^*)$ is also d-stabilizable. Finally, by Lemma 4.5.3, the same holds for $(\tilde{A}, \tilde{B})$. □

Now we are ready to state and prove the main comparison theorem between equations (13.1.1) and (13.3.4).

Theorem 13.3.5. *Assume that (A, B) is d-stabilizable, that the inequality (13.3.5) holds, and that $R > 0$, $\tilde{R} > 0$. If there is a hermitian matrix X such that $\tilde{\mathcal{R}}(X) \geq 0$ and $\tilde{R} + \tilde{B}^*X\tilde{B} > 0$, then the DAREs (13.1.1) and (13.3.4) have maximal hermitian solutions X_+ and $\tilde{X}_+$, respectively, and*

$$X_+ \geq \tilde{X}_+. \tag{13.3.7}$$

Proof. By Lemma 13.3.4 the pair $(\tilde{A}, \tilde{B})$ is d-stabilizable. We can now apply Theorem 13.1.1, and the existence of $\tilde{X}_+$ follows. Moreover, by Lemma 13.3.3, if $\tilde{\mathcal{R}}(X) \geq 0$, and $\tilde{R} + \tilde{B}^*X\tilde{B} > 0$ then also $\mathcal{R}(X) \geq 0$ and $R + B^*XB > 0$. By the same Theorem 13.1.1, the equation (13.1.1) has the maximal hermitian solution X_+, which also majorizes solutions of $\mathcal{R}(X) \geq 0$. Finally, the maximal hermitian solution $\tilde{X}_+$ of (13.3.4) is such that $\tilde{R} + \tilde{B}^*\tilde{X}_+\tilde{B} > 0$. Therefore, using Lemma 13.3.3 again we have $\mathcal{R}(\tilde{X}_+) \geq 0$. Hence (13.3.7) holds. □

There is a connection between the inequalities (13.3.2) and (13.3.5). For convenience, let us denote

$$T_1 = \begin{bmatrix} Q & C^* \\ C & R \end{bmatrix}, \quad T_2 = \begin{bmatrix} Q - C^*R^{-1}C & -C^*R^{-1}B^* \\ -BR^{-1}C & -BR^{-1}B^* \end{bmatrix},$$

$$T_3 = \begin{bmatrix} Q - C^*R^{-1}C & -C^*R^{-1} \\ -R^{-1}C & -R^{-1} \end{bmatrix}, \tag{13.3.8}$$

and let $\tilde{T}_1$, $\tilde{T}_2$ and $\tilde{T}_3$ be the corresponding matrices defined for the equation (13.3.4). Observe that A (resp. $\tilde{A}$) does not appear in the definition of T_1, T_2 and T_3 (resp. $\tilde{T}_1$, $\tilde{T}_2$ and $\tilde{T}_3$).

Lemma 13.3.6. *Assume $B = \tilde{B}$, $R > 0$ and $\tilde{R} > 0$. Then the inequality $T_1 \geq \tilde{T}_1$ is equivalent to $T_3 \geq \tilde{T}_3$, and each one of these inequalities implies $T_2 \geq \tilde{T}_2$.*

The reader is referred to Wimmer (1992b) for the proof.

13.4 Inequalities for partitioned matrices

This section is devoted to proving Lemma 13.3.3. The proof is rather long and involves some ideas which are more widely useful. In particular, the first lemma is already well-known from other problem areas.

Lemma 13.4.1. *Let $G, \tilde{G} \in \mathbb{C}^{n \times m}$ and satisfy $\tilde{G}\tilde{G}^* = I$, $GG^* \leq I$. Then there is $K \in \mathbb{C}^{m \times m}$ such that $\|K\| \leq 1$ and $G = \tilde{G}K$.*

As usual $\|\cdot\|$ denotes the spectral matrix norm in this statement.

Proof. Since $GG^* \leq \tilde{G}\tilde{G}^* = I$, $x^*\tilde{G} = 0$ implies $x^*G = 0$. It follows that Im $\tilde{G} \subseteq$ Im G and hence there exists a $K \in \mathbb{C}^{m \times m}$ such that $G = \tilde{G}K$ although, in general, K is not unique. Consider the orthogonal decomposition

$$\mathbb{C}^m = (\text{Ker } \tilde{G})^\perp \oplus (\text{Ker } \tilde{G}),$$

and K can be chosen so that, with respect to this decomposition, $K = \begin{bmatrix} K_0 & 0 \\ 0 & 0 \end{bmatrix}$.

Thus K_0 is a linear transformation on $(\text{Ker } \tilde{G})^\perp$. In the same decomposition, $\tilde{G} = [\tilde{G}_0 \ 0]$ and, since $\tilde{G}\tilde{G}^* = I$, $\tilde{G}_0$ is unitary. Thus, we have $G = [\tilde{G}_0 K_0 \ 0]$.

Using properties of the spectral norm,

$$\|K\|^2 = \|KK^*\| = \|K_0 K_0^*\| = \|GG^*\| \leq 1. \qquad \square$$

The next lemma is technical and its proof is facilitated by using the concept of the *Moore–Penrose (generalized) inverse*. Details can be found in Chapter 12 of Lancaster and Tismenetsky (1985) but, for completeness, the necessary minimum properties are summarized here. The Moore–Penrose inverse of a $p \times q$ matrix Z is the unique $q \times p$ matrix Z^+ satisfying the conditions:

(i) $Z^+ Z Z^+ = Z^+$, $\quad ZZ^+ Z = Z$,

(ii) $(Z^+ Z)^* = Z^+ Z$, $\quad (ZZ^+)^* = ZZ^+$.

It follows easily from these conditions that ZZ^+ and Z^+Z are the orthogonal projections onto Im Z and $(\text{Ker } Z)^\perp$, respectively. Furthermore, if $Z^* = Z$ then $(\text{Ker } Z)^\perp = $ Im Z and the projections are equal. Hence Z and Z^+ commute. Note also that

$$(Z^+)^+ = Z \qquad (13.4.1)$$

Returning to the notation of Lemma 13.4.1, let $n \times m$ matrices G and $\tilde{G}$ be given and write

$$\Gamma = GG^*, \quad \tilde{\Gamma} = \tilde{G}\tilde{G}^*, \quad \Delta = \Gamma - \tilde{\Gamma} \qquad (13.4.2)$$

$$P = \Delta^+ \Delta = \Delta \Delta^+. \qquad (13.4.3)$$

Now let X be a hermitian matrix such that $I + \Gamma X$ and $I + \tilde{\Gamma} X$ are nonsingular and define

$$M = X(I + \Gamma X)^{-1} = (I + X\Gamma)^{-1}X, \quad \tilde{M} = X(I + \tilde{\Gamma} X)^{-1} = (I + X\tilde{\Gamma})^{-1}X.$$

$$(13.4.4)$$

Lemma 13.4.2. *With the notations of equations* (13.4.2)–(13.4.4) *the following results hold:*

(1) $M - \tilde{M} = -M \triangle \tilde{M} = \tilde{M}(-\triangle + \triangle M \triangle)\tilde{M}$.

(2) *If* $L := -\triangle + \triangle M \triangle$ *and* $N = -\triangle^+ - \tilde{M}$ *then* $L = (PNP)^+$ *and*

$$PNL = LNP = P \quad \text{and} \quad LNL = L. \qquad (13.4.5)$$

(3) *If* $\tilde{\Gamma} \geq \Gamma$ *then* $I + \tilde{G}^* X \tilde{G} > 0$ *implies*

$$I + G^* X G > 0 \quad \text{and} \quad L \geq 0. \qquad (13.4.6)$$

Proof. The proof of (1) is obtained from the following calculation:

$$\begin{aligned} M - \tilde{M} &= (I + X\Gamma)^{-1} X - X(I + \tilde{\Gamma}X)^{-1} \\ &= (I + X\Gamma)^{-1}[X(I + \tilde{\Gamma}X) - (I + X\Gamma)X](I + \tilde{\Gamma}X)^{-1} \\ &= -X(I + \tilde{\Gamma}X)^{-1}(I + \tilde{\Gamma}X)(I + \Gamma X)^{-1} \triangle \tilde{M} \\ &= -\tilde{M}[(I + \Gamma X) + (\tilde{\Gamma} - \Gamma)X](I + \Gamma X)^{-1} \triangle \tilde{M} \\ &= \tilde{M}(-\triangle + \triangle M \triangle)\tilde{M}. \end{aligned} \qquad (13.4.7)$$

We now pass to the proof of (2). First of all observe that a calculation similar to (13.4.7) shows that

$$L = -(I + \tilde{\Gamma}X)(I + \Gamma X)^{-1} \triangle. \qquad (13.4.8)$$

Now

$$\begin{aligned} PN &= -P(\triangle^+ + P\tilde{M}) = -P \triangle^+ (I + \triangle \tilde{M}) \\ &= -P \triangle^+ [(I + \tilde{\Gamma}X)(I + \tilde{\Gamma}X)^{-1} + \triangle X(I + \tilde{\Gamma}X)^{-1}] \\ &= -P \triangle^+ (I + \Gamma X)(I + \tilde{\Gamma}X)^{-1}. \end{aligned} \qquad (13.4.9)$$

Thus, using (13.4.8) and (13.4.9) we obtain

$$PNL = P \triangle^+ \triangle = P \qquad (13.4.10)$$

and since $L = LP$ the equality $LNL = L$ follows from (13.4.10) as well. Finally, observe that $\triangle$, M and $\tilde{M}$ are hermitian and, therefore, so are N and L. Thus,

$$(LNP)^* = PNL = P, \qquad (13.4.11)$$

and the proof of (13.4.5) is complete.

Next, we verify that $L = (PNP)^+$. The definition of L together with (13.4.3) imply $L = PL$. So
$$L(PNP)L = L(PNL) = LP = L$$
and
$$(PNP)L(PNP) = (PNL)PNP = P(PNP) = PNP,$$
and the matrices $L(PNP) = LNP = P$ (we have used (13.4.11) here) and $(PNP)L = PNL = P$ are clearly hermitian. By the uniqueness of the Moore-Penrose inverse it now follows that $L = (PNP)^+$.

It remains to prove part (3) of the lemma. Given an invertible $n \times n$ matrix S, the transformation
$$G \to SG, \quad \tilde{G} \to S\tilde{G}, \quad M \to S^{*-1}MS^{-1}, \quad X \to S^{*-1}XS^{-1} \quad (13.4.12)$$
preserves the hypotheses and conclusion of the statement (3). Note also that Γ transforms to $S\Gamma S^*$. Therefore, applying the transformation (13.4.12) with a suitable S, we can assume without loss of generality that
$$\tilde{\Gamma} = \begin{bmatrix} I & 0 \\ 0 & 0 \end{bmatrix}, \quad (13.4.13)$$
where I is $p \times p$, and p is the rank of $\tilde{\Gamma}$. The hypothesis $\tilde{\Gamma} \geq \Gamma \geq 0$ now implies that
$$\Gamma = \begin{bmatrix} \Gamma_1 & 0 \\ 0 & 0 \end{bmatrix}, \quad (13.4.14)$$
where $\Gamma_1 = \Gamma_1^*$ is $p \times p$ and $0 \leq \Gamma_1 \leq I$. Partition
$$G = \begin{bmatrix} G_1 \\ G_2 \end{bmatrix}, \quad \tilde{G} = \begin{bmatrix} \tilde{G}_1 \\ \tilde{G}_2 \end{bmatrix}$$
where G_1 and $\tilde{G}_1$ are $p \times m$. In view of (13.4.13) and (13.4.14) we obtain $G_2 = 0$, $\tilde{G}_2 = 0$. So the proof of (3) reduces to the case when $\tilde{\Gamma} = I_n$. We can therefore assume without loss of generality that
$$\Gamma = \begin{bmatrix} \Gamma_2 & 0 \\ 0 & I_{n-q} \end{bmatrix} \quad (13.4.15)$$
where $0 \leq \Gamma_2 < I_q$ (for some q, $0 \leq q \leq n$).

Denote by $\lambda_{\min}(Z)$ the smallest eigenvalue of a hermitian matrix Z. The inequality $I + \tilde{G}^* X \tilde{G} > 0$ is equivalent to $\lambda_{\min}(\tilde{G}^* X \tilde{G}) > -1$. We will prove that this inequality implies
$$\lambda_{\min}(G^* X G) > -1. \quad (13.4.16)$$

By Lemma 13.4.1, we have $G = \tilde{G}K$ for some $m \times m$ matrix K with $\|K\| \leq 1$. Put $\lambda = \lambda_{\min}(\tilde{G}^* X \tilde{G})$. If $\lambda \geq 0$, then $\tilde{G}^* X \tilde{G} \geq 0$; consequently, $G^* X G =$

$K^*\tilde{G}^*X\tilde{G}K \geq 0$, and (13.4.16) is obvious. So assume $-1 < \lambda < 0$. We have $\tilde{G}^*X\tilde{G} - \lambda I \geq 0$, therefore

$$G^*XG - \lambda K^*K \geq 0. \qquad (13.4.17)$$

But $K^*K \leq I$, so $\lambda K^*K - \lambda I \geq 0$. Combining with (13.4.17), we conclude that $G^*XG \geq \lambda I$, i.e. $\lambda \leq \lambda_{\min}(G^*XG)$. Combining with $\lambda > -1$, the inequality (13.4.16) follows. This establishes the first inequality of (13.4.6).

Finally, it is verified that $L \geq 0$. From $I + \tilde{G}^*X\tilde{G} > 0$ it is deduced that $\tilde{G}\tilde{G}^* + \tilde{G}\tilde{G}^*X\tilde{G}\tilde{G}^* \geq 0$, and because of $\tilde{G}\tilde{G}^* = \tilde{\Gamma} = I$ we have $I + X \geq 0$. Since $I+X$ is nonsingular we have, in fact, $I+X > 0$. A simple algebraic manipulation shows that

$$(I + \Gamma X)(I + X)^{-1} = (I - \Gamma)(I + X)^{-1} + \Gamma.$$

Thus, using the formula (13.4.8) for L,

$$\begin{aligned} L &= -(I + X)(I + \Gamma X)^{-1} \Delta \\ &= -[(I + \Gamma X)(I + X)^{-1}]^{-1} \Delta \\ &= [(I - \Gamma)(I + X)^{-1} + \Gamma](-\Delta). \end{aligned} \qquad (13.4.18)$$

Since $L = -\Delta + \Delta M \Delta$, L has the form $\begin{bmatrix} L_0 & 0 \\ 0 & 0 \end{bmatrix}$, where by L_0 we denote a $q \times q$ block. Comparing with (13.4.18) it is seen that

$$L_0 = (WS + I - W)^{-1}W \qquad (13.4.19)$$

where S is the top left $q \times q$ corner of $(I + X)^{-1}$, and $W = I - \Gamma_2 > 0$. The equality (13.4.19) can be rewritten in the form $L_0 = S + W^{-1} - I$. Since $W^{-1} \geq I$ and $S > 0$ we conclude that $L_0 > 0$ and hence $L \geq 0$, as claimed. $\square$

Proof of Lemma 13.3.3. We use the notation and assume the hypotheses of Lemma 13.3.3. First of all, observe that

$$\mathcal{R}(X) = -X + F^*X(I + BR^{-1}B^*X)^{-1}F + Q - C^*R^{-1}C,$$

where $F = A - BR^{-1}C$, and a similar formula holds for $\tilde{\mathcal{R}}(X)$ (see Proposition 12.1.1). Therefore, $\mathcal{R}(X) \geq \tilde{\mathcal{R}}(X)$ is equivalent to

$$\begin{aligned} F^*X(I + BR^{-1}B^*X)^{-1}F + Q - C^*R^{-1}C \geq \\ \tilde{F}^*X(I + \tilde{B}\tilde{R}^{-1}\tilde{B}^*X)^{-1}\tilde{F} + \tilde{Q} - \tilde{C}^*\tilde{R}^{-1}\tilde{C}, \end{aligned} \qquad (13.4.20)$$

where $\tilde{F} = \tilde{A} - \tilde{B}\tilde{R}^{-1}\tilde{C}$. So we shall prove (13.4.20) rather than $\mathcal{R}(X) \geq \tilde{\mathcal{R}}(X)$.

INEQUALITIES FOR PARTITIONED MATRICES 323

Define $Q_d = (Q - C^*R^{-1}C) - (\tilde{Q} - \tilde{C}^*\tilde{R}^{-1}\tilde{C})$ and $F_d = F - \tilde{F}$. Define also (as in (13.4.4))

$$M = X(I + BR^{-1}B^*X)^{-1}, \quad \tilde{M} = X(I + \tilde{B}\tilde{R}^{-1}\tilde{B}X)^{-1}.$$

Now (13.4.20) can be rewritten in the form

$$\begin{aligned} 0 &\leq F^*MF + Q_d - \tilde{F}^*\tilde{M}\tilde{F} \\ &= F^*MF + Q_d - (F - F_d)^*\tilde{M}(F - F_d) \\ &= Q_d - F_d^*\tilde{M}F_d + F^*\tilde{M}F_d + F_d^*\tilde{M}F + F^*(M - \tilde{M})F \\ &= [I \ F^*\tilde{M}] \begin{bmatrix} Q_d - F_d^*\tilde{M}F_d & F_d^* \\ F_d & -\Delta + \Delta M\Delta \end{bmatrix} \begin{bmatrix} I \\ \tilde{M}F \end{bmatrix}, \end{aligned}$$

where $\Delta = BR^{-1}B^* - \tilde{B}\tilde{R}^{-1}\tilde{B}^* \leq 0$ (and we have used statement (1) of Lemma 13.4.2).

Observe that $\tilde{M}$ is hermitian (cf. (13.4.4)). It follows that the proof of Lemma 13.3.3 will be completed once we show that

$$\begin{bmatrix} Q_d - F_d^*\tilde{M}F_d & F_d^* \\ F_d & -\Delta + \Delta M\Delta \end{bmatrix} \geq 0. \tag{13.4.21}$$

To do this the hypothesis (13.3.5) is needed, which can be rewritten in the form

$$\begin{bmatrix} Q_d & F_d^* \\ F_d & -\Delta \end{bmatrix} \geq 0. \tag{13.4.22}$$

The equality (recall that Δ^+ stands for the Moore–Penrose inverse of Δ)

$$\begin{bmatrix} I & F_d^*\Delta^+ \\ 0 & I \end{bmatrix} \begin{bmatrix} Q_d & F_d^* \\ F_d & -\Delta \end{bmatrix} \begin{bmatrix} I & 0 \\ \Delta^+F_d & I \end{bmatrix} = \begin{bmatrix} Q_d + F_d^*\Delta^+F_d & F_d^* - F_d^*\Delta^+\Delta \\ F_d - \Delta\Delta^+F_d & -\Delta \end{bmatrix}$$

shows that $Q_d + F_d^*\Delta^+F_d \geq 0$. Also, (13.4.22) implies $\operatorname{Im} F_d \subseteq \operatorname{Im} \Delta$, and, consequently,

$$F_d = PF_d, \tag{13.4.23}$$

where $P = \Delta\Delta^+ = \Delta^+\Delta$ is the orthogonal projection on $\operatorname{Im} \Delta$. By part (3) of Lemma 13.4.2 $L := -\Delta + \Delta M\Delta \geq 0$ and therefore, using the equality $L(-\Delta^+ - \tilde{M})P = P$ (see (13.4.5)), we obtain

$$\begin{bmatrix} P(-\Delta^+ -\tilde{M})P & P \\ P & L \end{bmatrix} =$$

$$\begin{bmatrix} I & P(-\Delta^+ -\tilde{M}) \\ 0 & I \end{bmatrix} \begin{bmatrix} 0 & 0 \\ 0 & L \end{bmatrix} \begin{bmatrix} I & 0 \\ (-\Delta^+ -\tilde{M})P & I \end{bmatrix} \geq 0.$$

Thus, using (13.4.23),

$$0 \leq \begin{bmatrix} F_d^* & 0 \\ 0 & I \end{bmatrix} \begin{bmatrix} P(-\Delta^+ -\tilde{M})P & P \\ P & L \end{bmatrix} \begin{bmatrix} F_d & 0 \\ 0 & I \end{bmatrix} = \begin{bmatrix} F_d^*(-\Delta^+ -\tilde{M})F_d & F_d^* \\ F_d & L \end{bmatrix},$$

and since

$$\begin{bmatrix} Q_d - F_d^* \tilde{M} F_d & F_d^* \\ F_d & -\Delta + \Delta M \Delta \end{bmatrix} =$$
$$\begin{bmatrix} F_d^*(-\Delta^+ - \tilde{M})F_d & F_d^* \\ F_d & L \end{bmatrix} + \begin{bmatrix} Q_d + F_d^* \Delta^+ F_d & 0 \\ 0 & 0 \end{bmatrix},$$

where both summands on the right-hand side are positive semidefinite, the desired inequality (13.4.21) follows. □

13.5 Stabilizing and positive semidefinite solutions

A hermitian solution X of the DARE (13.1.1) is called *stabilizing* (resp. *almost stabilizing*) if all eigenvalues of

$$A - B(R + B^*XB)^{-1}(C + B^*XA) \qquad (13.5.1)$$

are in the open (resp. closed) unit disk. In this section we derive several results, many of them well-known in the literature, concerning existence and uniqueness of stabilizing and almost stabilizing solutions, especially in connection with positive semidefiniteness of such solutions. We draw extensively on the results obtained and the techniques used so far in this and the previous chapter.

Our first result is very general:

Proposition 13.5.1. *If the DARE* (13.1.1) *has a stabilizing solution, then it is unique in the set of almost stabilizing solutions.*

Proof. Let X_1 and X_2 be almost stabilizing solutions of (13.1.1), and at least one of them be stabilizing. For $j = 1, 2$, define

$$Z_j = -(R + B^*X_jB)^{-1}(C + B^*X_jA).$$

Use equation (13.1.1) to write:

$$\left. \begin{array}{l} X_1 = Q + A^*X_1A + (C^* + A^*X_1B)Z_1 \\ X_2 = Q + A^*X_2A + Z_2^*(C + B^*Z_2A) \end{array} \right\}. \qquad (13.5.2)$$

First note that

$$(R+B^*X_1B)^{-1}-(R+B^*X_2B)^{-1} = (R+B^*X_2B)^{-1}B^*(X_2-X_1)B(R+B^*X_1B)^{-1}.$$

Multiply on the left and right by $C^* + A^*X_2B$ and $C + B^*X_1A$, respectively, to obtain

$$-(C^* + A^*X_2B)Z_1 + Z_2^*(C + B^*X_1A) = Z_2^*B^*(X_2 - X_1)BZ_1.$$

Expand parentheses below and use the last identity to get

$$(A^* + Z_2^*B^*)(X_1 - X_2)(A + BZ_1)$$
$$= \{A^*X_1A + (C^* + A^*X_1B)Z_1\} - \{A^*X_2A + Z_2^*(C + B^*X_2A)\}.$$

But now equations (13.5.2) give

$$(A^* + Z_2^*B^*)(X_1 - X_2)(A + BZ_1) - (X_1 - X_2) = 0, \qquad (13.5.3)$$

and this is just a symmetric Stein equation for $X_1 - X_2$.

As the eigenvalues of both $A + BZ_1$ and $A + BZ_2$ are in the closed unit circle, and at least one of those matrices is d-stable, it follows from Theorem 5.2.3 that the unique solution of (13.5.3) is $X_1 - X_2 = 0$. □

Recall first of all, the general results of Theorem 13.1.3 which give sufficient conditions for a stabilizing and (semi) definite maximal solution of the DARE.

For the purposes of the remainder of this section it will be assumed that $R > 0$, and $Q \geq 0$. These assumptions are standard in the literature (especially in connection with the discrete LQR problem, Section 16.5), and they allow us to simplify the DARE considerably: in view of Proposition 12.1.1 we can (and do) assume that $C = 0$. Furthermore, replacing B by $BR^{-1/2}$ has the effect of assuming $R = I$ in (13.1.1). Finally, we write $Q = D^*D$ where D is a $p \times n$ matrix. The DARE takes the form

$$X = A^*XA + D^*D - A^*XB(I + B^*XB)^{-1}B^*XA, \qquad (13.5.4)$$

and the matrix (13.5.1) is now $A - B(I + B^*XB)^{-1}B^*XA$.

Theorem 13.5.2. *Assume that A is invertible.*

(i) *If (A, B) is d-stabilizable, then there exists a unique almost stabilizing solution of (13.5.4), and this almost stabilizing solution is positive semidefinite and maximal (in the set of all hermitian solutions).*

(ii) *If (D, A) is detectable (i.e. (A^*, D^*) is d-stabilizable), then every positive semidefinite solution of (13.5.4) is stabilizing.*

Proof. (i) Corollary 13.1.2 implies the existence of an almost stabilizing solution which is positive semidefinite and maximal. If (A, B) is controllable, then Theorem 12.2.3 guarantees the uniqueness of an almost stabilizing solution. If the pair (A, B) is merely d-stabilizable, then partition (A, B) as in (12.5.3), and replace (13.5.4) by (12.5.5)–(12.5.7), thereby reducing the proof of uniqueness of an almost stabilizing solution to the controllable case.

(ii) Let $X \geq 0$ be a solution of (13.5.4) and arguing by contradiction, assume that $Kx = \lambda x$, $x \neq 0$, $|\lambda| \geq 1$, where

$$K = A - B(I + B^*XB)^{-1}B^*XA. \tag{13.5.5}$$

The T-invariance of the graph of X (Proposition 12.2.2) implies (cf. formula (12.3.6))

$$A + BB^*A^{*-1}D^*D - BB^*A^{*-1}X = K, \quad -A^{*-1}D^*D + A^{*-1}X = XK.$$

Premultiplying the second equation by BB^* and adding to the first equation gives
$$K = A - BB^*XK; \quad A^*XK = -D^*D + X. \tag{13.5.6}$$
Therefore, $KXA = KXBB^*XK + KXK$, and we have

$$\bar{\lambda}x^*XAx = |\lambda|^2 x^*XBB^*Xx + |\lambda|^2 x^*Xx,$$

$$\lambda x^*A^*Xx = -x^*D^*Dx + x^*Xx,$$

where the second equality follows from the second equality in (13.5.6). It follows that

$$-|\lambda|^2 x^*XB^*BXx - x^*D^*Dx = (|\lambda|^2 - 1)x^*Xx. \tag{13.5.7}$$

Since $|\lambda|^2 \geq 1$, both sides in (13.5.7) must be zero and hence $Dx = 0$, $B^*Xx = 0$. But now $Kx = \lambda x$ implies (by (13.5.6))

$$\lambda x = Ax - \lambda BB^*Xx = Ax.$$

Since $x \neq 0$, the equalities $Ax = \lambda x$, $Dx = 0$ contradict the detectability of (D, A). □

The assumption that A is invertible is not essential in Theorem 13.5.2; these results are valid without this assumption (see DeSouza et al. (1986)).

Corollary 13.5.3. *Assume that A is invertible, (A, B) is d-stabilizable, and (D, A) is detectable. Then the DARE (13.5.4) has a stabilizing positive semidefinite solution which is unique in the set of almost stabilizing solutions, and also is unique in the set of positive semidefinite solutions.*

For the proof simply combine the two parts of Theorem 13.5.2.

13.6 Exercises

13.6.1. Prove that the weaker hypothesis $I+\tilde{G}^*X\tilde{G} \geq 0$ in place of $I+\tilde{G}^*X\tilde{G} > 0$ in Lemma 13.4.2(3) leads to the weaker conclusion $I + \tilde{G}XG \geq 0$. [Hint: Follow the proof of part (3) of Lemma 13.4.2.]

13.6.2. Prove that every positive definite solution of (13.5.4) is almost stabilizing (without further assumptions on d-stabilizability or detectability). [Hint: Follow the approach of the proof of Theorem 13.5.2(ii).]

13.6.3. A number $\lambda \in \mathbb{C}$ is called an *undetectable eigenvalue* for the pair (D, A) if there is an $x \neq 0$ such that $Dx = 0$ and $Ax = \lambda x$. Under the hypotheses of Theorem 13.5.2(i), show that the DARE (13.5.4) has a stabilizing solution (necessarily unique) if and only if (D, A) has no undetectable eigenvalues on the unit circle. [Hint: Let X be the unique almost stabilizing solution. If K (given by (13.5.5)) has eigenvalue λ on the unit circle, then the proof of Theorem 13.5.2(ii) shows that λ is undetectable for (D, A). Conversely, if $Ax = \lambda x$, $Dx = 0$, $x \neq 0$, $|\lambda| = 1$, then the equation (13.5.4) implies $B^*Xx = 0$, and therefore $Kx = \lambda x$.]

13.6.4. Show by example that the DARE (13.1.1) may have more than one almost stabilizing solution.

13.6.5. Define $\mathcal{R}(X)$ as in Section 13.1 (i.e. as the left-hand side of inequality (13.1.2)). Show that the recursion of equations (13.1.7) and (13.1.6) is obtained by applying the Newton–Kantorovich method to the equation $\mathcal{R}(X) = 0$ (cf. the discussion of Section 9.2).

13.6.6. Prove that the equation (13.1.1) is equivalent to (13.1.16).

13.7 Notes

The iterative approach used in the proof of Theorem 13.1.1 originates with Hewer (1971), and has been further extended and developed in a series of papers including Chan et al. (1984), Bitmead et al. (1985), Lancaster et al. (1987), and Ran and Vreugdenhil (1988). The review paper of Bitmead and Gevers (1991) contains an exposition of the convergence of Riccati difference equations to the correponding DARE, as well as additional bibliography on this subject. The iteration process for DARE in the framework of infinite dimensional spaces and operators has been studied by Hager and Horowitz (1976) and Gurvits (1990).

Example 13.2.1 was communicated to us by Gurvits (personal communication).

The material of Sections 13.3 and 13.4 is based on Wimmer (1992b). Proposition 13.5.1 appears in a paper of Ionescu and Weiss (1992). The proof of Theorem 13.5.2(ii) follows Kučera (1972b).

A criterion for uniqueness of a positive semidefinite almost stabilizing solution of the DARE was obtained in Payne and Silverman (1973). There, a more general DARE was studied in which the invertibility of $R + B^*XB$ is not required of a solution X, and the inverse of $R + B^*XB$ is replaced by its Moore–Penrose inverse in the DARE (13.1.1). The relationships between positive semidefinite solutions and almost stabilizing solutions of the DARE are studied in Wimmer (1989) (the almost stabilizing solutions are called strong solutions in this paper).

14
PERTURBATION THEORY FOR DISCRETE ALGEBRAIC RICCATI EQUATIONS

Taking advantage of the geometric theory for the discrete algebraic Riccati equation developed in Chapter 12, as well as the comparison theorems of Chapter 13, we can now study the behaviour of hermitian solutions of discrete algebraic Riccati equations under small perturbations of the coefficients of these equations. The emphasis is on the behaviour of extremal (minimal and maximal) solutions. Their behaviour as functions of the coefficients is studied here as well.

The material of this chapter is, of course, the analogue of that in Chapter 11 and is developed here for the DARE of equation (12.2.1) rather than the CARE of equation (7.2.1).

14.1 Existence of hermitian solutions

Consider a sequence of discrete algebraic Riccati equations with the form of (12.2.1):

$$X = A_p^* X A_p + Q_p - A_p^* X B_p (R_p + B_p^* X B_p)^{-1} B_p^* X A_p \qquad (14.1.1)$$

for $p = 1, 2, \ldots$, and assume that the limits

$$A = \lim_{p \to \infty} A_p, \quad Q = \lim_{p \to \infty} Q_p, \quad B = \lim_{p \to \infty} B_p, \quad R = \lim_{p \to \infty} R_p \qquad (14.1.2)$$

exist. The main result of this section is that, under suitable hypotheses, existence of hermitian solutions of every equation (14.1.1) implies the same property for the limiting equation

$$X = A^* X A + Q - A^* X B (R + B^* X B)^{-1} B^* X A. \qquad (14.1.3)$$

Theorem 14.1.1. *Assume that Q_p and R_p are hermitian matrices of sizes $n \times n$ and $m \times m$, respectively, for all p. Assume further that the matrices A and R are invertible, the pair (A, B) is d-sign controllable, and*

$$R + B^*(\eta^{-1} - A^*)^{-1} Q (\eta - A)^{-1} B > 0 \qquad (14.1.4)$$

for some η on the unit circle. Then, if each equation (14.1.1) admits a hermitian solution, then the limiting equation (14.1.3) admits a hermitian solution as well.

The proof of this theorem will be patterned after that of Theorem 11.1.1. First note that the d-sign controllability persists under small perturbations:

Lemma 14.1.2. *If a pair of matrices (A, B) is d-sign controllable, then there is an $\varepsilon > 0$ such that all pairs of matrices (A', B') satisfying $\|A-A'\|+\|B-B'\| < \varepsilon$ are d-sign controllable as well.*

We omit the proof of Lemma 14.1.2; it can be obtained as in the proof of Lemma 11.1.2 after an obvious modification (replacing the symmetry with respect to the real axis by symmetry with respect to the unit circle).

Proof of Theorem 14.1.1. In view of (14.1.2), the matrices A_p and R_p are invertible and
$$R_p + B_p^*(\eta^{-1} - A_p^*)^{-1}Q_p(\eta - A_p)B_p > 0$$
for sufficiently large p. Also, by Lemma 14.1.2 the pair (A_p, B_p) is d-sign controllable (again, for p large enough). Define $J = \begin{bmatrix} 0 & -iI \\ iI & 0 \end{bmatrix}$ and

$$T_p = \begin{bmatrix} A_p + B_p R_p^{-1} B_p^* A_p^{*-1} Q_p & -B_p R_p^{-1} B_p^* A_p^{*-1} \\ -A_p^{*-1} Q_p & A_p^{*-1} \end{bmatrix}.$$

(cf. equation (12.2.5)) and it follows from Theorem 12.5.1 that there exists a T_p-invariant J-Lagrangian subspace $\mathcal{M}_p$ in $\mathbf{C}^{2n}$.

Let $\mathcal{M}$ be the limit of a subsequence of $\{\mathcal{M}_p\}_{p=1}^\infty$ (the limit exists because of the compactness of the metric space of subspaces in $\mathbf{C}^{2n}$ (Theorem 7.A.4)). By Corollaries 7.A.6 and 7.A.7 the subspace $\mathcal{M}$ is n-dimensional, J-neutral and T-invariant, where

$$T = \begin{bmatrix} A + BR^{-1}B^* A^{*-1} Q & -BR^{-1} B^* A^{*-1} \\ -A^{*-1} Q & A^{*-1} \end{bmatrix}.$$

It remains to apply Theorem 12.5.1 to the equation (14.1.3). □

14.2 Extremal solutions

We consider the DARE in the form (14.1.3) but now the coefficients A, B, R and Q are allowed to vary. If $Q = Q^*$, $R = R^*$ and is invertible, the pair (A, B) is d-stabilizable, and (14.1.4) holds for some η on the unit circle then, by Theorem 13.1.1, the equation (14.1.3) has the maximal hermitian solution X_+, provided it has hermitian solutions at all.

Assuming, in addition, that A is invertible then, by Theorem 12.5.1, the matrix

$$T = \begin{bmatrix} A + BR^{-1}B^*A^{*-1}Q & -BR^{-1}B^*A^{*-1} \\ -A^{*-1}Q & A^{*-1} \end{bmatrix} \quad (14.2.1)$$

has all even multiplicities corresponding to its unimodular eigenvalues, and the sign characteristic of $\left(T, \begin{bmatrix} 0 & -iI \\ iI & 0 \end{bmatrix}\right)$ consists of -1's only. Also, the graph subspace of X_+,

$$\mathcal{M} = \mathrm{Im}\begin{bmatrix} I \\ X_+ \end{bmatrix},$$

is T-invariant and J-Lagrangian, and $T|\mathcal{M}$ is similar to

$$A - B(R + B^*X_+B)^{-1}B^*X_+A$$

(cf. formula (12.3.6)), and therefore all the eigenvalues of $T|\mathcal{M}$ are in the closed unit disk (Theorem 13.1.1).

Now results on continuous and analytic dependence of the maximal solution X_+ on the coefficients of the equation can be proved in exactly the same way as in Sections 11.2, 11.3, i.e. by reduction to the corresponding continuity and analyticity properties of the T-invariant J-Lagrangian subspace $\mathcal{M}$ such that all eigenvalues of $T|\mathcal{M}$ lie in the closed unit disk. We state the results without proof.

Theorem 14.2.1. *Let Z be the set of all quadruples of matrices (A, B, R, Q) with the following properties:*

(i) *the sizes of A, B, R and Q are $n \times n$, $n \times m$, $m \times m$ and $n \times n$, respectively;*

(ii) $Q = Q^*$, $R = R^*$;

(iii) *A and R are invertible;*

(iv) *(A, B) is d-stabilizable;*

(v) *(14.1.4) holds for some unimodular η;*

(vi) *the DARE (14.1.3) has a hermitian solution.*

Then the maximal hermitian solution $X_+ = X_+(A, B, R, Q)$ is a continuous function of $(A, B, R, Q) \in Z$.

Theorem 14.2.2. *Let $(A(t), B(t), R(t), Q(t)) \in Z$ be analytic functions of t on a real interval (α, β). Assume that the number of unimodular eigenvalues (counted with multiplicites) of the matrix $T = T(t)$ (given by (14.2.1)) is independent of t. Then the maximal solution $X_+(t)$ of*

$$X(t) = A(t)^*X(t)A(t) + Q(t) - A(t)^*X(t)B(t)$$
$$\cdot (R(t) + B(t)^*X(t)B(t))^{-1}B(t)^*X(t)A(t) \quad (14.2.2)$$

is an analytic function of $t \in (\alpha, \beta)$. Conversely, if $X_+(t)$ is an analytic function of $t \in (\alpha, \beta)$, then the number of unimodular eigenvalues of $T(t)$ is constant.

The proof of the converse statement, as well as the proof of the direct statement in case $T(t)$ has no unimodular eigenvalues, can be easily adapted from the proof of Theorem 11.3.1. The direct statement in the general case can be proved by the method of proof for the continuous Riccati equation given by Ran and Rodman (1988a). We omit the details.

Finally, observe that by using the methods of Section 12.6, results analogous to Theorems 14.1.1, 14.2.1 and 14.2.2 can be obtained for the more general DARE of (12.6.1).

14.3 Notes

Analyticity of the maximal (stabilizing) solution of the DARE is proved in Delchamps (1980, 1983). The paper by Kamen and Khargonekar (1984) contains analysis of the DARE and its stabilizing solutions in the framework of linear systems whose coefficients are continuous functions of real parameters. This approach was put in the abstract framework of matrices over commutative Banach *-algebras in Green and Kamen (1985). Robust hermitian solutions of the CARE (in the sense of persistence after small perturbations of the coefficients) have been identified in Ran and Rodman (1992b). Basic results on the sensitivity of the positive semidefinite solution of the DARE with respect to perturbations of the coefficients are found in Konstantinov et al. (1993).

15
DISCRETE ALGEBRAIC RICCATI EQUATIONS AND MATRIX PENCILS

The discussion of discrete algebraic Riccati equations of the form

$$X = A^*XA - A^*XB(R + B^*XB)^{-1}B^*XA + Q$$

has been closely connected to the classical eigenvalue problem for the matrix

$$T = \begin{bmatrix} A + BR^{-1}B^*A^{*-1}Q & -BR^{-1}B^*A^{*-1} \\ -A^{*-1}Q & A^{*-1} \end{bmatrix}.$$

The symmetries of T and its spectral properties play a vital part in the examination of solutions of the Riccati equation. In particular, the J-unitary property of T, where $J = \begin{bmatrix} 0 & -iI \\ iI & 0 \end{bmatrix}$, is most significant.

The use of the matrix T makes it necessary, of course, to assume that A is invertible. This assumption can be avoided by relating the DARE to a J-unitary pencil (see Section 2.9) of the form

$$\lambda \begin{bmatrix} I & BR^{-1}B^* \\ 0 & A^* \end{bmatrix} - \begin{bmatrix} A & 0 \\ -Q & I \end{bmatrix}$$

rather than the pencil $\lambda I - T$. When A is not invertible this pencil has an eigenvalue at infinity, a phenomenon which cannot arise in a classical eigenvalue problem. There are other advantages in the matrix pencil technique which will be mentioned in the sequel.

Some of the basic ideas and results are described in this chapter but, in the authors' view, the theory cannot be described as complete, and is still a developing area of research.

15.1 The DARE and a symplectic matrix pencil

Consider a discrete algebraic Riccati equation in the form of equation (12.1.2), which we recast in the following minor way:

$$X = \hat{A}^* X \hat{A} - \hat{A}^* X B (R + B^* X B)^{-1} B^* X \hat{A} + \hat{Q}, \qquad (15.1.1)$$

where

$$\hat{A} = A - B R^{-1} C, \qquad \hat{Q} = Q - C^* R^{-1} C \qquad (15.1.2)$$

$Q^* = Q$, $R^* = R$, and R is invertible. Recall that equation (15.1.1) is equivalent to equation (12.1.1) and also to equation (12.1.3) (with A_1, Q_1, replaced by $\hat{A}$ and $\hat{Q}$, respectively) in the sense that all three equations have the same set of hermitian solutions X.

Let us define $D = B R^{-1} B^*$ and consider the $2n \times 2n$ matrix pencil $\lambda F - G$ where

$$F = \begin{bmatrix} I & D \\ 0 & \hat{A}^* \end{bmatrix}, \qquad G = \begin{bmatrix} \hat{A} & 0 \\ -\hat{Q} & I \end{bmatrix}. \qquad (15.1.3)$$

Observe first that the pencil has a symmetry property with respect to $\hat{H} := \begin{bmatrix} 0 & iI \\ -iI & 0 \end{bmatrix}$. Namely,

$$F \hat{H} F^* = G \hat{H} G^* = \begin{bmatrix} 0 & i\hat{A} \\ -i\hat{A}^* & 0 \end{bmatrix}. \qquad (15.1.4)$$

Thus, in the terminology of Section 2.9 the pencil $\lambda F^* - G^*$ is H-unitary. It will be convenient to call a matrix pencil $\lambda L - M$ H-symplectic if $\lambda L^* - M^*$ is H-unitary. The properties established in Section 2.9 for H-unitary pencils can obviously be extended to H-symplectic pencils. Note that, when $\hat{A}$ is invertible, $\lambda F - G = F(\lambda I - F^{-1} G)$ and $F^{-1} G = T$, the fundamental matrix of (12.2.5) used in the analysis of Chapter 12. Thus, analysis of the pencil $\lambda F - G$ admits some relaxation of the condition that $\hat{A}$ be invertible which was used extensively in Chapter 12. Also, the symmetry property (15.1.4) yields

$$(F^{-1} G) \hat{H} (F^{-1} G)^* = \hat{H};$$

the $\hat{H}$-unitary property of matrix T^* (or, what is equivalent, of matrix T, see Exercise 2.10.3).

An extension of Proposition 12.2.2 can now be formulated in which invertibility of A is replaced by the hypothesis that $\lambda F - G$ be regular.

Theorem 15.1.1. *Let $\lambda F - G$ be a regular pencil (with F, G defined by (15.1.3)). Then an $n \times n$ hermitian matrix X is a solution of (15.1.1) if and only if the graph of X is $\hat{H}$-neutral and deflating for the pair (F, G).*

Proof. Note first that

$$\begin{bmatrix} I \\ X \end{bmatrix}^* \hat{H} \begin{bmatrix} I \\ X \end{bmatrix} = i(X^* - X) \qquad (15.1.5)$$

and so X is hermitian if and only if the graph of X is $\hat{H}$-neutral.

Thus, if X is a hermitian solution of (15.1.1) then the graph of X is $\hat{H}$-neutral and, by Proposition 12.1.1

$$\hat{A}^* X (I + DX)^{-1} \hat{A} = X - \hat{Q}. \tag{15.1.6}$$

Define $K := (I + DX)^{-1} \hat{A}$ so that $(I + DX)K = \hat{A}$ and, from (15.1.6), $\hat{A}^* XK = X - \hat{Q}$. The last two equations can be written in the form

$$\begin{bmatrix} I & D \\ 0 & \hat{A}^* \end{bmatrix} \begin{bmatrix} I \\ X \end{bmatrix} K = \begin{bmatrix} \hat{A} & 0 \\ -\hat{Q} & I \end{bmatrix} \begin{bmatrix} I \\ X \end{bmatrix} \tag{15.1.7}$$

and, by Lemma 1.6.1, the graph of X is deflating for (F, G).

Conversely, given the last statement (and $X^* = X$) then, by Lemma 1.6.1, equation (15.1.7) holds for some K and equation (15.1.6) is easily deduced. Finally, Proposition 12.1.1 shows that X is a hermitian solution of (15.1.1). □

Similarly, we take advantage of the H-symplectic pencil $\lambda F - G$ in proving an analogue of Theorem 7.2.8 which has some important applications.

Theorem 15.1.2. *Assume that $R > 0$, $\hat{Q} \geq 0$, $(\hat{A}, B)$ is d-stabilizable and $(\hat{Q}, \hat{A})$ is d-detectable. Then the pencil $\lambda F - G$ is regular and has no eigenvalues on the unit circle,* **T**.

Proof. Assume that $\lambda \in \mathbf{T}$ and

$$(\lambda F - G) \begin{bmatrix} x_1 \\ x_2 \end{bmatrix} = 0,$$

where $x_1, x_2 \in \mathbf{C}^n$. Then

$$(\lambda I - \hat{A}) x_1 + \lambda D x_2 = 0, \quad \hat{Q} x_1 + (\lambda \hat{A}^* - I) x_2 = 0 \tag{15.1.8}$$

Multiply the first equation on the left by x_2^* to obtain

$$\lambda x_2^* D x_2 = -\lambda x_2^* x_1 + x_2^* \hat{A} x_1$$

and, since $\lambda^{-1} = \bar{\lambda}$,

$$x_2^* D x_2 = -x_2^* x_1 + \bar{\lambda} x_2^* \hat{A} x_1.$$

Furthermore, as $x_2^* D x_2$ is real,

$$x_2^* D x_2 = -x_1^* x_2 + \lambda x_1^* \hat{A}^* x_2.$$

But the second of equations (15.1.8) yields

$$-x_1^* \hat{Q} x_1 = -x_1^* x_2 + \lambda x_1^* \hat{A}^* x_2$$

so that $x_2^* D x_2 = -x_1^* \hat{Q} x_1$. Now $R > 0$ implies that $D = BR^{-1}B^* \geq 0$. But $\hat{Q} \geq 0$ as well, and it follows that $x_2^* D x_2 = x_1^* \hat{Q} x_1 = 0$ and, hence, that $D x_2 = \hat{Q} x_1 = 0$.

The first equation of (15.1.8) now gives

$$x_1^*(\bar{\lambda} I - \hat{A}^*) = 0 \quad \text{and} \quad x_1^* \hat{Q} = 0.$$

Thus, $x_1 \neq 0$ implies rank $[\bar{\lambda} I - \hat{A}^* \; \hat{Q}] < n$ and, as $\bar{\lambda} \in \mathbf{T}$ the detectability of $(\hat{Q}, \hat{A})$ is contradicted (see Theorem 4.5.6). Hence $x_1 = 0$.

Similarly, since $D = BR^{-1}B^*$, Lemma 4.5.3 shows that $(\hat{A}, D)$ is d-stabilizable and it follows from

$$(\lambda \hat{A}^* - I)x_2 = 0, \quad Dx_2 = 0, \quad \lambda \in \mathbf{T}$$

and Theorem 4.5.6 that $x_2 = 0$. Thus, $\lambda \in \mathbf{T}$ implies that λ is not an eigenvalue of $\lambda F - G$. Also, $\det(\lambda F - G) \neq 0$ for $\lambda \in \mathbf{T}$ shows that $\lambda F - G$ is a regular pencil. □

Observe that we did not use the full strength of our hypotheses in the proof of Theorem 15.1.2. Indeed, the only properties used in place of the d-stability of $(\hat{A}, B)$ and the d-detectability of $(\hat{Q}, \hat{A})$ are

$$\text{rank}\,[\lambda I - \hat{A}^* \; \hat{Q}] < n, \quad \text{rank}\, \begin{bmatrix} \lambda I - \hat{A}^* \\ D \end{bmatrix} < n \quad \text{for } \lambda \in \mathbf{T}.$$

15.2 The DARE and a dilated matrix pencil

For numerical purposes the presence of R^{-1} in the terms $\hat{A}$, $\hat{Q}$, D of equation (15.1.1) is unfortunate. As in Section 7.10, a dilation of the appropriate pencil ($\lambda F - G$ in this case) can be used to avoid explicit inversion of R and admit the possibility of treating problems in which R is not invertible. The dilation chosen is strictly equivalent to a pencil of size $2n + m$ of the form

$$\lambda \begin{bmatrix} 0 & * \\ 0 & F \end{bmatrix} - \begin{bmatrix} R & 0 \\ 0 & G \end{bmatrix}$$

where F, G are defined in equations (15.1.3) and $*$ denotes a suitably chosen matrix. Clearly, invertibility of R ensures that the finite spectrum of the H-symplectic pencil $\lambda F - G$ is preserved in this dilation. Explicitly, the reader will readily verify the following statement:

Proposition 15.2.1. *Let*

$$F_e = \begin{bmatrix} I & 0 & 0 \\ 0 & A^* & 0 \\ 0 & -B^* & 0 \end{bmatrix}, \quad G_e = \begin{bmatrix} A & 0 & B \\ -Q & I & -C^* \\ C & 0 & R \end{bmatrix} \qquad (15.2.1)$$

and define nonsingular matrices

$$P_1 = \begin{bmatrix} BR^{-1} & I & 0 \\ -C^*R^{-1} & 0 & I \\ I & 0 & 0 \end{bmatrix}, \quad P_2 = \begin{bmatrix} R^{-1}C & 0 & I \\ I & 0 & 0 \\ 0 & I & 0 \end{bmatrix}, \quad (15.2.2)$$

then

$$P_1 \left(\lambda \begin{bmatrix} 0 & 0 & -B^* \\ 0 & I & D \\ 0 & 0 & \hat{A}^* \end{bmatrix} - \begin{bmatrix} R & 0 & 0 \\ 0 & \hat{A} & 0 \\ 0 & -\hat{Q} & I \end{bmatrix} \right) P_2 = \lambda F_e - G_e \quad (15.2.3)$$

Remark 1. It is clear that $\lambda F_e - G_e$ is regular if and only if R is invertible and $\lambda F - G$ is regular.

Remark 2. The $\hat{H}$-symplectic pencil $\lambda F - G$ has eigenvalues at infinity and at $\lambda = 0$ if and only if $\hat{A}$ is not invertible. The pencil $\lambda F_e - G_e$ defined by (15.2.1) always has spectrum at infinity because F_e is singular. Indeed, the eigenvalue at infinity has algebraic multiplicity not less than m.

Remark 3. The finite spectrum of $\lambda F_e - G_e$ is symmetric with respect to the unit circle because the relation (15.2.3) is a strict equivalence and $\lambda F - G$ is H-symplectic. However, neither the dilation of $\lambda F - G$, nor the pencil $\lambda F_e - G_e$ is H-symplectic for any H. This is because the eigenvalues at infinity and at the origin fail to have the necessary symmetry (see Theorem 2.9.5).

Remark 4. Our discussion is deliberately algebraic. However, the pencil $\lambda F_e - G_e$ arises naturally on applying the Lagrange multiplier technique to the LQR problem of Section 16.5 (see Bender and Laub (1987b), for example).

The next theorem is the basis of some useful algorithms for the numerical solution of discrete algebraic Riccati equations with the form (15.1.1) (cf. Theorem 7.10.1 for the CARE). It is formulated in terms of the "extended graph subspace" idea introduced in Section 7.10.

Theorem 15.2.2. Let the pencil $\lambda F_e - G_e$ (of equations (15.2.1)) be regular. Then equation (15.1.1) has a solution if and only if there is a deflating subspace for (F_e, G_e) of the form

$$\mathcal{S}_e = \operatorname{Im} \begin{bmatrix} X_1 \\ X_2 \\ X_3 \end{bmatrix} \quad (15.2.4)$$

which is an extended graph subspace (i.e. X_1 is invertible), and $R + B^*(X_2 X_1^{-1})B$ is invertible. When this is the case, $X := X_2 X_1^{-1}$ is a solution of (15.1.1) and

$$X_3 X_1^{-1} = -(R + B^* X B)^{-1}(B^* X A + C). \quad (15.2.5)$$

Proof. If (15.1.1) has a solution X put $X_1 = I$, $X_2 = X$, define X_3 by (15.2.5) and define $\mathcal{S}_e$ as in (15.2.4). Then a little manipulation using equations (15.1.1) and (15.2.5) yields:

$$G_e \begin{bmatrix} I \\ X \\ X_3 \end{bmatrix} = \begin{bmatrix} A + BX_3 \\ -Q + X - C^*X_3 \\ C + RX_3 \end{bmatrix}$$

$$= \begin{bmatrix} I \\ A^*X \\ -B^*X \end{bmatrix} (A + BX_3) = F_e \begin{bmatrix} I \\ X \\ X_3 \end{bmatrix} (A + BX_3).$$

Then $\mathcal{S}_e$ obviously has the required form and, by Lemma 1.6.1, $\mathcal{S}_e$ is deflating for the pair (F_e, G_e).

Conversely, let $\mathcal{S}_e$ have the form (15.2.4) with X_1 invertible and define $X = X_2 X_1^{-1}$ and $Z = X_3 X_1^{-1}$. Then

$$\mathcal{S}_e = \text{Im} \begin{bmatrix} I \\ X \\ Z \end{bmatrix}$$

and $(\text{Ker } F_e) \cap \mathcal{S}_e = \{0\}$. Applying Corollary 1.6.3 there is an $n \times n$ matrix K such that

$$F_e \begin{bmatrix} I \\ X \\ Z \end{bmatrix} K = G_e \begin{bmatrix} I \\ X \\ Z \end{bmatrix}$$

and, using the definitions (15.2.1),

$$A + BZ = K \qquad (15.2.6)$$
$$-Q + X - C^*Z = A^*XK \qquad (15.2.7)$$
$$C + RZ = -B^*XK. \qquad (15.2.8)$$

Equations (15.2.6) and (15.2.8) yield (15.2.5) and, as the required inverse is assumed to exist,

$$K = A - B(R + B^*XB)^{-1}(B^*XA + C). \qquad (15.2.9)$$

Then (15.2.7) can be used to show that X satisfies (12.1.1), and hence (15.1.1). $\square$

Note also that the spectrum of K is just that of $\lambda F_e - G_e$ associated with the deflating subspace $\mathcal{S}_e$ (see the remarks following Lemma 1.6.1, for example).

Now some sufficient conditions are formulated for solution matrices X of (15.1.1) to be hermitian.

Corollary 15.2.3. *If the deflating subspace S_e of Theorem 15.2.2 has the property that the spectrum of $(\lambda F_e - G_e)|_{S_e}$ has no pairs of eigenvalues λ_j, λ_k which are symmetric with respect to $\mathbf{T}$ (i.e. $\lambda_j \bar{\lambda}_k \neq 1$), then the corresponding solution X of (15.1.1) is hermitian.*

Proof. Multiply equation (15.2.8) on the left by Z^* and use equations (15.2.6) and (15.2.7) in turn to obtain

$$Z^*C + Z^*RZ = -K^*XK + A^*XK$$
$$= -K^*XK - Q + X - C^*Z.$$

Thus,
$$X - K^*XK = Q + C^*Z + Z^*C + Z^*RZ, \qquad (15.2.10)$$

which we interpret as a symmetric Stein equation for X.

Now the spectra of $(\lambda F_e - G_e)|_{S_e}$ and matrix K coincide, so the result follows from Theorem 5.2.3. □

As might be expected, properties of deflating subspaces for the pair (F_e, G_e) are easily interpreted in terms of the embedded H-symplectic pair (F, G) of (15.1.3). From the strict equivalence of equation (15.2.3) and the definition of a deflating subspace it immediately follows that S_e is deflating for (F_e, G_e) if and only if $P_2 S_e$ is deflating for the pair

$$\left[\begin{array}{c|cc} 0 & 0 & -B^* \\ \hline 0 & I & D \\ 0 & 0 & \hat{A}^* \end{array}\right], \quad \left[\begin{array}{c|cc} R & 0 & 0 \\ \hline 0 & \hat{A} & 0 \\ 0 & -\hat{Q} & I \end{array}\right]$$

where P_2 is defined in (15.2.2). Using this fact (or Lemma 1.6.1) the next Proposition is easily verified.

Proposition 15.2.4. *Let $\lambda F_e - G_e$ be regular. Then (F_e, G_e) has a deflating subspace of the form (15.2.4) which is also an extended graph subspace if and only if $\mathrm{Im}\begin{bmatrix} X_1 \\ X_2 \end{bmatrix}$ is a graph subspace which is deflating for (F, G) and $X_3 = -R^{-1}(B^*X_2 + CX_1)$.*

15.3 Stabilizing solutions

Consider the equation (12.1.1):

$$X = A^*XA + Q - (C + B^*XA)^*(R + B^*XB)^{-1}(C + B^*XA) \qquad (15.3.1)$$

where, as in Proposition 12.1.1, it is assumed that $Q^* = Q$, $R^* = R$ and that R is invertible. The condition that R be positive definite will not be required in this section. By using the matrix pencil methods of this chapter it is also possible to avoid the assumption that A is invertible. A solution of equation (15.3.1) is a matrix X for which $(R + B^*XB)^{-1}$ exists and (15.3.1) is satisfied. A hermitian solution X is said to be *stabilizing* if all eigenvalues of

$$A - B(R + B^*XB)^{-1}(B^*XA + C)$$

are inside the unit circle, $\mathbf{T}$. In the case that $R > 0$ (among other hypotheses) the reader may recognize the feedback associated with the optimal solution of the LQR problem (cf. Theorems 13.1.1 and 16.6.4).

Judging by the experience of Chapters 12–14 it might be expected that, by focussing attention on stabilizing solutions, difficulties associated with eigenvalues of $\lambda F_e - G_e$ on the unit circle can be avoided. This is the case. Furthermore, it turns out that by assuming the existence of a stabilizing solution, in which case (A, B) is certainly a stabilizable pair, other conditions on the coefficients of the DARE can be relaxed.

First recall Proposition 13.5.1 which asserts under general hypotheses (not even requiring invertibility of R) that, when a stabilizing solution for the DARE (15.3.1) exists, it is necessarily unique.

Proposition 15.3.1. *Let $\lambda F_e - G_e$ be regular. Then there are no eigenvalues of $\lambda F_e - G_e$ on the unit circle $\mathbf{T}$ if and only if the spectral deflating subspace $\mathcal{S}_e$ for (F_e, G_e) corresponding to all eigenvalues inside $\mathbf{T}$ has dimension n.*

Furthermore, when such a subspace $\mathcal{S}_e$ has the form (15.2.4) with X_1 and $R + B^*(X_2 X_1^{-1})B$ invertible, then $X = X_2 X_1^{-1}$ is the stabilizing solution of (15.3.1).

Note that different sufficient conditions have already been formulated in Theorem 15.1.2 which ensure that $\lambda F_e - G_e$ has no eigenvalues on $\mathbf{T}$.

Proof. It is apparent from equation (15.1.3) and the strict equivalence of equation (15.2.3) that the finite eigenvalues of $\lambda F_e - G_e$, and their partial multiplicities, are just those of $\lambda F - G$ and, furthermore, $\lambda F - G$ is a regular $\hat{H}$-symplectic pencil (see equations (15.1.4)). It follows from the known symmetries of the spectrum of such a pencil (see Corollary 2.9.4 and Theorem 2.9.5, for example) that $\lambda F - G$ has no eigenvalues on $\mathbf{T}$ if and only if there are n eigenvalues (with algebraic multiplicites) of $\lambda F - G$, and hence $\lambda F_e - G_e$, inside $\mathbf{T}$. Hence the first statement of the proposition follows.

When $\mathcal{S}_e$ has the form specified in the second statement, it follows from Theorem 15.2.2 that $X = X_2 X_1^{-1}$ is a solution of (15.3.1) and from Corollary 15.2.3 that X is hermitian. Finally, the matrix K of equation (15.2.9) has the spectrum of $(\lambda F_e - G_e)|_{\mathcal{S}_e}$ and is therefore d-stable, i.e. the solution X is stabilizing. □

The main theorem of this section is now:

Theorem 15.3.2. *The DARE (15.3.1) has a stabilizing solution if and only if the following conditions are satisfied:*

(a) $\lambda F_e - G_e$ *has no eigenvalues on* **T**;

(b) *the spectral subspace of $\lambda F_e - G_e$ corresponding to its eigenvalues inside* **T** *is an extended graph subspace.*

Notice that condition (a) implies, in particular, that $\det(\lambda F_e - G_e) \not\equiv 0$, i.e. that $\lambda F_e - G_e$ is regular.

Proof. Given conditions (a) and (b) the existence of a solution $X = X_2 X_1^{-1}$ follows from Theorem 15.2.2 where $\mathcal{S}_e$ of equation (15.2.4) is the spectral subspace of $\lambda F_e - G_e$ associated with its eigenvalues inside **T**, provided it can be verified that $R + B^* X B$ is invertible. Furthermore, by Proposition 15.3.1 this solution X will be hermitian and stabilizing. The necessary verification involves some computation.

Observe first that, with $Z := X_3 X_1^{-1}$, equations (15.2.6)–(15.2.8) hold. Equations (15.2.6) and (15.2.7) yield

$$A^* X A - X + (A^* X B + C^*) Z + Q = 0,$$

and (15.2.6) and (15.2.8) give

$$(R + B^* X B) Z = -B^* X A - C. \tag{15.3.2}$$

Combine these two equations to obtain:

$$A^* X A - X - Z^* (R + B^* X B) Z + Q = 0 \tag{15.3.3}$$

Define the rational matrix function

$$W(\lambda) := [B^*(\lambda^{-1} I - A^*)\ \ I] \begin{bmatrix} Q & C^* \\ C & R \end{bmatrix} \begin{bmatrix} (\lambda I - A)^{-1} B \\ I \end{bmatrix}. \tag{15.3.4}$$

Then a computation with equations (15.3.2) and (15.3.3) serves to verify that

$$\{I - B^*(\lambda^{-1} I - A^*)^{-1} Z\}(R + B^* X B)\{I - Z(\lambda I - A)^{-1} B\} = W(\lambda). \tag{15.3.5}$$

But another computation, with determinants and Schur complements this time (see p. 46 of Lancaster and Tismenetsky (1985), for example), and using the definition (15.2.1) shows that

$$\det(\lambda F_e - G_e) = (-1)^{n+m} \lambda^n \det(\lambda I - A) \det(\lambda^{-1} I - A^*) \det W(\lambda). \tag{15.3.6}$$

Comparing this with (15.3.5) it is clear that the regularity of $\lambda F_e - G_e$ implies the invertibility of $R + B^* X B$.

Conversely, let X be a stabilizing solution of equation (15.3.1). Then $R+B^*XB$ is invertible and, as in the proof of Theorem 15.2.2, equations (15.2.6)–(15.2.8) hold. Consequently, equation (15.3.5) also holds. Then it follows from (15.3.6) that $\lambda F_e - G_e$ is regular. Since X is stabilizing the extended graph subspace

$$\mathcal{S} = \operatorname{Im} \begin{bmatrix} I \\ X \\ Z \end{bmatrix}$$

is associated with eigenvalues of $\lambda F_e - G_e$ inside **T** and has dimension n. Conditions (a) and (b) now follow from Proposition 15.3.1. □

15.4 The real case

The discrete algebraic Riccati equation (15.1.1) with real coefficients takes the form
$$X = \hat{A}^T X \hat{A} - \hat{A}^T XB(R + B^T XB)^{-1} B^T X \hat{A} + \hat{Q} \qquad (15.4.1)$$
where
$$\hat{A} = A - BR^{-1}C, \qquad \hat{Q} = Q - C^T R^{-1} C,$$
A, B, C, Q, R are real matrices, $Q^T = Q$ is $n \times n$, $R^T = R$ is $m \times m$ and R is invertible. The associated real pencil $\lambda F - G$ is defined by setting $D = BR^{-1}B^T$ and
$$F = \begin{bmatrix} I & D \\ 0 & \hat{A}^T \end{bmatrix}, \qquad G = \begin{bmatrix} \hat{A} & 0 \\ -\hat{Q} & I \end{bmatrix}. \qquad (15.4.2)$$

If we also define the real, invertible, skew-symmetric matrix $\hat{H}_r = \begin{bmatrix} 0 & I \\ -I & 0 \end{bmatrix}$, then equation (15.1.4) takes the form

$$F\hat{H}_r F^T = G\hat{H}_r G^T = \begin{bmatrix} 0 & \hat{A} \\ -\hat{A}^T & 0 \end{bmatrix}. \qquad (15.4.3)$$

Thus, in the terminology of Section 3.5, $\lambda F^T - G^T$ is an $\hat{H}_r$-orthogonal pencil. It follows that the spectrum of $\lambda F - G$ is symmetric with respect to both the unit circle and the real axis, and more detailed properties of the spectrum and eigenvectors can be deduced from Theorem 3.5.2.

Theorem 15.1.1 is readily modified to yield:

Theorem 15.4.1. *Let the pencil $\lambda F - G$ defined by (15.4.2) be real and regular. Then an $n \times n$ real symmetric matrix X is a solution of equation (15.4.1) if and only if the graph of X is $\hat{H}_r$-neutral and deflating for the pair (F, G).*

Then Theorem 15.1.2 holds verbatim.

The dilated pencil $\lambda F_e - G_e$ introduced in Section 15.2 is now defined by

$$F_e = \begin{bmatrix} I & 0 & 0 \\ 0 & A^T & 0 \\ 0 & -B^T & 0 \end{bmatrix}, \quad G_e = \begin{bmatrix} A & 0 & B \\ -Q & I & -C^T \\ C & 0 & R \end{bmatrix} \quad (15.4.4)$$

and Proposition 15.2.1 holds without modification.

Theorem 15.2.2 must be modified to admit only real matrices X_1, X_2, X_3 in equation (15.2.4). Thus:

Theorem 15.4.2. *Let the pencil $\lambda F_e - G_e$ defined by (15.4.4) be real and regular. Then equation (15.4.1) has a real solution if and only if there is a deflating subspace for (F_e, G_e) of the form*

$$\mathcal{S}_e = \text{Im} \begin{bmatrix} X_1 \\ X_2 \\ X_3 \end{bmatrix} \quad (15.4.5)$$

where $X_1, X_2 \in \mathbb{R}^{n \times n}$, $X_3 \in \mathbb{R}^{m \times n}$, $\mathcal{S}_e$ is an extended graph subspace, and $R + B^T(X_2 X_1^{-1})B$ is invertible. When this is the case, $X = X_2 X_1^{-1}$ is a solution of (15.4.1) and

$$X_3 X_1^{-1} = -(R + B^T X B)^{-1}(B^T X A + C). \quad (15.4.6)$$

Corollary 15.4.3. *If the deflating subspace $\mathcal{S}_e$ of Theorem 15.4.2 has the property that the spectrum of $(\lambda F_e - G_e)|\mathcal{S}_e$ has no pairs of eigenvalues λ_j, λ_k which are symmetric with respect to* **T** *(i.e. $\lambda_j \bar{\lambda}_k \neq 1$), then the corresponding solution X of the DARE (15.4.1) is real and symmetric.*

Proof. The existence of a hermitian solution X here is established in Corollary 15.2.3. The fact that X must also be real follows because the symmetric Stein equation (15.2.10) satisfied by X has real coefficients. □

Proposition 15.2.4 requires no modification. Theorem 15.3.2 also holds in the real case if the reference to (15.2.4) is replaced by (15.4.5). Finally, Theorem 15.3.2 takes the form:

Theorem 15.4.4. *Equation (15.4.1) has a real symmetric stabilizing solution if and only if the following conditions are satisfied:*

(a) *$\lambda F_e - G_e$ has no eigenvalues on* **T**;

(b) *the spectral subspace of $\lambda F_e - G_e$ corresponding to its eigenvalues inside* **T** *is an extended graph subspace.*

Proof. The proof is just that of Theorem 15.3.2; it is only necessary to confirm that the spectral subspace of statement (b) can be expressed in the form (15.4.5) with real matrices X_1, X_2, X_3. This is the case because the spectrum of $\lambda F_e - G_e$ (and particularly that inside **T**) is symmetric with respect to the real line, and the matrices X_1, X_2, X_3 can be constructed from the real spectral projector associated with the interior of **T** (see Example 1.4.2). □

15.5 Exercises

15.5.1. Consider the equation (15.3.1) with

$$A = \begin{bmatrix} 0 & 1 \\ 0 & 0.5 \end{bmatrix}, \quad B = \begin{bmatrix} 0 \\ 1 \end{bmatrix}, \quad Q = \begin{bmatrix} 0 & 3 \\ 3 & -0.86 \end{bmatrix}$$

$$C = [0.8 \ -4], \quad R = -1.$$

(a) Compute the function (15.3.4).

(b) Verify that $\lambda F_e - G_e$ has no eigenvalues on the unit circle.

(c) Show that $X = \begin{bmatrix} -0.64 & 3 \\ 3 & 2 \end{bmatrix}$ is the stabilizing solution.

15.5.2. Consider equation (15.1.1) with $C = 0$,

$$A = \begin{bmatrix} 2 & -1 \\ 1 & 0 \end{bmatrix}, \quad G = \begin{bmatrix} 1 \\ 0 \end{bmatrix}, \quad Q = \begin{bmatrix} 0 & 0 \\ 0 & 1 \end{bmatrix}, \quad R = 0.$$

(a) Construct the pencil $\lambda F_e - G_e$ and find its deflating subspaces.

(b) Using Corollary 15.2.3, find all hermitian solutions of (15.1.1).

(c) Using Theorem 15.3.2, find the stabilizing solution of (15.1.1).

15.5.3. Verify Proposition 15.2.4.

15.5.4. [A more general pencil.] The quadratic regulator problem in discrete singular linear systems leads to the pencil $\lambda \hat{F}_e - \hat{G}_e$, where

$$\hat{F}_e = \begin{bmatrix} E & 0 & 0 \\ 0 & A^* & 0 \\ 0 & -B^* & 0 \end{bmatrix}, \quad \hat{G}_e = \begin{bmatrix} A & 0 & B \\ -Q & E^* & -C^* \\ C & 0 & R \end{bmatrix}.$$

((15.2.1) is a particular case when $E = I$). Prove that the finite spectrum (including the partial multiplicities) of $\lambda \hat{F}_e - \hat{G}_e$ is symmetric relative to the unit circle.

15.6 Notes

The use of matrix pencils for analysis of the DARE seems to originate with Pappas *et al.* (1980). More recently, this approach has been investigated by Wimmer (1991). The extension of the pencil $\lambda F - G$ of Section 15.1 to the dilated pencil $\lambda F_e - G_e$ follows the pattern set by Van Dooren (1981) in the context of the CARE. For discussion of numerical methods, extensions to singular Riccati equations, and other connections see Mehrmann (1991). The analysis of Section 15.3 is based on the paper of Ionescu and Weiss (1992). The crux of the proof of Theorem 15.3.2 is the use made of the function $W(\lambda)$ defined in equation (15.3.5). This function is frequently known as the Popov function (see Popov (1973)). Exercise 15.5.1 is from Molinari (1975). Exercise 15.5.2 is from Van Dooren (1981) (and was suggested by A. Amami–Naeini). For more information on the singular systems of Exercise 15.5.4 see Mehrmann (1991), for example.

Part IV

APPLICATIONS AND CONNECTIONS

In this part several topics are developed in which solutions of the CARE and DARE play an important, even central, part. There is a wide range of choice of suitable topics and, for reasons of space, and our own limitations, only a few topics have been selected for presentation here. Our selection was guided by widespread recognition of the importance and usefulness of topics on the one hand, and by relative accessibility of the material on the other hand. Inevitably, the choice of topics was also influenced by our personal taste.

We did not strive for depth of exposition in this part. Rather, we have opted to expose the guiding ideas, and present the basic, easily accessible results. Suggestions for further in-depth reading are found in the notes at the end of chapters.

Part IV

APPLICATIONS AND CONNECTIONS

16
LINEAR-QUADRATIC REGULATOR PROBLEMS

The "linear-quadratic regulator" (LQR) problem of optimal control has probably provided the greatest single stimulus for investigation of matrix Riccati equations in differential, difference, and algebraic forms. In this chapter the continuous and discrete LQR problems are to be outlined and then the solutions of these problems using Riccati equations are developed. Consistent with the spirit of this book attention is confined to time-invariant (i.e., constant coefficient) problems, even though significant parts of the theory are readily extended to linear time-varying problems (see, for example, Brockett (1970), Kalman, Falb and Arbib (1969), or Russell (1979)).

In the first four sections the LQR problem for time-invariant, continuous systems is described. We do not use the line of argument initiated by Kalman which involves the technique of dynamic programming. In view of our limited objectives a more direct line of argument seems appropriate even though it is rather delicate. This seems to originate with Brockett (1970), and has been refined by Hijab (1987). Here, we adapt the Brockett–Hijab development to our needs. This line of argument also has the great advantage that it is readily adapted to the *discrete* LQR problem. This is done in Sections 16.5 and 16.6.

16.1 The optimization problem

Consider a primitive time-invariant linear system of the form

$$\dot{x}(t) = Ax(t) + Bu(t), \quad x(0) = x^0 \qquad (16.1.1)$$

in which $A \in \mathbb{C}^{n \times n}$ and $B \in \mathbb{C}^{n \times m}$. The vector functions $u(t)$ and $x(t)$ are known as the *control* (or *input*) and the *state* vectors, respectively. Observe that $x(t)$ is uniquely defined by a given integrable control function $u(t)$ and an initial vector x^0 in the form

$$x(t) = e^{At}x^0 + \int_0^t e^{A(t-s)}Bu(s)\,ds. \qquad (16.1.2)$$

We write $x(t) = x^u(t; x^0)$ when these dependences are to be emphasized.

It will be assumed throughout that control functions $u(t)$ are defined for $0 \le t < \infty$ and have the property that $u \in L^2_m(0,T)$ for all $T > 0$. Let $\mathcal{U}$ denote this class of functions. Thus $\mathcal{U} = \bigcap_{T>0} L^2_m(0,T)$. In particular, functions in $L^2_m(0,\infty)$ are admissible controls. The space $L^2_k(a,b)$ is, as usual, the Hilbert space of vector functions y, z on (a,b) with values in $\mathbb{C}^k$ and with inner product

$$(y,z) = \int_a^b z(t)^* y(t)\, dt.$$

To define a *cost* associated with controls and initial vectors suppose first that a positive semidefinite matrix $\hat{R}$ of size $n+m$ is given with

$$\hat{R} = \begin{bmatrix} Q & S \\ S^* & R \end{bmatrix} \ge 0 \tag{16.1.3}$$

where Q is $n \times n$ and R is $m \times m$. It is assumed throughout that $R > 0$ and, necessarily, $Q \ge 0$. We say that $\hat{R}$ is *nondegenerate* if rank $\hat{R}$ = rank Q+ rank R. Now define the quadratic *cost functional* by

$$J^u(x^0) = \int_0^\infty [x(t)^* \ u(t)^*] \hat{R} \begin{bmatrix} x(t) \\ u(t) \end{bmatrix} dt = \int_0^\infty \left\| \begin{bmatrix} x(t) \\ u(t) \end{bmatrix} \right\|^2_{\hat{R}} dt. \tag{16.1.4}$$

Note that $x(t)$ is defined by u and x^0 as in equation (16.1.2). Obviously, $0 \le J^u(x^0) \le \infty$. The *optimal cost* at x^0 is defined by

$$\hat{J}(x^0) = \inf_u J^u(x^0), \tag{16.1.5}$$

and an *optimal control* is then a control function u for which this infimum is attained.

It may seem more natural to formulate the cost functional in terms of norms of the control and an output vector $y(t) = Cx(t) + Du(t)$; say $\|u\|^2_{R_1} + \|y\|^2_{R_2}$. But if we write

$$\hat{R} = \begin{bmatrix} 0 & 0 \\ 0 & R_1 \end{bmatrix} + \begin{bmatrix} C^* \\ D^* \end{bmatrix} R_2 [C \ D]$$

then $\|u\|^2_{R_1} + \|y\|^2_{R_2} = \left\| \begin{bmatrix} x \\ u \end{bmatrix} \right\|^2_{\hat{R}}$, as in (16.1.4).

The objective of the LQR problem is to determine conditions on the system (16.1.1) and $\hat{R}$ which will ensure that a unique optimal control exists for any choice of x^0, and to specify such a control. It will be seen that, under suitable conditions, an optimal control can be specified by a *feedback* mechanism. That is, by assuming the control to be coupled with the state by $u(t) = -Fx(t)$ for some $F \in \mathbb{C}^{m \times n}$. The unique solution of the resulting initial value problem

$$\dot{x}(t) = (A - BF)x(t), \quad x(0) = x^0$$

then determines $u(t)$ in the form

$$u(t) = -Fe^{(A-BF)t}x^0, \quad t \geq 0. \tag{16.1.6}$$

It will be shown that, under suitable conditions, the optimal control is determined by a feedback matrix of the form $F = -B^*X$ where X is an extremal solution of a related algebraic Riccati equation.

16.2 Properties of the cost functional

Looking ahead, a critical step in our analysis is transformation of the cost at finite time suggested by (16.1.4) in terms of the $\hat{R}$-seminorm to a form involving the R-norm (recalling that $R > 0$). This computation will appear as Lemma 16.3.1, but some preliminary investigations are required to tell us when the optimal cost exists and is nontrivial.

A classical result asserts that bounded sequences in L_m^2 have a weakly convergent subsequence. The first task is to extend this result to suitably bounded sequences of controls from $\mathcal{U}$.

Lemma 16.2.1. *Assume that, for each $T > 0$, the sequence $\{u_k(t)\}_{k=1}^\infty \subseteq \mathcal{U}$ is bounded in $L_m^2(0,T)$. Then there is a subsequence $\{v_k\}$ of $\{u_k\}$ and a $u \in \mathcal{U}$ such that*

$$\int_0^T v(t)^*(v_k(t) - u(t))\, dt \to 0 \tag{16.2.1}$$

as $k \to \infty$ for all $T > 0$ and all $v \in \mathcal{U}$.

Proof. First consider a fixed $T > 0$. Since $\{u_k\}$ is a bounded sequence in $L_m^2(0,T)$ there is a weakly convergent subsequence $\{v_{k,T}\} \subseteq \{u_k\}$. That is, there is a $u_T \in L_m^2(0,T)$ such that

$$\int_0^T v(t)^*(v_{k,T}(t) - u_T(t))\, dt \to 0$$

as $k \to \infty$, for all $v \in L_m^2(0,T)$ and hence all $v \in \mathcal{U}$. More briefly, $v_{k,T} \xrightarrow{w} u_T$ in $L_m^2(0,T)$.

If $T_0 < T$ it follows immediately from (16.2.1) that also $v_{k,T} \xrightarrow{w} u_T$ in $L_m^2(0,T_0)$. (Just consider those functions $v(t)$ which are identically zero on (T_0, T).) Now consider the restriction of $\{u_k(t)\}$ to $(T, T+1)$ and construct a subsequence $\{\hat{v}_k\}$ of $\{v_{k,T}\}$ such that $\hat{v}_k \xrightarrow{w} \hat{u}$ in $L_m^2(T, T+1)$. In the obvious way, $v_{k,T}$ and $\hat{v}_k$ define a function $v_{k,T+1} \in L^2_m(0, T+1)$ and u_T, $\hat{u}$ define a function u_{T+1} in $L^2_m(0, T+1)$ such that $v_{k,T+1} \xrightarrow{w} u_{T+1}$ in $L^2_m(0, T+1)$. Now extend a subsequence of $\{v_{k,T+1}\}$ to $\{v_{k,T+2}\}$ and u_{T+1} to u_{T+2} in the same way. Continuing this construction a subsequence $\{v_k\}$ of $\{u_k\}$ and a $u \in \mathcal{U}$ are obtained for which (16.2.1) holds for all $T > 0$. □

Now it is to be shown that if there is a sequence of initial values $\{x_k^0\}$ such that $x_k^0 \to x^0$ and (as in the lemma above) there is a sequence of controls v_k such that $v_k \xrightarrow{w} u$ in $\mathcal{U}$, then the corresponding states $x^{v_k}(t; x_k^0)$ are weakly convergent and $x^{v_k}(t, x_k^0) \xrightarrow{w} x^u(t; x^0)$.

Lemma 16.2.2. *Assume that, for each $T > 0$, the sequence $\{u_k(t)\} \subseteq \mathcal{U}$ is bounded in $L^2_m(0, T)$ and that $\{x_k^0\}$ is a sequence in $\mathbb{C}^n$ for which $x_k^0 \to x^0$ as $k \to \infty$. Then there is a subsequence $\{v_k\}$ of $\{u_k\}$ and a $u \in \mathcal{U}$ such that (16.2.1) holds and also*

$$\int_0^T w(t)^* (x^{v_k}(t; x_k^0) - x^u(t; x^0)) \, dt \to 0 \qquad (16.2.2)$$

as $k \to \infty$ for all $T > 0$ and all $w \in L^2_n(0, T)$.

Proof. Define the subsequence $\{v_k\}$ and $u \in \mathcal{U}$ as in the preceding lemma so that (16.2.1) holds. For any $w \in L^2_n(0, T)$ write

$$z(t) = \int_t^T B^* e^{(s-t)A^*} w(s) \, ds \in L^2_m(0, T).$$

Using (16.1.2) we have

$$x^{v_k}(t; x_k^0) - x^u(t; x^0) = e^{At} \int_0^t e^{-sA} B(v_k(s) - u(s)) \, ds + e^{At}(x_k^0 - x^0).$$

Thus, for any $w \in L^2_n(0, T)$

$$\int_0^T w(t)^* (x^{v_k}(t; x_k^0) - x^u(t; x^0)) \, dt$$

$$= \int_0^T w(t)^* e^{At} \int_0^t e^{-sA} B(v_k(s) - u(s)) \, ds + \int_0^T w(t)^* e^{At} \, dt (x_k^0 - x^0)$$

$$= \int_0^T \left(\int_s^T w(t)^* e^{At} \, dt \right) e^{-sA} B(v_k(s) - u(s)) \, ds$$

$$+ \int_0^T w(t)^* e^{At} \, dt (x_k^0 - x^0)$$

$$= \int_0^T z(s)^* (v_k(s) - u(s)) \, ds + \int_0^T w(t)^* e^{At} \, dt (x_k^0 - x^0).$$

Now from $x_k^0 \to x^0$ and (16.2.1) we obtain (16.2.2). □

PROPERTIES OF THE COST FUNCTIONAL

These lemmas are now used to obtain an important continuity property for the optimal cost $\hat{J}$ of (16.1.5). The more difficult part of the argument is contained in the next lemma.

Lemma 16.2.3. *Let $\{x_k^0\}$ be a sequence in $\mathbb{C}^n$ with $x_k^0 \to x^0$ as $k \to \infty$ and let $\{u_k\}$ be a sequence in $\mathcal{U}$ for which*

$$\lim_{k \to \infty} J^{u_k}(x_k^0) \tag{16.2.3}$$

exists. Then there is a $u \in \mathcal{U}$ such that

$$J^u(x^0) \leq \lim_{k \to \infty} J^{u_k}(x_k^0). \tag{16.2.4}$$

Proof. Define the cost at time T by

$$J_T^u(x^0) = \int_0^T \left\| \begin{bmatrix} x(t) \\ u(t) \end{bmatrix} \right\|_{\hat{R}}^2 dt$$

(see equation (16.1.4)). Given the existence of the limit (16.2.3) it may be assumed that $\{u_k\}$ is bounded in $L_m^2(0,T)$ for each $T > 0$ and define a subsequence $\{v_k\}$ of $\{u_k\}$ and $u \in \mathcal{U}$ as in Lemma 16.2.1. Then

$$J_T^{v_k - u}(x_k^0 - x^0) = \int_0^T \left\| \begin{bmatrix} x^{v_k}(t; x_k^0) - x^u(t; x^0) \\ v_k(t) - u(t) \end{bmatrix} \right\|_{\hat{R}}^2 dt. \tag{16.2.5}$$

Defining

$$[w(t)^* \; v(t)^*] = [x^u(t; x^0)^* \; u(t)^*]\hat{R}$$

and

$$E_k^T(v, w) = \int_0^T [w(t)^* \; v(t)^*] \begin{bmatrix} x^{v_k}(t; x_k^0) - x^u(t; x^0) \\ v_k(t) - u(t) \end{bmatrix} dt$$

a little calculation with (16.2.5) shows that

$$0 \leq J_T^{v_k - u}(x_k^0 - x^0) = J_T^{v_k}(x_k^0) - J_T^u(x^0) - 2 \operatorname{Re} E_k^T(v, w).$$

Lemmas 16.2.1 and 16.2.2 show that $E_k^T(v, w) \to 0$ as $k \to \infty$. Hence, if we first take the limit as $k \to \infty$ and then as $T \to \infty$ (16.2.4) is obtained. □

Theorem 16.2.4. *For each $x^0 \in \mathbb{C}^n$ there is a $\hat{u} \in \mathcal{U}$ at which the optimal cost is attained: $\hat{J}(x^0) = J^{\hat{u}}(x^0)$. Furthermore, for any sequence $\{x_k^0\}$ in $\mathbb{C}^n$ with $x_k^0 \to x^0$*

$$\hat{J}(x^0) \leq \lim_{k \to \infty} \hat{J}(x_k^0) \tag{16.2.6}$$

whenever the limit exists (i.e., $\hat{J}$ is lower semicontinuous).

Proof. From the definition of $\hat{J}$, there is a sequence of controls $\{u_k\}$ such that

$$J^{u_k}(x^0) \leq \hat{J}(x^0) + k^{-1}, \quad k = 1, 2, \ldots.$$

Thus, $\lim_{k \to \infty} J^{u_k}(x^0)$ exists and Lemma 16.2.3 aplies to give the first statement.
  Now use this statement to obtain $\{u_k\}$ for which $\hat{J}(x_k^0) = J^{u_k}(x_k^0)$, $k = 1, 2, \ldots$, and (16.2.6) follows from (16.2.4). □

Now let us consider some conditions which will ensure that the optimal costs are nontrivial, in the sense that they are finite and nonzero. Not surprisingly, the finiteness is ensured by a stabilizable hypothesis on the system (16.1.1). The nonzero property is ensured by a suitable observability condition.

Proposition 16.2.5. *If the pair (A, B) is stabilizable then $\hat{J}(x^0) < \infty$ for any initial vector $x^0 \in \mathbb{C}^n$.*

Proof. By definition of a stabilizable pair, there is a matrix F such that $A - BF$ is stable. Consider the feedback control $u(t) = -Fx(t)$. Thus, $x(t)$ satisfies $\dot{x} = (A - BF)x$ and $x(0) = x^0$. So

$$\begin{bmatrix} x(t) \\ u(t) \end{bmatrix} = \begin{bmatrix} I \\ -F \end{bmatrix} x(t) = \begin{bmatrix} I \\ -F \end{bmatrix} e^{(A-BF)t} x^0.$$

Since $\|e^{(A-BF)t} x^0\| \leq M e^{-kt}$ for some positive M and k it follows that

$$J^u(x^0) = \int_0^\infty \left\| \begin{bmatrix} x(t) \\ u(t) \end{bmatrix} \right\|_{\hat{R}}^2 dt < \infty$$

and so an optimal cost is also finite. □

Now a simple lemma concerning the cost functional (see (16.1.3) and (16.1.4)):

Lemma 16.2.6. *If $\hat{R} \geq 0$ and $R > 0$ then $\mathrm{Im}\, S \subseteq \mathrm{Im}\, Q$.*

Proof. If the result is false then there is a nonzero $x \in \mathrm{Im}\, S \cap (\mathrm{Im}\, Q)^\perp = \mathrm{Im}\, S \cap \mathrm{Ker}\, Q$. Then

$$0 \leq [x^* \quad -x^* S R^{-1}] \begin{bmatrix} Q & S \\ S^* & R \end{bmatrix} \begin{bmatrix} x \\ -R^{-1} S^* x \end{bmatrix} = -x^* S R^{-1} S^* x.$$

As $x \in \mathrm{Im}\, S$ implies $x \notin \mathrm{Ker}\, S^*$ we have $x^* S R^{-1} S^* x < 0$ a contradiction with $R > 0$. □

PROPERTIES OF THE COST FUNCTIONAL

Lemma 16.2.7. *If $\hat{R}$ is nondegenerate and (Q, A) is an observable pair, then*

$$(Q - SR^{-1}S^*, A - BR^{-1}S^*)$$

is also an observable pair.

Note first of all that the nondegeneracy of $\hat{R}$ is essential, for if

$$\hat{R} = \begin{bmatrix} 0 & 0 & \vdots & 0 \\ 0 & 1 & \vdots & 1 \\ \hdotsfor{4} \\ 0 & 1 & \vdots & 1 \end{bmatrix},$$

then $\operatorname{rank} \hat{R} < \operatorname{rank} Q + \operatorname{rank} R$, and $Q - SR^{-1}S^* = 0$.

Proof. By the preceding lemma we have $\operatorname{Im} S \subseteq \operatorname{Im} Q$ and hence

$$\operatorname{Im}(Q - SR^{-1}S^*) \subseteq \operatorname{Im} Q + \operatorname{Im} S = \operatorname{Im} Q \tag{16.2.7}$$

But $\begin{bmatrix} Q & S \\ S^* & R \end{bmatrix}$ and $\begin{bmatrix} Q - SR^{-1}S^* & 0 \\ 0 & R \end{bmatrix}$ are congruent and, since $\hat{R}$ is nondegenerate, Q and $Q - SR^{-1}S^*$ have the same rank. Thus equality holds in (16.2.7) and

$$\operatorname{Ker}(Q - SR^{-1}S^*) = \operatorname{Ker} Q$$

as well. For any $\lambda \in \mathbb{C}$ let

$$x \in \operatorname{Ker}(\lambda I - (A - BR^{-1}S^*)) \cap \operatorname{Ker}(Q - SR^{-1}S^*) \tag{16.2.8}$$

Then $x \in \operatorname{Ker} Q$. Also, $\operatorname{Im} S \subseteq \operatorname{Im} Q$ implies $\operatorname{Ker} S^* \supseteq \operatorname{Ker} Q$, so $S^*x = 0$. Hence $x \in \operatorname{Ker}(\lambda I - A)$ also. Thus,

$$x \in \operatorname{Ker}(\lambda I - A) \cap \operatorname{Ker} Q.$$

As (Q, A) is observable it follows from Corollary 4.3.4 that $x = 0$ and the result follows from (16.2.8) and the same corollary. $\square$

Proposition 16.2.8. *If $\hat{R}$ is nondegenerate and (Q, A) is observable then $\hat{J}(x^0) > 0$ whenever $x^0 \neq 0$.*

Proof. We first transform the problem to one in which the matrix $\hat{R}$ becomes block-diagonal. Thus, if we define

$$v(t) = R^{-1}S^* x(t) + u(t), \qquad (16.2.9)$$

then, since

$$\begin{bmatrix} I & -SR^{-1} \\ 0 & I \end{bmatrix} \begin{bmatrix} Q & S \\ S^* & R \end{bmatrix} \begin{bmatrix} I & 0 \\ -R^{-1}S^* & I \end{bmatrix} = \begin{bmatrix} Q - SR^{-1}S^* & 0 \\ 0 & R \end{bmatrix}$$

we have

$$J^u(x^0) = \int_0^\infty (x(t)^* \hat{Q} x(t) + v(t)^* R v(t))\, dt, \qquad (16.2.10)$$

where $\hat{Q} = Q - SR^{-1}S^*$, and also

$$\dot{x}(t) = \hat{A} x(t) + B v(t), \quad x(0) = x^0, \qquad (16.2.11)$$

where $\hat{A} = A - BR^{-1}S^*$.

From Theorem 16.2.4, $\hat{J}(x^0) = 0$ for some $x^0 \neq 0$ implies $J^u(x^0) = 0$ for some $u \in \mathcal{U}$. But then (using the transformed system) $R > 0$ implies $v \equiv 0$ and, since $\hat{Q} \geq 0$ we have $\hat{Q} x(t) \equiv 0$. Now we have $x(t) = e^{\hat{A}t} x^0$ and so

$$\hat{Q} e^{\hat{A}t} x^0 \equiv 0.$$

Differentiating repeatedly with respect to t and then setting $t = 0$ shows that $(Q - SR^{-1}S^*, A - BR^{-1}S^*)$ is not an observable pair. But this contradicts the result of the preceding lemma. Hence $\hat{J}(x^0) > 0$ for all $x^0 \neq 0$. $\square$

The next result gives useful sufficient conditions for a control to be "stabilizing"; particularly when combined with Proposition 16.2.5.

Proposition 16.2.9. *Let $\hat{R}$ be nondegenerate and the pair (Q, A) be observable. Then $J^u(x^0) < \infty$ implies that $x^u(t; x^0) \to 0$ as $t \to \infty$.*

Proof. Suppose, without loss of generality, that $\|x^0\| = 1$. The hypothesis $J^u(x^0) < \infty$ implies $\hat{J}(x^0) < \infty$, and the last proposition gives $\hat{J}(x^0) > 0$. Furthermore, by Theorem 16.2.4 and Proposition 16.2.8 there is a finite $m > 0$ for which $\min_{\|x\|=1} \hat{J}(x) = m$. Consequently, for any sequence $\{x_k\} \subseteq \mathbb{C}^n$ we have

$$\hat{J}(x_k) = \|x_k\|^2 \hat{J}\left(\frac{1}{\|x_k\|} x_k\right) \geq \|x_k\|^2 m \qquad (16.2.12)$$

for any k. In particular, if $\|x_k\| \to \infty$ as $k \to \infty$ then $\hat{J}(x_k) \to \infty$ as $k \to \infty$.

Now let $v(t) = u(t+T)$, $t \geq 0$, and it is easily verified that $x^v(t; x^u(T)) = x^u(t+T; x^0)$. Hence (abbreviating $x^u(T; x^0)$ to $x^u(T)$),

$$\hat{J}(x^u(T)) \leq J^v(x^u(T)) = \int_0^\infty \left\| \begin{bmatrix} x^v(t; x^u(T)) \\ v(t) \end{bmatrix} \right\|_{\hat{R}}^2 dt$$

$$= \int_0^\infty \left\| \begin{bmatrix} x^u(t+T; x^0) \\ u(t+T) \end{bmatrix} \right\|_{\hat{R}}^2 dt = \int_T^\infty \left\| \begin{bmatrix} x^u(t; x^0) \\ u(t) \end{bmatrix} \right\|_{\hat{R}}^2 dt.$$

Thus, as $T \to \infty$,

$$\hat{J}(x^u(T)) \to 0. \tag{16.2.13}$$

Using this property and applying (16.2.12) to $\{x^u(T) | T \geq 0\}$ this set is seen to be bounded. Then there is a sequence $\{T_k\}$ such that $T_k \to \infty$ and $\{x^u(T_k)\}$ is convergent; say $x^u(T_k) \to x$ as $k \to \infty$.

The lower semicontinuity of $\hat{J}$ (Theorem 16.2.4) implies that $\hat{J}(x) \leq \hat{J}(x^u(T_k))$ and (16.2.13) gives $\hat{J}(x) = 0$. But this contradicts Proposition 16.2.8 unless $x = 0$. This completes the proof. □

16.3 Finding an optimal control

We now make the connection between an optimal control for the LQR problem and a solution of a CARE. The matrices appearing in (16.1.1) and (16.1.3) determine the coefficients of the Riccati equation:

$$XBR^{-1}B^*X - X(A - BR^{-1}S^*) - (A - BR^{-1}S^*)^*X - (Q - SR^{-1}S^*) = 0. \tag{16.3.1}$$

It will be convenient to work with the equivalent problem of (16.2.9), (16.2.10), (16.2.11). There, we defined $\hat{A} = A - BR^{-1}S^*$, $\hat{Q} = Q - SR^{-1}S^*$ and so our equation can also be written

$$XBR^{-1}B^*X - X\hat{A} - \hat{A}^*X - \hat{Q} = 0. \tag{16.3.2}$$

The following computation for the cost at a finite time is crucial. It is a remarkable fact that the cost at time T with control $u \in \mathcal{U}$, and initial state x^0 can be expressed in terms of any hermitian solution X of (16.3.1). The cost at time T, $J_T^u(x^0)$, is defined as in the proof of Lemma 16.2.3.

Lemma 16.3.1. *If X is any hermitian solution of (16.3.1), then for any $u \in \mathcal{U}$ and $x_0 \in \mathbb{C}^n$,*

$$J_T^u(x^0) = -x(T)^*Xx(T) + x^{0*}Xx^0 + \int_0^T \|u(t) + R^{-1}(S^* + B^*X)x(t)\|_R^2 \, dt \tag{16.3.3}$$

where $\|\cdot\|_R$ denotes the R-norm, i.e. $\|y\|_R^2 = y^*Ry$ for any y (and we abbreviate $x^u(t; x^0) = x(t)$).

Proof. First consider the cost expressed in the form of (16.2.10). We have, using (16.2.11) and then (16.3.2),

$$\frac{d}{dt}x(t)^* X x(t) = \dot{x}(t)^* X x(t) + x(t)^* X \dot{x}(t)$$
$$= (v^* B^* + x^* \hat{A}^*) X x + x^* X (\hat{A} x + B v)$$
$$= x^* (X \hat{A} + \hat{A}^* X) x + v^* B^* X x + x^* X B v$$
$$= x^* X B R^{-1} B^* X x + v^* B^* X x + x^* X B v + v^* R v - v^* R v - x^* \hat{Q} x$$
$$= (v^* + x^* X B R^{-1}) R (v + R^{-1} B^* X x) - (x^* \hat{Q} x + v^* R v).$$

Integrating with respect to t from 0 to T we obtain (cf. (16.2.10)),

$$J_T^u(x^0) = -x(T)^* X x(T) + x^{0*} X x^0 + \int_0^T \|v + R^{-1} B^* X x\|_R^2 \, dt,$$

and the result follows from (16.2.9). □

Lemma 16.3.2. *Let $\hat{R}$ be nondegenerate and (Q, A) be observable. If X is an hermitian solution of (16.3.1) then $J^u(x^0) < \infty$ implies*

$$J^u(x^0) = x^{0*} X x^0 + \int_0^\infty \|u + R^{-1}(S^* + B^* X) x\|_R^2 \, dt. \tag{16.3.4}$$

Proof. By Proposition 16.2.9, $x(T) \to 0$ as $T \to \infty$, so the result follows immediately from (16.3.3). □

Now several results can be brought together to formulate the main theorem.

Theorem 16.3.3. *Let the pair (A, B) be c-stabilizable, $\hat{R}$ be nondegenerate, and (Q, A) be observable. Then there is a unique solution $X \geq 0$ of (16.3.1). Moreover*

(a) $\hat{J}(x^0) = x^{0*} X x^0$ *for all $x^0 \in \mathbb{C}^n$.*

(b) *For each x^0 there is a unique optimal control. This control is determined by the feedback matrix $F = R^{-1}(S^* + B^* X)$ and then $u(t) = -F e^{(A - BF)t} x^0$ for $t \geq 0$.*

(c) $A - B R^{-1}(S^* + B^* X)$ *is c-stable.*

(d) $X > 0$.

(e) X *is the maximal hermitian solution of the CARE (16.3.1).*

THE DIFFERENTIAL AND ALGEBRAIC RICCATI EQUATIONS 359

Proof. It is easily seen that (A, B) stabilizable implies the same property for $(A - BF, B)$ for any feedback matrix F. Thus, in equation (16.3.2) we have $(\hat{A}, B)$ stabilizable, $BR^{-1}B^* \geq 0$, $\hat{Q} \geq 0$ and the existence of a solution $X \geq 0$ of (16.3.2) (and hence (6.3.1)) follows by combining Theorems 7.2.8 and 9.3.3. The latter theorem also shows that (c) is satisfied for some solution $X \geq 0$. (The existence statement will also be proved independently in the next section.) Also, it follows from Proposition 16.2.5 that $\hat{J}(x^0) < \infty$ for all $x^0 \in \mathbb{C}^n$.

Now let $X \geq 0$ be a solution of (16.3.1), and Lemma 16.3.2 applies and yields $J^u(x^0) \geq x^{0*}Xx^0$ for all x^0 and all controls u. Hence $\hat{J}(x^0) \geq x^{0*}Xx^0$. Let u_c be the control at x^0 determined by the feedback matrix $F = R^{-1}(S^* + B^*X)$ (i.e., $u_c = -Fx$). Then it follows from (16.3.3) that $J^{u_c}(x^0) \leq x^{0*}Xx^0$. Consequently, $\hat{J}(x^0) = x^{0*}Xx^0$. This proves the uniqueness of X. Also, u_c is optimal and, in addition, part (a) is established.

The uniqueness of the optimal control also follows from Lemma 16.3.2, and statement (b) is easily obtained.

As $\hat{J}(x^0) > 0$ for all nonzero x^0 (Proposition 16.2.8), (d) follows from (a). Finally, statement (e) follows from Theorem 9.3.1. □

16.4 The differential and algebraic Riccati equations

In preceding sections the argument has been developed leading to the results of Theorem 16.3.3 without reference to Riccati differential equations. This makes for a more direct argument than most expositions (see, for example, Kwakernaak and Sivan (1972), or Russell (1979)), in which the differential equation makes an early appearance and has to be solved subject to a terminal boundary condition.

Here, following the argument of Hijab (1987), we show that *forward* solvability of the differential equation is readily established, and leads to another proof of existence of a maximal solution for the *algebraic* Riccati equation.

A fundamental lemma for time-dependent systems will be needed:

Lemma 16.4.1. *Let $A(t) \in \mathbb{C}^{n \times n}$ for $t \in [0, T]$ and assume that*

$$\int_0^T \|A(t)\| \, dt < \infty.$$

Then for each $x^0 \in \mathbb{C}^n$ the problem

$$\dot{x}(t) = A(t)x(t), \quad x(0) = x^0,$$

has a unique solution $x(t)$ on $[0, T]$.

Proof. See Lemma 2.4.1 of Hijab (1987), for example.

For our purposes, the Riccati differential equation takes the form

$$\dot{M}(t) = \hat{A}^* M(t) + M(t)\hat{A} - M(t)BR^{-1}B^* M(t) + \hat{Q} \tag{16.4.1}$$

where $\hat{A} = A - BR^{-1}S^*$, $\hat{Q} = Q - SR^{-1}S^*$. As in equations (16.3.1), (16.3.2) it is assumed that $R > 0$ and $\hat{Q} \geq 0$. We consider the initial value problem posed by (16.4.1) with $M(0) = 0$, $t \geq 0$, and show that when $M(T)$ exists it is intimately connected with the cost functional at time T for the LQR problem of (16.2.10) and (16.2.11).

Lemma 16.4.2. *If* (16.4.1) *has a solution* $M(t)$ *for* $t \in [0,T]$ *then* $M(t) \geq 0$ *on* $[0,T]$ *and for any* $x^0 \in \mathbb{C}^n$,

$$(x^0)^* M(T) x^0 = \min_{u \in \mathcal{U}} J_T^u(x^0). \tag{16.4.2}$$

Proof. Observe first that if $M(t)$ satisfies the initial value problem then so does $M(t)^*$. But the solution is well-known to be unique, and hence $M(t)^* = M(t)$. Let $x(t)$ be a solution of

$$\dot{x}(t) = Ax(t) + Bu(t) = \hat{A}x(t) + Bv(t), \quad x(0) = x^0$$

where $u \in \mathcal{U}$ and $v(t) = R^{-1}S^* x(t) + u(t)$. Make the abbreviations $M = M(T-t)$, $x = x^u(t)$ etc., and we have:

$$\begin{aligned}
x^{0*} M(T) x^0 &= x^{0*} M(T) x^0 - x(T)^* M(0) x(T) \\
&= \int_0^T -\frac{d}{dt}[x(t)^* M(T-t) x(t)]\, dt \\
&= \int_0^T (-\dot{x}^* M x + x^* \dot{M} x - x^* M \dot{x})\, dt \\
&= \int_0^T \{-(\hat{A}x + Bv)^* M x + x^* \dot{M} x - x^* M(\hat{A}x + Bv)\}\, dt \\
&= \int_0^T \{x^*(\dot{M} - \hat{A}^* M - M\hat{A}) x - v^* B^* M x - x^* M B v\}\, dt \\
&= \int_0^T x^* \{(-MBR^{-1}B^* M + \hat{Q}) x - v^* B^* M x - x^* M B v\}\, dt.
\end{aligned}$$

Now

$$\begin{aligned}
\int_0^T \|v + R^{-1} B^* M x\|_R^2\, dt &= \int_0^T (v^* + x^* M B R^{-1}) R (R^{-1} B^* M x + v)\, dt, \\
&= \int_0^T (\|v\|_R^2 + x^* M B R^{-1} B^* M x + v^* B^* M x + x^* M B v)\, dt,
\end{aligned}$$

so that

$$x^{0*}M(T)x^0 = \int_0^T \{x^*\hat{Q}x + \|v\|_R^2\}\,dt - \int_0^T \|v + R^{-1}B^*Mx\|_R^2\,dt,$$

$$= J_T^u(x^0) - \int_0^T \|v + R^{-1}B^*Mx\|_R^2\,dt$$

$$\leq J_T^u(x^0). \tag{16.4.3}$$

Now consider the time-dependent problem

$$\dot{x}(t) = G(t)x(t), \quad x(0) = x^0$$

on $[0,T]$ where $G(t) = A - BB^*M(T-t)$. By Lemma 16.4.1, there is a unique solution $x(t)$ and we define $w(t) = -B^*M(T-t)x(t)$. Then

$$\dot{x}(t) = Ax(t) + Bw(t), \quad x(0) = x^0$$

and $x^{0*}M(T)x^0 = J_T^w(x^0)$. Combining this with (16.4.3) the result (16.4.2) is obtained. Since this holds for all x^0 we also have $M(t) \geq 0$. $\square$

Theorem 16.4.3. *There exists a unique solution of* (16.4.1) *with* $M(0) = 0$, $t \geq 0$ *and it is well-defined for all* $t \in [0,\infty)$. *Also, for any* $x^0 \in \mathbb{C}^n$, $x^{0*}M(t)x^0$ *is an increasing function of* t *and for all* $t \geq 0$,

$$x^{0*}M(t)x^0 \leq \hat{J}(x^0) \tag{16.4.4}$$

Proof. The theory of ordinary differential equations ensures that a solution of the initial value problem exists locally, i.e. on an interval $[0,\tau]$ for some $\tau > 0$. Suppose there is a largest such number $\tau < \infty$.

Clearly, $J_t^u(x^0)$ is an increasing function of t for fixed u and x^0. Hence, by (16.4.2), $x^{0*}M(t)x^0$ is also an increasing function of t on $[0,\tau)$ and is bounded above by $J_\tau^0(x^0) < \infty$. Thus, $\lim_{t \to \tau^-} x^{0*}M(t)x^0$ exists for all choices of x^0. It is easily deduced from this that, because $M(t)^* = M(t)$, we have the existence of

$$\lim_{t \to \tau^-} M(t)_{jk}$$

for every element $M(t)_{jk}$ of $M(t)$ and we may define $M(\tau) = \lim_{t \to \tau^-} M(t)$. Now the initial value problem starting at time τ has a unique solution on an interval $[\tau, \tau_1)$ extending the solution given on $[0,\tau)$. This contradicts the maximality of τ, and so the solution $M(t)$ exists on $[0,\infty)$.

Since $J_T^u(x^0) \leq J^u(x^0)$, (16.4.4) follows from (16.4.2) and the definition (16.1.5). $\square$

Finally, we show that if, in addition to the hypotheses of the last theorem, the existence of $\lim_{t \to \infty} M(t)$ can be guaranteed, then this limit determines a semidefinite solution X of the algebraic equation (16.3.1) (or (16.3.2)).

Theorem 16.4.4. *If $\hat{J}(x^0) < \infty$ for all x^0 then there is a solution $X \geq 0$ of the algebraic Riccati equation (16.3.1).*

Proof. Let $t \to \infty$ in (16.4.4) and we find that
$$\lim_{t \to \infty} x^{0*} M(t) x^0$$
exists for each x^0. As in the proof of the preceding theorem it follows that, because $M(t)^* = M(t)$, $\lim_{t \to \infty} M(t)$ also exists. We may define
$$\lim_{t \to \infty} M(t) = X \geq 0.$$

Now we calculate, using (16.4.1),

$$\begin{aligned}
\hat{A}^* X + X \hat{A} &= \int_0^1 (\hat{A}^* X + X \hat{A}) \, dt \\
&= \int_0^1 \lim_{T \to \infty} (\hat{A}^* M(t+T) + M(t+T) \hat{A}) \, dt \\
&= \int_0^1 \lim_{T \to \infty} \{M(t+T) B R^{-1} B^* M(t+T) - \hat{Q} + \dot{M}(t+T)\} \, dt \\
&= X B R^{-1} B^* X - \hat{Q} + \lim_{T \to \infty} \int_0^1 \dot{M}(t+T) \, dt \\
&= X B R^{-1} B^* X - \hat{Q} + \lim_{T \to \infty} (M(T+1) - M(T)).
\end{aligned}$$

The result follows as $\lim_{T \to \infty}(M(T+1) - M(T)) = X - X = 0$. □

Combining this with Proposition 16.2.5 we obtain our existence result.

Corollary 16.4.5. *If (A, B) is stabilizable then the Riccati equation (16.3.1) has at least one positive semidefinite solution $X \geq 0$.*

On comparing this with Theorem 9.1.2, for example, keep in mind the general hypotheses of this section that $R > 0$ and $\hat{Q} \geq 0$.

16.5 The cost functional for the discrete LQR problem

Consider now a "discrete" system:
$$x_{k+1} = A x_k + B u_k, \quad k = 0, 1, 2, \ldots \tag{16.5.1}$$
where $A \in \mathbb{C}^{n \times n}$, $B \in \mathbb{C}^{n \times m}$, x_0 is given, and $\{u_k\}_{k=0}^{\infty}$ is a given sequence of control vectors in $\mathbb{C}^m$. The solution of this system, or recurrence relation, is
$$x_k = A^k x_0 + \sum_{r=0}^{k-1} A^{k-r-1} B u_r, \quad k = 1, 2, 3, \ldots. \tag{16.5.2}$$

For the purposes of this section it is convenient to have an abbreviated notation for sequences, whether of finite or infinite length. Thus, we may write

$\tilde{u} = \{u_k\}_{k=0}^{\infty}$, $\tilde{v} = \{v_k\}_{k=0}^{K}$, for example. Furthermore, for the solution sequence of (16.5.2) we may write $\tilde{x} = \{x_k\}_{k=0}^{\infty} = \tilde{x}(\tilde{u}, x_0) = \{x_k(\tilde{u}, x_0)\}_{k=0}^{\infty}$ and note that

$$\tilde{x}(\tilde{v} - \tilde{u}, x_0 - y_0) = \tilde{x}(\tilde{v}, x_0) - \tilde{x}(\tilde{u}, y_0). \tag{16.5.3}$$

Admissible control sequences are simply infinite sequences of vectors from $\mathbb{C}^m$. Thus, the space of controls is

$$\mathcal{U} = \{\tilde{u} = \{u_k\}_{k=0}^{\infty} : u_k \in \mathbb{C}^m\},$$

and we write $\mathbb{C}^{m,K} = \{\{u_k\}_{k=0}^{K} : u_k \in \mathbb{C}^m\}$. An inner product is defined on $\mathbb{C}^{m,K}$ in the usual way:

$$(\tilde{u}, \tilde{v})_K = \sum_{j=1}^{K} v_j^* u_j,$$

and then $\|\tilde{u}\|_K = ((\tilde{u}, \tilde{u})_K)^{1/2}$.

The cost functional associated with (16.5.1) is defined in terms of a positive semidefinite matrix $\hat{R}$ of size $n + m$ of the form

$$\hat{R} = \begin{bmatrix} Q & S \\ S^* & R \end{bmatrix} \geq 0$$

where $R > 0$ and is of size $m \times m$. Then, necessarily, $Q \geq 0$ and is $n \times n$. The cost associated with control sequence $\tilde{u} \in \mathcal{U}$ and initial vector $x_0 \in \mathbb{C}^m$ is then

$$J(\tilde{u}, x_0) = \sum_{k=0}^{\infty} [x_k^*, u_k^*] \begin{bmatrix} Q & S \\ S^* & R \end{bmatrix} \begin{bmatrix} x_k \\ u_k \end{bmatrix} = \sum_{k=0}^{\infty} \left\| \begin{bmatrix} x_k \\ u_k \end{bmatrix} \right\|_{\hat{R}}^2. \tag{16.5.4}$$

The *optimal cost* at x_0 is defined by

$$\hat{J}(x_0) = \inf_{\tilde{u} \in \mathcal{U}} J(\tilde{u}, x_0). \tag{16.5.5}$$

These definitions should be compared with definitions (16.1.4) and (16.1.5) of the continuous problem. Investigation of the properties of J and $\hat{J}$ follows the same direction taken in Section 16.2. An analogue of Lemmas 16.2.1 and 16.2.2 is developed more easily in this context:

Lemma 16.5.1. *Let there be given a sequence (of sequences) $\{\tilde{u}_k\} \subseteq \mathcal{U}$ and a sequence $\{x_{k,0}\} \subseteq \mathbb{C}^n$ such that $x_{k,0} \to x_0$ as $k \to \infty$. Then there is a subsequence $\{\tilde{v}_k\} \subseteq \{\tilde{u}_k\}$ and a $\tilde{u} \in U$ such that*

$$(\tilde{w}, \tilde{v}_k - \tilde{u}_k)_K \to 0 \quad \text{and} \quad (\tilde{w}, \tilde{x}(\tilde{v}_k, x_{k,0}) - \tilde{x}(\tilde{u}, x_0))_K \to 0$$

as $k \to \infty$ for all $K > 0$ and all $\tilde{w} \in \mathcal{U}$.

The arguments used to establish Lemmas 16.2.1 and 16.2.2 are easily adapted to this situation. The details are omitted.

Lemma 16.5.2. *Let $\{x_{r,0}\}_{r=0}^{\infty} \subseteq \mathbf{C}^n$ with $x_{r,0} \to x_0$ as $r \to \infty$ and $\{\tilde{u}_r\} \subseteq \mathcal{U}$ be a sequence for which*
$$\lim_{r \to \infty} J(\tilde{u}_r, x_{r,0})$$
exists. Then there is a $\tilde{v} \in \mathcal{U}$ such that
$$J(\tilde{v}, x_0) \leq \lim_{r \to \infty} J(\tilde{u}_r, x_{r,0}). \tag{16.5.6}$$

Proof. Define the cost at step K by
$$J_K(\tilde{u}, x_0) = \sum_{k=0}^{K} \left\| \begin{bmatrix} x_k \\ u_k \end{bmatrix} \right\|_{\hat{R}}^2$$

and define a subsequence $\{\tilde{v}_k\}$ of $\{\tilde{u}_k\}$ and $\tilde{u} \in \mathcal{U}$, with the properties of Lemma 16.5.1. Then, taking advantage of (16.5.3),

$$J_K(\tilde{v}_k - \tilde{u}, x_{k,0} - x_0)$$
$$= \sum_{r=0}^{K} [x_r(\tilde{v}_k, x_{k,0})^* - x_r(\tilde{u}, x_0)^*, v_{k,r}^* - u_r^*] \hat{R} \begin{bmatrix} x_r(\tilde{v}_k, x_{k,0}) - x_r(\tilde{u}, x_0) \\ v_{k,r} - u_r \end{bmatrix}. \tag{16.5.7}$$

Define
$$[y_r^*, z_r^*] = [x_r(\tilde{u}, x_0)^*, u_r^*] \hat{R}$$

and then
$$E_k^K(\tilde{y}, \tilde{z}) = \sum_{r=0}^{K} [y_r^*, z_r^*] \begin{bmatrix} x_r(\tilde{v}_k, x_{k,0}) - x_r(\tilde{u}, x_0) \\ v_{k,r} - u_r \end{bmatrix}$$
$$= \overline{(\tilde{z}, \tilde{v}_k - \tilde{u})}_K + \overline{(\tilde{y}, \tilde{x}(\tilde{v}_k, x_{k,0}) - \tilde{x}(\tilde{u}, x_0))}_K,$$

and it follows from Lemma 16.5.1 that $E_k^K(\tilde{y}, \tilde{z}) \to 0$ as $k \to \infty$.

A little calculation with (16.5.7) shows that
$$0 \leq J_K(\tilde{v}_k - \tilde{u}, x_{k,0} - x_0) = J_K(\tilde{v}_k, x_k^0) - J_K(\tilde{u}, x_0) - 2 \operatorname{Re} E_k^K(\tilde{y}, \tilde{z})$$

and so, taking the limits as $k \to \infty$ and $K \to \infty$, in this order, (16.5.6) is obtained. □

The lower semicontinuity of $\hat{J}$ is now easily established by adapting the arguments used to prove Theorem 16.2.4.

Theorem 16.5.3. *For each $x_0 \in \mathbb{C}^n$ there is a $\tilde{u} \in \mathcal{U}$ at which the optimal cost is attained: $\hat{J}(x_0) = J(\tilde{u}, x_0)$. Furthermore, for any $\{x_{k,0}\} \subseteq \mathbb{C}^n$ with $x_{k,0} \to x_0$ as $k \to \infty$,*

$$\hat{J}(x_0) \leq \lim_{k \to \infty} \hat{J}(x_{k,0})$$

whenever this limit exists.

Proof. Replace Lemma 16.2.3 in the proof of Theorem 16.2.4 by Lemma 16.5.2. □

Lemma 16.5.4. *If the pair (A, B) is d-stabilizable then $\hat{J}(x_0) < \infty$ for any initial vector $x_0 \in \mathbb{C}^n$.*

Proof. There is an $m \times n$ matrix F such that $A - BF$ is d-stable. Consider the feedback control $u_k = -Fx_k$, so that $\{x_k\}$ is generated by $x_{k+1} = (A - BF)x_k$ for $k = 0, 1, 2, \ldots$. Then

$$\begin{bmatrix} x_k \\ u_k \end{bmatrix} = \begin{bmatrix} I \\ -F \end{bmatrix} x_k = \begin{bmatrix} I \\ -F \end{bmatrix} (A - BF)^k x_0.$$

Since $A - BF$ is d-stable there is a norm in which $\|A - BF\| = \alpha < 1$ (see Lemma 5.6.10 of Horn and Johnson (1985), for example). Thus,

$$\left\| \begin{bmatrix} x_k \\ u_k \end{bmatrix} \right\| \leq (\text{const.}) \alpha^k,$$

and it follows easily that $J(\tilde{u}, x_0) < \infty$, whence $\hat{J}(x_0) < \infty$. □

Proposition 16.5.5. *If $\hat{R}$ is nondegenerate (i.e. $\text{rank } \hat{R} = \text{rank } Q + \text{rank } R$) and (Q, A) is an observable pair then $\hat{J}(x_0) > 0$ whenever $x_0 \neq 0$.*

Proof. Modify the formulation of the problem by defining $\tilde{v} \in \mathcal{U}$ in terms of $\tilde{u} \in \mathcal{U}$ by

$$v_k = R^{-1}S^* x_k + u_k \tag{16.5.8}$$

(cf. equation (16.2.9)). Then it is found that

$$J(\tilde{u}, x_0) = \sum_{k=0}^{\infty} (x_k^* \hat{Q} x_k + v_k^* R v_k) \tag{16.5.9}$$

where $\hat{Q} = Q - SR^{-1}S^*$ (as usual) and also

$$x_{k+1} = \hat{A} x_k + B v_k \tag{16.5.10}$$

where $\hat{A} = A - BR^{-1}S^*$.

From Theorem 16.5.3, if $\hat{J}(x_0) = 0$ for some $x_0 \neq 0$, there is a $\tilde{u} \in \mathcal{U}$ such that $J(\tilde{u}, x_0) = 0$. Since $R > 0$ and $\hat{Q} \geq 0$ it follows from (16.5.9) that $\tilde{v} = \tilde{0}$ and $\hat{Q} x_k = 0$ for each k. Consequently, $\hat{Q}\hat{A}^k x_0 = 0$ for $k = 0, 1, 2, \ldots$, and $(\hat{Q}, \hat{A})$ is not observable. But this contradicts Lemma 16.2.7 and so $\hat{J}(x_0) > 0$. □

Proposition 16.5.6. *Let $\hat{R}$ be nondegenerate and (Q, A) be an observable pair. Then $J(\tilde{u}, x_0) < \infty$ implies that $x_k(\tilde{u}, x_0) \to 0$ as $k \to \infty$.*

Proof. Assume $\|x_0\| = 1$. Then $J(\tilde{u}, x_0) < \infty$ implies $\hat{J}(x_0) < \infty$ and Proposition 16.5.5 gives $\hat{J}(x_0) > 0$. Furthermore, Theorem 16.5.3 implies the existence of a finite $m > 0$ for which $\min_{\|x\|=1} \hat{J}(x) = m$. Thus, for any $y \in \mathbb{C}^n$

$$\hat{J}(y) = \|y\|^2 \hat{J}\left(\frac{1}{\|y\|}y\right) \geq \|y\|^2 m. \qquad (16.5.11)$$

Let $v_k = u_{k+K}$ and, using (16.5.2), it is easily verified that

$$x_k(\tilde{v}, x_K(\tilde{u}, x_0)) = x_{k+K}(\tilde{u}, x_0).$$

Hence

$$\hat{J}(x_K(\tilde{u}, x_0)) \leq J(\tilde{v}, x_K(\tilde{u}, x_0)) = \sum_{k=0}^{\infty} \left\| \begin{bmatrix} x_k(\tilde{v}, x_K(\tilde{u}, x_0)) \\ v_k \end{bmatrix} \right\|_{\hat{R}}^2$$

$$= \sum_{k=0}^{\infty} \left\| \begin{bmatrix} x_{k+K}(\tilde{u}, x_0) \\ u_{k+K} \end{bmatrix} \right\|_{\hat{R}}^2 = \sum_{r=K}^{\infty} \left\| \begin{bmatrix} x_r(\tilde{u}, x_0) \\ u_r \end{bmatrix} \right\|_{\hat{R}}^2.$$

Consequently, as $K \to \infty$

$$\hat{J}(x_K(\tilde{u}, x_0)) \to 0. \qquad (16.5.12)$$

Using this property and applying (16.5.11) to the members of $\{x_K(\tilde{u}, x_0) : K \geq 0\}$ this set is seen to be bounded. Hence there is a convergent subsequence; say $x_s(\tilde{u}, x_0) \to x$ as $s \to \infty$.

Now Theorem 16.5.3 yields $\hat{J}(x) \leq \hat{J}(x_s(\tilde{u}, x_0))$ and so, from (16.5.12), $\hat{J}(x) = 0$. Finally, the observability condition and Proposition 16.5.5 imply that $x = 0$. □

16.6 Solution of the discrete LQR problem

Having made the preparations of the preceding section a direct attack can now be made on the solution of the discrete LQR problem, i.e., the problem of finding a control $\tilde{u} \in \mathcal{U}$ for the system (16.5.1) which, starting from an arbitrary $x_0 \in \mathbb{C}^n$ attains the optimal (minimal) cost of (16.5.5). This optimal control will be expressed in terms of a solution of the DARE of equation (12.1.1) (with C replaced by S^*):

$$X = A^* X A + Q - (S^* + B^* X A)^* (R + B^* X B)^{-1} (S^* + B^* X A), \qquad (16.6.1)$$

and we recall that Q, R, S are defined as in (16.1.3). Furthermore, since $R > 0$ and $\hat{R} \geq 0$ we have $Q \geq 0$.

Note also that, in view of Lemma 16.5.4, the stabilizability of (A, B) is a natural hypothesis. But these are just the conditions of Corollary 13.1.2, which ensures the existence of a positive semidefinite solution $\tilde{X}$ for (16.6.1). Then it is obvious that $R + B^*\tilde{X}B > 0$ and, furthermore, it follows from Theorem 12.2.1 that $R + B^*XB > 0$ for all hermitian solutions X of (16.6.1). We have:

Lemma 16.6.1. *If $R > 0$, $\hat{Q} \geq 0$, and (A, B) is d-stabilizable then equation (16.6.1) has a positive semidefinite solution. Moreover, $R + B^*XB > 0$ for all hermitian solutions X of equation (16.6.1).*

Now an analogue of Lemma 16.3.1 can be established in which the cost after K steps can be expressed in terms of a norm, rather than the $\hat{R}$-seminorm. Recall that the basic recursion is

$$x_{k+1} = Ax_k + Bu_k, \quad k = 0, 1, 2, \ldots, \tag{16.6.2}$$

and the cost after K steps is (cf. (16.5.4)):

$$J_K(\tilde{u}, x_0) = \sum_{k=0}^{K} \left\| \begin{bmatrix} x_k \\ u_k \end{bmatrix} \right\|_{\hat{R}}^2. \tag{16.6.3}$$

Lemma 16.6.2. *If X is any hermitian solution of (16.6.1) then, for any $\tilde{u} \in U$ and $x_0 \in \mathbb{C}^n$*

$$J_K(\tilde{u}, x_0) = -x_K^* X x_K + x_0^* X x_0^* + \sum_{k=0}^{K-1} \| R_X^{-1}(S^* + B^*XA)x_k + u_k \|_{R_X}^2 \tag{16.6.4}$$

where $R_X = R + B^*XB$.

Note that conditions under which $\|\cdot\|_{R_X}$ is a genuine norm are apparent from Lemma 16.6.1.

Proof. Using (16.6.2) followed by (16.6.1) it is found that

$x_{k+1}^* X x_{k+1} - x_k^* X x_k$
$= (x_k^* A^* + u_k^* B)X(Ax_k + Bu_k) - x_k^* X x_k$
$= x_k^*(A^*XA - X)x_k + x_k^* A^*XBu_k + u_k^* BXAx_k + u_k^* R_X u_k - u_k^* R u_k.$
$= x_k^*(S + A^*XB)R_X^{-1}(S^* + B^*XA)x_k + x_k^*(S + A^*XB)u_k$
$\quad + u_k^*(S^* + B^*XA)x_k + u_k^* R_X u_k - (x_k^* Q x_k + x_k^* S u_k + u_k^* S^* x_k + u_k^* R u_k)$
$= \{x_k^*(S + A^*XB)R_X^{-1} + u_k^*\} R_X \{R_X^{-1}(S^* + B^*XA)x_k + u_k\} - \left\| \begin{bmatrix} x_k \\ u_k \end{bmatrix} \right\|_{\hat{R}}^2$
$= \| R_X^{-1}(S^* + B^*XA)x_k + u_k \|_{R_X}^2 - \left\| \begin{bmatrix} x_k \\ u_k \end{bmatrix} \right\|_{\hat{R}}^2$

Sum with respect to k from 0 to $K - 1$ and (16.6.4) is obtained. □

Lemma 16.6.3. *Let $\hat{R}$ be nondegenerate and (Q, A) be an observable pair. If X is a hermitian solution of (16.6.1) then $J(\tilde{u}, x_0) < \infty$ implies*

$$J(\tilde{u}, x_0) = x_0^* X x_0 + \sum_{k=0}^{\infty} \|R_X^{-1}(S^* + B^* X A)x_k + u_k\|_{R_X}^2 \qquad (16.6.5)$$

Proof. By Proposition 16.5.6, $x_k(\tilde{u}, x_0) \to 0$ as $k \to 0$ so the result follows from Lemma 16.6.2. □

Now we are ready to complete the proof of the main theorem concerning the discrete LQR problem and the DARE (16.6.1).

Theorem 16.6.4. *Let the pair (A, B) be d-stabilizable, $\hat{R}$ be nondegenerate, and (Q, A) be observable. Then there is a unique solution $X \geq 0$ of (16.6.1) and*

(a) *$\hat{J}(x_0) = x_0^* X x_0$ for all $x_0 \in \mathbb{C}^n$.*

(b) *For each x_0 there is a unique optimal control. The control is determined by the feedback matrix*

$$F = (R + BXB)^{-1}(S^* + B^* X A) \qquad (16.6.6)$$

and then $u_k = -F(A - BF)^k x_0$, $k = 0, 1, 2, \ldots$.

(c) *$A - BF$ is d-stable.*

(d) *$X > 0$.*

(e) *X is the maximal hermitian solution of the DARE (16.6.1).*

Proof. It has been shown in Lemma 16.6.1 that equation (16.6.1) has a unique positive semidefinite solution X and that $R + B^* X B > 0$. Thus, $\|\cdot\|_{R_X}$ in equations (16.6.4) and (16.6.5) is a genuine norm. Also, Lemma 16.5.4 shows that $\hat{J}(x_0) < \infty$. Then Lemmas 16.6.2 and 16.6.3 show that $J(\tilde{u}, x_0) \geq x_0^* X x_0$.
Consider the control $\tilde{u}^c$ at x_0 given by $u_k^c = -Fx_k$ and F is given by (16.6.6). Then equation (16.6.4) shows that $J(\tilde{u}^c, x_0) \leq x_0^* X x_0$. Consequently, $\hat{J}(x_0) = x_0^* X x_0$. Since this equality holds for any positive semidefinite solution X and any $x_0 \in \mathbb{C}^n$, we conclude that (16.6.1) has a unique positive semidefinite solution. Also, $J(\tilde{u}^c, x_0) = x_0^* X x_0$ and $\tilde{u}_k^c = -F(A - BF)^k x_0$ determines the optimal control. The uniqueness of the optimal control obviously follows from Lemma 16.6.2. Thus, (a) and (b) are established.
It follows from Proposition 16.5.6 that $x_k(\tilde{u}^c, x_0) \to 0$ as $k \to \infty$ for any choice of x_0, i.e. $(A - BF)^k x_0 \to 0$ as $k \to \infty$. Hence part (c) follows.
Finally, $\hat{J}(x_0) > 0$ for any x_0 follows from Proposition 16.5.5 and (d) is obtained. □

16.7 Exercises

16.7.1. Prove that, under the hypotheses of Theorem 16.6.4, there is a unique solution $X = X^*$ of (16.6.1) such that the eigenvalues of $A - BF$ are in the closed unit disk; moreover, $X \geq 0$ in this case.

16.7.2. Prove that, under the hypotheses of Theorem 16.3.3, there is a unique solution $X = X^*$ of (16.3.1) such that all eigenvalues of $A - BR^{-1}(S^* + B^*X)$ are in the closed left half-plane; moreover, $X \geq 0$.

16.7.3. Consider the time-invariant system (16.1.1) with $A = \begin{bmatrix} 0 & 1 \\ -1 & 0 \end{bmatrix}$, $B = \begin{bmatrix} 0 \\ 1 \end{bmatrix}$, and with the cost functional (16.1.4) given by

$$\hat{R} = \begin{bmatrix} \alpha + \beta & 0 & -\beta \\ 0 & \alpha & 0 \\ -\beta & 0 & \beta + \gamma \end{bmatrix},$$

where α, β, γ are positive parameters.

(1) Solve the Riccati equation (16.3.1).

(2) Find the optimal control.

16.8 Notes

The linear quadratic regulator problem was treated first in the seminal paper Kalman (1960b). Since then, the topic has generated an immense literature. The line of attack followed in this chapter has been developed by Brockett (1970) and Hijab (1987). Among many texts on the LQR problem, we mention here the graduate level textbooks of Anderson and Moore (1990), and Kwakernaak and Sivan (1972); both oriented towards an engineering audience. For LQR problems for systems with delays see Yanushevsky (1993) and references given there. Exercise 16.7.3 is taken from the text of Russell (1979).

17
THE DISCRETE KALMAN FILTER

One of the more important motivations for the study of algebraic Riccati equations was recognized in the early 1960s after publication of the works of Kalman (1960a) and Kalman and Bucy (1961) on filtering and prediction problems of discrete and continuous types, respectively. A full development of the theory behind these processes is beyond the scope of this book. However, this chapter is devoted to an intuitive development of the ideas behind discrete Kalman filtering and, in Section 17.5, an account of the basic convergence theorem for the Riccati difference equation. This may be all that the reader needs to know about the problem, or it may lead the way to more thorough treatments such as the careful mathematical development of Catlin (1989) or the graduate engineering text of Anderson and Moore (1979).

The kind of problem to be resolved is easily described in the context of the time-invariant system $x_{k+1} = Fx_k$, $k = 0, 1, 2, \ldots$, where x_0 is given. This system is supposed to be corrupted by a noise process determined by a sequence of random vectors $\{w_k\}$ in such a way that the original recursion is replaced by

$$x_{k+1} = Fx_k + Gw_k, \quad k = 0, 1, 2, \ldots,$$

where G is a matrix of compatible size. Observations of the system should be of the form $z_k = Hx_k$, but are also corrupted by a noise sequence, say $\{v_k\}$, so that

$$y_k = Hx_k + v_k, \quad k = 0, 1, 2, \ldots.$$

The question then is: given the corrupted observations $\{y_r\}_{r=0}^k$ (described in statistical terms) and the statistics of x_0 and the noise processes $\{w_r\}_{r=0}^k$, $\{v_r\}_{r=0}^k$, describe the statistics of the state vector sequence $x_1, x_2, \ldots, x_{k+1}$.

Our presentation begins with a quick review of some necessary ideas from probability theory, presented in as accessible a format as possible. More detail can be found in the text of Anderson (1958), for example. The time-dependent problem is formulated in Section 17.2 and its resolution is the topic of the next two sections. The role played by the algebraic Riccati equation in time-invariant systems is the topic of Section 17.5.

17.1 Some concepts and results concerning random vectors

For our purposes, a *probability distribution function* (or *pdf*) is a function p defined on $\mathbb{R}^k$, taking nonnegative real values, and for which

$$\int_{-\infty}^{\infty} \cdots \int_{-\infty}^{\infty} p(x_1, \ldots, x_k) \, dx_1 \ldots dx_k = 1.$$

If $k > 1$ this is a *joint* pdf for the *random variables* $x_1, x_2, \ldots, x_k$. Given any k-tuple $(a_1, \ldots, a_k) \in \mathbb{R}^k$, the probability that $x_j < a_j$ for $j = 1, 2, \ldots, k$ is written

$$P(x_1 < a_1, x_2 < a_2, \ldots, x_k < a_k) = \int_{-\infty}^{a_k} \cdots \int_{-\infty}^{a_1} p(x_1, \ldots, x_k) \, dx_1 \ldots dx_k.$$

Note that the probability distribution for x_1, called a *marginal pdf*, is

$$p_1(x_1) = \int_{-\infty}^{\infty} \cdots \int_{-\infty}^{\infty} p(x_1, x_2, \ldots, x_k) \, dx_2 \ldots dx_k.$$

Also, the joint pdf for x_1 and x_2 is

$$p_{12}(x_1, x_2) = \int_{-\infty}^{\infty} \cdots \int_{-\infty}^{\infty} p(x_1, x_2, \ldots, x_k) \, dx_3 \ldots dx_k \qquad (17.1.1)$$

and similarly for other subsets of $x_1, x_2, \ldots, x_k$.

The *expected value* of a function $h(x_1, \ldots, x_k)$ defined on $\mathbb{R}^k$ is

$$E(h(x_1, \ldots, x_k)) = \int_{-\infty}^{\infty} \cdots \int_{-\infty}^{\infty} h(x_1, \ldots, x_k) p(x_1, \ldots, x_k) \, dx_1 \ldots dx_k$$

when this integral exists. Observe that E is a linear transformation on the linear space of functions h which are integrable over $\mathbb{R}^k$ with respect to p.

In particular, with the notation of equation (17.1.1),

$$E(x_1 x_2) = \int_{-\infty}^{\infty} \int_{-\infty}^{\infty} x_1 x_2 p_{12}(x_1, x_2) \, dx_1 dx_2, \qquad (17.1.2)$$

and similarly for other products of the x_j's.

Let x, y denote column vectors of jointly distributed random variables $x_1, \ldots, x_k$ and $y_1, \ldots, y_\ell$, respectively. We call x and y *random vectors*. Given joint probability distributions $p_{rs}(x_r, y_s)$, $(1 \le r \le k, 1 \le s \le \ell)$, we may define the $k \times \ell$ matrix of expected values (calculated elementwise as in equation (17.1.2)),

$$E(xy^T) = [E(x_r y_s)]_{r,s=1}^{k,\ell}.$$

For a scalar random variable there are the familiar definitions

$$\bar{x} := E(x) = \int_{-\infty}^{\infty} x p(x) \cdot dx, \quad \sigma^2 := E(x - \bar{x})^2 \ge 0;$$

the *mean* and *variance* of the random variable x, respectively. For a random vector x the mean $\bar{x}$ is defined elementwise and the *covariance* matrix of x is

$$\Sigma_x := E((x - \bar{x})(x - \bar{x})^T).$$

For a pair of random vectors x and y we write

$$\Sigma_{xy} := E((x - \bar{x})(y - \bar{y})^T), \quad \Sigma_{yx} = \Sigma_{xy}^T.$$

Thus $\Sigma_x = \Sigma_{xx}$ and we note that, if x has n components then, for any $u \in \mathbb{R}^n$, $u^T \Sigma_x u = E(\mu^2)$ where $\mu = \Sigma_{j=1}^n u_j (x_j - \bar{x}_j)$ is again a random variable. Thus $u^T \Sigma_x u \ge 0$ for all u, i.e. $\Sigma_x \ge 0$.

If random variables $x_1, \ldots, x_k$ have a joint pdf $p(x_1, \ldots, x_k)$, the *conditional pdf* for variables $x_1, \ldots, x_j$ for given $x_{j+1}, \ldots, x_k$ is

$$p(x_1, \ldots, x_j | x_{j+1}, \ldots, x_k) = \frac{p(x_1, \ldots, x_k)}{q(x_{j+1}, \ldots, x_k)}$$

where

$$q(x_{j+1}, \ldots, x_k) = \int_{-\infty}^{\infty} \cdots \int_{-\infty}^{\infty} p(x_1, \ldots, x_k) \, dx_1 \ldots dx_j.$$

It is easily verified from this definition that conditional probabilities can be calculated recursively. For example, if scalar random variables u, v, w have a joint pdf $p(u, v, w)$ then $p(u|v, w)$ can be calculated as the conditional probability of $p(u, v|w)$ for given v.

Scalar random variables x and y with joint pdf $p(x, y)$ are said to be *statistically independent* if

$$p(x, y) = \int_{-\infty}^{\infty} p(x, u) \, du \int_{-\infty}^{\infty} p(u, y) \, du \quad (17.1.3)$$

for all $x, y \in \mathbb{R}$. In this case the conditional pdfs become

$$p(x|y) = \int_{-\infty}^{\infty} p(x, y) \, dy, \quad p(y|x) = \int_{-\infty}^{\infty} p(x, y) \, dx.$$

In particular, $E(xy) = E(x)E(y) = \bar{x}\bar{y}$. Thus, the covariance of x and y is zero, because

$$E((x-\bar{x})(y-\bar{y})) = E(xy) - E(x)\bar{y} - \bar{x}E(y) + E(\bar{x}\bar{y})$$
$$= E(xy) - \bar{x}\bar{y} = 0.$$

However, when the covariance is zero we obtain $E(xy) = \bar{x}\bar{y}$, but this need not imply condition (17.1.3), i.e. x and y are not necessarily statistically independent.

We say that random *vectors* x and y are statistically independent if all pairs of components (x_i, y_j) are statistically independent. Thus, when this is the case, $\Sigma_{xy} = E((x-\bar{x})(y-\bar{y})^T) = 0$, although the converse statement does not hold.

The components $x_1, \ldots, x_k$ of a random vector x are said to be (jointly) normally distributed with mean $\bar{x}$ and covariance $\Sigma_x > 0$ if the joint pdf has the form

$$p(x_1, \ldots, x_k) = \frac{1}{(2\pi)^{k/2}(\det \Sigma_x)^{1/2}} \exp\left(-\frac{1}{2}(x-\bar{x})^T \Sigma_x^{-1}(x-\bar{x})\right).$$

A simple manipulation serves to establish the first part of the following lemma. The last part is a basic result for normal distributions.

Lemma 17.1.1. *If x and y are random vectors with n components and F, G are $m \times n$ constant matrices, then $Fx + Gy$ is a random vector with mean $F\bar{x} + G\bar{y}$ and covariance*

$$F\Sigma_x F^T + F\Sigma_{xy} G^T + G\Sigma_{yx} F^T + G\Sigma_y G^T$$

and if x, y are statistically independent random vectors then the covariance of $Fx + Gy$ is

$$F\Sigma_x F^T + G\Sigma_y G^T.$$

Furthermore, if x and y are normally distributed, so is $Fx + Gy$.

An important result requiring more proof is:

Theorem 17.1.2. *Let x and y be random n- and m-vectors, respectively, and let the $(n+m)$-vector $\begin{bmatrix} x \\ y \end{bmatrix}$ be normally distributed with mean $\begin{bmatrix} \bar{x} \\ \bar{y} \end{bmatrix}$ and covariance*

$$\begin{bmatrix} \Sigma_x & \Sigma_{xy} \\ \Sigma_{yx} & \Sigma_y \end{bmatrix},$$

where $\Sigma_y > 0$. Then the conditional pdf for x given y is normal with mean $\bar{x} + \Sigma_{xy}\Sigma_y^{-1}(y-\bar{y})$ and covariance $\Sigma_x - \Sigma_{xy}\Sigma_y^{-1}\Sigma_{yx}$.

The proofs of Lemma 17.1.1 and Theorem 17.1.2 can be found in Chapter 2 of Anderson (1958), for example.

A final set of definitions begins with the notion of a *random process*, which simply means a sequence of random vectors, say $\{x_r\}_{r=0}^{\infty}$, where each x_r has (say) n component random variables. A random process is said to be a *white noise process* if $r \neq s$ implies that the vectors x_r and x_s are statistically independent. In particular, this implies that $\Sigma_{x_r x_s} = 0$ (the $n \times n$ zero matrix, if x_r, x_s have n components). More generally, if the sequence is white then for $r, s = 0, 1, 2, \ldots$

$$\Sigma_{x_r x_s} = \Sigma_{x_r} \delta_{rs}.$$

A *white noise normal process* is a white noise process as just defined and, for each r, the components of vector x_r are jointly normally distributed.

A random process $\{x_r\}_{r=0}^{\infty}$ is said to have *zero mean* if $\bar{x}_r = 0 \ (\in \mathbb{R}^n)$ for every r.

17.2 Statement of the problem

We describe a general abstract model which has found many useful applications. Our discussion is largely heuristic and serves to set up the model and useful terminology.

Consider a system which can be described in terms of n scalar coordinates represented in a single vector $x \in \mathbb{R}^n$ known as the state vector. The system evolves in time and can be observed, at least partially, at equally spaced moments in time at which the state vector takes values $x_0, x_1, x_2, \ldots$.

This evolution in time is determined by the linear recursion

$$x_{k+1} = F_k x_k + G_k w_k, \quad k = 0, 1, 2, \ldots \qquad (17.2.1)$$

where $F_k \in \mathbb{R}^{n \times n}$, $G_k \in \mathbb{R}^{n \times m}$, and w_k is a squence of vectors in $\mathbb{R}^n$ which represent disturbances in the otherwise homogeneous process $x_{k+1} = F_k x_k$.

It is generally the case that the state variables themselves cannot be observed directly and there is access to (generally only a few) linear combinations of the state variables. Thus, at each moment k there is a vector $y_k \in \mathbb{R}^p$ of observations given by

$$y_k = H_k x_k + v_k, \quad k = 0, 1, 2, \ldots, \qquad (17.2.2)$$

where $H_k \in \mathbb{R}^{p \times n}$ and, again, errors or disturbances in the measurement process are lumped together in the inhomogeneous term v_k.

The whole process in deterministic form is then as follows: given x_0, and sequences $\{A_k\}$, $\{G_k\}$, $\{H_k\}$, $\{w_k\}$, $\{v_k\}$, the state and observed vectors evolve as described in equations (17.2.1) and (17.2.2). The block diagram of Figure 17.2.1 can be helpful in visualizing the process.

The point of view now changes rather dramatically and we assume that the data concerning x_0 and the disturbance and error sequences $\{w_k\}$, $\{v_k\}$ are

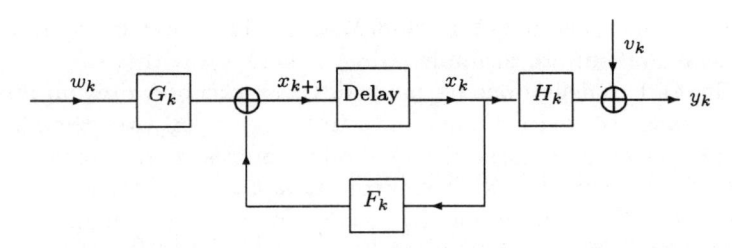

Fig. 17.2.1. The linear system in block form.

described only in probabilistic terms. (The coefficient sequences $\{F_k\}$, $\{G_k\}$, $\{H_k\}$ are given in numerical form, as before.) The following hypotheses are made using concepts discussed in Section 17.1:

(1) $\{w_k\}$ and $\{v_k\}$ are normally distributed, zero mean, white noise processes with given covariance sequences

$$E(w_k w_k^T) = Q_k, \quad E(v_k v_k^T) = R_k. \qquad (17.2.3)$$

(2) For each j and k the vectors w_j and v_k are statistically independent (so that $E(w_j v_k^T) = 0$).

(3) x_0 is normally distributed and the mean $\bar{x}_0$ and covariance $\Sigma_0 := \Sigma_x$ are given. Furthermore, for any j, k (x_0, w_j) and (x_0, v_k) are statistically independent.

First observe that, because of the linearity of the relations (17.2.1) and (17.2.2) and Lemma 17.1.1 the processes $\{x_k\}$ and $\{y_k\}$ will also be normally distributed (and, in general, x_j, y_k will not be statistically independent). Apart from the data described in items (1), (2), and (3) it is supposed that at time step k information about observations $y_0, y_1, \ldots, y_{k-1}$ and possibly y_k, is available. Using this information predictions are to be made about the state of the system at time k, i.e. about x_k.

In probabilistic terms we are to predict the conditional probability for x_k given $y_0, y_1, \ldots, y_{k-1}$ which we write as $x_{k|k-1}$ and then, updating with information about y_k, we are interested in the probability of x_k given $y_0, y_1, \ldots, y_k$, which we write as $x_{k|k}$. We seek only the means $\bar{x}_{k|k-1}$ and $\bar{x}_{k|k}$ and the covariances, which it is convenient to write in the form

$$\Sigma_{k|k-1} := E\{(x_k - \bar{x}_{k|k-1})(x_k - \bar{x}_{k|k-1})^T | y_0, y_1, \ldots, y_{k-1}\} \qquad (17.2.4)$$

$$\Sigma_{k|k} := E\{(x_k - \bar{x}_{k|k})(x_k - \bar{x}_{k|k})^T | y_0, y_1, \ldots, y_k\}. \qquad (17.2.5)$$

A remarkable feature of the subsequent argument is the fact that these quantities can be formed recursively, i.e. allowing stepping forward in time from step k to $k+1$ without explicit reference to steps $0, 1, \ldots, k-1$. This is sometimes referred to as a *Markovian* property.

Why would such a process be called a "filter"? The intuition is that the state is predicted from the observations in such a way that effects of the "noise", represented by $\{w_k\}$ and $\{v_k\}$, are minimized, or filtered out, in the process.

17.3 The recursive process

In this section it will be shown how the conditional expectations $\bar{x}_{k|k-1}$, $\bar{x}_{k|k}$, and covariances $\Sigma_{k|k-1}$, $\Sigma_{k|k}$ defined at the end of Section 17.2 (see equations (17.2.4), and (17.2.5)) can be calculated recursively. The line of argument is that presented by Anderson and Moore (1979) (see Chapter 2).

Step 1 Since $y_0 = H_0 x_0 + v_0$ and $\bar{v}_0 = 0$, $\bar{y}_0 = H_0 \bar{x}_0$ and

$$E((x_0 - \bar{x}_0)(y_0 - \bar{y}_0)^T) = E((x_0 - \bar{x}_0)(x_0 - \bar{x}_0)^T H_0^T) + E((x_0 - \bar{x}_0)v_0^T)$$
$$= \Sigma_0 H_0^T$$

since x_0, v_0 are statistically independent. It follows from Lemma 17.1.1 that y_0 has covariance $H_0 \Sigma_0 H_0^T + R_0$ and hence $\begin{bmatrix} x_0 \\ y_0 \end{bmatrix}$ has mean and covariance

$$\begin{bmatrix} \bar{x}_0 \\ H_0 \bar{x}_0 \end{bmatrix}, \quad \begin{bmatrix} \Sigma_0 & \Sigma_0 H_0^T \\ H_0 \Sigma_0 & H_0 \Sigma_0 H_0^T + R_0 \end{bmatrix},$$

respectively. It follows from Theorem 17.1.2 that, using the notation of equations (17.2.4) and (17.2.5), and assuming $H_0 \Sigma_0 H_0^T + R_0$ to be invertible,

$$\bar{x}_{0|0} = \bar{x}_0 + \Sigma_0 H_0^T (H_0 \Sigma_0 H_0^T + R_0)^{-1}(y_0 - H_0 \bar{x}_0) \quad (17.3.1)$$

$$\Sigma_{0|0} = \Sigma_0 - \Sigma_0 H_0^T (H_0 \Sigma_0 H_0^T + R_0)^{-1} H_0 \Sigma_0. \quad (17.3.2)$$

Step 2 The predictive step is made now from time 0 to time 1. Since (x_0, w_0) and (v_0, w_0) are statistically independent pairs and since $y_0 = Hx_0 + v_0$, (y_0, w_0) are also statistically independent. Hence, $x_1 = F_0 x_0 + G_0 w_0$ implies

$$x_{1|0} = F_0 x_{0|0} + G_0 w_0.$$

Applying Lemma 17.1.1 we obtain

$$\bar{x}_{1|0} = F_0 \bar{x}_{0|0}, \quad (17.3.3)$$

$$\Sigma_{1|0} = F_0 \Sigma_{0|0} F_0^T + G_0 Q_0 G_0^T. \quad (17.3.4)$$

Step 3 This step consists of calculation of the mean and covariance of the conditional distribution of $\begin{bmatrix} x_1 \\ y_1 \end{bmatrix}$ given y_0, which we write $\begin{bmatrix} x_{1|0} \\ y_{1|0} \end{bmatrix}$.

By hypothesis (x_0, v_1) and (v_0, v_1) are statistically independent pairs and, since $y_0 = H_0 x_0 + v_0$, the same is true of the pair (y_0, v_1). So it follows from $y_1 = H_1 x_1 + v_1$ that

$$y_{1|0} = H_1 x_{1|0} + v_1.$$

Then Lemma 17.1.1 shows that $y_{1|0}$ has mean $H_1 \bar{x}_{1|0}$ and covariance $H_1 \Sigma_{1|0} H_1^T + R_1$.

We also have

$$(x_1 - \bar{x}_{1|0})(y_1 - \bar{y}_{1|0})^T = (x_1 - \bar{x}_{1|0})(x_1 - \bar{x}_{1|0})^T H_1^T + (x_1 - \bar{x}_{1|0})v_1^T.$$

But $x_1 = A x_0 + G_0 w_0$, and the independence of pairs (x_0, v_1), (w_0, v_1) shows that (x_1, v_1) are also independent. Thus

$$E\{(x_1 - \bar{x}_{1|0})(y_1 - \bar{y}_{1|0})^T | y_0\} = \Sigma_{1|0} H_1^T.$$

Bringing these results together, it has been shown that the conditional pdf for $\begin{bmatrix} x_1 \\ y_1 \end{bmatrix}$ given y_0 has mean and covariance given by

$$\begin{bmatrix} \bar{x}_{1|0} \\ H_1 \bar{x}_{1|0} \end{bmatrix}, \quad \begin{bmatrix} \Sigma_{1|0} & \Sigma_{1|0} H_1^T \\ H_1 \Sigma_{1|0} & H_1 \Sigma_{1|0} H_1^T + R_1 \end{bmatrix}$$

Step 4 Using a remark of Section 17.2 concerning recursive calculation of conditional probabilities, and assuming that $H_1 \Sigma_{1|0} H_1^T + R_1$ is invertible, an application of Theorem 17.1.2 now shows that the conditional pdf for $\begin{bmatrix} x_1 \\ y_1 \end{bmatrix}$ given y_0 and y_1 has mean

$$\bar{x}_{1|1} = \bar{x}_{1|0} + \Sigma_{1|0} H_1^T (H_1 \Sigma_{1|0} H_1^T + R_1)^{-1} (y_1 - H_1 \bar{x}_{1|0}) \tag{17.3.5}$$

and covariance

$$\Sigma_{1|1} = \Sigma_{1|0} - \Sigma_{1|0} H_1^T (H_1 \Sigma_{1|0} H_1^T + R_1)^{-1} H_1 \Sigma_{1|0}. \tag{17.3.6}$$

Observe now that a cycle of computation has been completed. Starting with the initial data $\bar{x}_{0|0}$ and $\Sigma_{0|0}$ given by equations (17.3.1) and (17.3.2) we arrive at the equations (17.3.5) and (17.3.6) for $\bar{x}_{11}$ and $\Sigma_{1|1}$ via the intermediate calculation of $\bar{x}_{1|0}$ and $\Sigma_{1|0}$ in equations (17.3.3) and (17.3.4). Application of equations (17.3.3) and (17.3.4) is the *predictive*, or *time-update* step and then the incorporation of information on y_1, in the form of equations (17.3.5) and (17.3.6), is the *measurement-update* step.

Step 5 consists simply of the observation that the calculations of Steps 2, 3, 4 can be extended in a recursive way. It is assumed that the matrices $H_k \Sigma_{k|k-1} H_k^T + R_k$ are invertible for $k = 0, 1, 2, \ldots$. Note that, since $\Sigma_{k|k-1} \geq 0$, this is certainly the case if the covariance matrices R_k of the noise process $\{v_k\}$ affecting the observations are positive definite. The recursion can then be summarized in the following way:

(a) Define $\bar{x}_{0|-1} = \bar{x}_0$, $\Sigma_{0|-1} = \Sigma_0$, and then:

For $k = 0, 1, 2, \ldots$, set

(b) $\bar{x}_{k|k} = \bar{x}_{k|k-1} + \Sigma_{k|k-1} H_k^T (H_k \Sigma_{k|k-1} H_k^T + R_k)^{-1} (y_k - H_k \bar{x}_{k|k-1})$, (17.3.7)

$$\Sigma_{k|k} = \Sigma_{k|k-1} - \Sigma_{k|k-1} H_k^T (H_k \Sigma_{k|k-1} H_k^T + R_k)^{-1} H_k \Sigma_{k|k-1}. \quad (17.3.8)$$

These equations are derived from (17.3.5) and (17.3.6) and from the measurement update equations. They are followed by the predictive equations:

(c) $$\bar{x}_{k+1|k} = F_k \bar{x}_{k|k}, \qquad (17.3.9)$$

$$\Sigma_{k+1|k} = F_k \Sigma_{k|k} F_k^T + G_k Q_k G_k^T, \qquad (17.3.10)$$

derived from equations (17.3.3) and (17.3.4).

17.4 Discussion

1. Note that, as predicted in Section 17.2, a Markov process has been obtained. That is, the values of $\bar{x}_{k+1|k}$ and $\Sigma_{k+1|k}$ (i.e. the mean and covariance of the conditional pdf for x_{k+1} given $y_0, y_1, \ldots, y_k$) depend only on the data at time k and the preceding values $\bar{x}_{k|k-1}$ and $\Sigma_{k|k-1}$.

2. Observe that the evolution of the covariance matrices through (17.3.8) and (17.3.10) is independent of the $\bar{x}_{k+1|k}$ sequence, and can be computed independently.

3. Combining equations (17.3.7) and (17.3.9) we obtain a recurrence relation for $\bar{x}_{k+1|k}$, $k = 0, 1, 2, \ldots$. Defining the matrix

$$K_k = F_k \Sigma_{k|k-1} H_k^T (H_k \Sigma_{k|k-1} H_k^T + R_k)^{-1} \qquad (17.4.1)$$

and setting $\bar{x}_{0|-1} = \bar{x}_0$ the relation becomes

$$\bar{x}_{k+1|k} = (F_k - K_k H_k) \bar{x}_{k|k-1} + K_k y_k, \quad k = 0, 1, 2, \ldots. \qquad (17.4.2)$$

Here, K_k is known as the "gain matrix" and is generated via the sequence of covariance matrices.

4. Combination of the equations (17.3.8) and (17.3.10) gives the recursion for the sequence $\{\Sigma_{k+1|k}\}$. Set $\Sigma_{0|-1} = \Sigma_0$ and then for $k = 0, 1, 2, \ldots$ set

$$\Sigma_{k+1|k} = F_k \Sigma_{k|k-1} F_k^T + G_k Q_k G_k^T$$
$$- F_k \Sigma_{k|k-1} H_k^T (H_k \Sigma_{k|k-1} H_k^T + R_k)^{-1} H_k \Sigma_{k|k-1} F_k^T, \qquad (17.4.3)$$

and here a Riccati function can be recognized on the right-hand side. This function is time dependent in the sense that the coefficients A_k, G_k, H_k, Q_k, R_k are dependent on the time-step k. However, time-invariant systems are also of great importance and are the subject of the next section.

17.5 Time-invariant filters and a Riccati equation

Time-invariant systems (17.2.1) and (17.2.2) determine what is sometimes called the "standard" Kalman filter. Thus, the dependence of F, G, H, Q, R on k disappears. Recall that $F \in \mathbb{R}^{n \times n}$, $G \in \mathbb{R}^{n \times m}$ and $H \in \mathbb{R}^{p \times n}$. Matrix Q is the covariance matrix for the white noise process entering into equation (17.2.1) so that $Q \geq 0$. Similarly, R is the covariance for the noise process entering into (17.2.2), and it is convenient to assume throughout the discussion that $R > 0$.

Writing X_k for $\Sigma_{k|k-1}$ in equations (17.4.1)–(17.4.3), the standard Kalman filter takes the form:

Given $X_0 \geq 0$, the observations $y_0, y_1, \ldots$, and $x_0 = \bar{x}_0$, for $k = 0, 1, 2, \ldots$, define the gain matrices

$$K(X_k) = F X_k H^T (H X_k H^T + R)^{-1}, \qquad (17.5.1)$$

and then

$$x_{k+1} = (F - K(X_k)H) x_k + K(X_k) y_k. \qquad (17.5.2)$$

$$X_{k+1} = F X_k F^T + G Q G^T - F X_k H^T (H X_k H^T + R)^{-1} H X_k F^T. \qquad (17.5.3)$$

A straightforward computation, which is left to the reader, verifies that we may also write

$$X_{k+1} = (F - K(X_k)H) X_k (F - K(X_k)H)^T + G Q G^T + K(X_k) R K(X_k)^T. \qquad (17.5.4)$$

Now for an arbitrary sequence of $n \times m$ gain matrices $\{K_k\}_{k=0}^{\infty}$ and a given Y_0 define $\{Y_k\}_{k=1}^{\infty}$ by

$$Y_{k+1} = (F - K_k H) Y_k (F - K_k H)^T + G Q G^T + K_k R K_k^T. \qquad (17.5.5)$$

We are to compare the sequences $\{X_k\}$ and $\{Y_k\}$ generated by (17.5.4) and (17.5.5) and show that, if $0 \leq X_0 \leq Y_0$, then the sequence $\{X_k\}$, generated by the gains $K(X_k)$, is optimal in a sense to be made precise in the next theorem.

Theorem 17.5.1. *Let an arbitrary sequence* $\{K_k\}_{k=0}^{\infty} \subset \mathbb{R}^{n \times m}$ *be given and also a matrix* $Y_0 \geq 0$, *and let a sequence* $\{Y_k\}_{k=0}^{\infty}$ *be defined using equation* (17.5.5). *Let* X_0 *be a matrix for which* $0 \leq X_0 \leq Y_0$ *and define a sequence* $\{X_k\}_{k=0}^{\infty}$ *using* (17.5.4). *Then* $0 \leq X_k \leq Y_k$ *for* $k = 0, 1, 2, \ldots$.

Proof. Let us introduce the abbreviated notation

$$Y_{k+1} = f(Y_k, K_k) \qquad (17.5.6)$$

for equation (17.5.5). Then we may also write $X_{k+1} = f(X_k, K(X_k))$.

TIME-INVARIANT FILTERS AND A RICCATI EQUATION

The proof is by induction. The relation $0 \leq X_0 \leq Y_0$ is given and we assume that $0 \leq X_k \leq Y_k$. Then define

$$\hat{X}_{k+1} = f(Y_k, K(Y_k))$$

and it will first be proved that $\hat{X}_{k+1} \leq Y_{k+1}$. For brevity, write $K = K(Y_k)$ and $\hat{R} = HY_k H^T + R$ in the following calculation. Write $K_k = K + (K_k - K)$ in equation (17.5.5) and it is easily checked that

$$Y_{k+1} = FY_k F^T + GQG^T + (K\hat{R} - FY_k H^T)K^T + (K_k - K)(\hat{R}K^T - HY_k F^T)$$
$$+ (K\hat{R} - FY_k H^T)(K_k - K)^T - KHY_k F^T + (K_k - K)\hat{R}(K_k - K)^T.$$

But equation (17.5.1) now gives $K\hat{R} = FY_k H^T$ and so

$$Y_{k+1} = FY_k F^T + GQG^T - FY_k H^T \hat{R}^{-1} HY_k F^T + (K_k - K)\hat{R}(K_k - K)^T.$$

As in (17.5.3), (17.5.4) we may therefore write

$$Y_{k+1} = f(Y_k, K(Y_k)) + (K_k - K)\hat{R}(K_k - K)^T.$$

Since $R > 0$ we obtain $Y_{k+1} \geq \hat{X}_{k+1}$, or $f(Y_k, K_k) \geq f(Y_k, K(Y_k))$.

Thus, the choice of the gain $K(Y_k)$ produces a lower bound $\hat{X}_{k+1}$ for Y_{k+1}. Applying the same argument to the $\{X_k\}$ sequence shows that

$$f(X_k, K(Y_k)) \geq f(X_k, K(X_k)) = X_{k+1}. \tag{17.5.7}$$

However, the induction hypothesis $X_k \leq Y_k$ yields

$$f(X_k, K(Y_k)) = (F - K(Y_k)H)X_k(F - K(Y_k)H)^T + GQG^T + K(Y_k)RK(Y_k)^T$$
$$\leq (F - K(Y_k)H)Y_k(F - K(Y_k)H)^T + GQG^T + K(Y_k)RK(Y_k)^T$$
$$= f(Y_k, K(Y_k)) = \hat{X}_{k+1}.$$

Combining this with (17.5.7) we obtain $X_{k+1} \leq \hat{X}_{k+1}$.

It follows from (17.5.4) that $X_k \geq 0$ implies $X_{k+1} \geq 0$ and so we have $0 \leq X_{k+1} \leq \hat{X}_{k+1} \leq Y_{k+1}$ and the induction is complete. □

Lemma 17.5.2. *Let the sequence $\{X_k\}_{k=0}^{\infty}$ be defined as in Theorem 17.5.1. If (H, F) is a d-detectable pair, then there is a matrix Y such that $0 \leq X_k \leq Y$ for $k = 0, 1, 2, \ldots$.*

Proof. Since (H, F) is d-detectable there is a matrix K such that $F - KH$ is d-stable (see the remarks following Theorem 4.4.2). Generate a sequence $\{Z_k\}$ from $Z_0 = X_0$ and the recurrence relation $Z_{k+1} = f(Z_k, K)$, i.e. set $K_k \equiv K$ in equation (17.5.5). Then Theorem 17.5.1 can be applied and gives $0 \leq X_k \leq Z_k$, $k = 0, 1, 2, \ldots$, and for $j = 1, 2, \ldots$

$$Z_{j+1} - Z_j = -6ptf(Z_j, K) - f(Z_{j-1}, K) = (F - KH)(Z_j - Z_{j-1})(F - KH)^T.$$

Thus,
$$Z_j - Z_{j-1} = (F - KH)^{j-1}(Z_1 - Z_0)((F - KH)^T)^{j-1},$$

and hence
$$Z_n = Z_0 + \sum_{j=1}^{n}(Z_j - Z_{j-1}),$$
$$= Z_0 + \sum_{j=1}^{n}(F - KH)^{j-1}(Z_1 - Z_0)((F - KH)^T)^{j-1}.$$

As $F - KH$ is d-stable there is a matrix norm in which $\kappa := \|F - KH\| < 1$ and it follows that

$$\|Z_n\| \leq \|Z_0\| + \|Z_1 - Z_0\| \left(\sum_{j=1}^{\infty} \kappa^{2j-2}\right) =: \kappa_0$$

and κ_0 is independent of n. Thus, we may take $Y = \kappa_0 I$ and obtain $X_n \leq Z_n \leq \|Z_n\|I \leq \kappa_0 I = Y$. (Note that Lemma 5.6.10 and Theorem 5.6.9 of Horn and Johnson (1985) are used here.) □

Now let us state the main theorem.

Theorem 17.5.3. *Assume that $R > 0$, $Q \geq 0$, (H, F) is a d-detectable pair and (F, G) is d-stabilizable. Define a sequence $\{X_k\}_{k=0}^{\infty}$ by equation (17.5.3) and an $X_0 \geq 0$. Then there is a unique $X_+ \geq 0$ such that $X_k \to X_+$ as $k \to \infty$. The limit X_+ is the maximal solution of the DARE*

$$X = FXF^T + GQG^T - FXH^T(HXH^T + R)^{-1}HXF^T. \tag{17.5.8}$$

Furthermore, we have $X_+ \geq 0$ and the matrix

$$F - K(X_+)H = F - FX_+H^T(HX_+H^T + R)^{-1}H \tag{17.5.9}$$

is d-stable. Moreover, if (F, G) is controllable, then $X_+ > 0$.

It is an interesting fact that, if $X_0 = 0$ in this theorem then $X_k \leq X_{k+1}$ for $k = 0, 1, 2, \ldots$, and the existence of the bound proved in Lemma 17.5.2 leads to a straightforward proof that the $\{X_k\}$ is a convergent sequence. However, when $X_0 \geq 0$ and $X_0 \neq 0$, the sequence $\{X_k\}$ need not display monotonic behaviour (see Bitmead et al. (1985)) and the proof of the theorem takes some more work. Indeed, it is intuitively clear that, if equation (17.5.8) has a solution $X_+ \geq 0$ then the behaviour of the sequence $\{X_k\}$ will undoubtedly depend on the relationship of X_0 to X_+.

Let us first resolve the easier case in which $X_0 = 0$.

TIME-INVARIANT FILTERS AND A RICCATI EQUATION 383

Lemma 17.5.4. *Assume that $R > 0$, $Q \geq 0$, and that (H, F) is a d-detectable pair. Define a sequence $\{X_k\}_{k=0}^{\infty}$ by equation (17.5.3) using initial matrix $X_0 = 0$. Then $\{X_k\}$ is a nondecreasing sequence and converges to a matrix $X_+ \geq 0$ which satisfies the Riccati equation (17.5.8) and for which the matrix $F - K(X_+)H$ of equation (17.5.9) has all its eigenvalues in the closed unit disk.*

Furthermore, if (F, G) is a d-stabilizable pair then $X_+ \geq 0$ and the matrix of (17.5.9) is d-stable. Moreover, if (F, G) is controllable, then $X_+ > 0$.

Proof. Starting with $X_0 = W_0 = 0$ generate two sequences of matrices $\{X_k\}$ and $\{W_k\}$ by

$$X_{k+1} = f(X_k, K(X_k)), \quad W_{k+1} = f(W_k, K(X_{k+1}))$$

(see equations (17.5.5) and (17.5.6)). Then Theorem 17.5.1 implies that $X_{k+1} \leq W_{k+1}$ and (writing $K_k = K(X_k)$)

$$X_{k+1} - W_k = (F - K_k H)(X_k - W_{k-1})(F - K_k H)^T. \tag{17.5.10}$$

Now $X_0 = 0$ implies $X_1 = GQG^T \geq 0$ so that $X_1 - W_0 \geq 0$. Make the induction hypothesis $X_k - W_{k-1} \geq 0$ and (17.5.10) yields $X_{k+1} \geq W_k$. Consequently, $X_k \leq W_k \leq X_{k+1}$ and, by Lemma 17.5.2, $\{X_k\}$ is a nondecreasing sequence with an upper bound. It therefore has a limit X_+ and, taking limits in equation (17.5.3), it is found that X_+ is a solution of (17.5.8). Then use Theorem 13.1.1 and its corollary to show that all eigenvalues of $F - K(X_+)H$ are in the closed unit disk.

With the further hypothesis of stabilizability or controllability for the pair (F, G), Theorem 13.1.3 applies and yields the last part of the lemma. $\square$

Proof of Theorem 17.5.3. Let $\{X_k\}_{k=0}^{\infty}$ be defined as in the preceding lemma, and generate $\{\tilde{X}_k\}_{k=0}^{\infty}$ in the same way, but from an initial matrix $\tilde{X}_0 \geq 0$. We are to show that $\tilde{X}_k \to X_+$, the limit of the X_k-sequence.

By Lemma 17.5.2, $\{\tilde{X}_k\}$ is a bounded sequence in $\mathbb{R}^{n \times n}$ and we are to show that it has the unique limit point X_+.

Writing $K_+ = K(X_+)$ in equation (17.5.1), we have (from 17.5.4)

$$X_+ = (F - K_+ H)X_+(F - K_+ H)^T + GQG^T + K_+ R K_+^T,$$

and, by Theorem 17.5.1,

$$\tilde{X}_{k+1} = f(\tilde{X}_k, K(\tilde{X}_k)) \leq f(\tilde{X}_k, K_+),$$
$$= (F - K_+ H)\tilde{X}_k(F - K_+ H)^T + GQG^T + K_+ R K_+^T.$$

Thus,

$$\tilde{X}_{k+1} - X_+ \leq (F - K_+ H)(\tilde{X}_k - X_+)(F - K_+ H)^T$$

and hence, for $n = 1, 2, \ldots$,

$$\tilde{X}_{k+n} - X_+ \leq (F - K_+ H)^n (\tilde{X}_k - X_+)((F - K_+ H)^T)^n. \tag{17.5.11}$$

By Lemma 17.5.4, $F - K_+ H$ is d-stable and, if $\tilde{X}$ is the limit of any subsequence of $\{\tilde{X}_k\}$, it follows on taking the limit in (17.5.11) that $\tilde{X} \leq X_+$.

On the other hand, by Theorem 17.5.1,

$$X_{n+1} = f(X_n, K(X_n)) \leq f(X_n, K(\tilde{X}_n))$$
$$= (F - \tilde{K}_n H) X_n (F - \tilde{K}_n H)^T + GQG^T + \tilde{K}_n R \tilde{K}_n^T$$

and we write $\tilde{K}_n$ for $K(\tilde{X}_n)$. Making the induction hypothesis $X_n \leq \tilde{X}_n$ (which is certainly satisfied at $n = 0$) we obtain

$$X_{n+1} \leq (F - \tilde{K}_n H)^T \tilde{X}_n (F - \tilde{K}_n H)^T + GQG^T + \tilde{K}_n R \tilde{K}_n^T$$
$$= f(\tilde{X}_n, K(\tilde{X}_n)) = \tilde{X}_{n+1}.$$

It follows that $X_k \leq \tilde{X}_k$ for $k = 0, 1, 2, \ldots$ and hence that $X_+ \leq \tilde{X}$. Thus X_+ is the only limit point of $\{\tilde{X}_k\}_{k=0}^{\infty}$.

As in Lemma 17.5.4, the last statement of the theorem follows from Theorem 13.1.3. □

17.6 Properties of the observer system

The chapter concludes with some remarks concerning the "steady state" of the recurrence (17.5.3) achieved under the conditions of Theorem 17.5.3, for example. Thus, we take it that $X_k \to X_+ \geq 0$ as $k \to \infty$ and X_+ is the maximal solution of the DARE (17.5.8). Define the limiting Kalman gain:

$$K_+ = F X_+ H^T (H X_+ H^T + R)^{-1}.$$

The time-invariant linear system

$$\hat{x}_{k+1} = (F - K_+ H)\hat{x}_k + K_+ z_k \qquad (17.6.1)$$

(cf. (17.4.2)) is called an *observer* for the uncorrupted system

$$\xi_{k+1} = F\xi_k, \quad z_k = H\xi_k,$$

whatever the initial vector ξ_0 may be. The reason is that

$$\xi_{k+1} - \hat{x}_{k+1} = F(\xi_k - \hat{x}_k) + K_+ H \hat{x}_k - K_+ z_k$$
$$= F(\xi_k - \hat{x}_k) + K_+ H (\hat{x}_k - \xi_k)$$
$$= (F - K_+ H)(\xi_k - \hat{x}_k).$$

Thus, when $F - K_+ H$ is d-stable, $\|\xi_k - \hat{x}_k\| \to 0$ as $k \to \infty$. The vectors $\hat{x}_k$ obtained from (17.6.1) with the help of the maximal solution of (17.5.8) give approximations for the state vectors ξ_k when k is large; and this is the case whatever the initial vectors ξ_0 and $\hat{x}_0$ may be.

Theorem 17.5.3 summarizes our main conclusions concerning the standard Kalman filter and hence the observer system. But let us note also the weaker conclusions which follow when the stabilizability condition on (F, G) is dropped. The following statement follows from Theorem 13.1.1 and its Corollary 13.1.2.

Theorem 17.6.1. *If $R > 0$, $Q \geq 0$ and (H, F) is a d-detectable pair, then equation (17.5.8) has a unique maximal solution X_+. Furthermore $X_+ \geq 0$ and, if*

$$K_+ := FX_+H^T(HX_+H^T + R)^{-1},$$

then all eigenvalues of $F - K_+H$ lie in the closed unit disk.

17.7 Exercises

17.7.1. Verify the equivalence of equations (17.5.3) and (17.5.4).

17.7.2. (This exercise shows that the sequence $\{X_k\}$ determined by (17.5.3) need not be monotonic.)

Let $R = 1$, $H = [0\ 1]$, $F = \begin{bmatrix} 0 & 1 \\ 1 & 0 \end{bmatrix}$; $GQG^T = I_2$; $X_0 = \begin{bmatrix} 100 & 0 \\ 0 & 10 \end{bmatrix}$.

Compute X_1, from (17.5.3) and the maximal solution X_+ of (17.5.8) and verify that $X_0 > X_+$, but $X_0 \not> X_1$.

17.8 Notes

The Kalman filter (introduced originally in Kalman (1960a) and Kalman and Bucy (1961)) is one of the most influential developments in mathematics and engineering of the past several decades. Today, many expositions of this theory and its applications are available. We mention here Davis (1977), Kwakernaak and Sivan (1972), Anderson and Moore (1971), the last two books being oriented towards engineering. Rigorous mathematical developments of Kalman filter theory can be found in Ruymgaart and Soong (1988) and Catlin (1981). For a general overview, perspective, and historical background of the Kalman filter see the volume of Antoulas (1991).

The proof of Theorems 17.5.1 and 17.5.2 given here is due to Caines and Mayne (1970 and 1971). For a further strengthening of the important Theorem 17.5.3 see Wimmer (1994b). Exercise 17.7.2 is taken from the paper of Bitmead et al. (1985).

18
THE TOTAL LEAST SQUARES TECHNIQUE

Consider the problem of finding an $n \times p$ matrix X, if any, satisfying the linear equation $AX = B$, where A is $m \times n$ and B is $m \times p$ and let us focus attention on inconsistent equations. Thus, A is not invertible and $\text{Im } B \not\subseteq \text{Im } A$. It is well-known that, if A^+ denotes the Moore–Penrose inverse of A, then the matrix $X_0 = A^+ B$ has the following property: it is the matrix of least Frobenius (euclidean) norm (say $\|M\|_F = (\sum |m_{ij}|^2)^{1/2}$) in the manifold of matrices for which the minimum of $\|AX - B\|_F^2$ (over $n \times p$ matrices X) is attained. (See Lancaster and Tismenetsky (1985), for example.) It is classically described as the "best least squares solution" of $AX = B$.

Notice that $AX_0 = (AA^+)B$ and, as AA^+ is just the orthogonal projector onto $\text{Im } A$, X_0 is the minimum norm solution of the consistent system $AX = (AA^+)B$ obtained by projecting $\text{Im } B$ onto $\text{Im } A$. Thus, in this process, the subspace $\text{Im } A$ plays the pivotal role and the "right-hand side" matrix B is "adjusted" to produce a solvable system.

In many applications inconsistent systems arise in an attempt to model consistent equations which are corrupted by observation, modelling, or computational errors. The argument given above can be criticized in this context for holding A (or at least $\text{Im } A$) inviolate and adjusting only B. If A and B are real and are equally subject to error it can be argued that one should try to minimize

$$\|A - P\|_F^2 + \|B - Q\|_F^2 = \|[A \ B] - [P \ Q]\|_F^2$$

over pairs of matrices $P \in \mathbb{R}^{m \times n}$ and $Q \in \mathbb{R}^{m \times p}$ subject to the constraint that the equation $PX = Q$ is solvable. Any $X \in \mathbb{R}^{n \times p}$ for which this is true is said to solve the "total least squares" (TLS) problem for $[A, B]$. This problem is the subject of Section 18.1. If the further condition is made that X be symmetric (so that $p = n$) the problem will be described as the TLS problem for $[A \ B]$ with symmetry. This is the subject of Section 18.2.

The best least squares solution of linear systems remains a very important and widely useful technique after a 200-year history. The total least squares idea can be traced back for about 100 years, but has come into prominence more recently with investigations of numerical techniques (notably Golub and Van Loan

(1980)). The strong connection of solutions of the total least squares problems with solutions of algebraic Riccati equations (of CARE type) was revealed by De Moor and David (1992), and their line of argument is followed here. Our concern is with the algebraic connections between the two problems, rather than the issues raised by numerical analysis.

18.1 The total least squares problem

As described above, matrices $A \in \mathbb{R}^{m \times n}$ and $B \in \mathbb{R}^{m \times p}$ are given and matrices $P = [p_{ij}]$, $Q = [q_{ij}]$ of corresponding sizes are sought for which the error functional

$$E(P, Q) := \|A - P\|_F^2 + \|B - Q\|_F^2 \qquad (18.1.1)$$

is minimized subject to the constraint that $PX = Q$ for some $n \times p$ matrix X. For any real matrix $M = [m_{ij}]$ we have $\|M\|_F^2 = \text{tr}(M^T M) = \sum_{i,j} m_{ij}^2$, where tr denotes the *trace* of a matrix and recall that, if MN and NM are well-defined matrix products, then they have the same trace.

Our minimization problem can be tackled using the familiar method of Lagrange multipliers, one multiplier for each scalar equation of constraint. Thus, a Lagrangian quadratic form is introduced as follows:

$$\mathcal{L}(P, Q, X) := E(P, Q) + 2 \sum_{i=1}^{m} \sum_{j=1}^{p} \ell_{ij}(PX - Q)_{ij} \qquad (18.1.2)$$

(and the "2" is introduced for convenience). Write

$$\mathcal{L}(P, Q, X) = \text{tr}\{(A^T - P^T)(A - P) + (B^T - Q^T)(B - Q)\} + 2 \sum_{i,j} \ell_{ij}(PX - Q)_{ij}.$$

Equating all partial derivatives of $\mathcal{L}$ with respect to p_{ij} ($1 \le i \le m$, $1 \le j \le n$) to zero leads to the equation

$$P - A + LX^T = 0, \qquad (18.1.3)$$

where $L = [\ell_{ij}]$ is the matrix of Lagrange multipliers. Equating the derivatives with respect to the q_{ij} to zero gives

$$Q - B - L = 0. \qquad (18.1.4)$$

The derivatives with respect to x_{ij} give

$$P^T L = 0 \qquad (18.1.5)$$

and, finally, there are the constraint equations themselves

$$PX - Q = 0. \qquad (18.1.6)$$

It is now a matter of computation with equations (18.1.3)–(18.1.6) to show that X satisfies a Riccati equation. The second and third equations give $P^T Q = P^T B$ and combining this with (18.1.6) gives

$$P^T P X = P^T B. \tag{18.1.7}$$

Now equations (18.1.4) and (18.1.6) give $L = PX - B$ and substituting in (18.1.3) gives $P(I_n + XX^T) = A + BX^T$. As $I_n + XX^T$ is positive definite, we have

$$P = (A + BX^T)(I_n + XX^T)^{-1}. \tag{18.1.8}$$

Substituting this expression for P in (18.1.7):

$$(A^T + XB^T)(A + BX^T)(I_n + XX^T)^{-1}X = (A^T + XB^T)B.$$

However, it is easily seen that $(I_n + XX^T)^{-1}X = X(I_p + X^T X)^{-1}$ and so

$$(A^T + XB^T)(A + BX^T)X - (A^T + XB^T)B(I_p + X^T X) = 0,$$

and hence

$$X(B^T A)X + (A^T A)X - X(B^T B) - A^T B = 0. \tag{18.1.9}$$

We have proved:

Theorem 18.1.1. *Let $A \in \mathbb{R}^{m \times n}$ and $B \in \mathbb{R}^{m \times p}$. Then all solutions of the TLS problem for $[A \ B]$ (if any) satisfy the Riccati equation (18.1.9).*

Referring to the theory of Section 7.1 we see that solutions of (18.1.9) correspond to the p-dimensional T-invariant graph subspaces of the $(p+n) \times (p+n)$ matrix

$$T = \begin{bmatrix} -B^T B & B^T A \\ A^T B & -A^T A \end{bmatrix} = -\begin{bmatrix} B^T \\ -A^T \end{bmatrix} [B \ -A], \tag{18.1.10}$$

and T is clearly negative semidefinite, $T \leq 0$. Let the eigenvalues of T be

$$\lambda_{p+n} \leq \lambda_{p+n-1} \leq \cdots \leq \lambda_1 \leq 0. \tag{18.1.11}$$

Our next task is to show that, if the T-invariant subspace $\mathcal{S}_0$ corresponding to $\lambda_1, \ldots, \lambda_p$ is a graph subspace, then it determines a total least squares solution X for $[A \ B]$.

Let $\mathcal{S}$ be any p-dimensional invariant subspace of T. Since G is real and symmetric we may assume without loss of generality that $\mathcal{S}$ is the span of orthonormal eigenvectors $v_1, \ldots, v_p$. Let $\lambda_{i_1}, \ldots, \lambda_{i_p}$ be the corresponding eigenvalues and write

$$\Lambda = \text{diag}\,[\lambda_{i_1}, \ldots, \lambda_{i_p}], \quad [v_1\ v_2\ \ldots\ v_p] = \begin{bmatrix} Y \\ Z \end{bmatrix},$$

(cf. Theorem 7.1.2) so that Y is $p \times p$ and Z is $m \times p$. We have

$$\begin{bmatrix} -B^T B & B^T A \\ A^T B & -A^T A \end{bmatrix} \begin{bmatrix} Y \\ Z \end{bmatrix} = \begin{bmatrix} Y \\ Z \end{bmatrix} \Lambda, \tag{18.1.12}$$

and $\mathcal{S}$ is a graph subspace if and only if Y^{-1} exists. Furthermore, when this is the case, $X = ZY^{-1}$ and (as in equation (7.1.3)):

$$\begin{bmatrix} -B^T B & B^T A \\ A^T B & -A^T A \end{bmatrix} \begin{bmatrix} I \\ X \end{bmatrix} = \begin{bmatrix} I \\ X \end{bmatrix} (Y \Lambda Y^{-1}).$$

The crux of the following argument is the simple observation that the relation

$$\text{tr}\,(-B^T B + B^T A X) = \text{tr}\,(\Lambda) \tag{18.1.13}$$

follows from the preceding equation.

Set $Q = PX$ (equation (18.1.6)) in the expression for the error functional and we have

$$\begin{aligned} E(P, PX) &= \|A - P\|_F^2 + \|B - PX\|_F^2, \\ &= \text{tr}\,(A^T A + B^T B) + \text{tr}\,(P^T P - 2A^T P + X^T P^T P X - 2B^T P X), \\ &= \text{tr}\,(A^T A + B^T B) + \text{tr}\,\{-2(A^T + XB^T)P + P^T P(I + XX^T)\}. \end{aligned}$$

Using (18.1.8) write $P(I + XX^T) = A + BX^T$, and

$$\begin{aligned} E(P, PX) &= \text{tr}\,(A^T A + B^T B) + \text{tr}\,\{-2(A^T + XB^T)P + P^T(A + BX^T)\}, \\ &= \text{tr}\,(A^T A + B^T B) - \text{tr}\,\{(A^T + XB^T)P\}, \\ &= \text{tr}\,(A^T A + B^T B) - \text{tr}\,\{(A^T + XB^T)(A + BX^T)(I + XX^T)^{-1}\}, \end{aligned} \tag{18.1.14}$$

using (18.1.8) again. But the Riccati equation (18.1.9) gives

$$(A^T + XB^T)B = (A^T + XB^T)AX$$

and it follows that

$$(A^T + XB^T)(A + BX^T)(I + XX^T)^{-1} = (A^T + XB^T)A.$$

Thus, from equation (18.1.14),

$$E(P, PX) = \operatorname{tr}(A^T A + B^T B) - \operatorname{tr}(A^T A + X B^T A),$$
$$= \operatorname{tr}(B^T B - X B^T A) \qquad (18.1.15)$$
$$= -\operatorname{tr}\Lambda$$

by (18.1.13). It is apparent that $E(P, PX)$ is minimized by chosing $\lambda_{i_j} = \lambda_j$ for $j = 1, 2, \ldots, p$.

Theorem 18.1.2. *Let* $A \in \mathbb{R}^{m \times n}$, $B \in \mathbb{R}^{n \times p}$, *and* T *be defined as in equation* (18.1.10) *with eigenvalues as in* (18.1.11). *If* $\lambda_{p+1} < \lambda_p$ *and the* T-*invariant subspace corresponding to* $\{\lambda_1, \ldots, \lambda_p\}$ *is a graph subspace, then it determines the unique TLS solution* X *for* $[A \ B]$ *(and* $X = ZY^{-1}$ *as constructed from equation* (18.1.12)*).*

Notice that if $\lambda_{p+1} = \lambda_p$ the situation is indeterminate. There may be no solution, one solution, or infinitely many solutions all achieving the same minimum for E. Note that this situation is to be expected from the more familiar and intimately connected analysis via the singular-value decomposition of $[A \ B]$. In the papers of De Moor and David (1992) and Golub and Van Loan (1980) the case $\lambda_{p+1} < \lambda_p$ is said to be "generic". Algorithms for the "nongeneric" case, $\lambda_{p+1} = \lambda_p$, have been studied by Van Huffel and Vandewalle (1988).

18.2 Total least squares with a symmetry constraint

As in the preceding section, our topic concerns the (generally inconsistent) linear matrix equation $AX = B$, but now it is required that the optimal choice of X be symmetric, $X^T = X$. This implies that A and B must now have the same size, say $m \times n$, and $X \in \mathbb{R}^{n \times n}$. A matrix X which minimizes the error functional $E(P, Q)$ of equation (18.1.1) subject to the constraints $PX = Q$ and $X^T = X$ will be said to solve the TLS problem for $[A \ B]$ with symmetry.

The Lagrange method is applied again to the error functional $E(P, Q)$, but now there are two matrix equations of constraint: $PX = Q$ and $X^T = X$. Accordingly, we consider a modified Lagrangian (see equation (18.1.2)),

$$\mathcal{L}_s(P, Q, X) = \mathcal{L}(P, Q, X) + 2 \sum_{i,j=1}^{n} m_{ij}(x_{ij} - x_{ji}),$$

where the m_{ij} are Lagrange multipliers.

The necessary conditions for a critical point take the form:

$$P - A + LX^T = 0, \quad Q - B - L = 0,$$
$$P^T L + M - M^T = 0,$$
$$PX = Q, \quad X^T = X.$$

Eliminating L and Q, and writing $Z = M^T - M$, one easily obtains

$$P = (A + BX)(I + X^2)^{-1}, \tag{18.2.1}$$
$$(P^T P)X = P^T B + Z.$$

Substitute from the first equation into the second to obtain

$$(A^T + XB^T)(A + BX)(I + X^2)^{-1}X = (A^T + XB^T)B + (I + X^2)Z$$

and hence

$$(A^T + XB^T)(A + BX)X - (A^T + XB^T)B(I + X^2) = (I + X^2)Z(I + X^2).$$

Multiplying out one obtains (cf. equation (18.1.9))

$$X(B^T A)X + (A^T A)X - X(B^T B) - A^T B = (I + X^2)Z(I + X^2).$$

Taking the transpose and recalling $Z^T = -Z$,

$$X(A^T B)X - (B^T B)X + X(A^T A) - B^T A = -(I + X^2)Z(I + X^2).$$

Finally, add the last two equations to obtain a Riccati equation for X:

$$X(A^T B + B^T A)X + (A^T A - B^T B)X + X(A^T A - B^T B) - (A^T B + B^T A) = 0. \tag{18.2.2}$$

Theorem 18.2.1. *Let $A, B \in \mathbb{R}^{m \times n}$. Then all solutions of the TLS problem for $[A \; B]$ with symmetry (if any) satisfy the Riccati equation (18.2.2).*

Introducing the abbreviations

$$R = A^T A - B^T B, \quad S = A^T B + B^T A \tag{18.2.3}$$

the Riccati equation becomes

$$XSX + RX + XR - S = 0$$

and the solutions are connected to the invariant subspaces of the $2n \times 2n$ matrix

$$T := \begin{bmatrix} R & S \\ S & -R \end{bmatrix} \tag{18.2.4}$$

(see Section 7.1). Note that R, S and T are all real symmetric matrices.

Let us define the skew-symmetric matrix $K = \begin{bmatrix} 0 & I_n \\ -I_n & 0 \end{bmatrix}$ and observe that $KT = -TK$. Thus, T is K-skew-symmetric in the sense of Section 3.2. Because T is also real and symmetric, Theorem 3.2.4 yields simple canonical forms for T and K. Indeed, that theorem shows that there is a real nonsingular matrix W such that

$$W^{-1}TW = \text{diag}\,[\lambda_k, -\lambda_k, \ldots, \lambda_1, -\lambda_1, 0, 0, \ldots, 0] \quad (18.2.5)$$

where $\lambda_k \geq \lambda_{k-1} \geq \cdots \geq \lambda_1 > 0$ are all the positive eigenvalues of T, and

$$W^T K W = \text{diag}\,\left[\begin{bmatrix} 0 & 1 \\ -1 & 0 \end{bmatrix}, \ldots, \begin{bmatrix} 0 & 1 \\ -1 & 0 \end{bmatrix}\right] \quad (18.2.6)$$

with n diagonal blocks of size 2×2.

It is apparent that odd-numbered columns of W generate an invariant subspace $\mathcal{S}_0$ of T which is n-dimensional and K-neutral. Furthermore, if all eigenvalues of T are distinct (implying $k = n$), it is clear that there are as many as 2^n n-dimensional, T-invariant, and K-neutral subspaces. It follows from part (β) of Proposition 7.2.1 that, subject to the graph-subspace property, each of these subspaces may generate a real symmetric solution of equation (18.2.2). We will show that, under suitable conditions, the subspace $\mathcal{S}_0$ determines a solution X which solves the TLS problem with symmetry.

Lemma 18.2.2. *If X solves the TLS problem for $[A\ B]$ with symmetry, then*

$$E(P, PX) = \text{tr}\,(B^T B - X B^T A). \quad (18.2.7)$$

It is rather surprising that this formula is the same as that for the TLS problem without symmetry (equation (18.1.15)), although it applies to a more restrictive class of matrices P and X.

Proof. Equation (18.2.2) can be written in the form

$$(A^T + XB^T)(B - AX) = -(B^T - XA^T)(A + BX)$$

and expresses the fact that the matrix on the left is skew-symmetric. Using this equation write

$$(A^T + XB^T)B = (A^T + XB^T)AX - (B^T - XA^T)(A + BX)$$

and hence

$$(A^T + XB^T)(A + BX) = (A^T + XB^T)A(I + X^2) - (B^T - XA^T)(A + BX)X. \quad (18.2.8)$$

Using equation (18.2.1) instead of (18.1.8) it is found that equation (18.1.14) holds in this context too, and so

$$E(P, PX) = \operatorname{tr}(A^T A + B^T B) - \operatorname{tr}\{(A^T + XB^T)(A + BX)(I + X^2)^{-1}\}$$
$$= \operatorname{tr}(A^T A + B^T B) - \operatorname{tr}\{(A^T + XB^T)A\}$$
$$+ \operatorname{tr}\{(B^T - XA^T)(A + BX)X(I + X^2)^{-1}\},$$

using (18.2.8). But the last term is zero, as it is the trace of the product of the skew-symmetric matrix $(B^T - XA^T)(A + BX)$ and the symmetric matrix $X(I + X^2)^{-1}$. Equation (18.2.7) follows. □

Theorem 18.2.3. *Let $A, B \in \mathbb{R}^{m \times n}$ and T be defined by equations (18.2.3) and (18.2.4), with eigenvalues as described in equation (18.2.5). Then there is an n-dimensional, T-invariant, and K-neutral subspace S_0 such that $\sigma(T|S_0) = \{\lambda_k, \ldots, \lambda_1, 0\}$ (or $\sigma(T|S_0) = \{\lambda_k, \ldots, \lambda_1\}$ if $k = n$).*

If S_0 is a graph subspace it determines a real symmetric solution X of equation (18.2.2) which solves the TLS problem for $[A \ B]$ with symmetry. In this case, the solution is unique if, in addition, $k = n$.

Proof. The existence of S_0 with the properties described in the first statement was established in the discussion of the canonical forms (18.2.5) and (18.2.6) (it is generated by the odd-numbered columns of matrix W).

For any symmetric solution X of (18.2.2) we have

$$\begin{bmatrix} R & S \\ S & -R \end{bmatrix} \begin{bmatrix} I \\ X \end{bmatrix} = \begin{bmatrix} I \\ X \end{bmatrix} (U \Lambda U^{-1}) \qquad (18.2.9)$$

where Λ is the $n \times n$ diagonal matrix of eigenvalues of $T|S$ and S is the T-invariant subspace corresponding to X. Hence

$$\operatorname{tr}(R + SX) = \operatorname{tr}\{A^T A - B^T B + (A^T B + B^T A)X\} = \operatorname{tr}(\Lambda). \qquad (18.2.10)$$

It follows from Lemma 18.2.2 that X must be chosen to maximize $\operatorname{tr}(XB^T A)$. Equation (18.2.10) shows that this is done by maximizing $\operatorname{tr}(\Lambda)$. Thus, the choice of S_0 of the theorem does determine the optimal X - provided S_0 is a graph subspace.

Finally, note that when $k = n$ the subspace S_0 is uniquely defined. □

Useful conditions under which the subspace S_0 of Theorem 18.2.3 is a graph subspace are not known to the authors at the time of writing. However, one (rather restrictive) condition is presented in:

Proposition 18.2.4. *If, in Theorem 18.2.3 we have $k = n$ (so that $\Lambda > 0$) and $R = A^T A - B^T B \geq 0$, then S_0 is a graph subspace and determines the unique TLS solution for $[A \ B]$ with symmetry.*

Proof. Let the columns of a matrix $\begin{bmatrix} U \\ V \end{bmatrix}$ be determined by real orthogonal eigenvectors of T corresponding to $\lambda_n, \lambda_{n-1}, \ldots, \lambda_1$, where U, V are both $n \times n$ matrices. Then
$$U^T U + V^T V = I$$
and
$$\begin{bmatrix} R & S \\ S & -R \end{bmatrix} \begin{bmatrix} U \\ V \end{bmatrix} = \begin{bmatrix} U \\ V \end{bmatrix} \Lambda.$$
Hence
$$U^T RU + U^T SV + V^T SU - V^T RV = \Lambda > 0.$$
Let $x \in \operatorname{Ker} U$ and multiply on the right and left by x, x^T, respectively, to obtain
$$-(Vx)^T R(Vx) = x^T \Lambda x.$$
Since $R \geq 0$ there is a contradiction unless $x = 0$. Thus, U must be invertible. Hence $\mathcal{S}_0$ is a graph subspace and the proposition is established. $\square$

18.3 Notes

The presentation of this chapter has been guided by the original work of De Moor and David (1992), however our presentation reveals that there are some interesting open questions. In particular, the determination of useful sufficient conditions on A and B ensuring the existence of TLS solutions with and without symmetry, and the resolution of the "indeterminacy" mentioned after the statement of Theorem 18.1.2.

Computational aspects of the total least squares problem are discussed in the recent book of Van Huffel and Vandewalle (1991).

19
CANONICAL FACTORIZATION

Let Γ be a simple closed rectifiable contour in the extended complex plane $\overline{\mathbb{C}} = \mathbb{C} \cup \{\infty\}$, which partitions $\overline{\mathbb{C}} \backslash \Gamma$ into two connected components Γ_+ and Γ_-. Let an $n \times n$ matrix function $W(\lambda)$ be given whose domain includes Γ. We say that $W(\lambda)$ admits a (right) *canonical factorization* (relative to Γ) if $W(\lambda)$ can be factored as

$$W(\lambda) = W_-(\lambda)W_+(\lambda)_1, \quad \lambda \in \Gamma$$

where $W_+(\lambda)$ (resp. $W_-(\lambda)$) is an $n \times n$ matrix function which is analytic and invertible in Γ_+ (resp. Γ_-) and continuous and invertible on $\Gamma_+ \cup \Gamma$ (resp. $\Gamma_- \cup \Gamma$).

The concept of canonical factorization plays a crucial role in numerous applications, and therefore has been thoroughly studied, especially in recent years. The conclusion of this chapter and its main objective concern canonical factorizations of *rational* matrix functions $W(\lambda)$ with respect to the imaginary axis. In this case we choose $\Gamma = \{i\alpha : \alpha \in \mathbb{R}\}$, and we choose Γ_+ to be the open right half-plane. This choice is motivated firstly by the importance of canonical factorization with respect to the imaginary axis in various applications (notably H^∞ control; see Chapter 20) and, secondly, because solutions to certain algebraic Riccati equations play an essential role in such factorizations.

The chapter consists of three sections. In the first section a criterion is derived for existence of a canonical factorization in the geometric terms of spectral invariant subspaces and this is done for the general case of any simple closed rectifiable contour Γ. In section 19.2 the corresponding result for real factors of real rational matrix functions is considered. In the last section a particular class of canonical factorizations with respect to the imaginary axis — spectral factorizations — is studied, and the connection with algebraic matrix Riccati equations is established.

19.1 Canonical factorization in geometric terms

Throughout this section Γ is a closed simple rectifiable contour in $\overline{\mathbb{C}}$, and $W(\lambda)$ is an $n \times n$ rational matrix function having a (not necessarily minimal) realization

$$W(\lambda) = D + C(\lambda I - A)^{-1} B. \tag{19.1.1}$$

For $W(\lambda)$ to admit a canonical factorization, it is obviously necessary that $W(\lambda)$ have no poles on Γ. This will be ensured if we assume that A has no eigenvalues on Γ. Furthermore, if the realization (19.1.1) is minimal then, by Theorem 6.2.2, $W(\lambda)$ has no poles on Γ precisely when A has no eigenvalues on Γ.

Another obvious necessary condition for a canonical factorization of $W(\lambda)$ is that $W(\lambda)$ has no zeros on Γ or, equivalently, $W(\lambda)^{-1}$ has no poles on Γ. This condition can be expressed in terms of the realization (19.1.1) as follows.

Lemma 19.1.1. *Assume that $\sigma(A) \cap \Gamma = \emptyset$. Then:*

(i) *If $\infty \in \Gamma$, then $W(\lambda)$ has no zeros on Γ precisely when $\det D \neq 0$ and $\sigma(A - BD^{-1}C) \cap \Gamma = \emptyset$.*

(ii) *Suppose $\infty \notin \Gamma$ and assume that $\det D \neq 0$. Then $W(\lambda)$ has no zeros on Γ precisely when $\sigma(A - BD^{-1}C) \cap \Gamma = \emptyset$.*

Proof. We prove part (i) only. The proof of (ii) is similar. Assume that $W(\lambda)$ has no zeros on Γ. Then, as $\infty \in \Gamma$, D must be invertible. By formula (6.3.14) we have

$$W(\lambda)^{-1} = D^{-1} - D^{-1}C(\lambda I - (A - BD^{-1}C))^{-1}BD^{-1}. \qquad (19.1.2)$$

Next, we prove that $\sigma(A - BD^{-1}C) \cap \Gamma = \emptyset$. Assume that $(A - BD^{-1}C)x = \lambda_0 x$ for some $\lambda_0 \in \Gamma$. We have to prove that $x = 0$. Now compute

$$\begin{aligned}W(\lambda_0)D^{-1}Cx &= Cx + C(\lambda_0 I - A)^{-1}BD^{-1}Cx \\ &= Cx + (\lambda_0 I - A)^{-1}[Ax - (A - BD^{-1}C)x] \\ &= Cx + C(\lambda_0 I - A)^{-1}(A - \lambda_0 I)x = 0,\end{aligned}$$

and since $\det W(\lambda_0) \neq 0$, we obtain $Cx = 0$. But then $Ax = \lambda_0 x$, and in view of the hypothesis $\sigma(A) \cap \Gamma = \emptyset$, we must have $x = 0$.

Conversely, assume $\det D \neq 0$ and $\sigma(A - BD^{-1}C) \cap \Gamma = \emptyset$. Then formula (19.1.2) shows that $W(\lambda)^{-1}$ has no poles on Γ, i.e. $W(\lambda)$ has no zeros on Γ. □

The necessary conditions for canonical factorization mentioned in Lemma 19.1.1 are generally not sufficient. We need an extra condition which "matches" certain spectral invariant subspaces of A and $A - BD^{-1}C$. Denote by $\mathcal{R}_+(X)$ (resp. $\mathcal{R}_-(X)$) the sum of root subspaces of X corresponding to the eigenvalues of X in Γ_+ (resp. Γ_-).

Theorem 19.1.2. *Let $W(\lambda)$ be an $n \times n$ rational matrix function without poles and zeros on Γ and having realization (19.1.1), where $\sigma(A) \cap \Gamma = \emptyset$ and D is invertible. Then $W(\lambda)$ admits a canonical factorization if and only if*

$$\mathbb{C}^m = \mathcal{R}_-(A - BD^{-1}C) \dotplus \mathcal{R}_+(A), \qquad (19.1.3)$$

where A is an $m \times m$ matrix. When this equation holds a canonical factorization of W is constructed in the following way:

(1) *Choose bases* $x_1, \ldots, x_r$ *for* $\mathcal{R}_-(A - BD^{-1}C)$ *and* $x_{r+1}, \ldots, x_n$ *for* $\mathcal{R}_+(A)$, *and put*
$$T = [x_1 \ldots x_r | x_{r+1} \ldots x_n]. \qquad (19.1.4)$$
Then T is nonsingular.

(2) *Make the following partitions:*
$$T^{-1}AT = \begin{bmatrix} A_{11} & A_{12} \\ A_{21} & A_{22} \end{bmatrix}, \quad T^{-1}B = \begin{bmatrix} B_1 \\ B_2 \end{bmatrix}, \quad CT = (C_1 \; C_2),$$
conformally with the partitioning of T in (19.1.4).

(3) *Put*
$$W_+(\lambda) = I + D^{-1}C_1(\lambda I - A_{11})^{-1}B_1, \qquad (19.1.5)$$
$$W_-(\lambda) = D + C_2(\lambda I - A_{22})^{-1}B_2. \qquad (19.1.6)$$
Then $W = W_-W_+$ *is a canonical factorization.*

For the proof of Theorem 19.1.2 we need auxilliary results on block triangular matrices which are also of independent interest (especially Theorem 19.1.4 below).

Lemma 19.1.3. *Let*
$$X = \begin{bmatrix} X_1 & X_{12} \\ 0 & X_2 \end{bmatrix}$$
be a block triangular $m \times m$ matrix, where X_1 is $m_1 \times m_1$ and X_2 is $m_2 \times m_2$. Then for any set $\Omega \subset \mathbb{C}$ the spectral X_2-invariant subspace $\mathcal{R}_\Omega(X_2)$ corresponding to the eigenvalues of X_2 in Ω is given by
$$\mathcal{R}_\Omega(X_2) = \left\{ x_2 \in \mathbb{C}^{m_2} \;\middle|\; \begin{bmatrix} x_1 \\ x_2 \end{bmatrix} \in \mathcal{R}_\Omega(X) \text{ for some } x_1 \in \mathbb{C}^{m_1} \right\},$$
where $\mathcal{R}_\Omega(X)$ is the analogous subspace for the matrix X.

Proof. Let $p(\lambda)$ be the characteristic polynomial of the restriction $X|\mathcal{R}_\Omega(X)$. Then $\mathcal{R}_\Omega(X) = \operatorname{Ker} p(X)$. For a vector $x \in \mathbb{C}^m$ the second component of $p(X)x$, which we denote by $(p(X)x)_2$, is equal to $p(X_{22})x_2$. So if $x \in \mathcal{R}_\Omega(X)$, then $p(X_{22})x_2 = 0$. This proves that
$$\left\{ x_2 \;\middle|\; \begin{bmatrix} x_1 \\ x_2 \end{bmatrix} \in \mathcal{R}_\Omega(X) \text{ for some } x_1 \right\} \subseteq \mathcal{R}_\Omega(X_{22}).$$

To prove the converse inclusion choose $x \in \mathbb{C}^m$ such that $x_2 \in \mathcal{R}_\Omega(X_{22})$ and construct a vector u such that $u_2 = x_2$ and $u \in \mathcal{R}_\Omega(X)$. Put $q(\lambda) = \prod_{i=1}^r (\lambda - \lambda_i)^m$, where $\lambda_1, \ldots, \lambda_r$ are the eigenvalues of X which do not lie in

Ω. Since the polynomials $p(\lambda)$ and $q(\lambda)$ do not have a common zero, there are polynomials $g(\lambda)$ and $h(\lambda)$, such that $g(\lambda)p(\lambda) + h(\lambda)q(\lambda) = 1$. Note that $(p(X)x)_2 = p(X_{22})x_2 = 0$. So we get $(g(X)p(X)x)_2 = 0$. Use the fact that $p(\lambda)q(\lambda)$ is a multiple of the characteristic polynomial of X to deduce from the Cayley–Hamilton Theorem that $p(X)h(X)q(X)x = 0$. Choose $u = h(X)q(X)x$. Then $u \in \mathcal{R}_\Omega(X)$. Moreover $x = g(X)p(X)x + u$, and since $(G(X)p(X)x)_2 = 0$ it follows that $(u - x)_2 = 0$. This proves that $u_2 = x_2$. □

In the next theorem we use a disjoint partition $\mathbb{C} = \mathbb{C}_+ \cup \mathbb{C}_-$ of the complex plane, and denote by $\mathcal{R}_+(Z)$ (resp. $\mathcal{R}_-(Z)$) the spectral Z-invariant subspace corresponding to the eigenvalues of a matrix Z lying in $\mathbb{C}_+$ (resp. in $\mathbb{C}_-$).

Theorem 19.1.4. *Let*

$$X = \begin{bmatrix} X_1 & * & \cdots & * \\ 0 & X_2 & \cdots & * \\ \vdots & \vdots & \ddots & \vdots \\ & & & * \\ 0 & 0 & \cdots 0 & X_k \end{bmatrix}, \quad Y = \begin{bmatrix} Y_1 & * & \cdots & * \\ 0 & Y_2 & \cdots & * \\ \vdots & \vdots & \ddots & \vdots \\ & & & * \\ 0 & 0 & \cdots 0 & Y_k \end{bmatrix}$$

be block upper triangular matrices, where X_i and Y_i are $m_i \times m_i$ ($i = 1, \ldots, k$). Assume that

$$\mathbb{C}^{m_j} = \mathcal{R}_+(X_j) \dotplus \mathcal{R}_-(Y_j) \quad (j = 1, \ldots, k). \tag{19.1.7}$$

Then

$$\mathbb{C}^m = \mathcal{R}_+(X) \dotplus \mathcal{R}_-(Y), \tag{19.1.8}$$

where $m = m_1 + \cdots + m_k$.

Proof. Using induction on k, it suffices to consider the case $k = 2$. Let

$$\begin{bmatrix} y_1 \\ y_2 \end{bmatrix} \in \mathcal{R}_+(X) \cap \mathcal{R}_-(Y), \quad \text{where } y_1 \in \mathbb{C}^{m_1}, \ y_2 \in \mathbb{C}^{m_2}.$$

By Lemma 19.1.3, $y_2 \in \mathcal{R}_+(X_2) \cap \mathcal{R}_-(Y_2) = \{0\}$ in view of (19.1.7). So $y_2 = 0$, and therefore $y_1 \in \mathcal{R}_+(X_1) \cap \mathcal{R}_-(Y_1)$. Using (19.1.7) again we get $y_1 = 0$ also. We have proved that

$$\mathcal{R}_+(X) \cap \mathcal{R}_-(Y) = \{0\}. \tag{19.1.9}$$

Now

$$\dim \mathcal{R}_+(X) = \dim \mathcal{R}_+(X_1) + \dim \mathcal{R}_+(X_2)$$
$$\dim \mathcal{R}_-(Y) = \dim \mathcal{R}_-(Y_1) + \dim \mathcal{R}_-(Y_2).$$

Adding these equalities and, since

$$\dim \mathcal{R}_+(X_j) + \dim \mathcal{R}_-(Y_j) = m_j$$

for $j = 1, 2$, we obtain

$$\dim \mathcal{R}_+(X) + \dim \mathcal{R}_-(Y) = m.$$

In view of (19.1.9) the equality (19.1.8) follows. □

CANONICAL FACTORIZATION IN GEOMETRIC TERMS

Theorem 19.1.4 holds also for real matrices X and Y (with $\mathbb{C}$ replaced by $\mathbb{R}$ in (19.1.7) and (19.1.8)) provided the sets $\mathbb{C}_+$ and $\mathbb{C}_-$ are symmetric with respect to the real axis.

Proof of Theorem 19.1.2. First note that in view of Lemma 19.1.1 we have $\sigma(A - BD^{-1}C) \cap \Gamma = \emptyset$.

Assume that W admits a canonical factorization $W = W_- W_+$, where (without loss of generality) we can assume $W_+(\infty) = I$. In view of Theorems 6.1.3 and 6.1.5 we can write

$$A = \begin{bmatrix} A_{11} & A_{12} & A_{13} \\ 0 & A_0 & A_{23} \\ 0 & 0 & A_{33} \end{bmatrix}, \quad B = \begin{bmatrix} B_1 \\ B_0 \\ 0 \end{bmatrix}, \quad C = [0 \ C_0 \ C_1],$$

where
$$W(\lambda) = D + C_0(\lambda I - A_0)^{-1} B_0,$$

and this realization is minimal. Since, in particular, a canonical factorization is minimal, by Theorem 6.4.2 (applied to the matrix function $D^{-1}W(\lambda)$) we have

$$W_-(\lambda) = D + C_0 \pi_{\mathcal{L}} (\lambda I - A_0)^{-1} \pi_{\mathcal{L}} B_0,$$

$$W_+(\lambda) = I + D^{-1} C_0 \pi_{\mathcal{N}} (\lambda I - A_0)^{-1} \pi_{\mathcal{N}} B_0.$$

Here $\mathbb{C}^{m_0} = \mathcal{L} \dotplus \mathcal{N}$ is a supporting decomposition for $D^{-1}W(\lambda)$ (so that $\mathcal{L}$ is A_0-invariant and $\mathcal{N}$ is $(A_0 - B_0 D^{-1} C_0)$-invariant), and $\pi_{\mathcal{L}} = I - \pi_{\mathcal{N}}$ is the projector on $\mathcal{L}$ along $\mathcal{N}$. Furthermore, by Theorem 6.4.3 and by location of the poles and zeros of $W_+(\lambda)$ and $W_-(\lambda)$, we see that, in fact, $\mathcal{L} = \mathcal{R}_+(A_0)$ and $\mathcal{N} = \mathcal{R}_-(A_0 - B_0 D^{-1} C_0)$. Now

$$A = \begin{bmatrix} A_{11} & * & * \\ 0 & A_0 & * \\ 0 & 0 & A_{33} \end{bmatrix}, \quad A - BD^{-1}C = \begin{bmatrix} A_{11} & * & * \\ 0 & A_0 - B_0 D^{-1} C_0 & * \\ 0 & 0 & A_{33} \end{bmatrix},$$

and by Theorem 19.1.4 the equality (19.1.3) follows.

Conversely, assume that (19.1.3) holds. We then verify that $W_-(\lambda)$ and $W_+(\lambda)$ given by (19.1.5) and (19.1.6) satisfy all the properties required by the definition of canonical factorization $W = W_+ W_-$. The following properties are needed:

(a) $A_{12} = 0$;

(b) $\sigma(A_{22}) \subset \Gamma_+$;

(c) $\sigma(A_{11}) \subset \Gamma_-$;

(α) $A_{21} - B_2 D^{-1} C_1 = 0$;

(β) $\sigma(A_{11} - B_1 D^{-1} C_1) \subset \Gamma_-$;

(γ) $\sigma(A_{22} - B_2 D^{-1} C_2) \subset \Gamma_+$.

Proof of (a). Write $T = [T_1 \mid T_2]$ according to the partitioning in equation (19.1.4). We know that $A\mathcal{R}_+(A) \subseteq \mathcal{R}_+(A)$. Thus $A \operatorname{Im} T_2 \subseteq \operatorname{Im} T_2$, and hence for an arbitrary vector $y \in \mathbb{C}^{n-r}$ we have $AT\binom{0}{y} \subseteq \operatorname{Im} T_2$. On the other hand

$$AT \begin{bmatrix} 0 \\ y \end{bmatrix} = T \begin{bmatrix} A_{11} & A_{12} \\ A_{21} & A_{22} \end{bmatrix} \begin{bmatrix} 0 \\ y \end{bmatrix}$$

$$= T \begin{bmatrix} A_{12} y \\ A_{22} \end{bmatrix} = T_1 A_{12} y + T_2 A_{22} y,$$

and thus $T_1 A_{12} y = 0$. Since T_1 has full column rank, $A_{12} y = 0$. But y is arbitrary, and so $A_{12} = 0$.

Proof of (b). Let $E : \mathbb{C}^{n-r} \to \mathcal{R}_+(A)$ be defined by $Ey = T_2 y$. By the proof of (a),

$$EA_{22} y = T_2 A_{22} y = AT_2 y = (A|\mathcal{R}_+(A)) Ey.$$

Thus
$$A_{22} = E^{-1}(A|\mathcal{R}_+(A))E, \qquad (19.1.10)$$

and hence $\sigma(A_{22}) \subset \Gamma_+$.

Proof of (c). Write $\alpha(\lambda) = \det(\lambda I - A)$ as $\alpha(\lambda) = \alpha_-(\lambda) \alpha_+(\lambda)$, where α_+ has all its zeros in Γ_+ and α_- in Γ_-. Furthermore, put

$$\alpha_1(\lambda) = \det(\lambda I - A_{11}), \quad \alpha_2(\lambda) = \det(\lambda I - A_{22}).$$

From (19.1.10) it follows that $\alpha_2 = \alpha_+$. On the other hand, by (a),

$$\alpha(\lambda) = \det \begin{bmatrix} \lambda I - A_{11} & 0 \\ -A_{21} & \lambda I - A_{22} \end{bmatrix} = \alpha_1(\lambda) \alpha_2(\lambda).$$

Thus $\alpha_1 = \alpha_-$, and hence $\sigma(A_{11}) \subset \Gamma_-$.

Proof of (α)–(γ). Check that

$$T^{-1}(A - BD^{-1}C)T = \begin{bmatrix} A_{11} - B_1 D^{-1} C_1 & -B_1 D^{-1} C_1 \\ A_{21} - B_2 D^{-1} C_1 & A_{22} - B_2 D^{-1} C_2 \end{bmatrix}, \qquad (19.1.11)$$

and repeat the arguments of the proofs of (a)–(c).

From (a)–(c) it follows that $W_+(\lambda)$ (resp. $W_-(\lambda)$) has no poles in Γ_+ (resp. Γ_-). Next, note that

$$W_+(\lambda)^{-1} = I - D^{-1}C_1[\lambda I - (A_{11} - B_1 D^{-1}C_1)]^{-1}B_1,$$

$$W_-(\lambda)^{-1} = D^{-1} - D^{-1}C_2[\lambda I - (A_{22} - B_2 D^{-1}C_2)]^{-1}B_2 D^{-1}.$$

Thus, by (α)–(γ), $W_+(\lambda)^{-1}$ (resp. $W_-(\lambda)^{-1}$) also has no poles in Γ_+ (resp. Γ_-). Finally, a straightforward calculation shows that

$$W_-(\lambda)W_+(\lambda) = D + [C_1 \ C_2] \begin{bmatrix} \lambda I - A_{11} & 0 \\ -B_2 D^{-1}C_1 & \lambda I - A_{22} \end{bmatrix} \begin{bmatrix} B_1 \\ B_2 \end{bmatrix}.$$

By (a) and (α) this implies $W_-W_+ = W$, and the canonical factorization is established. □

19.2 The real case

The concept of canonical factorization and Theorem 19.1.2 are also valid in the framework of real rational matrix functions. Recall that an $m \times n$ rational matrix function $W(\lambda)$ is called *real* if $W(\lambda_0)$ is a real matrix for all real λ_0 which are not poles of $W(\lambda)$. Recall also that a rational matrix functions $W(\lambda)$ with finite value at infinity is real if and only if it admits a realization

$$W(\lambda) = D + C(\lambda I - A)^{-1}B, \qquad (19.2.1)$$

where all four matrices A, B, C and D are real (see Section 6.7).

A *real canonical factorization* of a real rational matrix function $W(\lambda)$ has the form described in the introduction to this chapter, but with the additional requirement that the factors $W_-(\lambda)$ and $W_+(\lambda)$ be real. In this context it is natural to assume that Γ is symmetric with respect to the real axis, i.e. that $\lambda_0 \in \Gamma$ implies $\bar{\lambda}_0 \in \Gamma$.

The real analogue of Theorem 19.1.2 is then:

Theorem 19.2.1. *Assume that Γ is symmetric relative to the real axis. Let $W(\lambda)$ be a real rational $n \times n$ matrix function without poles and zeros on Γ, and having a realization as described in equation (19.2.1) where $\sigma(A) \cap \Gamma = \emptyset$, D is invertible, and all four matrices A, B, C and D are real. Then $W(\lambda)$ admits a real canonical factorization if and only if*

$$\mathbb{R}^m = \mathcal{R}_-(A - BD^{-1}C) \dotplus \mathcal{R}_+(A),$$

where A is of size $m \times m$.

Furthermore, in this case the canonical factorization $W = W_-W_+$ is given by the formulas (19.1.5) and (19.1.6).

Observe that, under the hypotheses of Theorem 19.2.1, the sets Γ_+ and Γ_- are symmetric relative to the real axis as well, and therefore

$$\mathcal{R}_+(A) := \sum \mathcal{R}_{\lambda \pm i\mu}(A),$$

where the sum is taken over all $\lambda \pm i\mu \in \sigma(A) \cap \Gamma_+$, is a well-defined real subspace. Similarly, $\mathcal{R}_-(A - BD^{-1}C)$ is a well-defined real subspace.

The proof of Theorem 19.2.1 is obtained in the same way as the proof of Theorem 19.1.2. The details are omitted.

19.3 Spectral factorization

In this section $\Gamma = \{i\lambda : \lambda \in \mathbb{R}\} \cup \{\infty\}$, and we let Γ_+ (resp. Γ_-) be the open right (resp. left) half-plane. An $n \times n$ rational matrix function $W(\lambda)$ is said to admit a *spectral factorization* if

$$W(\lambda) = (V(-\bar\lambda))^* V(\lambda) \tag{19.3.1}$$

for some $n \times n$ rational matrix function $V(\lambda)$ such that all poles and zeros of $V(\lambda)$ are in Γ_- and this relation holds for all complex numbers λ which are not poles of any of the functions involved. If (19.3.1) holds, then clearly $V(\lambda)$ (resp. $(V(-\bar\lambda))^*$) is analytic and invertible on $\Gamma_+ \cup \Gamma$ (resp. $\Gamma_- \cup \Gamma$), so spectral factorization is a particular case of canonical factorization.

As we have seen in Section 19.1, a necessary condition for existence of a spectral factorization is that $W(\lambda)$ have no poles and zeros on Γ. In addition, $W(\lambda)$ must clearly be positive definite for $\lambda \in \Gamma$. It turns out that these conditions are also sufficient and, moreover, the spectral factorization is given in terms of the maximal solution of a certain algebraic Riccati equation.

Theorem 19.3.1. *Let $W(\lambda)$ be an $n \times n$ rational matrix function having no poles and zeros on Γ and which is positive definite for all $\lambda \in \Gamma$. Then $W(\lambda)$ admits a spectral factorization with respect to Γ.*

Furthermore, a spectral factorization of $W(\lambda)$ is constructed as follows. Write (using elementary fractions in every entry of $W(\lambda)$)

$$W(\lambda) = W_1(\lambda) + W_2(\lambda),$$

where $W_1(\lambda)$ has all its poles in Γ_- and $W_1(\infty) = 0$, while $W_2(\lambda)$ has all its poles in Γ_+. Let

$$W_1(\lambda) = C_1(\lambda I - A_1)^{-1} B_1 \tag{19.3.2}$$

be a realization of $W_1(\lambda)$ such that A_1 has no eigenvalues on the imaginary axis and the pair (A_1, B_1) is c-stabilizable. Then the algebraic Riccati equation

$$XB_1 D^{-1} B_1^* X - X(A_1 - B_1 D^{-1} C_1) - (A_1 - B_1 D^{-1} C_1)^* X + C_1^* D^{-1} C_1 = 0, \tag{19.3.3}$$

where $D = W(\infty)$, has a maximal hermitian solution X_0, and the function $V(\lambda)$ of equation (19.3.1) is given by

$$V(\lambda) = D^{1/2} + D^{-1/2}(C_1 + B_1^* X_0)(\lambda I - A_1)^{-1} B_1. \tag{19.3.4}$$

If $W(\lambda)$ is real, then C_1, A_1 and B_1 in (19.3.2) can be chosen real as well, the matrix X_0 is real, and consequently the function $V(\lambda)$ of (19.3.4) is also real.

Observe that a realization (19.3.2) with $\sigma(A_1) \subset \Gamma_-$ and (A_1, B_1) c-stabilizable certainly exists. For example, a minimal realization of $W_1(\lambda)$ has this property (see Theorem 6.1.5).

Proof. Consider the function

$$Q(\lambda) = C_1(\lambda I - A_1)^{-1} B_1 + B_1^*(-\lambda I - A_1^*)^{-1} C_1^*.$$

By the construction of $W_1(\lambda)$ we see that $Q(\lambda) - W(\lambda)$ is analytic on Γ_-. On the other hand, the function $W(\lambda)$ satisfies $(W(-\bar{\lambda}))^* = W(\lambda)$ for $\lambda \in i\mathbb{R}$, and since $(W(-\bar{\lambda}))^*$ is a rational matrix function as well, we must have $(W(-\bar{\lambda}))^* = W(\lambda)$ for all complex λ which are not poles of $W(\lambda)$. Consequently,

$$W(\lambda) = (W_1(-\bar{\lambda}))^* + (W_2(-\bar{\lambda}))^*,$$

and therefore $Q(\lambda) - W(\lambda)$ is analytic in Γ_+. Since both $Q(\lambda)$ and $W(\lambda)$ has no poles on Γ, we conclude that the rational matrix function $Q(\lambda) - W(\lambda)$ is analytic on $\mathbb{C} \cup \{\infty\}$, and therefore must be a constant. As a result, we obtain the following realization for $W(\lambda)$:

$$W(\lambda) = D + [C_1 \ B_1^*] \left(\lambda I - \begin{bmatrix} A_1 & 0 \\ 0 & -A_1^* \end{bmatrix} \right)^{-1} \begin{bmatrix} B_1 \\ -C_1^* \end{bmatrix}.$$

Next, we use Theorem 19.1.2 to prove that $W(\lambda)$ admits a canonical factorization. Let

$$A = \begin{bmatrix} A_1 & 0 \\ 0 & -A_1^* \end{bmatrix}, \quad B = \begin{bmatrix} B_1 \\ -C_1^* \end{bmatrix}, \quad C = [C_1 \ B_1^*],$$

$$A - BD^{-1}C = \begin{bmatrix} A_1 - B_1 D^{-1} C_1 & -B_1 D^{-1} B_1^* \\ C_1^* D^{-1} C_1 & -A_1^* + C_1^* D^{-1} B_1^* \end{bmatrix}.$$

Clearly,

$$\mathcal{R}_+(A) = \text{Im} \begin{bmatrix} 0 \\ I \end{bmatrix}.$$

In order to find $\mathcal{R}_-(A - BD^{-1}C)$, we verify first that the Riccati equation (19.3.3) does indeed have the maximal hermitian solution X_0. We start with the observation that, since (A_1, B_1) is c-stabilizable, so is $(A_1 - B_1 D^{-1} C_1, B_1 D^{-1} B_1^*)$

(see Lemma 4.5.3). Let $\mathcal{C}$ be the controllable subspace of $(A_1 - B_1 D^{-1} C_1, B_1 D^{-1} B_1^*)$. With respect to the orthogonal decomposition $\mathbb{C}^p = \mathcal{C} \oplus \mathcal{C}^\perp$ (where $p \times p$ is the size of A_1) write

$$A_1 - B_1 D^{-1} C_1 = \begin{bmatrix} A_{11} & A_{12} \\ 0 & A_{22} \end{bmatrix}, \quad B_1 D^{-1} B_1^* = \begin{bmatrix} D_{11} & 0 \\ 0 & 0 \end{bmatrix}, \quad (19.3.5)$$

$$-C_1^* D^{-1} C_1 = \begin{bmatrix} C_{11} & C_{12} \\ C_{12}^* & C_{22} \end{bmatrix}, \quad (19.3.6)$$

and let

$$M_1 = \begin{bmatrix} -A_{11} & D_{11} \\ C_{11} & A_{11}^* \end{bmatrix}.$$

By Lemma 19.1.1, the matrix $A - BD^{-1}C$ has no eigenvalues on the imaginary axis. Therefore the same is true for $-(A - BD^{-1}C)$. Now the definition of M_1, together with the block forms of (19.3.5) and (19.3.6) imply that

$$\sigma(-(A - BD^{-1}C)) = \sigma(M_1) \cup \sigma(-A_{22}) \cup \sigma(A_{22}^*),$$

and consequently M_1 has no eigenvalues on the imaginary axis. Theorems 7.9.3 and 7.9.4 are now applicable, and so (19.3.3) has indeed the maximal hermitian solution X_0; moreover, $\sigma(A_1 - B_1 D^{-1} C_1 - B_1 D^{-1} B_1^* X_0)$ lies in the open left half-plane. The equality

$$(A - BD^{-1}C) \begin{bmatrix} I \\ X_0 \end{bmatrix} = \begin{bmatrix} I \\ X_0 \end{bmatrix} (A_1 - B_1 D^{-1} C_1 - B_1 D^{-1} B_1^* X_0)$$

shows that $\operatorname{Im} \begin{bmatrix} I \\ X_0 \end{bmatrix} \subseteq \mathcal{R}_-(A - BD^{-1}C)$. Dimensional considerations (taking into account that $i(A - BD^{-1}C)$ is $\begin{bmatrix} 0 & iI \\ -iI & 0 \end{bmatrix}$-self-adjoint and therefore $\dim \mathcal{R}_+(A - BD^{-1}C) = \dim \mathcal{R}_-(A - BD^{-1}C)$) show that in fact

$$\operatorname{Im} \begin{bmatrix} I \\ X_0 \end{bmatrix} = \mathcal{R}_-(A - BD^{-1}C).$$

As we obviously have

$$\mathbb{C}^{2p} = \operatorname{Im} \begin{bmatrix} 0 \\ I \end{bmatrix} \dotplus \operatorname{Im} \begin{bmatrix} I \\ X_0 \end{bmatrix},$$

it follows from Theorem 19.1.2 that $W(\lambda)$ admits a canonical factorization $W = W_- W_+$ given by the (slightly modified) formulas (19.1.5) and (19.1.6):

$$W_+(\lambda) = D^{1/2} + D^{-1/2} \tilde{C}_1 (\lambda I - \tilde{A}_{11})^{-1} \tilde{B}_1,$$

$$W_-(\lambda) = D^{1/2} + \tilde{C}_2(\lambda I - \tilde{A}_{22})^{-1}\tilde{B}_2 D^{-1/2},$$

where

$$\begin{bmatrix} \tilde{A}_{11} & \tilde{A}_{12} \\ \tilde{A}_{21} & \tilde{A}_{22} \end{bmatrix} = \begin{bmatrix} I & 0 \\ X_0 & I \end{bmatrix}^{-1} \begin{bmatrix} A_1 & 0 \\ 0 & -A_1^* \end{bmatrix} \begin{bmatrix} I & 0 \\ X_0 & I \end{bmatrix},$$

$$\begin{bmatrix} \tilde{B}_1 \\ \tilde{B}_2 \end{bmatrix} = \begin{bmatrix} I & 0 \\ X_0 & I \end{bmatrix}^{-1} \begin{bmatrix} B_1 \\ -C_1^* \end{bmatrix}; \quad [\tilde{C}_1 \ \tilde{C}_2] = [C_1 \ B_1^*] \begin{bmatrix} I & 0 \\ X_0 & I \end{bmatrix}.$$

Thus, $\tilde{A}_{11} = A_1$, $\tilde{A}_{22} = -A_1^*$, $\tilde{B}_1 = B_1$,

$$\tilde{C}_2 = B_1^* = \tilde{B}_1^*, \quad \tilde{C}_1 = C_1 + B_1^* X_0, \quad \tilde{B}_2 = -X_0 B_1 - C_1^* = -\tilde{C}_1^*.$$

It follows that $W_+(\lambda)$ coincides with $V(\lambda)$ given by (19.3.4), and that $W_-(\lambda) = (V(-\bar{\lambda}))^*$. So our canonical factorization is in fact spectral.

Finally, the assertion of Theorem 19.3.1 concerning the case when $W(\lambda)$ is real are easily verified; the details are omitted. □

Note that the hypotheses of Theorem 19.3.1 can be restated in a somewhat different form. It suffices to require that $W(\lambda)$ has no poles and zeros on Γ, satisfies $W(\lambda) = (W(\lambda))^*$ for $\lambda \in i\mathbb{R}$, and $W(\infty)$ is positive definite. Then $W(\lambda)$ is positive definite for all $\lambda \in i\mathbb{R}$. Indeed, the eigenvalue functions $\mu_1(\lambda), \ldots, \mu_n(\lambda)$ of $W(\lambda)$ ($\lambda \in i\mathbb{R}$) are real continuous functions of λ which are never zero (because otherwise $W(\lambda)$ would have a zero on the imaginary axis). In view of the positive semidefiniteness of $W(\infty)$ the functions $\mu_j(\lambda)$ ($j = 1, \ldots, n$) are positive for $\lambda = i\alpha$, $\alpha > 0$ sufficiently large. Therefore $\mu_j(\lambda)$ ($j = 1, \ldots, n$) must be positive for all $\lambda \in i\mathbb{R}$.

19.4 Notes

The exposition of Theorem 19.1.2 and its proof are based on the work of Kaashoek and Ran (1991), and Lemma 19.1.3 and its proof are taken from Gohberg et al. (1994). In the latter book, the lemma is stated and proved only for the case when Ω is a singleton.

For the general theory of canonical (and more general) factorizations of matrix valued functions, as well as several applications of this theory, see the books of Clancey and Gohberg (1981), and Gohberg and Kaashoek (1986). There is extensive literature on the solution of H^∞ control problems via canonical factorization. The interested reader is referred to Ball and Cohen (1987), Ball and Ran (1987), Green et al. (1990), as well as Chapter 7 in Francis (1987).

20
H^∞ CONTROL PROBLEMS

A typical H^∞ control problem involves minimizing the expression

$$\max_{\lambda \in \mathbb{R}} \|G_K(i\lambda)\|, \tag{1}$$

where $G_K(\lambda)$ is the transfer function of a feedback system, with an admissibile feedback K. The minimization is over all possible matrices K. It is often required that $G_K(\lambda)$ also be stable in the sense that all of its poles lie in the open left half-plane. Since the expression (1) represents the norm of $G_K(\lambda)$ as a member of the space (denoted H^∞) of bounded functions which are analytic in the open right half-plane, the terminology "H^∞ control" is used for problems of this type.

In applications, $\|G_K(i\lambda)\|$ often measures the amplitude of a disturbance or an error or, more generally, of a "bad" signal at the frequency λ. Thus, (1) is interpreted as a "worst case" error, and a typical H^∞ control problem becomes a "worst case" minimization problem; in other words, a min-max problem of the type well-known in optimization theory.

An important and more "robust" variant of this problem runs as follows. Given a tolerance level of $\gamma > 0$, find (if possible) an admissible feedback K such that

$$\max_{\lambda \in \mathbb{R}} \|G_K(i\lambda)\| < \gamma. \tag{2}$$

Clearly, the minimum of (1) coincides with the greatest lower bound of the set of positive numbers γ for which (2) is satisfied for some admissible feedback K. Control problems in the form (2) are the subject of this chapter.

In recent years, H^∞ control problems in various guises have become the subjects of intensive research, mainly by electrical engineers. Hundreds of research papers (and some books) using various approaches to the subject have appeared. We mention the sources Francis (1987), and Kaashoek and Ran (1991), though it cannot be claimed that they cover more than a fraction of the significant work in the area. Other important sources are mentioned in the notes of Section 20.4.

It turns out that one successful approach to H^∞ control problems involves algebraic Riccati equations of continuous type. Here, we treat two simple cases of the so-called "standard problem" of H^∞ control, and use the Riccati equation as the main tool for their solution. The first case is that of full state information, i.e. state feedback. The second case involves state estimation. Our exposition is based on the informal lecture notes of Kaashoek and Ran (1991).

20.1 The bounded real lemma

In this section we prove a result which is now popularly known as the *bounded real lemma*. It is a significant result in its own right, but appears here as a preparation for the solution of H^∞ control problems to be given in later sections.

Theorem 20.1.1. *Let an $n \times n$ rational matrix function $W(\lambda)$ be given in realized form:*
$$W(\lambda) = C(\lambda I - A)^{-1}B, \qquad (20.1.1)$$
where the matrix A has no eigenvalues on the imaginary axis and the pair (A, B) is c-stabilizable. Then
$$\|W(\lambda)\| < 1 \quad \text{for all } \lambda \in i\mathbb{R} \qquad (20.1.2)$$
if and only if the algebraic Riccati equation
$$XBB^*X - XA - A^*X + C^*C = 0 \qquad (20.1.3)$$
has a hermitian solution X_0 such that $\sigma(A - BB^*X_0)$ lies in the open left half-plane. When such a hermitian solution X_0 exists it is unique and is the maximal hermitian solution of equation (20.1.3).

We emphasize that the realization (20.1.1) need not be minimal (which is the case when (A, B) is a controllable pair and (C, A) is an observable pair). The hypotheses of Theorem 20.1.1 are satisfied, in particular, when all eigenvalues of A are in the open left half-plane.

Proof. Consider the rational matrix function
$$G(\lambda) = I - (W(-\bar{\lambda}))^*W(\lambda).$$
An easy calculation, using the realization (20.1.1), shows that
$$G(\lambda) = I + [0 \ B^*] \begin{bmatrix} \lambda I - A & 0 \\ -C^*C & \lambda I - (-A^*) \end{bmatrix}^{-1} \begin{bmatrix} B \\ 0 \end{bmatrix}. \qquad (20.1.4)$$
Since A has no pure imaginary or zero eigenvalues, $G(\lambda)$ has no poles on $i\mathbb{R}$. Furthermore, applying formula (6.3.2) to the realization (20.1.4), we obtain
$$G(\lambda)^{-1} = I - [0 \ B^*](\lambda I - N)^{-1} \begin{bmatrix} B \\ 0 \end{bmatrix}, \qquad (20.1.5)$$
where N is the matrix associated with the CARE of (20.1.3), i.e.,
$$N = \begin{bmatrix} A & -BB^* \\ C^*C & -A^* \end{bmatrix} = iM, \qquad (20.1.6)$$
where M is defined in Section 7.2.

If inequality (20.1.2) holds, then for $\lambda = i\omega \in i\mathbb{R}$ and x a nonzero vector we have

$$\begin{aligned}\langle G(\lambda)x, x\rangle &= \|x\|^2 - \langle W(i\omega)^*W(i\omega)x, x\rangle \\ &= \|x\|^2 - \langle W(i\omega)x, W(i\omega)x\rangle \\ &= \|x\|^2 - \|W(i\omega x)\|^2 > 0.\end{aligned} \qquad (20.1.7)$$

Hence $G(i\omega) > 0$. In particular, $G(\lambda)$ has no zeros on $i\mathbb{R}$.

Next we verify that N has no eigenvalues on $i\mathbb{R}$. Indeed, let

$$\begin{bmatrix} A & -BB^* \\ C^*C & -A^* \end{bmatrix} \begin{bmatrix} x \\ y \end{bmatrix} = \lambda_0 \begin{bmatrix} x \\ y \end{bmatrix} \qquad (20.1.8)$$

for some $\lambda_0 \in i\mathbb{R}$. Then

$$\begin{aligned}G(\lambda)B^*y &= G(\lambda)[0 \ B^*]\begin{bmatrix} x \\ y \end{bmatrix} \\ &= \left\{ I + [0 \ B^*]\begin{bmatrix} \lambda I - A & 0 \\ -C^*C & \lambda I - (-A^*) \end{bmatrix}^{-1} \begin{bmatrix} B \\ 0 \end{bmatrix} \right\} [0 \ B^*]\begin{bmatrix} x \\ y \end{bmatrix} \\ &= [0 \ B^*]\left\{ I + \begin{bmatrix} \lambda I - A & 0 \\ -C^*C & \lambda I - (-A^*) \end{bmatrix}^{-1} \begin{bmatrix} 0 & BB^* \\ 0 & 0 \end{bmatrix} \right\}\begin{bmatrix} x \\ y \end{bmatrix} \\ &= [0 \ B^*]\begin{bmatrix} \lambda I - A & 0 \\ -C^*C & \lambda I - (-A^*) \end{bmatrix}^{-1}\left\{\begin{bmatrix} \lambda I - A & 0 \\ -C^*C & \lambda I - (-A^*) \end{bmatrix} \right. \\ &\qquad\qquad \left. + \begin{bmatrix} 0 & BB^* \\ 0 & 0 \end{bmatrix} \right\}\begin{bmatrix} x \\ y \end{bmatrix} \\ &= [0 \ B^*]\begin{bmatrix} \lambda I - A & 0 \\ -C^*C & \lambda I - (-A^*) \end{bmatrix}^{-1}(\lambda I - N)\begin{bmatrix} x \\ y \end{bmatrix} \\ &= (\lambda - \lambda_0)[0 \ B^*]\begin{bmatrix} \lambda I - A & 0 \\ -C^*C & \lambda I - (-A^*) \end{bmatrix}^{-1}\begin{bmatrix} x \\ y \end{bmatrix},\end{aligned}$$

and thus $G(\lambda_0)B^*y = 0$. But $G(\lambda_0) > 0$ and therefore $B^*y = 0$. Now (20.1.8) takes the form

$$\begin{bmatrix} A & 0 \\ C^*C & -A^* \end{bmatrix}\begin{bmatrix} x \\ y \end{bmatrix} = \lambda_0 \begin{bmatrix} x \\ y \end{bmatrix}.$$

Since A has no eigenvalues on the imaginary axis, we conclude that $x = y = 0$, and thus λ_0 cannot be an eigenvalue of N. So $\sigma(N) \cap i\mathbb{R} = \emptyset$, and Theorem 7.9.1 can be applied to show that (20.1.3) has hermitian solutions. Furthermore, by Theorem 7.9.4 the equation (20.1.3) admits a unique hermitian solution X_0 with the required properties.

Conversely, assume that (20.1.3) has a hermitian solution X_0 for which $\sigma(A - BB^*X_0)$ lies in the open left half-plane. Then by Theorem 7.9.4 the matrix N has no eigenvalues on the imaginary axis, and it follows from equation (20.1.5) that $G(\lambda)$ has no zeros on $i\mathbb{R}$. Obviously, $G(\lambda)$ is hermitian for $\lambda \in i\mathbb{R}$ and $G(\infty)$ is positive definite. It follows from Theorem 19.3.1 that $G(\lambda)$ admits a spectral factorization. In particular, $G(\lambda) > 0$ for $\lambda \in i\mathbb{R}$, and therefore (20.1.2) follows from (20.1.7). □

20.2 An H^∞ control problem with state feedback

Consider the following linear control system:

$$\begin{cases} \dot{x} = Ax + B_1 w + B_2 u; \quad x(0) = 0 \\ z = Cx + Du. \end{cases} \quad (20.2.1)$$

Here A, B_1, B_2, C and D are constant matrices of sizes $n \times n$, $n \times m_1$, $n \times m_2$, $p \times n$ and $p \times m_2$, respectively. The vectors $x = x(t)$, $w = w(t)$, $u = u(t)$ and $z = z(t)$ of appropriate sizes are time dependent, and satisfy the system (20.2.1). They have the following heuristic descriptions: u is the input, w is disturbance, z is output, and x is the state variable. Our goal is to devise a static state feedback (i.e., independent of time) when it exists, so that the closed loop system is stable and the influence of the disturbance on the output is smaller than a preassigned tolerance level for all frequencies.

Mathematically, the problem is formulated as follows: The state feedback law $u = Kx$, where K is a constant $m_2 \times n$ matrix, brings the system (20.2.1) into the closed loop form

$$\begin{cases} \dot{x} = (A + B_2 K)x + B_1 w; \quad x(0) = 0 \\ z = (C + DK)x. \end{cases} \quad (20.2.2)$$

Let $\hat{w}$ and $\hat{z}$ be the Laplace transforms of w and z, respectively:

$$\hat{w}(\lambda) = \int_0^\infty e^{-\lambda t} w(t)\, dt,$$

and similarly for $\hat{z}$. Taking Laplace transforms in (20.2.2), and solving for $\hat{z}$ in terms of $\hat{w}$, it is found that $\hat{z} = G_K(\lambda)\hat{w}$, where

$$G_K(\lambda) = (C + DK)(\lambda I - (A + B_2 K))^{-1} B_1. \quad (20.2.3)$$

If possible, a matrix K is to be found such that $G_K(\lambda)$ has no poles in the closed right half-plane and

$$\max_{\lambda \in i\mathbb{R}} \|G_K(\lambda)\| < \gamma, \quad (20.2.4)$$

where $\gamma > 0$ is a preassigned tolerance level. In fact, we will require more; namely, that the matrix $A + B_2 K$ is c-stable (i.e., all its eigenvalues are in the

AN H^∞ CONTROL PROBLEM WITH STATE FEEDBACK

open left half-plane). If the realization (20.2.3) is minimal, then the c-stability of $A + B_2 K$ is equivalent to $G_K(\lambda)$ having no poles in the closed right half-plane. In general, the c-stability of $A + B_2 K$ is a stronger requirement.

At this point we make the following hypotheses which make the subsequent analysis considerably more tractable:

(i) the pair (C, A) is observable;

(ii) the pairs (A, B_1) and (A, B_2) are c-stabilizable;

(iii) the equality $D^*[C \ D] = [0 \ I]$ holds.

We shall prove the following theorem.

Theorem 20.2.1. *Suppose the system (20.2.1) satisfies the conditions (i)—(iii). Then there exists a state feedback K such that $A + B_2 K$ is c-stable and the closed loop transfer function $G_K(\lambda)$, given by (20.2.3), satisfies (20.2.4) if and only if there is a positive definite solution X_∞ of the algebraic Riccati equation*

$$0 = C^*C + A^*X_\infty + X_\infty A + X_\infty(\gamma^{-2}B_1 B_1^* - B_2 B_2^*)X_\infty \qquad (20.2.5)$$

such that $A + (\gamma^{-2}B_1 B_1^* - B_2 B_2^*)X_\infty$ is c-stable. In that case one such state feedback is given by $K = -B_2^* X_\infty$.

The rest of this section is devoted to the proof of Theorem 20.2.1. The proof relies heavily on the bounded real lemma (Theorem 20.1.1).

Proof. Let us first suppose that K is a feedback for which $A + B_2 K$ is c-stable and (20.2.4) holds. The function

$$\frac{1}{\gamma}G_K(\lambda) = (C + DK)(\lambda I - (A + B_2 K))^{-1}\frac{1}{\gamma}B_1$$

has the property that

$$\max_{\lambda \in i\mathbb{R}} \|\frac{1}{\gamma}G_K(\lambda)\| < 1.$$

According to Theorem 20.1.1 this holds if and only if the algebraic Riccati equation

$$-(A + B_2 K)^* X - X(A + B_2 K) + \gamma^{-2} X B_1 B_1^* X + (C + DK)^*(C + DK) = 0 \qquad (20.2.6)$$

has a hermitian solution X_+ such that $A + B_2 K - \gamma^{-2} B_1 B_1^* X_+$ is c-stable (such an X_+ is unique).

We show first that X_+ is invertible. Suppose $X_+ x = 0$ for some $x \in \mathbb{C}^n$. Multiplying (20.2.6) (with X replaced by X_+) on the left with x^* and on the right with x we see that
$$x^*(C + DK)^*(C + DK)x = 0$$
and, hence, $(C + DK)x = 0$. Now use condition (iii) to obtain
$$D^*(C + DK)x = Kx = 0.$$
Thus, $Cx = 0$ as well. We have proved that $\operatorname{Ker} X_+ \subseteq \operatorname{Ker} C$. Now multiply (20.2.6) on the right by x and it is found that $X_+ Ax = 0$. So $\operatorname{Ker} X_+$ is A-invariant. Therefore,
$$\operatorname{Ker} X_+ \subseteq \bigcap_{j=0}^{\infty} \operatorname{Ker}(CA^j) = \{0\},$$
because (C, A) is observable. Hence X_+ is invertible.

The next step is to show that, in fact, $X_+ < 0$. Because of the invertibility of X_+ it suffices to show that $X_+ \leq 0$. Rewrite equation (20.2.6) (with X replaced by X_+) in the form
$$\begin{aligned}(A + B_2 K)^* X_+ + X_+(A + B_2 K) &= \gamma^{-2} X_+ B_1 B_1^* X_+ + (C + DK)^*(C + DK) \\ &= [(C + DK)^* \quad \gamma^{-1} X_+ B_1] \begin{bmatrix} C + DK \\ \gamma^{-1} B_1^* X_+ \end{bmatrix}.\end{aligned}$$
(20.2.7)

It turns out that the pair
$$\left(\begin{bmatrix} C + DK \\ \gamma^{-1} B_1^* X_+ \end{bmatrix}, A + B_2 K \right) \tag{20.2.8}$$
is observable. Indeed, in view of the Hautus test (Theorem 4.3.3) we have to check that the equalities
$$(C + DK)x = 0, \quad \gamma^{-1} B_1^* X_+ x = 0, \quad (A + B_2 K)x = \lambda x \tag{20.2.9}$$
for some $\lambda \in \mathbb{C}$ and $x \in \mathbb{C}^n$ are possible only when $x = 0$. Using (iii), it follows from (20.2.9) that $0 = D^*(C + DK)x = Kx$. So $Cx = 0$ and $Ax = \lambda x$. Since the pair (C, A) is observable, x must be the zero vector, and the observability of the pair (20.2.8) is verified.

Now Theorem 5.3.1(b) can be applied to equation (20.2.7) and shows that $X_+ < 0$.

Let us denote $-X_+$ by X_-. Rewriting equation (20.2.6) this matrix satisfies
$$0 = \gamma^{-2} X_- B_1 B_1^* X_- + (A + B_2 K)^* X_- + X_-(A + B_2 K) + (C + DK)^*(C + DK). \tag{20.2.10}$$
Let us rewrite this relation again (using condition (iii) once more):

$$\gamma^{-2} X_- B_1 B_1^* X_- - X_- B_2 B_2^* X_- + A^* X_- + X_- A + C^* C$$
$$= -K^* B_2^* X_- - X_- B_2 K - K^* K - X_- B_2 B_2^* X_-$$
$$= -(B_2^* X_- + K)^*(B_2^* X_- + K). \tag{20.2.11}$$

It follows from this equation that there is a matrix $X_- > 0$ with

$$C^* C + X_- A + A^* X_- + X_-(\gamma^{-2} B_1 B_1^* - B_2 B_2^*) X_- \leq 0. \tag{20.2.12}$$

Now introduce

$$R(X) = -XC^* CX - XA^* - AX - \gamma^{-2} B_1 B_1^* + B_2 B_2^*$$

and also

$$S(X) = (A + B_2 K)^* X + X(A + B_2 K) - \gamma^{-2} X B_1 B_1^* X - C^* C - K^* K.$$

Because equation (20.2.6) has a stabilizing solution X_+, and using the condition (iii) again, it follows from Theorem 9.1.3 that there is a hermitian matrix X_0 with $S(X_0) > 0$. Using a perturbation argument (for example, replacing X_0 by $X_0 + \varepsilon I$ for sufficiently small real nonzero ε) it is found that X_0 can be assumed to be invertible. Now compute:

$$X_0 R(-X_0^{-1}) X_0 = -C^* C + A^* X_0 + X_0 A - X_0(\gamma^{-2} B_1 B_1^* - B_2 B_2^*) X_0$$
$$= S(X_0) + K^* K - K^* B_2^* X_0 - X_0 B_2 K + X_0 B_2 B_2^* X_0$$
$$= S(X_0) + (B_2^* X_0 - K)^*(B_2^* X_0 - K) > 0.$$

As X_0 is invertible we have $R(-X_0^{-1}) > 0$. Applying Theorem 9.1.3 again it is found that there is a hermitian solution X_m of

$$XC^* CX + XA^* + AX + \gamma^{-2} B_1 B_1^* - B_2 B_2^* = 0, \tag{20.2.13}$$

such that $\sigma(A^* + C^* C X_m)$ lies in the open right half-plane. Applying Theorem 9.1.1 to inequality (20.2.12) we obtain $X_m \geq X_1^{-1} > 0$, and therefore X_m is positive definite. Denoting $X_\infty = X_m^{-1}$, it follows that X_∞ solves the equation

$$0 = C^* C + A^* X_\infty + X_\infty A + X_\infty(\gamma^{-2} B_1 B_1^* - B_2 B_2^*) X_\infty. \tag{20.2.14}$$

Furthermore, this equation yields

$$-X_\infty^{-1}(A^* + C^* C X_\infty^{-1}) X_\infty = -X_\infty^{-1}(A^* X_\infty + C^* C)$$
$$= A + (\gamma^{-2} B_1 B_1^* - B_2 B_2^*) X_\infty,$$

and since $\sigma(A^* + C^* C X_\infty^{-1})$ lies in the open right half-plane, the matrix $A + (\gamma^{-2} B_1 B_1^* - B_2 B_2^*) X_\infty$ is c-stable. This proves Theorem 20.2.1 in one direction.

For the converse, assume that $X_\infty > 0$ satisfies (20.2.5) and is such that $A + (\gamma^{-2} B_1 B_1^* - B_2 B_2^*) X_\infty$ is c-stable. First, we show that $A + B_2 K = A - B_2 B_2^* X_\infty$ is c-stable, where $K = -B_2^* X_\infty$. A little calculation with equation (20.2.5) shows that

$$X_\infty(A + B_2 K) + (A + B_2 K)^* X_\infty = -C^* C - X_\infty \gamma^{-2} B_1 B_1^* X_\infty - X_\infty B_2 B_2^* X_\infty. \tag{20.2.15}$$

Now suppose for some $x \neq 0$ and λ with $\operatorname{Re} \lambda \geq 0$ we have $(A + B_2 K) x = \lambda x$, i.e. $A + B_2 K$ is not c-stable. Then

$$x^* \{ X_\infty(A + B_2 K) + (A + B_2 K)^* X_\infty \} x = 2(\operatorname{Re} \lambda) x^* X_\infty x \geq 0.$$

On the other hand, by (20.2.15)

$$x^* \{ X_\infty(A + B_2 K) + (A + B_2 K)^* X_\infty \} x \leq 0. \tag{20.2.16}$$

Because $X_\infty > 0$ it follows that $\operatorname{Re} \lambda = 0$ and that equality holds in (20.2.16). Thus,

$$x^* \{ C^* C + X_\infty \gamma^{-2} B_1 B_1^* X_\infty + X_\infty B_2 B_2^* X_\infty \} x = 0.$$

Hence $C x = 0$, $B_1^* X_\infty x = 0$, $B_2^* X_\infty x = 0$. But then

$$(A + B_2 K) x = (A - B_2 B_2^* X_\infty) x = A x.$$

So $A x = \lambda x$ and $C x = 0$. As (C, A) is observable this yields $x = 0$, which is a contradiction. Hence $A - B_2 B_2^* X_\infty$ is c-stable.

Next, we show that with $K = -B_2^* X_\infty$ the function $G_K(\lambda)$ satisfies

$$\max_{\lambda \in i\mathbb{R}} \| G_K(\lambda) \| < \gamma.$$

This is another application of Theorem 20.1.1. Indeed, if we substitute $X = -X_\infty$, $K = -B_2^* X_\infty$ in (20.2.6), and use condition (iii), we find that (20.2.6) is satisfied. As

$$A + B_2 K - \gamma^{-2} B_1 B_1^* (-X_\infty) = A - B_2 B_2^* X_\infty + \gamma^{-2} B_1 B_1^* X_\infty$$

is c-stable it follows that, for this choice of K, equation (20.2.6) has a stabilizing hermitian solution; namely $-X_\infty$. □

20.3 H^∞ filtering using state estimation

Consider a linear time-invariant system with (vector) input $w = w(t)$, (vector) state $x = x(t)$, and (vector) outputs $z = z(t)$ and $y = y(t)$:

$$\begin{cases} \dot{x} = Ax + Bw; & x(0) = 0 \\ z = C_1 x; & y = C_2 x + Dw. \end{cases} \quad (20.3.1)$$

Here A, B, C_1, C_2 and D are constant matrices of suitable sizes. For this system we are looking for a linear time-invariant filter F producing an estimate $\tilde{z}$ of z based on y, and such that the transfer function $W(\lambda)$ from the input w to the error $e = z - \tilde{z}$ is stable and satisfies

$$\max_{\lambda \in i\mathbb{R}} \|W(\lambda)\| < \gamma. \quad (20.3.2)$$

Here again $\gamma > 0$ is a preassigned tolerance level.

To make the problem tractable the class of admissible filters must be specified. They are to have the form of a state estimator:

$$\begin{cases} \dot{\tilde{x}} = A\tilde{x} + K(y - C_2\tilde{x}), & \tilde{x}(0) = 0 \\ \tilde{z} = C_1 \tilde{x}. \end{cases} \quad (20.3.3)$$

In other words, $\tilde{x}$ is the estimate (which is produced using the constant matrix K) of the state x based on the output y. The state estimator $\tilde{x}$ is then transformed to the estimator $\tilde{z}$ using the same transformation (given by the matrix C_1) that transforms x to z. Now the problem is reduced to finding the matrix K (when it exists) which gives the requisite properties of $W(\lambda)$.

Introduce a new state $X = x - \tilde{x}$. Then, combining equations (20.3.1) and (20.3.3), the state equations for the closed loop system become

$$\begin{cases} \dot{X} = (A - KC_2)X + (B - KD)w; & X(0) = 0 \\ e = z - \tilde{z} = C_1 X. \end{cases}$$

By taking Laplace transforms (as in Section 20.2), the closed loop transfer function $W(\lambda)$ from w to e is given by

$$W(\lambda) = C_1(\lambda I - (A - KC_2))^{-1}(B - KD). \quad (20.3.4)$$

Thus the problem we consider is to find K such that $A - KC_2$ is c-stable and

$$\max_{\lambda \in i\mathbb{R}} \|W(\lambda)\| < \gamma.$$

As in the preceding section, this problem is solved under simplifying assumptions:

Theorem 20.3.1. *Assume that the pairs (C_1, A) and (C_2, A) are detectable, the pair (A, B) is controllable, and the equality $D[B^* \ D^*] = [0 \ I]$ holds. Then there is a filter of the form (20.3.3) with $A - KC_2$ c-stable and such that the closed loop transfer function $W(\lambda)$ given by (20.3.4) satisfies*

$$\max_{\lambda \in i\mathbb{R}} \|W(\lambda)\| < \gamma$$

if and only if there is a positive definite solution Y_∞ of the algebraic Riccati equation

$$0 = BB^* + AY_\infty + Y_\infty A^* + Y_\infty(\gamma^{-2}C_1^*C_1 - C_2^*C_2)Y_\infty \qquad (20.3.5)$$

such that $A^* + (\gamma^{-2}C_1^*C_1 - C_2^*C_2)Y_\infty$ is stable. In that case, one such filter is given by (20.3.3) with $K = Y_\infty C_2^*$.

The proof is easily reduced to Theorem 20.2.1. Indeed, let A, B_1, B_2, C, D and K be as in formula (20.2.3), and let

$$\tilde{A} = A^*, \quad \tilde{C}_1 = B_1^*, \quad \tilde{C}_2 = B_2^*, \quad \tilde{K} = -K^*, \quad \tilde{D} = D^*, \quad \tilde{B} = C^*. \qquad (20.3.6)$$

Then if $\lambda \in i\mathbb{R}$ (20.2.3) gives

$$(G_K(-\lambda))^* = \tilde{C}_1(\lambda I - (\tilde{A} - \tilde{K}\tilde{C}_2))^{-1}(\tilde{B} - \tilde{K}\tilde{D}),$$

which is exactly the formula (20.3.4) with the tildas omitted. Also, the hypotheses of Theorem 20.2.1 correspond to those of Theorem 20.3.1 (under the transformation defined by (20.3.6)), and the Riccati equation (20.2.5) corresponds to (20.3.5). Since it is obvious that

$$\max_{\lambda \in i\mathbb{R}} \|(G_K(-\lambda))^*\| = \max_{\lambda \in i\mathbb{R}} \|G_K(\lambda)\|,$$

Theorem 20.3.1 follows immediately from Theorem 20.2.1.

20.4 Notes

In the literature, the term "bounded real lemma" (see Theorem 20.1.1) is used to identify several closely related results which have to do with characterizations of rational matrix functions (often assumed to have real coefficients) that are bounded on the imaginary axis. Such results are well-known, and have important applications in network synthesis (besides their theoretical significance); see, for example, Anderson and Vongpanitlerd (1973).

The problem of H_∞ optimal control was introduced by Zames (1981). Since then, many approaches have been developed in the engineering and mathematical literature to study various H_∞ control problems. Some of these, especially the frequency domain methods, are discussed in the mathematically oriented survey paper of Francis and Doyle (1987) and the book of Francis (1987). The approach based on the Riccati equations was developed in the late 1980s in the works Petersen (1987), Zhou and Khargenekar (1988), Khargonekar et al. (1988), Tadmor (1990), Bernstein and Haddad (1989), and Doyle et al. (1989). The last paper was particularly influential in our development. This approach was further extended to singular H_∞ control problems in Stoorvogel and Trentelman (1990); there, a matrix inequality (rather than equation) of Riccati type appears.

We also mention Glover et al. (1991) where the solutions of the so-called "four block" H_∞ control problem are described, and where the Riccati equations (of continuous type) play a crucial role; this paper also contains an extensive and useful bibliography.

21
CONTRACTIVE RATIONAL MATRIX FUNCTIONS

A rational matrix function $W(\lambda)$ (generally rectangular of size $p \times m$) is called a *contraction* if
$$W(\lambda)(W(\lambda))^* \leq I$$
for all real λ. This definition implies that a contraction has no poles on the real line and no pole at infinity. A contraction $W(\lambda)$ is called a *strict contraction* if the limiting condition $W(\infty)W(\infty)^* < I$ is satisfied.

In this chapter we study strict contractions. In particular, realizations of strict contractions are characterized, and their minimal unitary completions are described. It will be seen that a continuous algebraic Riccati equation (of the type studied in Part II) and its hermitian solutions play a crucial role in the analysis of strict contractions.

21.1 Realizations of strict contractions

Let $W(\lambda)$ be a $p \times m$ rational matrix function which has no pole at infinity (in other words, $W(\infty)$ is well-defined) and
$$I - W(\infty)W(\infty)^* > 0. \tag{21.1.1}$$

Every such function $W(\lambda)$ admits a minimal realization
$$W(\lambda) = D + C(\lambda I - A)^{-1}B, \tag{21.1.2}$$

where $D = W(\infty)$. Suppose that A is of size $n \times n$, so that C is $p \times n$ and B is $n \times m$. Observe that (21.1.1) implies $I - D^*D > 0$. In particular, the matrices $I - DD^*$ and $I - D^*D$ are invertible. Define
$$\alpha = A + BD^*(I - DD^*)^{-1}C \tag{21.1.3}$$
$$\beta = B(I - D^*D)^{-1}B^* \tag{21.1.4}$$

$$\gamma = C^*(I - DD^*)^{-1}C. \tag{21.1.5}$$

Note that α, β and γ are $n \times n$ and $\gamma \geq 0$. (For convenience, the convention of using only Roman capitals to denote matrices is broken in this chapter.) Using the matrices α, β and γ, form the Riccati equation

$$X\gamma X + (i\alpha)X + X(i\alpha)^* + \beta = 0. \tag{21.1.6}$$

This equation will play an important role. Its coeffiencits α, β and γ obviously depend on the coefficients A, B, C and D of the minimal realization (21.1.2). If one takes another minimal realization

$$W(\lambda) = \tilde{D} + \tilde{C}(\lambda I - \tilde{A})^{-1}\tilde{B} \tag{21.1.7}$$

in place of (21.1.2), the connections between the two Riccati equations are easily found:

Proposition 21.1.1. *Let* (21.1.2) *and* (21.1.7) *be minimal realizations of* $W(\lambda)$, *and define* $\tilde{\alpha}$, $\tilde{\beta}$, $\tilde{\gamma}$ *by analogy with* (21.1.3)–(21.1.5) *using the realization* (21.1.7) *instead of* (21.1.2). *Then for some invertible matrix* S,

$$\tilde{\alpha} = S^{-1}\alpha S, \quad \tilde{\beta} = S^{-1}\beta(S^{-1})^*, \quad \tilde{\gamma} = S^*\gamma S.$$

In particular, X is a solution of (21.1.6) *if and only if* $Y = S^{-1}X(S^*)^{-1}$ *is a solution of*
$$Y\tilde{\gamma}X + (i\tilde{\alpha})X + X(i\tilde{\alpha})^* + \tilde{\beta} = 0.$$

Proposition 21.1.1 follows immediately from the uniqueness (up to similarity) of a minimal realization of $W(\lambda)$ described in Theorem 6.1.4.

Following Gohberg and Rubinstein (1988), call the $2n \times 2n$ matrix $\begin{bmatrix} \alpha & \beta \\ \gamma & \alpha^* \end{bmatrix}$ the *state characteristic matrix* of $W(\lambda)$ associated with the minimal realization (21.1.2). Proposition 21.1.1 shows that state characteristic matrices of $W(\lambda)$ associated with different minimal realizations are similar:

$$\begin{bmatrix} \tilde{\alpha} & \tilde{\beta} \\ \tilde{\gamma} & \tilde{\alpha}^* \end{bmatrix} = \begin{bmatrix} S^{-1} & 0 \\ 0 & S^* \end{bmatrix} \begin{bmatrix} \alpha & \beta \\ \gamma & \alpha^* \end{bmatrix} \begin{bmatrix} S & 0 \\ 0 & S^{*-1} \end{bmatrix}.$$

The distinctive properties of state characteristic matrices are summarized in the following theorem.

Theorem 21.1.2. *Let $W(\lambda)$ be a $p \times m$ rational matrix function with no pole at infinity and satisfying* (21.1.1), *and let* $\begin{bmatrix} \alpha & \beta \\ \gamma & \alpha^* \end{bmatrix}$ *be its state characteristic matrix. Then*

(i) β and γ are positive semidefinite;

(ii) the pair (α, β) is controllable, and the pair (γ, α) is observable.

Conversely, any 2×2 block matrix $\begin{bmatrix} \alpha & \beta \\ \gamma & \alpha^* \end{bmatrix}$ with $n \times n$ blocks which has properties (i) and (ii) is the state characteristic matrix of a rational matrix function $W(\lambda)$ for which $I - W(\infty)W(\infty)^* > 0$.

The set of all such functions $W(\lambda)$ for which $\begin{bmatrix} \alpha & \beta \\ \gamma & \alpha^* \end{bmatrix}$ is the state characteristic matrix is given by the formula

$$W(\lambda) = D + (I - DD^*)^{1/2} S(\lambda I - (\alpha - TD^*S))^{-1} T(I - D^*D)^{1/2}, \quad (21.1.8)$$

where T and S are any $n \times m$ and $p \times n$ matrices, respectively, such that $\beta = TT^*$, $\gamma = S^*S$, and D is any $p \times m$ matrix satisfying $DD^* < I$.

In view of Proposition 21.1.1 the result of Theorem 21.1.2 is independent of the choice of minimal realization (21.1.2), as might be expected.

Proof. Part (i) follows immediately from the definition of the state characteristic matrix. The controllability of

$$(\alpha, \beta) = (A + BD^*(I - DD^*)^{-1}C, B(I - D^*D)^{-1}B^*)$$

follows from that of (A, B) in view of Corollary 4.1.3 and Lemma 4.4.1(iii). The observability of

$$(\gamma, \alpha) = (C^*(I - DD^*)^{-1}C, A + BD^*(I - DD^*)^{-1}C)$$

is verified in a similar way.

We now prove the converse part of this theorem. First of all, observe the equality

$$(I - Z^*Z)^{1/2} Z^* = Z^*(I - ZZ^*)^{1/2} \quad (21.1.9)$$

for any $p \times m$ matrix Z satisfying $ZZ^* \leq I$ or, equivalently, $Z^*Z \leq I$. Indeed, use the singular value decomposition $Z = UD_0V$, where U and V are unitary matrices, and D_0 is a $p \times m$ diagonal matrix with the diagonal entries in the interval $[0, 1]$. Then (21.1.9) takes the form

$$V^*(I - D_0^*D_0)^{1/2} D_0^* U^* = V^* D_0^* (I - D_0 D_0^*)^{1/2} U^*,$$

and the proof of (21.1.9) is reduced to the case when $Z = D_0$, in which case it can be easily verified.

The next observation is that the realization (21.1.8) is minimal, i.e., (see Theorem 6.1.5), the pair

$$(\alpha - TD^*S, T(I - D^*D)^{1/2}) \tag{21.1.10}$$

is controllable and the pair

$$((I - DD^*)^{1/2}S, \alpha - TD^*S) \tag{21.1.11}$$

is observable. We will show only the controllability of (21.1.10), since the observability of (21.1.11) can be shown in a similar way. To this end note that $(\alpha, \beta) = (\alpha, TT^*)$ is controllable by (ii) and therefore, by Corollary 4.1.3, the pair (α, T) is controllable as well. Now Lemma 4.4.1(iii) ensures the controllability of $(\alpha - TD^*S, T)$. Since $(I - D^*D)^{1/2}$ is invertible, the pair $(\alpha - TD^*S, T(I - D^*D)^{1/2})$ is controllable as well.

Now let $\begin{bmatrix} \alpha' & \beta' \\ \gamma' & \alpha'^* \end{bmatrix}$ be the state characteristic matrix of $W(\lambda)$ given by (21.1.8). Using (21.1.9), we have

$$\alpha' = \alpha - TD^*S + T(I - D^*D)^{1/2}D^*(I - DD^*)^{-1}(I - DD^*)^{1/2}S$$
$$= \alpha - TD^*S + TD^*(I - DD^*)^{1/2}(I - DD^*)^{-1}(I - DD^*)^{1/2}S = \alpha,$$

and the equalities $\beta' = \beta$ and $\gamma' = \gamma$ follow, since $\beta = TT^*$ and $\gamma = S^*S$. Thus, $\begin{bmatrix} \alpha & \beta \\ \gamma & \alpha^* \end{bmatrix}$ is indeed the state characteristic matrix of $W(\lambda)$ given by its minimal realization (21.1.8).

To complete the proof of Theorem 21.1.2, we have to show that every rational matrix function $W(\lambda)$ with $I - W(\infty)W(\infty)^* > 0$ and for which $\begin{bmatrix} \alpha & \beta \\ \gamma & \alpha^* \end{bmatrix}$ is the state characteristic matrix, has the form (21.1.8). Write a minimal realization for such a $W(\lambda)$:

$$W(\lambda) = D + C(\lambda I - A)^{-1}B. \tag{21.1.12}$$

As $D = W(\infty)$, we have $DD^* < I$. We also have

$$\alpha = A + BD^*(I - DD^*)^{-1}C$$
$$\beta = B(I - D^*D)^{-1}B^*, \qquad \gamma = C^*(I - DD^*)^{-1}C.$$

Put

$$T = B(I - D^*D)^{-1/2}, \qquad S = (I - DD^*)^{-1/2}C$$

to rewrite (21.1.12) in the form (21.1.8). For the matrix A we obtain

$$A = \alpha - BD^*(I - DD^*)^{-1}C$$
$$= \alpha - T(I - D^*D)^{1/2}D^*(I - DD^*)^{-1}(I - DD^*)^{1/2}S$$
$$= \alpha - TD^*(I - DD^*)^{1/2}(I - DD^*)^{-1}(I - DD^*)^{1/2}S = \alpha - TD^*S,$$

where the formula (21.1.9) has been used. □

REALIZATIONS OF STRICT CONTRACTIONS   425

We now return to the equation (21.1.6). The main result of this section expresses the strict contraction property of $W(\lambda)$ in terms of hermitian solutions of (21.1.6).

Theorem 21.1.3. *Let $W(\lambda)$ be a $p \times m$ rational matrix function satisfying (21.1.1) with a minimal realization (21.1.2). Then the following statements are equivalent:*

(i) $W(\lambda)$ *is a strict contraction;*

(ii) *the equation (21.1.6) admits hermitian solutions X;*

(iii) *all real eigenvalues (if any) of the state characteristic matrix* $\begin{bmatrix} \alpha & \beta \\ \gamma & \alpha^* \end{bmatrix}$ *of $W(\lambda)$ have only even partial multiplicities.*

Proof. We first prove the equivalence of (ii) and (iii). A calculation shows that

$$\begin{bmatrix} \alpha & \beta \\ \gamma & \alpha^* \end{bmatrix} = \begin{bmatrix} 0 & iI \\ I & 0 \end{bmatrix} M \begin{bmatrix} 0 & I \\ -iI & 0 \end{bmatrix},$$

where $M = i \begin{bmatrix} -i\alpha^* & \gamma \\ -\beta & -i\alpha \end{bmatrix}$ is the matrix associated with the equation (21.1.6) (see Section 7.2). Since the pair $(-i\alpha^*, \gamma)$ is controllable by Theorem 21.1.2(ii), Theorem 7.3.7 is applicable, and the equivalence of (ii) and (iii) follows.

For the proof of the equivalence of (i) and (iii) we use the results of Sections 6.3 and 6.5. Assume that $W(\lambda)$ is a $p \times m$ rational matrix function satisfying the hypotheses of Theorem 21.1.3. Let

$$\begin{bmatrix} \alpha & \beta \\ \gamma & \alpha^* \end{bmatrix}$$

be the state characteristic matrix of $W(\lambda)$, where α, β and γ are given by equations (21.1.3)–(21.1.5). First of all, note the following formula:

$$(I - W(\lambda)(W(\bar{\lambda}))^*)^{-1} = (I - DD^*)^{-1} + [(I - DD^*)^{-1}C, (I - DD^*)^{-1}DB^*]$$

$$\left(\lambda I - \begin{bmatrix} \alpha & \beta \\ \gamma & \alpha^* \end{bmatrix}\right)^{-1} \begin{bmatrix} BD^*(I - DD^*)^{-1} \\ C^*(I - DD^*)^{-1} \end{bmatrix}, \qquad (21.1.13)$$

which is easily checked by a direct verification, using the formula (6.5.16) for the product $W(\lambda)(W(\bar{\lambda}))^*$. By Theorem 6.5.5, the realization (21.1.13) is minimal on $\mathbb{R}$.

Assume now that $W(\lambda)$ is a strict contraction. Then $(I - W(\lambda)(W(\bar{\lambda}))^*)^{-1}$ is nonnegative (in the sense of Section 6.5). By Lemma 6.5.1, the partial multiplicities of $(I - W(\lambda)(W(\bar{\lambda}))^*)^{-1}$ at every $\lambda_0 \in \mathbb{R}$ are all even. By Theorem 6.3.3, the real eigenvalues of $\begin{bmatrix} \alpha & \beta \\ \gamma & \alpha^* \end{bmatrix}$ have only even partial multiplicities. We have proved that statement (i) implies (iii).

Conversely, assume that $\begin{bmatrix} \alpha & \beta \\ \gamma & \alpha^* \end{bmatrix}$ has only even partial multiplicities corresponding to its real eigenvalues. Since the realization (21.1.13) is minimal on $\mathbb{R}$ it follows from Theorem 6.3.3 that all positive partial multiplicities of $I - W(\lambda)(W(\bar{\lambda}))^*$ corresponding to its real zeros are even. On the other hand, by Lemma 6.5.1, the negative partial multiplicities of $-W(\lambda)(W(\bar{\lambda}))^*$ at every real pole λ_0 are also even. But the functions $-W(\lambda)(W(\bar{\lambda}))^*$ and $I - W(\lambda)(W(\bar{\lambda}))^*$ have the same poles, and for each pole they have the same negative partial multiplicities (this can be verified using Theorem 6.2.2, for example). Thus, all partial multiplicities of $I - W(\lambda)(W(\bar{\lambda}))^*$ (positive and negative) corresponding to any real pole and/or zero λ_0 are even.

At this point we invoke a theorem of Rellich (see equation (6.5.1)):

$$I - W(\lambda)(W(\bar{\lambda}))^* = (Q(\lambda))^* \operatorname{diag}[\mu_1(\lambda), \ldots, \mu_p(\lambda)] Q(\lambda), \qquad (21.1.14)$$

in a real neighborhood U of $\lambda_0 \in \mathbb{R}$, where $Q(\lambda)$ is analytic and unitary valued, and $\mu_1(\lambda), \ldots, \mu_p(\lambda)$ are real meromorphic functions of $\lambda \in U$. Write

$$\mu_j(\lambda) = (\lambda - \lambda_0)^{\alpha_j} \nu_j(\lambda)$$

where the functions $\nu_j(\lambda)$ are analytic at λ_0 and $\nu_j(\lambda_0) \neq 0$. As in the proof of Lemma 6.5.1, one verifies that the nonzero integers among the α_j's are exactly the partial multiplicities of $I - W(\lambda)(W(\bar{\lambda}))^*$ at λ_0. Thus, all α_j's are even integers. It follows now from (21.1.14) that the number of positive eigenvalues of the matrix $I - W(\lambda)(W(\bar{\lambda}))^*$ is constant for $\lambda \in U \setminus \{\lambda_0\}$. Since $\lambda_0 \in \mathbb{R}$ was arbitrary, we find that the number of positive eigenvalues of the matrix $I - W(\lambda)(W(\bar{\lambda}))^*$ is constant for all real λ which are not poles or zeros of $I - W(\lambda)(W(\bar{\lambda}))^*$. Since $I - W(\infty)W(\infty)^* > 0$ by hypothesis, it follows that $W(\lambda)(W(\lambda))^* < I$ for all such $\lambda \in \mathbb{R}$. In particular, $W(\lambda)$ has no poles on the real axis and, by continuity, we obtain

$$W(\lambda)(W(\lambda))^* \leq I$$

for all real λ. This proves the remaining implication (iii) $\Rightarrow$ (i) of Theorem 21.1.3 and concludes the proof. $\square$

21.2 Inertia of solutions of a "CARE" and poles of strict contractions

Theorem 21.1.3 shows that, under certain hypotheses, the strict contraction property of $W(\lambda)$ is equivalent to the existence of hermitian solutions of the corresponding Riccati equation

$$X\gamma X + (i\alpha)X + X(i\alpha)^* + \beta = 0. \tag{21.2.1}$$

Here, $\begin{bmatrix} \alpha & \beta \\ \gamma & \alpha^* \end{bmatrix}$ is the state characteristic matrix of $W(\lambda)$ associated with its minimal realization

$$W(\lambda) = D + C(\lambda I - A)^{-1}B. \tag{21.2.2}$$

It turns out that there are strong connections between the properties of hermitian solutions of (21.2.1) and the properties of $W(\lambda)$. This section, and the next, are devoted to the clarification of some of these connections.

We start with the concept of inertia (i.e., the numbers of positive and negative eigenvalues) of hermitian solutions of (21.2.1).

Theorem 21.2.1. *Let $W(\lambda)$ be a strict contraction having minimal realization (21.2.2). Then every hermitian solution X of (21.2.1) is invertible, and for any such X the number of negative (resp. positive) eigenvalues of X, counted with multiplicities, is equal to the number of poles (also counted with multiplicities) of $W(\lambda)$ in the lower (resp. upper) half-plane.*

Proof. We prove the first part of the theorem, and refer the reader to Gohberg and Rubinstein (1988) for a proof of the second part.

Arguing by contradiction, assume that there is a singular hermitian solution X of (21.2.1). There exists a unitary matrix U such that

$$UXU^* = \begin{bmatrix} 0 & 0 \\ 0 & \Lambda \end{bmatrix}$$

for some invertible diagonal matrix Λ. Partition the matrices $U\alpha U^*$, $U\beta U^*$ and $U\gamma U^*$ accordingly:

$$U\alpha U^* = \begin{bmatrix} \alpha_{11} & \alpha_{12} \\ \alpha_{21} & \alpha_{22} \end{bmatrix}, \quad U\beta U^* = \begin{bmatrix} \beta_{11} & \beta_{12} \\ \beta_{21} & \beta_{22} \end{bmatrix},$$

$$U\gamma U^* = \begin{bmatrix} \gamma_{11} & \gamma_{12} \\ \gamma_{21} & \gamma_{22} \end{bmatrix}.$$

Equation (21.2.1) takes the form

$$\begin{bmatrix} \beta_{11} & i\alpha_{12}\Lambda + \beta_{12} \\ -i\Lambda\alpha_{21}^* + \beta_{21} & \Lambda\gamma_{22}\Lambda - i\Lambda\alpha_{22}^* + i\alpha_{22}\Lambda + \beta_{22} \end{bmatrix} = \begin{bmatrix} 0 & 0 \\ 0 & 0 \end{bmatrix}. \tag{21.2.3}$$

We obtain $\beta_{11} = 0$. But $U\beta U^*$ is positive semidefinite, so $\beta_{12} = 0$ and $\beta_{21} = 0$. Now (21.2.3) implies $\alpha_{12} = 0$ and $\alpha_{21} = 0$. Hence $U\beta U^*$ and $U\alpha^k \beta U^*$ ($k = 1, 2, \ldots$) have the form

$$U\beta U^* = \begin{bmatrix} 0 & 0 \\ 0 & * \end{bmatrix}, \quad U\alpha^k \beta U^* = \begin{bmatrix} 0 & 0 \\ 0 & * \end{bmatrix}.$$

It follows that

$$U[\beta, \alpha\beta, \ldots, \alpha^{n-1}\beta]U^* = \begin{bmatrix} 0 & 0 & \cdots & 0 \\ & * & & \end{bmatrix},$$

which contradicts the controllability of (α, β) (see Theorem 21.1.2) □

21.3 Relaxing the minimality of realizations

The results of Sections 21.1 and 21.2 can be generalized by relaxing the minimality condition on the realization (21.1.2). Accordingly, throughout this section, $W(\lambda)$ is a $p \times m$ rational matrix function such that $I - (W(\infty))^* W(\infty) < I$, and having a realization

$$W(\lambda) = D + C(\lambda I - A)^{-1} B \tag{21.3.1}$$

which need no longer be minimal.

Our first observation follows simply by analysing the proof of Theorem 21.1.3.

Theorem 21.3.1. *Assume that the realization of equation (21.3.1) is minimal on* $\mathbb{R}$. *Then* $W(\lambda)$ *is a strict contraction if and only if the state characteristic matrix* $\begin{bmatrix} \alpha & \beta \\ \gamma & \alpha^* \end{bmatrix}$ *has only even partial multiplicities corresponding to its real eigenvalues (if any).*

To connect with the existence of a hermitian solution X of the equation

$$X \gamma X + (i\alpha) X + X(i\alpha)^* + \beta = 0, \tag{21.3.2}$$

we need additional hypotheses on the pairs (C, A) and (A, B). A pair of matrices (A, B), where A is $n \times n$ and B is $n \times m$, will be called *conjugate controllable* if for every $\lambda \in \mathbb{C}$ either $\mathcal{R}_\lambda(A) \subseteq \mathcal{C}_{A,B}$ or $\mathcal{R}_{\bar{\lambda}}(A) \subseteq \mathcal{C}_{A,B}$ (or both). Here, $\mathcal{C}_{A,B}$ denotes the controllable subspace of (A, B) and $\mathcal{R}_\lambda(A)$ is the root subspace of A corresponding to its eigenvalue λ (if λ is not an eigenvalue of A, we put $\mathcal{R}_\lambda(A) = \{0\}$).

Theorem 21.3.2. *Assume that the realization of equation (21.3.1) is minimal on* $\mathbb{R}$ *and assume, in addition, that the pair* (A^*, C^*) *is conjugate controllable. Then* $W(\lambda)$ *is a strict contraction if and only if the equation (21.3.2) has hermitian solutions.*

Proof. Let

$$A^* = \begin{bmatrix} X_{11} & X_{12} \\ 0 & X_{22} \end{bmatrix}, \quad C^* = \begin{bmatrix} Y \\ 0 \end{bmatrix}$$

be the control normal form of (A^*, C^*) (see Section 4.5). Since (A^*, C^*) is conjugate controllable, it follows that, for every $\lambda \in \mathbb{C}$, at least one of the numbers λ and $\bar{\lambda}$ is not an eigenvalue of X_{22} (in fact, this condition is equivalent to the conjugate controllability of (A^*, C^*)). As a consequence we immediately obtain that for every matrix F the pair $(A^* + C^* F, C^*)$ is conjugate controllable as well. In particular, in view of the definition of equation (21.1.3), the pair (α^*, C^*) is conjugate controllable.

Next, by Proposition 4.1.2, the controllable subspace $\mathcal{C}_{A,B}$ coincides with the minimal A-invariant subspace that contains Im B. Since Im γ = Im C^* (in view of formula (21.1.5)) it follows that

$$\mathcal{C}_{\alpha^*,C^*} = \mathcal{C}_{\alpha^*,\gamma},$$

and therefore the pair (α^*, γ) is conjugate controllable as well. But now the pair $(-i\alpha^*, \gamma)$ is sign controllable, and we can apply Theorem 7.2.5. So, if $W(\lambda)$ is a strict contraction, then by Theorem 21.3.1 $\begin{bmatrix} \alpha & \beta \\ \gamma & \alpha^* \end{bmatrix}$ has even partial multiplicities corresponding to its real eigenvalues. Therefore by Theorem 7.2.5 (taking into account the formula

$$\begin{bmatrix} \alpha & \beta \\ \gamma & \alpha^* \end{bmatrix} = \begin{bmatrix} 0 & iI \\ I & 0 \end{bmatrix} M \begin{bmatrix} 0 & I \\ -iI & 0 \end{bmatrix} \qquad (21.3.3)$$

where M is the matrix associated with the equation (21.3.2) as in Section 7.2), the equation (21.3.2) has hermitian solutions.

Conversely, if (21.3.2) has hermitian solutions, then by Theorem 7.3.7 the matrix M associated with equation (21.3.2) has only even multiplicities corresponding to its real eigenvalues. By formula (21.3.3) the same is true for $\begin{bmatrix} \alpha & \beta \\ \gamma & \alpha^* \end{bmatrix}$, and by Theorem 21.3.1 $W(\lambda)$ is a strict contraction. □

It is instructive to compare Theorem 21.3.2 with Theorem 20.1.1. Thus, let $D = 0$ in (21.3.1), and let $\tilde{W}(\lambda) = (-i)W(-i\lambda)$. So

$$\tilde{W}(\lambda) = C(\lambda I - iA)^{-1}B. \qquad (21.3.4)$$

Assume further that the realization (21.3.4) is minimal on the imaginary axis and that the pair (iA, B) is sign controllable. Applying Theorem 21.3.2 to $(W(\bar{\lambda}))^*$ it is found that $\tilde{W}(\lambda)$ is a contraction for all pure imaginary values of λ, i.e.

$$\tilde{W}(\lambda)(\tilde{W}(\lambda))^* \leq I, \quad \lambda \in i\mathbb{R},$$

if and only if the equation

$$XBB^*X + X(-iA) + (iA^*)X + C^*C = 0$$

admits a hermitian solution X. But this is exactly equation (20.1.3) applied to the realization (21.3.4). We obtain the following variation of Theorem 20.1.1:

Theorem 21.3.3. *Assume that the realization $W(\lambda) = C(\lambda I - A)^{-1}B$ is minimal on the imaginary axis, and that the pair (A, B) is sign controllable. Then*

$$\|W(\lambda)\| \leq 1, \quad \lambda \in i\mathbb{R} \qquad (21.3.5)$$

if and only if the equation

$$XBB^*X - XA - A^*X + C^*C = 0$$

has a hermitian solution X.

Observe that the hypotheses of Theorem 20.1.1 ($\sigma(A) \cap i\mathbb{R} = \emptyset$ and c-stabilizability of (A, B)) are stronger than those of Theorem 21.3.3. On the other hand, the conclusions of Theorem 20.1.1 are stronger as well (the inequality in (20.1.2) is strict in contrast with (21.3.5) and existence of a stabilizing hermitian solution is required in Theorem 20.1.1, in contrast with existence of any hermitian solution in Theorem 21.3.3.

21.4 Minimal unitary completions

Let a $p \times m$ rational matrix function $W(\lambda)$ be given. A *unitary completion* of $W(\lambda)$ is, by definition, a $(p+m) \times (p+m)$ rational matrix function $\tilde{W}(\lambda)$ of the form

$$\tilde{W}(\lambda) = \begin{bmatrix} * & W(\lambda) \\ * & * \end{bmatrix}$$

and such that

$$(\tilde{W}(\lambda))^* \tilde{W}(\lambda) = \tilde{W}(\lambda)(\tilde{W}(\lambda))^* = I \qquad (21.4.1)$$

for all real λ which are not poles of $\tilde{W}(\lambda)$.

Since the latter condition obviously implies $W(\lambda)(W(\lambda))^* \leq I$ for all real λ (not poles of $\tilde{W}(\lambda)$), a necessary condition for the existence of a unitary completion of $W(\lambda)$ is $W(\lambda)(W(\lambda))^* \leq I$, or, equivalently, $(W(\bar{\lambda}))^* W(\lambda) \leq I$ for all real λ which are not poles of $W(\lambda)$. In particular, $W(\lambda)$ has no poles on the real line and at infinity and therefore admits a minimal realization of the form

$$W(\lambda) = D + C(\lambda I - A)^{-1} B. \qquad (21.4.2)$$

The size of A (which is uniquely determined by $W(\lambda)$) is called the *McMillan degree* of $W(\lambda)$ and is denoted $\delta(W)$.

A unitary completion $\tilde{W}(\lambda)$ of $W(\lambda)$ is called *minimal* if $\delta(\tilde{W}) = \delta(W)$.

The following theorem describes all minimal unitary completions of a given strict contraction in terms of hermitian solutions of the corresponding Riccati equation.

Theorem 21.4.1. *Let $W(\lambda)$ be a strict contraction with a minimal realization (21.4.2) and with a corresponding Riccati equation*

$$X\gamma X + i\alpha X - iX\alpha^* + \beta = 0. \qquad (21.4.3)$$

(a) *For every hermitian solution X of (21.4.3) the function $\tilde{W}_X(\lambda)$ defined by*

$$\tilde{W}_X(\lambda) = \begin{bmatrix} (I - DD^*)^{1/2} & D \\ -D^* & (I - D^*D)^{1/2} \end{bmatrix} + \begin{bmatrix} C \\ (I - D^*D)^{-1/2}(iB^*X^{-1} - D^*C) \end{bmatrix}$$
$$(\lambda I - A)^{-1}[(iXC^* - BD^*)(I - DD^*)^{-1/2}, B] \qquad (21.4.4)$$

is a minimal unitary completion of $W(\lambda)$. (The invertibility of X is guaranteed by Theorem 21.2.1.)

(b) *Every minimal unitary completion* $\tilde{W}(\lambda)$ *of* $W(\lambda)$ *has the form*

$$\tilde{W}(\lambda) = \begin{bmatrix} I & 0 \\ 0 & T \end{bmatrix} \tilde{W}_X(\lambda) \begin{bmatrix} S & 0 \\ 0 & I \end{bmatrix} \quad (21.4.5)$$

where T and S are unitary matrices of sizes $m \times m$ and $p \times p$ respectively, and X is a hermitian solution of (21.4.3). The unitary matrices S and T are uniquely determined by the value of $\tilde{W}(\lambda)$ at infinity and are given by

$$S = (I - DD^*)^{-1/2}[\tilde{W}(\infty)]_{11} \quad (21.4.6a)$$

$$T = [\tilde{W}(\infty)]_{22}(I - D^*D)^{-1/2}. \quad (21.4.6b)$$

Here $[\tilde{W}(\infty)]_{11}$ and $[\tilde{W}(\infty)]_{22}$ stand for the top left and the bottom right blocks of $\tilde{W}(\infty)$, respectively.

(c) *The formula (21.4.4) provides a one-to-one correspondence between the set*

$$\{(X, T, S) : X = X^* \text{ is a solution of } (21.4.3); T \text{ and } S$$
$$\text{are } m \times m \text{ and } p \times p \text{ unitary matrices, respectively}\}$$

and the set of all minimal unitary completions of $W(\lambda)$.

An interesting observation following from formula (21.4.4) is that all minimal unitary completions of a given strict contraction $W(\lambda)$ have the same poles as $W(\lambda)$ with the same partial multiplicites, but their zeros are generally different from those of $W(\lambda)$. In particular, the theorem guarantees existence of a minimal unitary completion for every strict contraction.

A full proof of Theorem 21.4.1 can be found in Gohberg and Rubinstein (1988). We do not reproduce the proof here, but consider instead several illustrative examples.

Example 21.4.1. Let
$$W(\lambda) = \mu\lambda(\lambda - i)^{-1},$$
where μ is a parameter, $-1 < \mu < 1$, $\mu \neq 0$. A minimal realization for $W(\lambda)$ is easily found:
$$W(\lambda) = \mu + \mu(\lambda - i)^{-1}i.$$

The equation (21.4.3) takes the form

$$\frac{1}{1-\mu^2}\mu^2 x^2 - \frac{2}{1-\mu^2}x + \frac{1}{1-\mu^2} = 0.$$

This equation has two real solutions:

$$x_{1,2} = \mu^{-2} \pm \mu^{-2}\sqrt{1-\mu^2}.$$

According to Theorem 21.4.1, all minimal unitary completions $\tilde{W}(\lambda)$ of $W(\lambda)$ are given by the formula

$$\tilde{W}(\lambda) = \begin{bmatrix} 1 & 0 \\ 0 & e^{it} \end{bmatrix} \tilde{W}_x(\lambda) \begin{bmatrix} e^{is} & 0 \\ 0 & 1 \end{bmatrix},$$

where $0 \leq t, s < 2\pi$, and where

$$\tilde{W}_x(\lambda) = \begin{bmatrix} \sqrt{1-\mu^2} & \mu \\ -\mu & \sqrt{1-\mu^2} \end{bmatrix} + \begin{bmatrix} \mu \\ (1-\mu^2)^{-1/2}(\mu^2(1 \pm \sqrt{1-\mu^2})^{-1} - \mu^2) \end{bmatrix}$$
$$(\lambda - i)^{-1}[(i\mu^{-1}(1 \pm \sqrt{1-\mu^2}) - i\mu)(1-\mu^2)^{-1/2}, i]. \quad \square$$

Example 21.4.2. Let

$$W(\lambda) = \mu I + \begin{bmatrix} \lambda - (i-\mu)(1-\mu^2)^{-1} & -i(1-\mu^2)^{-1} \\ -i(1-\mu^2)^{-1} & \lambda + (i+\mu)(1-\mu^2)^{-1} \end{bmatrix}^{-1},$$

where μ, $0 < \mu < 1$ is a parameter. A minimal realization is found by inspection:

$$W(\lambda) = \mu I + I \left(\lambda I - \begin{bmatrix} (i-\mu)(1-\mu^2)^{-1} & i(1-\mu^2)^{-1} \\ i(1-\mu^2)^{-1} & -(i+\mu)(1-\mu^2)^{-1} \end{bmatrix} \right)^{-1} I.$$

Furthermore,

$$\alpha = \frac{1}{1-\mu^2}\begin{bmatrix} i & i \\ i & -i \end{bmatrix}; \quad \beta = \frac{1}{1-\mu^2}I; \quad \gamma = \frac{1}{1-\mu^2}I;$$

and the Riccati equation (21.4.3) takes the form

$$X^2 + \begin{bmatrix} -1 & -1 \\ -1 & 1 \end{bmatrix} X + X \begin{bmatrix} -1 & -1 \\ -1 & 1 \end{bmatrix} + I = 0. \tag{21.4.7}$$

The equation (21.4.7) has a hermitian solution $X = \begin{bmatrix} 1 & 0 \\ 0 & -1 \end{bmatrix}$. Thus, by Theorem 21.1.3, $W(\lambda)$ is a strict contraction. In fact, (21.4.7) admits a continuum of hermitian solutions (consisting of three connected components) given by the formulas

$$X = \begin{bmatrix} a & 1 \pm \sqrt{1-(a-1)^2} \\ 1 \pm \sqrt{1-(a-1)^2} & -a \end{bmatrix}, \quad 0 \leq a \leq 2$$

$$X = \begin{bmatrix} 0 & 1 \\ 1 & -2 \end{bmatrix}; \quad X = \begin{bmatrix} 2 & 1 \\ 1 & 0 \end{bmatrix}.$$

By Theorem 21.4.1., we obtain a continuum of minimal unitary completions $\tilde{W}(\lambda)$ of $W(\lambda)$, given by (21.4.5). $\square$

21.5 The real case

We present here real analogues of some of the results of previous sections. The definitions introduced earlier in this chapter will be used in this section without further explanation.

Consider a $p \times m$ strict contraction $W(\lambda)$ which is real, i.e., $W(\lambda)$ is a real matrix for every real λ. Any real rational matrix function $W(\lambda)$ which is analytic at infinity admits a minimal realization,

$$W(\lambda) = D + C(\lambda I - A)^{-1}B \qquad (21.5.1)$$

with real matrices A, B, C, and D. In the sequel, only real minimal realizations of $W(\lambda)$ will be used. According to Theorem 21.1.3, the associated Riccati equation,

$$X\gamma X - iX\alpha^T + i\alpha X + \beta = 0, \qquad (21.5.2)$$

has hermitian solutions. Note that, in view of formulas (21.1.3)–(21.1.5), the matrices α, β, and γ are real.

Being real, $W(\lambda)$ has equal numbers of poles in the open upper and in the open lower half-plane. Thus Theorem 21.2.1 gives the following result:

Theorem 21.5.1. *The hermitian solutions of* (21.5.2) *are invertible and have equal numbers of positive and negative eigenvalues.*

Now consider minimal unitary completions of $W(\lambda)$ which are also real.

Theorem 21.5.2. *Let X be a hermitian solution of* (21.5.2)*. Then the associated minimal unitary completion $\tilde{W}_X(\lambda)$ of $W(\lambda)$ (given by formula (21.4.4)) is real if and only if X is pure imaginary: $X = iX_0$ for some real skew-symmetric matrix X_0.*

Proof. The formula (21.4.4) shows that if X is pure imaginary, then $\tilde{W}_X(\lambda)$ is real. Conversely, let $\tilde{W}_X(\lambda)$ be real. Formula (21.4.4) shows that $C(\lambda I - A)^{-1} iXC^T$ is real for real λ. This implies easily that $CA^j iXC^T$ is real for $j = 0, 1, \ldots$, and from the observability of (C, A), it follows that iXC^T is real. The equation

$$X\gamma X - iX\alpha^T + i\alpha X + \beta = 0$$

can be rewritten in the form

$$XC^T(I - DD^T)^{-1}CX - iX(A^T + C^T(I - DD^T)^{-1}DB^T)$$
$$+ i(A + BD^T(I - DD^T)^{-1}C)X + B(I - DD^T)^{-1}B^T = 0,$$

and we see now that $-iXA^T + iAX$ is real. Write $X = Z_1 + iZ_2$, where Z_1 and Z_2 are real matrices; then we have $CZ_1 = 0$ and $-Z_1 A^T + AZ_1 = 0$. Premultiply the latter equation by C to obtain $CAZ_1 = 0$, and by induction, $CA^j Z_1 = 0$ for $j = 0, 1, \ldots$. Now by the observability of (C, A), $Z_1 = 0$, and X is pure imaginary. □

Corollary 21.5.3. *A real strict contraction $W(\lambda)$ admits a real minimal unitary completion if and only if the equation*

$$-Y\gamma Y + Y\alpha^T - \alpha Y + \beta = 0 \qquad (21.5.3)$$

has a real skew-symmetric solution Y.

Proof. If (21.5.3) has a real skew-symmetric solution Y, then (21.5.2) has a pure imaginary hermitian solutions $X = iY$ and, by Theorem 21.5.2, $\tilde{W}_X(\lambda)$ is a real minimal unitary completion of $W(\lambda)$.

Conversely, let $\tilde{W}(\lambda)$ be a real minimal unitary completion of $W(\lambda)$. By Theorem 21.4.1

$$\tilde{W}(\lambda) = \begin{bmatrix} I & 0 \\ 0 & T \end{bmatrix} \tilde{W}_X(\lambda) \begin{bmatrix} S & 0 \\ 0 & I \end{bmatrix}$$

for some hermitian solution X of (21.5.2). Formulas (21.4.6) show that S and T are real, and therefore $W_X(\lambda)$ is real as well. Now appeal to Theorem 21.5.2. □

All real minimal unitary completions of a real strict contraction $W(\lambda)$ are described by the formulas (21.4.4) and (21.4.5) in which X is replaced by iY, where Y is a real skew-symmetric solution of (21.5.3), and in which S and T are real unitary matrices of sizes $m \times m$ and $p \times p$, respectively.

In view of Corollary 21.5.3 it is natural to ask if there exist any pure imaginary hermitian solutions of (21.5.2) provided this equation has hermitian solutions at all. The following example shows that the answer is generally "no" and consequently, not every real strict contraction admits a minimal real unitary completion.

Example 21.5.1. Let

$$\alpha = \begin{bmatrix} 0 & 1 \\ -1 & 0 \end{bmatrix}; \quad \gamma = \begin{bmatrix} a & 0 \\ 0 & 0 \end{bmatrix}; \quad \beta = \begin{bmatrix} 0 & 0 \\ 0 & a \end{bmatrix},$$

where $0 < a < 1$. By Theorem 21.1.2 there is a rational matrix function $W(\lambda)$ with $I - W(\infty)W(\infty)^* > 0$ for which $Q = \begin{bmatrix} \alpha & \beta \\ \gamma & \alpha^T \end{bmatrix}$ is a state characteristic matrix. Formula (21.1.8) shows that $W(\lambda)$ may be chosen real. The eigenvalues of Q turn out to be $\pm i\sqrt{1-a}; \pm i\sqrt{1+a}$. Thus, by Theorem 21.1.3 $W(\lambda)$ is a strict contraction and the equation (21.5.2) admits hermitian solutions. However, a straightforward check shows that (21.5.3) has no real skew-symmetric solutions Y. By Corollary 21.5.3 $W(\lambda)$ has no real minimal unitary completions. □

21.6 Notes

The main results of Sections 21.1, 21.2 and 21.4, as well as the examples in Section 21.4, are taken from Gohberg and Rubinstein (1988). Our exposition of the proofs is also based on that paper. The real case (Section 21.5) was studied first by Lancaster and Rodman (1991).

Rational matrix functions $W(\lambda)$ which are contractions on the unit circle (i.e., satisfying $W(\lambda)(W(\lambda^{-1}))^* \leq I$ for $\|\lambda\| = 1$) and their unitary completions have been studied in Gohberg et al. (1992). There, the function $W(\lambda)$ is given by its realization in the form

$$W(\lambda) = D + (1-\lambda)C(\lambda G - A)^{-1}B.$$

Contractions on the unit circle, as well as minimal unitary (on the unit circle) completions, are described in terms of hermitian solutions of an algebraic Riccati equation of continuous type, with coefficients constructed from A, B, C, D and G.

22
THE MATRIX SIGN FUNCTION

Consider a symmetric algebraic Riccati equation of the form

$$XDX + AX + XA^* - C = 0$$

where $D^* = D$ and $C^* = C$ and all matrices have size $n \times n$. Consider also the related matrix

$$M = \begin{bmatrix} A & D \\ C & -A^* \end{bmatrix}.$$

It has been shown in Chapter 7 that (under suitable hypotheses) the hermitian solutions of this Riccati equation are intimately connected to certain n-dimensional invariant subspaces of M and, in particular, to the spectral subspace $\mathcal{M}_+$ of M corresponding to eigenvalues in the open right half-plane (see Theorem 7.6.2, for example).

The case in which M has no eigenvalues on the imaginary axis is important (see Theorem 7.9.4), and the notion of the matrix sign function $\operatorname{sgn}(M)$, to be discussed in Section 22.1, is a natural tool because it is a projector of the form

$$\operatorname{sgn}(M) = P_+ - (I - P_+),$$

where P_+ is the spectral (or Riesz) projector onto $\mathcal{M}_+$ (see Section 1.3). Of course, $I - P_+$ is then the complementary projector onto the spectral subspace of M corresponding to its eigenvalues in the open left half-plane.

Thus, knowledge of $\operatorname{sgn}(M)$ includes implicit information about P_+ and hence about an extremal solution X of the Riccati equation (in the sense of Section 7.9). In this chapter it is our objective to make this information explicit. In large measure, the motivation for doing this is the fact that there are efficient and well-developed algorithms for the computation of $\operatorname{sgn}(M)$ and hence for the solution of Riccati equations. We do not discuss these algorithms but take it as an axiom that $\operatorname{sgn}(M)$ can be readily and accurately computed. There are several other problems which can be resolved using the matrix sign function so an extensive literature on these algorithms has evolved since the matrix sign function was first introduced in 1971. See the extensive list of references in the review of Laub (1991) for an introduction to this literature.

In Section 22.1 there is a brief introduction to the matrix sign function. Theorem 22.2.1 is a key result which shows how, using sgn (M), the stabilizing solution of the Riccati equation can be found (when it exists). Then Theorem 22.2.2 gives necessary and sufficient conditions for the existence of such a solution, formulated in terms of sgn (M). Section 22.3 contains a discussion of the connection of the matrix sign function with a "DARE" and, finally, in Section 22.4 it is shown that equations in real matrices introduce no essentially new features.

22.1 Matrix sign function: definition and basic properties

Let M be an $n \times n$ complex matrix having no eigenvalues on the imaginary axis. Reducing M to its Jordan form, we can write

$$M = T(D+N)T^{-1} \qquad (22.1.1)$$

for some invertible matrix T, where D is a diagonal matrix of eigenvalues, say $D = \text{diag}[\lambda_1, \ldots, \lambda_n]$, and N is a nilpotent matrix commuting with D (see Section 1.1). Because of our assumption on M, $\text{Re}\,\lambda_j \neq 0$ for $j = 1, \ldots, n$. Define the *matrix sign* function of M by

$$\text{sgn}\,(M) = T \,\text{diag}\,[\text{sgn}\,(\text{Re}\,\lambda_1), \ldots, \text{sgn}\,(\text{Re}\,\lambda_n)]T^{-1},$$

where $\text{sgn}\,\alpha = 1$ if $\alpha > 0$ and $\text{sgn}\,\alpha = -1$ if $\alpha < 0$. Equivalently, $\text{sgn}\,(M)$ can be defined as a function of M in the sense of functional analysis as follows. Let $f(z)$ be a polynomial such that

$$\left. \begin{array}{l} f(\lambda_0) = \text{sgn}\,(\text{Re}\,\lambda_0) \\ f^{(j)}(\lambda_0) = 0, \quad j = 1, \ldots, m(\lambda_0) - 1 \end{array} \right\} \qquad (22.1.2)$$

for every eigenvalue λ_0 of M, where $m(\lambda_0)$ is the index of λ_0 (see Example 1.2.5). Then

$$\text{sgn}\,(M) = f(M)$$

(see Section 1.7). This second definition shows, in particular, that $\text{sgn}\,(M)$ is well-defined and does not depend on the choice of the representation (22.1.1).

We list several properties of the matrix sign function, which can be immediately verified using the definitions.

Proposition 22.1.1.
  (a) *The matrix* $\text{sgn}\,(M)$ *is diagonable and satisfies* $(\text{sgn}\,(M))^2 = I$ *(and therefore all eigenvalues of* $\text{sgn}\,(M)$ *are* ± 1*)*.
  (b) $\text{Im}\,(\text{sgn}\,(M)-I) = \text{Ker}\,(\text{sgn}\,(M)+I)$ *is the stable M-invariant subspace (i.e. the spectral M-invariant subspace corresponding to its eigenvalues in the open left half-plane)*.

(c) Im $(\mathrm{sgn}\,(M) + I) = \mathrm{Ker}\,(\mathrm{sgn}\,(M) - I)$ *is the antistable M-invariant subspace (corresponding to the eigenvalues of M in the open right half-plane).*

(d) *For any invertible matrix S*

$$\mathrm{sgn}\,(S^{-1}MS) = S^{-1}(\mathrm{sgn}\,M)S.$$

(e) *For any nonzero real number c,*

$$\mathrm{sgn}\,(cM) = \mathrm{sgn}\,(c)\,\mathrm{sgn}\,(M).$$

(f) $\mathrm{sgn}\,(M^*) = (\mathrm{sgn}\,(M))^*.$

In the next section the following property of the matrix sign function will be useful.

Lemma 22.1.2. *Let M be an $n \times n$ matrix without eigenvalues on the imaginary axis, and let k be the dimension of the spectral M-invariant subspace corresponding to the eigenvalues of M in the open right half-plane. Form the partition*

$$\mathrm{sgn}\,(M) + I = \begin{bmatrix} W_{11} & W_{12} \\ W_{21} & W_{22} \end{bmatrix},$$

where W_{11} is $(n-k) \times (n-k)$ and W_{22} is $k \times k$. If $\mathrm{rank} \begin{bmatrix} W_{12} \\ W_{22} \end{bmatrix} = k$, *then there exists a unique $k \times (n-k)$ matrix X satisfying*

$$\begin{bmatrix} W_{12} \\ W_{22} \end{bmatrix} X = - \begin{bmatrix} W_{11} \\ W_{21} \end{bmatrix}. \tag{22.1.3}$$

Proof. The uniqueness of X, when it exists, is clear because $\begin{bmatrix} W_{12} \\ W_{22} \end{bmatrix}$ has full column rank. To prove the existence of X observe that by part (c) of Proposition 22.1.1, and by hypothesis,

$$\mathrm{rank} \begin{bmatrix} W_{11} & W_{12} \\ W_{21} & W_{22} \end{bmatrix} = k = \mathrm{rank} \begin{bmatrix} W_{12} \\ W_{22} \end{bmatrix}.$$

Therefore the columns of $\begin{bmatrix} W_{11} \\ W_{21} \end{bmatrix}$ must be linear combinations of the columns of $\begin{bmatrix} W_{12} \\ W_{22} \end{bmatrix}$. In other words,

$$\begin{bmatrix} W_{11} \\ W_{21} \end{bmatrix} = \begin{bmatrix} W_{12} \\ W_{22} \end{bmatrix}(-X)$$

for some matrix X, which is exactly the equation (22.1.3). □

22.2 Connection with the Riccati equation

Consider the algebraic Riccati equation

$$XDX + XA + A^*X - C = 0, \tag{22.2.1}$$

where the matrices D and C are hermitian, and the matrices A, C and D have size $n \times n$. Form the matrix

$$M = \begin{bmatrix} A & D \\ C & -A^* \end{bmatrix}.$$

As in Section 7.9, a hermitian solution X of (22.2.1) is called *stabilizing* if $A+DX$ has all its eigenvalues in the open left half-plane.

Theorem 22.2.1. *Assume that the matrix M has no eigenvalues on the imaginary axis, and let*

$$Z = \begin{bmatrix} Z_{11} & Z_{12} \\ Z_{21} & Z_{22} \end{bmatrix} = \mathrm{sgn}\,(M)$$

where Z_{11} is $n \times n$. If X is a stabilizing solution of (22.2.1), then

$$\begin{bmatrix} Z_{12} \\ Z_{22} + I \end{bmatrix} X = -\begin{bmatrix} Z_{11} + I \\ Z_{21} \end{bmatrix}. \tag{22.2.2}$$

Proof. Since $A + DX$ is stable, an $n \times n$ matrix Y is defined uniquely by the equation

$$(A + DX)Y + Y(A + DX)^* = D, \tag{22.2.3}$$

(see Theorem 5.2.2). Now define

$$S = \begin{bmatrix} I & Y \\ 0 & I \end{bmatrix} \begin{bmatrix} I & 0 \\ -X & I \end{bmatrix} = \begin{bmatrix} I - YX & Y \\ -X & I \end{bmatrix}.$$

The inverse of S is easily computed:

$$S^{-1} = \begin{bmatrix} I & 0 \\ X & I \end{bmatrix} \begin{bmatrix} I & -Y \\ 0 & I \end{bmatrix} = \begin{bmatrix} I & -Y \\ X & I - XY \end{bmatrix}.$$

It turns out that the matrix S (when used as a similarity matrix) block diagonalizes M. Indeed,

$$S\begin{bmatrix} A & D \\ C & -A^* \end{bmatrix} S^{-1} = \begin{bmatrix} I-YX & Y \\ -X & I \end{bmatrix} \begin{bmatrix} A & D \\ C & -A^* \end{bmatrix} \begin{bmatrix} I & -Y \\ X & I-XY \end{bmatrix}$$
$$= \begin{bmatrix} W_{11} & W_{12} \\ W_{21} & W_{22} \end{bmatrix},$$

where

$W_{11} = A - YXA + YC + DX - YXDX - YA^*X$
$W_{12} = -AY + YXAY - YCY + D - DXY - YXD + YXDXY$
$\qquad\qquad\qquad\qquad\qquad\qquad\qquad\qquad -YA^* + YA^*XY$
$W_{21} = -XA + C - XDX - A^*X$
$W_{22} = XAY - CY - XD - A^* + XDXY + A^*XY.$

We see that, since X is a hermitian solution of (22.2.1), $W_{21} = 0$, $W_{11} = A + DX$ and $W_{22} = -(A + DX)^*$. The expression for W_{12} can be rewritten in the form

$$W_{12} = Y(XA - C + XDX + A^*X)Y - AY + D - DXY - YXD - YA^*,$$

which is zero in view of equation (22.2.3). Thus,

$$S\begin{bmatrix} A & D \\ C & -A^* \end{bmatrix} S^{-1} = \begin{bmatrix} A+DX & 0 \\ 0 & -(A+DX)^* \end{bmatrix}.$$

Now

$$\operatorname{sgn}(M) = S^{-1} \operatorname{sgn} \begin{bmatrix} A+DX & 0 \\ 0 & -(A+DX)^* \end{bmatrix} S$$
$$= S^{-1} \begin{bmatrix} -I & 0 \\ 0 & I \end{bmatrix} S = \begin{bmatrix} -I+2YX & -2Y \\ -2X+2XYX & I-2XY \end{bmatrix}.$$

So

$$\begin{bmatrix} Z_{11}+I & Z_{12} \\ Z_{21} & Z_{22}+I \end{bmatrix} = \operatorname{sgn}(M) + I = \begin{bmatrix} 2YX & -2Y \\ -2X+2XYX & 2I-2XY \end{bmatrix}$$
$$= \begin{bmatrix} \begin{bmatrix} 2Y \\ 2(XY-I) \end{bmatrix} X, & -\begin{bmatrix} 2Y \\ 2(XY-I) \end{bmatrix} \end{bmatrix}. \quad (22.2.4)$$

By comparing both sides of this equation, the equality (22.2.2) is obtained. □

A hermitian solution X of (22.2.1) is called *antistabilizing* if $A + DX$ has all its eigenvalues in the open right half-plane. For antistabilizing solutions a result analogous to Theorem 22.2.1 holds:

Theorem 22.2.2. Let M and $Z = \text{sgn}(M)$ be as in Theorem 22.2.1. If X is an antistabilizing solution of (22.2.1), then

$$\begin{bmatrix} Z_{12} \\ Z_{22} - I \end{bmatrix} X = - \begin{bmatrix} Z_{11} - I \\ Z_{21} \end{bmatrix}. \tag{22.2.5}$$

Proof. The proof is similar to that of Theorem 22.2.1, except that now $\text{sgn}(A + DX) = +I$ and $\text{sgn}(M) - I = ([\ * \ ] X, -[\ * \ ])$. □

The equalities (22.2.2) and (22.2.5) are used in practice to compute stabilizing and antistabilizing solutions for the Riccati equation (22.2.1). However, in order to use these equalities we must know if the stabilizing and antistabilizing solutions exist, and if they are unique. There are sufficient conditions for existence and uniqueness of these solutions. For example, standard and useful sufficient conditions are provided by Theorems 7.2.8 and 7.9.4 in the form: (A, D) is stabilizable, $D \geq 0$, $C \geq 0$, and (C, A) is detectable; in which case the stabilizing and antistabilizing solutions exist and are unique. It turns out, however, that a criterion for existence and uniqueness of the stabilizing and antistabilizing solutions can be given in terms of the matrix sign function:

Theorem 22.2.3. (a) *Equation* (22.2.1) *has a stabilizing solution if and only if the matrix M has no eigenvalues on the imaginary axis and*

$$\text{rank} \begin{bmatrix} Z_{12} \\ Z_{22} + I \end{bmatrix} = n \tag{22.2.6}$$

where

$$\text{sgn}(M) = \begin{bmatrix} Z_{11} & Z_{12} \\ Z_{21} & Z_{22} \end{bmatrix}.$$

(b) *Equation* (22.2.1) *has an antistabilizing solution if and only if M has no eigenvalues on the imaginary axis and*

$$\text{rank} \begin{bmatrix} Z_{12} \\ Z_{22} - I \end{bmatrix} = n.$$

Before considering the proof of this theorem note that Proposition 7.9.2 can be applied and shows that if a stabilizing solution exists it must be unique. Also, the proof of that proposition is easily adjusted to show that, also, when an antistabilizing solution exists it is unique.

Proof. We will prove part (a) only, leaving the proof of (b) as an exercise.

Suppose there exists a stabilizing solution $X = X^*$ of equation (22.2.1). Then $\sigma(A + DX) \subset \{\lambda \in \mathbb{C} : \operatorname{Re} \lambda < 0\}$ and

$$\begin{bmatrix} I & 0 \\ -X & I \end{bmatrix} \begin{bmatrix} A & D \\ C & -A^* \end{bmatrix} \begin{bmatrix} I & 0 \\ -X & I \end{bmatrix}^{-1} = \begin{bmatrix} A + DX & D \\ 0 & -(A + DX)^* \end{bmatrix},$$

and therefore M has no eigenvalues on the imaginary axis. Further, let

$$v \in \operatorname{Ker} \begin{bmatrix} Z_{12} \\ Z_{22} + I \end{bmatrix}.$$

Using the formula (22.2.4), we obtain

$$0 = \begin{bmatrix} Z_{11} + I & Z_{12} \\ Z_{21} & Z_{22} + I \end{bmatrix} \begin{bmatrix} 0 \\ v \end{bmatrix} = \begin{bmatrix} 2YX & -2Y \\ -2X + 2XYX & 2I - 2XY \end{bmatrix} \begin{bmatrix} 0 \\ v \end{bmatrix}$$

$$= \begin{bmatrix} -2Yv \\ -2XYv + 2v \end{bmatrix}.$$

Thus, $Yv = 0$ and $0 = -2XYv + 2v$, so that $v = 0$ as well and (22.2.6) is established.

Conversely, let (22.2.6) hold and define X_p by equation (22.2.2). That such an X_p indeed exists follows from Lemma 22.1.2. (Observe that iM is self-adjoint in the indefinite scalar product induced by the matrix $i\begin{bmatrix} 0 & I \\ -I & 0 \end{bmatrix}$, and therefore the spectral subspace of M corresponding to the eigenvalues in the open right half-plane has dimension n). Now equation (22.2.2) can be rewritten in the form

$$\begin{bmatrix} Z_{11} & Z_{12} \\ Z_{21} & Z_{22} \end{bmatrix} \begin{bmatrix} I \\ X_p \end{bmatrix} = -\begin{bmatrix} I \\ X_p \end{bmatrix}.$$

By Proposition 22.1.1(b), and taking into account the fact that the stable M-invariant subspace has dimension n, we find that $\operatorname{Im} \begin{bmatrix} I \\ X_p \end{bmatrix}$ coincides with the stable M-invariant subspace. By Proposition 7.1.1 X_p is a solution of the Riccati equation (22.2.1). Furthermore, iM is $i\begin{bmatrix} 0 & I \\ -I & 0 \end{bmatrix}$-self-adjoint, and therefore $\operatorname{Im} \begin{bmatrix} I \\ X_p \end{bmatrix}$ is the spectral subspace of iM corresponding to its eigenvalues in the open left half-plane, and is $i\begin{bmatrix} 0 & I \\ -I & 0 \end{bmatrix}$-neutral. So, by Propositon 7.2.1, $X_p = X_p^*$. Finally,

$$\begin{bmatrix} A & D \\ C & -A^* \end{bmatrix} \begin{bmatrix} I \\ X_p \end{bmatrix} = \begin{bmatrix} I \\ X_p \end{bmatrix} (A + DX_p),$$

so the restriction of M to $\operatorname{Im} \begin{bmatrix} I \\ X_p \end{bmatrix}$ is similar to $A + DX_p$, and therefore all eigenvalues of $A + DX_p$ are in the open left half-plane. □

22.3 The sign function and the DARE

A fundamental existence statement for the hermitian solutions of a discrete algebraic Riccati equation

$$X = A^*XA - A^*XB(R + B^*XB)^{-1}B^*XA + Q \qquad (22.3.1)$$

with $R > 0$ and $Q^* = Q$ is contained in Theorem 15.1.1. This is formulated in terms of deflating subspaces of the pencil $\lambda F - G$ where

$$F = \begin{bmatrix} I & D \\ 0 & A^* \end{bmatrix}, \quad G = \begin{bmatrix} A & 0 \\ -Q & I \end{bmatrix} \qquad (22.3.2)$$

and $D = BR^{-1}B^*$. Furthermore (see equation (15.1.4)) the pencil $\lambda F - G$ is $\hat{H}$-symplectic where

$$\hat{H} = \begin{bmatrix} 0 & iI \\ -iI & 0 \end{bmatrix}.$$

As the spectrum of $\lambda F - G$ is symmetric with respect to the unit circle $\mathbf{T}$, the sign function algorithm cannot be applied directly to isolate spectral properties associated with that part of the spectrum inside $\mathbf{T}$. The Cayley transform device can be used, however, to transform $\lambda F - G$ to an $\hat{H}$-self-adjoint pencil, while preserving the structure of the deflating subspaces, after which the sign function process can be applied.

To be specific, let us use Proposition 2.9.1 (part (ii)) with $\alpha = 1$ and $\omega = -i$. Then an $\hat{H}$-self-adjoint pencil $\lambda \hat{A} - i\hat{B}$ is obtained where

$$\hat{A} = F - G = \begin{bmatrix} I - A & D \\ Q & A^* - I \end{bmatrix}, \quad \hat{B} = F + G = \begin{bmatrix} I + A & D \\ -Q & I + A^* \end{bmatrix}. \qquad (22.3.3)$$

Note that we are using the transformation of equation (2.2.10) in the form

$$z = -i(\zeta + 1)/(\zeta - 1),$$

and the eigenvalues of $\lambda L - M$ inside and outside $\mathbf{T}$ are mapped to the lower and upper halves of the complex plane, respectively. Now make a transformation of parameter $\lambda \to i\lambda$ and observe that the pencil $\lambda \hat{A} - \hat{B}$ has the spectrum of $\lambda \hat{A} - i\hat{B}$ rotated anti-clockwise through a right-angle.

The properties of this transformation follow:

Lemma 22.3.1. *If $\lambda F - G$ has no eigenvalues on the unit circle, then:*

(a) *$\lambda \hat{A} - \hat{B}$ has no eigenvalues on the imaginary axis (and conversely);*

(b) *$\hat{A}$ and $\hat{B}$ are invertible;*

(c) *the eigenvalues of $\lambda F - G$ inside $\mathbf{T}$ correspond to the eigenvalues of $\lambda \hat{A} - \hat{B}$ in the open left half-plane.*

Proof. Parts (a) and (c) are clear. Part (a) shows, in particular, that $\hat{B}$ is invertible. If $\hat{A}$ is singular then $\lambda\hat{A} - \hat{B}$ has an eigenvalue at infinity. But this would imply that $\lambda F - G$ has an eigenvalue at $\zeta = 1$, on the unit circle. Hence $\hat{A}$ must also be invertible. □

On the assumption that $\lambda F - G$ has no eigenvalue on the unit circle (as, in Theorem 15.1.2, for example), we see that the pencil problem for $\lambda\hat{A} - \hat{B}$ is equivalent to the classical eigenvalue problem for the matrix

$$M := \hat{A}^{-1}\hat{B} = (F - G)^{-1}(F + G). \tag{22.3.4}$$

Furthermore, M has no eigenvalues on the imaginary axis and its spectrum is symmetric about this axis.

Since deflating subspaces are invariant under our Cayley transform (Proposition 2.9.2), it is clear that the maximal d-stable deflating subspace of $\lambda F - G$ is just the maximal c-stable M-invariant subspace. This subspace determines a hermitian d-stabilizing solution X of equation (22.3.1) which, by Proposition 13.5.1, is necessarily unique. Thus, X can now be determined by applying the sign-function algorithm to matrix M (see Gardiner and Laub (1986)).

22.4 The real case

In this section the matrix sign function for real matrices and its connection to the algebraic Riccati equation (22.2.1) are briefly discussed under the hypotheses that the coefficients A, D and C are all real and the solutions X are real symmetric matrices.

The concept of the matrix sign function is applicable, in particular, to real matrices M. If M is real without eigenvalues on the imaginary axis, then $\text{sgn}(M)$ is real as well. This is easily seen by representing $\text{sgn}(M) = f(M)$, where $f(\lambda)$ is a polynomial (which can be chosen to be real) satisfying conditions (22.1.2). All the properties of $\text{sgn}(M)$ given in Section 22.1 are valid for real matrices as well.

The real analogue of Theorem 22.2.3 reads as follows. We state explicitly part (a) of the theorem concerning stabilizing solutions; the statement of part (b) concerning antistabilizing solutions is analogous.

Theorem 22.4.1. *The equation*

$$XDX + XA + A^T X - C = 0, \tag{22.4.1}$$

where A, C, D are $n \times n$ real matrices and D, C are symmetric has a real, symmetric, stabilizing solution X if and only if the matrix

$$M = \begin{bmatrix} A & D \\ C & -A^T \end{bmatrix}$$

has no eigenvalues on the imaginary axis and (22.2.6) holds.

Proof. The "only if" statement is just a special case of Theorem 22.2.3. For the "if" statement, it follows from Theorem 22.2.3 that there exists a stabilizing solution X_0 of (22.4.1) (a priori X_0 is not necessarily real). Clearly, $\bar{X}_0$ is also a hermitian solution of (22.4.1), and since

$$\sigma(A + D\bar{X}_0) = \{\bar{\lambda} : \lambda \in \sigma(A + DX_0)\},$$

the solution $\bar{X}_0$ is also stabilizing. But by Proposition 7.9.2 the stabilizing solution is unique; so we must have $\bar{X}_0 = X_0$, i.e. X_0 is real. □

22.5 Exercises

22.5.1. Let M be an $n \times n$ matrix without eigenvalues on the imaginary axis, and let P be the spectral projector of M associated with the eigenvalues of M lying in the open right half-plane. Prove that $\operatorname{sgn}(M) = 2P - I$.

22.5.2. Let M and X be as in Theorem 22.2.1. Prove that

$$X[Z_{11} - I, \quad Z_{12}] = [Z_{21}, \quad Z_{22} - I].$$

Hint: Use the proof of Theorem 22.2.1 but work with $\operatorname{sgn}(M) - I$ instead of $\operatorname{sgn}(M) + I$.

22.5.3. Prove part (b) of Theorem 22.2.3.

22.6 Notes

The main result of this chapter (Theorem 22.2.1) is due to Roberts (1980). The proof presented here is based on the exposition of Laub (1991). Results closely connected to Theorem 22.2.3 were proved by Kenney et al. (1989).

Recently, algorithms have been developed for computation of the stabilizing solutions of the "CARE" using the matrix sign functions (see, e.g. the review paper of Laub (1991) and numerous references therein). In turn, computation of the matrix sign function is done via various iteration procedures; the simplest of these has the form

$$M_0 = M; \quad M_{k+1} = (2c_k)^{-1}(M_k + c_k^2 M_k^{-1}), \quad k = 1, 2, \ldots$$

where c_k are suitable constants. We refer the reader to Kenney and Laub (1991), where a general theory of matrix sign function iterations is developed. Scaling (i.e. the choice of c_k) is an important issue here, which can significantly improve convergence; see Kenney and Laub (1992); one of the early works in this direction is Balzer (1980). See Mehrmann (1991) for some discussion from the computational point of view.

The paper of Denman and Beavers (1976) contains applications of the matrix sign function to many invariant subspace computations that appear in control theory.

23
STRUCTURED STABILITY RADIUS

Let $\Omega \subseteq \mathbb{C}$ be an open set, and let A, B and C be given complex matrices of sizes $n \times n$, $n \times m$ and $p \times n$, respectively. We assume that $\sigma(A) \subset \Omega$. The *structured stability radius* of A (with respect to the set Ω and the perturbation structure determined by the matrices B and C) is defined by

$$r(A; B, C, \Omega) = \inf\{\|D\|\},$$

where the infimum is taken over all $m \times p$ complex matrices D such that $A+BDC$ has at least one eigenvalue outside Ω. The norm $\|\cdot\|$ used here is the operator norm induced by the standard euclidean norms in $\mathbb{C}^m$ and $\mathbb{C}^p$. If, for given A, B, C and Ω, there is no D for which $\sigma(A+BDC) \cap (\mathbb{C}\backslash\Omega) \neq \emptyset$, we put formally $r(A; B, C, \Omega) = 0$.

In applications, Ω is considered a "good" set, and a matrix X having all its eigenvalues in Ω is considered a "good" matrix. For example, if Ω is the open left half-plane, then a "good" matrix means that the matrix is c-stable. In general, $r(A; B, C, \Omega)$ can be interpreted informally as the "distance" from a given "good" matrix A to the set of "bad" matrices, where the distance is measured by the size of additive perturbation of the form BDC, with fixed B and C and variable D. The term *structured* refers to this particular structure of admissible perturbations.

In this chapter we study the structured stability radius, especially for the important cases when Ω is the open left half-plane or when Ω is the open unit disk. It turns out that in these cases the structured stability radius can be described in terms of the solvability of certain Riccati equations. This description is used for numerical computation of the structured stability radius (see Hinrichsen *et al.* (1989)).

We consider here the complex case because, although the real case (i.e. when the matrices A, B, C and D are assumed to be real) is especially important in applications, the complex case is simpler and, as far as the authors know at the time of writing, no direct connection to Riccati equations has been found in the real case. Thus, all matrices in this chapter will be assumed to be complex.

23.1 Basic properties of the structured stability radius

Let A, B, C be $n \times n$, $n \times m$ and $p \times n$ complex matrices, respectively, and let $\Omega \subseteq \mathbb{C}$ be a set of complex numbers for which $\sigma(A) \subseteq \Omega$. Recall the definition of the structured stability radius:

$$r(A; B, C, \Omega) = \inf\{\|D\| : D \text{ is such that } \sigma(A + BCD) \cap (\mathbb{C}\setminus\Omega) \neq \emptyset\}. \quad (23.1.1)$$

Denote by $\partial\Omega$ the boundary of the open set Ω. The continuity of eigenvalues of a matrix as functions of the entries of that matrix implies that

$$r(A; B, C, \Omega) = \inf\{\|D\| : D \text{ is such that } \sigma(A + BDC) \cap \partial\Omega \neq \emptyset\}. \quad (23.1.2)$$

To see this, observe that as Ω is open, $\partial\Omega \subseteq \mathbb{C}\setminus\Omega$. So

$$r(A; B, C, \Omega) \leq \inf\{\|D\| : D \text{ is such that } \sigma(A + BDC) \cap \partial\Omega \neq \emptyset\}. \quad (23.1.3)$$

On the other hand, let $\{D_\alpha\}_{\alpha=1}^\infty$ be a sequence of $m \times p$ matrices such that

$$\sigma(A + BD_\alpha C) \cap (\mathbb{C}\setminus\Omega) \neq \emptyset$$

and

$$r(A; B, C, \Omega) = \lim_{\alpha \to \infty} \|D_\alpha\|.$$

In particular, the sequence $\{D_\alpha\}_{\alpha=1}^\infty$ is bounded, and passing (if necessary) to a converging subsequence, we can assume that the sequence $\{D_\alpha\}_{\alpha=1}^\infty$ converges to an $m \times p$ matrix D_0. Obviously,

$$\|D_0\| = r(A; B, C, \Omega). \quad (23.1.4)$$

We claim that

$$\sigma(A + BD_0 C) \cap \partial\Omega \neq \emptyset,$$

thereby proving (in view of (23.1.3)) the equality (23.1.2). Assuming the contrary, i.e. that $\sigma(A + BD_0 C) \cap \partial\Omega = \emptyset$, by continuity of the eigenvalues of $A + BD_\alpha C$ (as functions of α) and by using the closedness of $\partial\Omega$, we find an eigenvalue of $A + BD_0 C$, say λ_0, in the interior of $\mathbb{C}\setminus\Omega$. Again, by continuity of eigenvalues, there exists an $\varepsilon > 0$ such that every matrix $A + BXC$ with $\|D_0 - X\| < \varepsilon$ will have an eigenvalue in the interior of $\mathbb{C}\setminus\Omega$. Choose X_0 such that $\|D_0 - X_0\| < \varepsilon$ and $\|X_0\| < \|D_0\|$.

Here is a concrete way of constructing such an X_0. Let

$$D_0 = U \operatorname{diag}[\sigma_1, \ldots, \sigma_r, 0, \ldots, 0] V$$

be a singular value decomposition of D_0, where U and V are unitary, and $\sigma_1 = \cdots = \sigma_s > \sigma_{s+1} \geq \cdots \geq \sigma_r > 0$. Observe that $D_0 \neq 0$ because all the eigenvalues of A are in Ω; thus $s \geq 1$. Now let $\partial = \min(\sigma_r, \varepsilon/2)$, and put

$$X_0 = U \operatorname{diag}[\sigma_1 - \partial, \ldots, \sigma_r - \partial, 0, \ldots, 0] V.$$

We find that $\sigma(A + BX_0 C) \cap (\mathbb{C}\setminus\Omega) \neq \emptyset$ but also $\|X_0\| < \|D_0\|$, a contradiction with the definition of $r(A; B, C, \Omega)$ and the equality (23.1.4). Hence (23.1.2) is proved.

BASIC PROPERTIES

The proof implies also that the infima in (23.1.1) and (23.1.2) are achieved, i.e. can be replaced by minima. We recapitulate these properties of the structural stability radius in the following proposition.

Proposition 23.1.1. *Let A, B and C be matrices of sizes $n \times n$, $n \times m$ and $p \times n$, respectively, such that $\sigma(A) \subset \Omega$, where Ω is an open set in the complex plane. Then there exists an $m \times p$ matrix D_0 such that $r(A;B,C,\Omega) = \|D_0\|$ and $A + BD_0C$ has at least one eigenvalue on the boundary $\partial\Omega$ of Ω.*

We now express the structured stability radius in terms of the norm of a rational matrix function.

Theorem 23.1.2. *Let A, B, C and Ω be as in Proposition 23.1.1, and let*

$$G(z) = C(zI - A)^{-1}B$$

be the $p \times m$ rational matrix function associated with the triple (A, B, C). Then

$$r(A;B,C,\Omega) = (\max_{z \in \partial\Omega} \|G(z)\|)^{-1}, \qquad (23.1.5)$$

where $\|\cdot\|$ is again the operator norm induced by the euclidean norms in $\mathbb{C}^p$ and $\mathbb{C}^m$, and $0^{-1} = \infty$.

Observe that
$$\lim_{|z| \to \infty} \|G(z)\| = 0,$$
and therefore the supremum of $\|G(z)\|$ when $z \in \partial\Omega$ is always achieved, even if Ω is unbounded: there is a $z_0 \in \partial\Omega$ such that

$$\|G(z_0)\| = \sup_{z \in \partial\Omega} \|G(z)\|.$$

Thus, the usage of "maximum" in (23.1.5) is justified.

Proof. Let D_0 be an $m \times p$ matrix that satisfies the properties indicated in Proposition 23.1.1. Let z_0 be the eigenvalue of $A + BD_0C$ in $\partial\Omega$, with the corresponding eigenvector x ($x \neq 0$). Then

$$(A + BD_0C)x = z_0 x.$$

Since $\sigma(A) \subset \Omega$, $z_0 I - A$ is invertible, and so

$$x = (z_0 I - A)^{-1} BD_0 C x. \qquad (23.1.6)$$

Now $D_0 C x \neq 0$ (otherwise $Ax = z_0 x$, a contradiction with $\sigma(A) \subset \Omega$). Letting $y = D_0 C x$, and premultiplying (23.1.6) by $D_0 C$, we obtain $y = D_0 G(z_0) y$. As $y \neq 0$, it follows that

$$\|D_0\| \cdot \|G(z_0)\| \geq \|D_0 G(z_0)\| \geq 1,$$

and therefore

$$r(A; B, C, \Omega) = \|D_0\| \geq \|G(z_0)\|^{-1} \geq (\max_{z \in \partial\Omega} \|G(z)\|)^{-1}. \qquad (23.1.7)$$

To prove the converse inequality, without loss of generality we may assume

$$\max_{z \in \partial\Omega} \|G(z)\| > 0$$

(otherwise (23.1.7) already implies Theorem 23.1.2). Let $z_0 \in \partial\Omega$ be such that

$$\|G(z_0)\| = \max_{z \in \partial\Omega} \|G(z)\|,$$

and choose $u \in \mathbb{C}^m$ satisfying $\|u\| = 1$ and $\|G(z_0)u\| = \|G(z_0)\|$. Let $v = G(z_0)u$ and $v' = \frac{v}{\|v\|}$. As $\|u\| = \|v'\| = 1$, it is clear that there exist unitary matrices U and V such that u is the first column in U and $Vv' = (1 \; 0 \ldots 0)^T$. Let

$$D = U \operatorname{diag}[\|v\|^{-1}, 0, \ldots, 0] V.$$

Clearly,

$$\|D\| = \|v\|^{-1} = \|G(z_0)\|^{-1} = \left(\max_{z \in \partial\Omega} \|G(z)\|\right)^{-1}. \qquad (23.1.8)$$

On the other hand,

$$Dv = \|v\| \cdot Dv' = \|v\| \cdot U \operatorname{diag}[\|v\|^{-1}, 0, \ldots, 0] V v'$$
$$= \|v\| \cdot U \cdot \operatorname{diag}[\|v\|^{-1}, 0, \ldots, 0][1, 0, \ldots, 0]^T = u,$$

i.e.,

$$DC(z_0 I - A)^{-1} Bu = u. \qquad (23.1.9)$$

The vector $w := (z_0 I - A)^{-1} Bu$ is clearly nonzero. Multiplying (23.1.9) on the left by B, we have $BDCw = Bu = (z_0 I - A)w$. In other words, z_0 is an eigenvalue of $A + BDC$ with the eigenvector w. Thus,

$$r(A; B, C, \Omega) \leq \|D\| = \left(\max_{z \in \partial\Omega} \|G(z)\|\right)^{-1}$$

by (23.1.8). □

23.2 The Hamiltonian matrix

In this section we assume that $\Omega = \{\lambda \in \mathbb{C} : \operatorname{Re}\lambda < 0\}$ is the open left half-plane. Given matrices A, B and C of sizes $n \times n$, $n \times m$ and $p \times n$ respectively, let

$$T(\rho) = \begin{bmatrix} A & BB^* \\ -\rho C^*C & -A^* \end{bmatrix} \qquad (23.2.1)$$

be the $2n \times 2n$ matrix depending on the nonnegative parameter ρ. (T also depends on A, B and C, of course, but we will consider A, B and C as fixed matrices and therefore suppress this dependence in the notation.)

From Chapter 7 we know that $T(\rho)$ is closely related to the Riccati equation

$$XBB^*X + XA + A^*X + \rho C^*C = 0; \qquad (23.2.2)$$

in fact, $M = iT(\rho)$, where M is the matrix introduced in Section 7.2. The matrix $T(\rho)$ will be called the *Hamiltonian matrix* associated with the triple of matrices (A, B, C).

The connection with the structured stability radius is revealed in the following theorem which is the main result of this section.

Theorem 23.2.1. *Let (A, B, C) be as above, let Ω denote the open left half of the complex plane and assume that $\sigma(A) \subset \Omega$. Then*

$$\rho \geq r(A; B, C, \Omega)^2$$

if and only if $T(\rho)$ has at least one eigenvalue which is pure imaginary or zero.

The proof of Theorem 23.2.1 is based on the following lemma:

Lemma 23.2.2. *Given the hypotheses of Theorem 23.2.1, a pure imaginary number $i\omega$, $\omega \in \mathbb{R}$, is an eigenvalue of $T(\rho)$, $\rho > 0$ if and only if ρ^{-1} is an eigenvalue of $(G(i\omega))^*G(i\omega)$, where $G(z) = C(zI - A)^{-1}B$ is the rational matrix function associated with (A, B, C).*

Proof. Fix $\rho > 0$. For any $z \in \mathbb{C}$ which is not an eigenvalue of A, we have

$$zI - T(\rho) = \begin{bmatrix} zI - A & 0 \\ \rho C^*C & I \end{bmatrix} \begin{bmatrix} I & -(zI-A)^{-1}BB^* \\ 0 & zI + A^* + \rho C^*C(zI-A)^{-1}BB^* \end{bmatrix},$$

and therefore

$$\det(zI - T(\rho)) = \det(zI - A) \cdot \det(zI + A^* + \rho C^*C(zI-A)^{-1}BB^*).$$

Assuming, in addition, that z is not an eigenvalue of $-A^*$, we can rewrite this equality in the form

$$\det(zI - T(\rho)) = \det(zI - A)\det(zI + A^*)\det(I + \rho(zI + A^*)^{-1}C^*G(z)B^*)$$
$$= \det(zI - A)\det(zI + A^*)\det[I - \rho B^*(-zI - A^*)^{-1}C^*G(z)]$$
$$= \rho^m \det(zI - A)\det(zI + A^*)\det[\rho^{-1}I - (G(-\bar{z}))^*G(z)],$$
$$(23.2.3)$$

where in the last but one equality we used the fact that $\det(I + XY) = \det(I + YX)$ for any matrices X and Y of sizes $p \times q$ and $q \times p$, respectively. Lemma 23.2.2 now follows by setting $z = i\omega$, $\omega \in \mathbb{R}$ in (23.2.3) and by observing that A and $-A^*$ have no eigenvalues on the imaginary axis. □

Proof of Theorem 23.2.1. Suppose $i\omega_0$ is an eigenvalue of $T(\rho)$, where $\omega_0 \in \mathbb{R}$. Then by Lemma 23.2.2 and Theorem 23.1.2 we have

$$\rho^{-1} \leq \|G(i\omega_0)\|^2 \leq (r(A, B, C; \Omega))^{-2},$$

or
$$\rho \geq (r(A, B, C; \Omega))^2. \qquad (23.2.4)$$

Conversely, if (23.2.4) holds, then (again by Theorem 23.1.2)

$$\rho^{-1/2} \leq (r(A, B, C; \Omega))^{-1} = \max_{\omega \in \mathbb{R}} \|G(i\omega)\|.$$

The function $f(\omega) = \|G(i\omega)\|$ is continuous (as a function of $\omega \in \mathbb{R}$) and $\lim_{|\omega| \to \infty} f(\omega) = 0$. Therefore, there exists an $\omega_0 \in \mathbb{R}$ such that

$$\rho^{-1/2} = \|G(i\omega_0)\|.$$

Now clearly $\rho^{-1} \in \sigma((G(i\omega_0))^*G(i\omega_0))$, and by Lemma 23.2.2 we conclude that $i\omega_0 \in T(\rho)$. □

Theorem 23.2.1 and its proof have several interesting consequences. We indicate a few of them:

Corollary 23.2.3. *Let A, B, C be matrices of sizes $n \times n$, $n \times m$, $p \times n$, respectively, and assume that A is c-stable. If the matrix*

$$\begin{bmatrix} A & BB^* \\ -\rho_0 C^*C & -A^* \end{bmatrix}$$

has pure imaginary or zero eigenvalues for some $\rho_0 > 0$, then any matrix

$$\begin{bmatrix} A & BB^* \\ -\rho C^*C & -A^* \end{bmatrix}$$

has pure imaginary or zero eigenvalues for any $\rho \geq \rho_0$.

This follows immediately from Theorem 23.2.1.

Corollary 23.2.4. *Let A, B, C be as in Corollary 23.2.3, and let*

$$G(z) = C(zI - A)^{-1}B.$$

Then $\omega_0 \in \mathbb{R}$ is such that

$$\|G(i\omega_0)\| = \max_{\omega \in \mathbb{R}} \|G(i\omega)\|$$

if and only if $i\omega_0$ is an eigenvalue of

$$T(r) = \begin{bmatrix} A & BB^* \\ -r^2 C^*C & -A^* \end{bmatrix},$$

where $r = r(A; B, C, \Omega)$.

This follows from the proof of Theorem 23.2.1 by specializing to the case $\rho = (r(A; B, C, \Omega))^2$.

The last corollary can be used to derive information about the function $f(\omega) = \|G(i\omega)\|$, $\omega \in \mathbb{R}$, using the eigenvalues of $T(r)$. For example: by Theorem 23.2.1, r is the least positive number for which $T(r)$ has pure imaginary or zero eigenvalues. As the eigenvalues of $T(\rho)$ are symmetric relative to the imaginary axis, it follows (by continuity of the eigenvalues) that $T(r)$ cannot have a simple eigenvalue (i.e. having algebraic multiplicity one) on the imaginary axis. As $T(r)$ has size $2n \times 2n$, the number of distinct eigenvalues of $T(r)$ on the imaginary axis cannot exceed n. Combining this deduction with Corollary 23.2.4, we obtain:

Corollary 23.2.5. *The number of points ω_0, at which $f(\omega) = \|C(i\omega - A)^{-1}B\|$, $\omega \in \mathbb{R}$, attains the global maximum, cannot exceed n, the size of the matrix A.*

23.3 Connection with the Riccati equation

We now focus on the connection between the structured stability radius and the algebraic Riccati equation (23.2.2). As in the previous section, we assume $\Omega = \{\lambda \in \mathbb{C} : \text{Re}\,\lambda < 0\}$, and that $\sigma(A) \subset \Omega$. The main result of this section is:

Theorem 23.3.1. *The positive number ρ satisfies*

$$\rho \leq r(A; B, C, \Omega)^2 \qquad (23.3.1)$$

if and only if the algebraic Riccati equation

$$XBB^*X + XA + A^*X + \rho C^*C = 0 \qquad (23.3.2)$$

has hermitian solutions X.

In other words, $r(A; B, C, \Omega)$ can be characterized as the square root of the infinum of all such numbers $\rho > 0$ for which (23.3.2) has no hermitian solutions; moreover, the equation (23.3.2) with $\rho = (r(A; B, C, \Omega))^2$ has hermitian solutions.

Proof of Theorem 23.3.1 (first part). Assume that equation (23.3.2) has a hermitian solution X for some $\rho > 0$. Then for every $\omega \in \mathbb{R}$

$$XBB^*X + X(A - i\omega I) + (A - i\omega I)^*X + \rho C^*C = 0$$

and therefore

$$\begin{aligned}
(B^*X(A &- i\omega I)^{-1}B + I)^*(B^*X(A - i\omega I)^{-1}B + I) \\
&= I + B^*X(A - i\omega I)^{-1}B + B^*(A - i\omega I)^{-1*}XB \\
&\quad + B^*(A - i\omega I)^{-1*}XBB^*X(A - i\omega I)^{-1}B \\
&= I + B^*X(A - i\omega I)^{-1}B + B^*(A - i\omega I)^{-1*}XB \\
&\quad + B^*[-(A - i\omega I)^{*-1}X - X(A - i\omega I)^{-1} \\
&\quad - \rho(A - i\omega I)^{-1*}C^*C(A - i\omega I)^{-1}]B \\
&= I - \rho B^*(i\omega I - A)^{-1*}C^*C(i\omega I - A)^{-1}B \\
&= I - \rho(G(i\omega))^*G(i\omega).
\end{aligned}$$

As the left-hand side of this equality is positive semidefinite, it follows that

$$\rho \leq (\max_{\omega \in \mathbb{R}} \|G(i\omega)\|^2)^{-1}$$

which, by Theorem 23.1.2, coincides with $r(A; B, C, \Omega)^2$. $\square$

We need a preliminary result before proving the other part of Theorem 23.3.1; namely, that if $\rho \leq r(A; B, C, \Omega)^2$, then equation (23.3.2) has hermitian solutions. Define $T(\rho)$ by (23.2.1) with $\rho \geq 0$. Define $\mathcal{L}$ to be the controllability subspace of the pair (A, B), i.e. the smallest A-invariant subspace that contains $\operatorname{Im} B$. With respect to the orthogonal decomposition $\mathbb{C}^n = \mathcal{L} \oplus \mathcal{L}^\perp$ we have

$$A = \begin{bmatrix} A_1 & A_2 \\ 0 & A_3 \end{bmatrix}, \quad B = \begin{bmatrix} B_1 \\ 0 \end{bmatrix}, \quad C = [C_1 \ C_2] \qquad (23.3.3)$$

and

$$T(\rho) = \begin{bmatrix} A_1 & A_2 & B_1 B_1^* & 0 \\ 0 & A_3 & 0 & 0 \\ -\rho C_1^* C_1 & -\rho C_1^* C_2 & -A_1^* & 0 \\ -\rho C_2^* C_1 & -\rho C_2^* C_2 & -A_2^* & -A_3^* \end{bmatrix}.$$

For example, here $A_1 : \mathcal{L} \to \mathcal{L}$ and $C_2 : \mathcal{L}^\perp \to \mathbb{C}^p$. From the choice of $\mathcal{L}$ it follows that the pair (A_1, B_1) is controllable. We observe also that

$$\sigma(T(\rho)) = \sigma(A_3) \cup \sigma(-A_3^*) \cup \sigma \begin{bmatrix} A_1 & B_1 B_1^* \\ -\rho C_1^* C_1 & -A_1^* \end{bmatrix}. \qquad (23.3.4)$$

Next, with respect to the same decomposition $\mathbb{C}^n = \mathcal{L} \oplus \mathcal{L}^\perp$ we write

$$X = \begin{bmatrix} X_1 & X_2 \\ X_2^* & X_3 \end{bmatrix} \qquad (23.3.5)$$

(so, for example, $X_2 : \mathcal{L}^\perp \to \mathcal{L}$) and write equation (23.3.2), where the solution X is assumed to be hermitian, in the form

$$\begin{bmatrix} X_1 & X_2 \\ X_2^* & X_3 \end{bmatrix} \begin{bmatrix} B_1 B_1^* & 0 \\ 0 & 0 \end{bmatrix} \begin{bmatrix} X_1 & X_2 \\ X_2^* & X_3 \end{bmatrix} + \begin{bmatrix} X_1 & X_2 \\ X_2^* & X_3 \end{bmatrix} \begin{bmatrix} A_1 & A_2 \\ 0 & A_3 \end{bmatrix}$$
$$+ \begin{bmatrix} A_1^* & 0 \\ A_2^* & A_3^* \end{bmatrix} \begin{bmatrix} X_1 & X_2 \\ X_2^* & X_3 \end{bmatrix} + \begin{bmatrix} \rho C_1^* C_1 & \rho C_1^* C_2 \\ \rho C_2^* C_1 & \rho C_2^* C_2 \end{bmatrix} = 0.$$

Here $X_1 = X_1^*$ and $X_3 = X_3^*$. This equation decomposes naturally as follows:

$$X_1 B_1 B_1^* X_1 + X_1 A_1 + A_1^* X_1 + \rho C_1^* C_1 = 0 \qquad (23.3.6)$$

$$X_1 B_1 B_1^* X_2 + X_1 A_2 + X_2 A_3 + A_1^* X_2 + \rho C_1^* C_2 = 0 \qquad (23.3.7)$$

$$X_2^* B_1 B_1^* X_2 + X_2^* A_2 + X_3 A_3 + A_2^* X_2 + A_3^* X_3 + \rho C_2^* C_2 = 0. \qquad (23.3.8)$$

We claim that equation (23.3.2) admits a hermitian solution X if and only if (23.3.6) admits a hermitian solution X_1. Indeed, let $X_1 = X_1^*$ be a solution of (23.3.6). Since the pair $(A_1, B_1 B_1^*)$ is controllable (see Corollary 4.1.3) there exists a hermitian solution $\tilde{X}_1$ of (23.3.6) such that $\sigma(A + BB^* \tilde{X}_1)$ lies in the closed left half-plane (by Theorem 7.5.1). Rewrite (23.3.7) with $X_1 = \tilde{X}_1$ in the form

$$(A_1 + B_1 B_1^* \tilde{X}_1)^* X_2 + X_2 A_3 = -\tilde{X}_1 A_2 - \rho C_1^* C_2. \qquad (23.3.9)$$

Because $\sigma(A_3)$ is in the *open* left half-plane, the equation (23.3.9) is uniquely solvable for X_2 (see Theorem 5.2.2). For the same reason, the equation (23.3.8) written in the form

$$X_3 A_3 + A_3^* X_3 = -\rho C_2^* C_2 - X_2^* A_2 - A_2^* X_2 - X_2^* B_1 B_1^* X_2$$

is uniquely solvable for X_3, and the solution X_3 turns out to be hermitian. We have therefore verified our claim.

Proof of Theorem 23.3.1 (second part). Suppose that $\rho \leq r(A; B, C, \Omega)^2$. By the claim above, we only have to prove that equation (23.3.6) admits a hermitian solution X_1. If $\rho < r(A; B, C, \Omega)^2$, then by Theorem 23.2.1 and formula (23.3.4) we conclude that the matrix

$$\begin{bmatrix} A_1 & B_1 B_1^* \\ -\rho C_1^* C_1 & -A_1^* \end{bmatrix}$$

has no pure imaginary eigenvalues. By Theorem 7.2.4 the equation (23.3.6) has hermitian solutions. If $\rho = r(A; B, C, \Omega)^2$, the existence of hermitian solutions of (23.3.6) follows from the already proved part of the theorem and from Theorem 11.1.1. □

In fact, we can say more about hermitian solutions of equation (23.3.2) (under the assumption (23.3.1)):

Theorem 23.3.2. *Let a triple of matrices A, B, C of sizes $n \times n$, $n \times m$ and $p \times n$, respectively, be given such that $\sigma(A) \subset \Omega = \{\lambda \in \mathbf{C} : \operatorname{Re} \lambda < 0\}$. If*

$$0 \leq \rho < r(A; B, C, \Omega)^2, \qquad (23.3.10)$$

then there exists a unique hermitian solution X_0 of equation (23.3.2) such that

$$\sigma(A + BB^* X_0) \subset \{\lambda \in \mathbf{C} : \operatorname{Re} \lambda \leq 0\}, \qquad (23.3.11)$$

and, in fact, $\sigma(A + BB^* X_0) \subset \Omega$. If $\rho = r(A; B, C, \Omega)^2$, then there is a unique hermitian solution X_0 of (23.3.2) with the property (23.3.11) and, in fact, $A + BB^* X_0$ does have eigenvalues on the imaginary axis.

Proof. We use the orthogonal decomposition $\mathbf{C}^n = \mathcal{L} \oplus \mathcal{L}^\perp$, where $\mathcal{L}$ is the controllability subspace for (A, B), and the corresponding partitions of A, B, C and X given in equations (23.3.3) and (23.3.5). Then

$$A + BB^* X = \begin{bmatrix} A_1 + B_1 B_1^* X_1 & A_2 + B_1 B_1^* X_2 \\ 0 & A_3 \end{bmatrix},$$

and therefore $\sigma(A + BB^* X)$ is contained in the closed (resp. open) left half-plane if and only if the same is true of $\sigma(A_1 + B_1 B_1^* X_1)$. We have seen above that (23.3.7) and (23.3.8) uniquely determine X_2 and X_3 provided $X_1 = X_1^*$ is chosen to satisfy (23.3.6). If (23.3.10) holds, then (as we have seen in the second part of the proof of Theorem 23.3.1) the matrix

$$\begin{bmatrix} A_1 & B_1 B_1^* \\ -\rho C_1^* C_1 & -A_1^* \end{bmatrix} \qquad (23.3.12)$$

has no pure imaginary eigenvalues. Therefore, by Theorem 7.5.1 the equation (23.3.6) has a unique hermitian solution X_1 such that

$$\sigma(A_1 + B_1 B_1^* X_1) \subseteq \{\lambda \in \mathbf{C} : \operatorname{Re} \lambda \leq 0\},$$

and for this solution X_1 in fact $\sigma(A_1 + B_1 B_1^* X_1) \subseteq \Omega$.

Assume now that $\rho = r(A; B, C, \Omega)^2$. Then the matrix (23.3.12) has eigenvalues on the imaginary axis and therefore, by Theorem 7.5.3 (taking into account Corollary 7.6.5), for every hermitian solution X_1 of (23.3.6) the matrix $A_1 + B_1 B_1^* X_1$ has eigenvalues on the imaginary axis. It remains to observe (by the same Theorem 7.5.1) that the hermitian solution X_1 of (23.3.6) with

$$\sigma(A_1 + B_1 B_1^* X_1) \subset \{\lambda \in \mathbf{C} : \operatorname{Re} \lambda \leq 0\} \qquad (23.3.13)$$

is unique. □

23.4 Exercises

23.4.1. Show that the structured stability radius is invariant under similarity of the ordered triples (A, B, C): $r(A; B, C, \Omega) = r(TAT^{-1'}TB, CT^{-1}, \Omega)$ where T is any invertible matrix.

23.4.2. A triple of matrices (A, B, C) of sizes $n \times n$, $n \times m$ and $p \times n$, respectively, is called a *dilation* of a triple of matrices (A_0, B_0, C_0) of sizes $n_0 \times n_0$, $n_0 \times m$ and $p \times n_0$, respectively, if there is a direct sum decomposition $\mathbb{C}^n = \mathcal{L} \dotplus \mathcal{M} \dotplus \mathcal{N}$ such that, with respect to this decomposition, and by choosing suitable bases in $\mathcal{L}$, $\mathcal{M}$ and $\mathcal{N}$, we have

$$A = \begin{bmatrix} * & * & * \\ 0 & A_0 & * \\ 0 & 0 & * \end{bmatrix}, \quad B = \begin{bmatrix} * \\ B_0 \\ 0 \end{bmatrix}, \quad C = \begin{bmatrix} 0 & C_0 & * \end{bmatrix},$$

where $*$ stands for entries of no immediate interest (in particular, $\dim \mathcal{M} = n_0$). Prove that if (A, B, C) is a dilation of A_0, B_0, C_0, then for every open set Ω containing $\sigma(A)$ the equality $r(A; B, C, \Omega) = r(A_0; B_0, C_0, \Omega)$ holds.

23.4.3. Let $G(z)$ be a $p \times m$ rational matrix function with value 0 at infinity. Assume that $G(z)$ has McMillan degree 1, and the only pole of $G(z)$ is not on the imaginary axis. Show that $f(\omega) = \|G(i\omega)\|$, $\omega \in \mathbb{R}$, has a unique maximum. Find the point where the maximum is attained in terms of a minimal realization of $G(z)$.

23.5 Notes

The main results of this chapter (Theorems 23.1.2, 23.3.1, 23.3.2) were obtained by Hinrichsen and Pritchard (1986b). Another paper by the same authors (1986a) contains further properties of the structured stability radius. Computational aspects of the structured stability radius are addressed in Hinrichsen et al. (1989) and Hinrichsen and Pritchard (1990), where reliable algorithms have been developed and applied. We mention also the paper of Hinrichsen et al. (1990), where further analysis of the stability radius based on the Riccati equations is given, as well as numerous references. In particular, the supremum of the stability radii that can be achieved by a linear state feedback is determined there.

Bibliography

1. Anderson, B.D.O. and Moore, J.B. (1971). *Linear optimal control*. Prentice Hall, Englewood Cliffs, N.J.

2. Anderson, B.D.O. and Moore, J.B. (1979). *Optimal filtering*. Prentice Hall, Englewood Cliffs, N.J.

3. Anderson, B.D.O. and Moore, J.B. (1990). *Optimal control: Linear quadratic problems*. Prentice Hall, Englewood Cliffs, N.J.

4. Anderson, B.D.O. and Vongpanitlerd, S. (1973). *Network analysis and synthesis*. Prentice Hall, Englewood Cliffs, N.J.

5. Anderson, T.W. (1958). *An introduction to multivariate statistical analysis*. John Wiley & Sons, New York.

6. Ando, T. (1988). *Matrix quadratic equations*. Hokkaido University, Sapporo, Japan.

7. Antoulas, A.C. (ed.) (1991). *Mathematical system theory. The influence of R.E. Kalman*. Springer Verlag, Berlin.

8. Ball, J.A. and Cohen, N. (1987). Sensitivity minimization in an H^∞ norm: Parametrization of all suboptimal solutions. *International Journal of Control*, **46**, 785–816.

9. Ball, J.A. and Ran, A.C.M. (1987). Optimal Hankel norm model reduction and Wiener–Hopf factorization, I. The canonical case. *SIAM J. Control and Optim.*, **25**, 362–383.

10. Ball, J.A., Gohberg, I., and Rodman, L. (1990). *Interpolation of rational matrix functions*. Birkhäuser Verlag, Basel.

11. Balzer, L.A. (1980). Accelerated convergence of the matrix sign function method of solving Lyapunov, Riccati and other matrix equations. *Intern. J. Control*, **32**, 1057–1078.

12. Bart, H., Gohberg, I., and Kaashoek, M.A. (1979). Minimal factorizations of matrix and operator functions. *Operator Theory*, **1**, Birkhäuser Verlag, Basel.

13. Bart, H., Gohberg, I., Kaashoek, M.A., and Van Dooren, P. (1980). Factorizations of transfer functions. *SIAM J. Control and Optim.*, **18**, 675–696.

14. Basar, T. and Bernhard, P. (1991). H^∞-optimal control and related minimax design problems. Birkhäuser, Boston.

15. Baumgartel, H. (1985). Analytic perturbation theorem for matrices and operators. *Operator Theory*, **OT15**. Birkhäuser, Basel.

16. Belevitch, V. (1968). *Classical network theory*. Holden-Day, San Francisco-Cambridge-Amsterdam.

17. Bender, D.J. and Laub, A.J. (1987a). The linear-quadratic optimal regulator problem for descriptor systems. *IEEE Trans. Autom. Control*, **AC–32**, 672–688.

18. Bender, D.J. and Laub, A.J. (1987b). The linear-quadratic optimal regulator problem for descriptor systems: Discrete-time case. *Automatica*, **23**, 71–85.

19. Bernstein, D.S. and Haddad, W.M. (1989). LQG control with an H_∞ performance bound: A Riccati equation approach. *IEEE Trans. Autom. Control*, **34**, 293–305.

20. Bitmead, R.R. and Gevers, M. (1991). Riccati difference and differential equations: Convergence, monotonicity and stability. In *The Riccati equation* (ed. S. Bittanti, A.J. Laub, and J.C. Willems), Springer Verlag, Berlin, p. 263–291.

21. Bitmead, R.R., Gevers, M.R., Petersen, I.R., and Kaye, R.J. (1985). Monotonicity and stabilizability properties of solutions of the Riccati difference equation: Propositions, lemmas, theorems, fallacious conjectures and counterexamples. *Systems and Control Letters*, **5**, 309–315.

22. Bittanti, S., Colaneri, P., and De Nicolao, G. (1991a). The periodic Riccati equation. Chapter 6 of *The Riccati equation* (ed. S. Bittanti, A.J. Laub, and J.C. Willems), Springer Verlag, Berlin.

23. Bittanti, S., Laub, A.J., and Willems, J.C. (ed.) (1991b). *The Riccati equation*. Springer Verlag, Berlin.

24. Brockett, R.W. (1970). *Finite dimensional linear systems*. Wiley, New York.

25. Byers, R. (1984). Numerical condition of the algebraic Riccati equation. Linear algebra and its role in systems theory. *Contemp. Math.*, **47**, 35–49, American Mathematical Society.

26. Caines, P.E. and Mayne, D.Q. (1970). On the discrete time matrix Riccati equation of optimal control. *International Journal of Control*, **12**, 785–794.

27. Caines, P.E. and Mayne, D.Q. (1971). On the discrete time matrix Riccati equation of optimal control – a correction. *International Journal of Control*, **14**, 205–207.

28. Catlin, D.E. (1989). *Estimation, control, and the discrete Kalman filter*. Springer Verlag, New York.

29. Chan, S.W., Goodwin, G.C., and Sin, K.S. (1984). Convergence properties of the Riccati difference equation in optimal filtering of nonstabilizable systems. *IEEE Trans. Autom. Control*, **29**, 110–118.

30. Chen, C.T. (1973). A generalization of the inertia theorem. *SIAM J. Appl. Math.*, **25**, 158–161.

31. Clancey, K.F. and Gohberg, I. (1981). Factorization of matrix functions and singular integral operators. *Operator Theory*, **3**, Birkhäuser Verlag, Basel.

32. Coppel, W.A. (1971). *Disconjugacy* (Lecture Notes in Math., Vol. 220). Springer Verlag, Berlin.

33. Coppel, W.A. (1974). Matrix quadratic equations. *Bulletin of the Australian Mathematics Society*, **10**, 377–401.

34. Cross, G.W. and Lancaster, P. (1974). Square roots of complex matrices. *Linear and Multilinear Algebra*, **1**, 289–293.

35. Curilov, A.N. (1978). On the solutions of quadratic matrix equations. *Nonlinear Vibrations and Control Theory* (Udmurt State University, Izhevsk), **2**, 24–33 (Russian).

36. Curilov, A.N. (1979). The frequency theorem and the Lur'e equation. *Sibirskii Matematicheskii Zhurnal*, **20**, 600–611 (in Russian).

37. Curilov, A.N. (1986). On solutions of a quadratic matrix equation encountered in investigation of discrete control systems. *Izv. Vyssh. Uchebn. Zaved. Mat.*, **11**, 59–65 (Russian).

38. Daleckii, Ju.L. and Krein, M.G. (1974). *Stability of solutions of differential equations in Banach space*. Amer. Math. Society, Providence, Rhode Island. (Translation of Russian edition of 1970.)

39. Davis, M.H.A. (1977). *Linear estimation and stochastic control*. Chapman and Hall, London.

40. Delchamps, D.F. (1980). A note on the analyticity of the Riccati metric. In *Lectures in applied mathematics*, Vol. 18 (ed. C.I. Byrnes and C.F. Martin), pp. 37–41. AMS, Providence, Rhode Island.

41. Delchamps, D.F. (1983). Analytic stabilization and the algebraic Riccati equation. In *Proceedings of the 22nd IEEE Conference on Decision and Control*, pp. 1396–1401.

42. Delchamps, D.F. (1984). Analytic feedback control and the algebraic Riccati equation. *IEEE Trans. Autom. Control*, **29**, 1032–1033.

43. De Moor, B. and David, J. (1992). Total linear least squares and the algebraic Riccati equation. *Systems and Control Letters*, **18**, 329–337.

44. Denman, E.D. and Beavers, A.N. (1976). The matrix sign function and computation in systems. *Appl. Math. Comput.*, **2**, 63–94.

45. De Souza, C.E., Gevers, M.R., and Goodwin, G.C. (1986). Riccati equations in optimal filtering of nonstabilizable systems having singular state transition matrices. *IEEE Trans. Autom. Control*, **31**, 831–838.

46. Dewilde, P. and Vandewalle, J.P. (1975). On the factorization of a nonsingular rational matrix. *IEEE Trans. Circuits and Systems*, **CAS-22**, 637–645.

47. Dorato, P. (1983). Theoretical developments in discrete- time control. *Automatica*, **19**, 395–400.

48. Dorato, P. and Levis, A.H. (1971). Optimal linear regulators: The discrete-time case. *IEEE Trans. Autom. Control*, **16**, 613–620.

49. Doyle, J.C., Glover, K., Khargonekar, P.P., and Francis, F. (1989). State-space solutions to standard H_2 and H_∞ control problems. *IEEE Trans. Autom. Control*, **34**, 831–847.

50. Djokovic, D.Z., Patera, J., Winternitz, P., and Zassenhaus, H. (1983). Normal forms of elements of classical real and complex Lie and Jordan algebras. *Journal of Mathematical Physics*, **24**, 1363–1374.

51. Elsner, L., He, C., and Mehrmann, V. (1994). Minimization of the norm, the norm of the inverse and the condition number of a matrix by completion. *Linear Algebra and its Applications* (to appear).

52. Faibusovich, L.E. (1986). Algebraic Riccati equation and symplectic algebra. *International Journal of Control*, **43**, 781–792.

53. Faibusovich, L.E. (1987). Matrix Riccati inequality: Existence of solutions. *Systems and Control Letters*, **9**, 59–64.

54. Finesso, L. and Picci, G. (1982). A characterization of minimal square spectral factors. *IEEE Trans. Automatic Control*, **27**, 122–127.

55. Finkbeiner, D.T. (1978). *Introduction to matrices and linear transformations* (3rd edn). W.H. Freeman, San Francisco.

56. Francis, B.A. (1987). *A course in H_∞ control theory*. Lecture Notes in Control and Information Science, Vol. 88. Springer Verlag, Berlin.

57. Francis, B.A. and Doyle, J.C. (1987). Linear control theory with an H_∞ optimality criterion. *SIAM J. Control and Optim.*, **25**, 815–844.

58. Fuhrmann, P.A. (1985). The algebraic Riccati equation – a polynomial approach. *Systems and Control Letters*, **5**, 369–376.

59. Gahinet, P. and Laub, A.J. (1990). Computable bounds for the sensitivity of the algebraic Riccati equation. *SIAM J. Control and Optim.*, **28**, 1461–1480.

60. Gantmacher, F.R. (1959). *The theory of matrices*, Vols. 1 and 2. Chelsea, New York.

61. Gardiner, J.D. and Laub, A.J. (1986). A generalization of the matrix-sign-function solution for algebraic Riccati equations. *International Journal of Control*, **44**, 823–832.

62. Glover, K., Green, J., Limebeer, D., and Doyle, J. (1990). A J-spectral factorization approach of H_∞ control. *Progress Systems and Control Theory*, **4**, 277–284, Birkhäuser, Boston.

63. Glover, K., Limebeer, D.J.N., Doyle, J.C., Kasenally, E.M., and Safonov, M.G. (1991). A characterization of all solutions to the four block general distance problem. *SIAM J. Control and Optim.*, **29**, 283–324.

64. Gohberg, I. and Kaashoek, M.A. (eds.) (1986). Constructive methods of Wiener–Hopf factorization. *Operator Theory*, **21**, Birkhäuser Verlag, Basel.

65. Gohberg, I. and Kaashoek, M.A. (1987). An inverse spectral problem for rational matrix functions and minimal divisibility. *Integral Equations and Operator Theory*, **10**, 437–465.

66. Gohberg, I. and Rubinstein, S. (1988). Proper contractions and their unitary minimal completions. *Operator Theory: Advances and Applications*, **33** (ed. I. Gohberg), 233–247.

67. Gohberg, I., Lancaster, P., and Rodman, L. (1979). *Spectral analysis of selfadjoint matrix polynomials*, Research paper 419, 126pp. Dept. of Mathematics & Statistics, University of Calgary.

68. Gohberg, I., Lancaster, P., and Rodman, L. (1980). Spectral analysis of selfadjoint matrix polynomials. *Annals of Math.*, **112**, 33–71.

69. Gohberg, I., Lancaster, P., and Rodman, L. (1982a). *Matrix polynomials*. Academic Press, New York.

70. Gohberg, I., Lancaster, P., and Rodman, L. (1982b). Perturbations of H-selfadjoint matrices, with applications to differential equations. *Integral Equations and Operator Theory*, **5**, 718–757.

71. Gohberg, I., Lancaster, P., and Rodman, L. (1983). Matrices and indefinite scalar products. *Operator Theory*, **8**, Birkhäuser Verlag, Basel.

72. Gohberg, I., Lancaster, P., and Rodman, L. (1986a). *Invariant subspaces of matrices with applications*. Wiley, New York.

73. Gohberg, I., Lancaster, P., and Rodman, L. (1986b). On hermitian solutions of the symmetric algebraic Riccati equations. *SIAM J. Control and Optim.*, **24**, 1323–1334.

74. Gohberg, I., Goldberg, S. and Kaashoek, M.A. (1990). Classes of linear operators, Vol. 1. *Operator Theory*, **49**, Birkhäuser Verlag, Basel, Boston, Berlin.

75. Gohberg, I., Kaashoek, M.A., and Ran, A.C.M. (1992). Factorization of and extension to J-unitary rational matrix functions on the unit circle. *Integral Equations and Operator Theory*, **15**, 262–300.

76. Gohberg, I., Kaashoek, M.A., and van Schagen, F. (1994). Partially specified matrices and eigenvalue completion problems. *Operator Theory: Advances and Applications*, Birkhäuser Verlag (to appear).

77. Golub, G.H. and Van Loan, C. (1980). An analysis of the total least squares problem. *SIAM J. Numer. Anal.*, **17**, 883–893.

78. Golub, G.H. and Van Loan, C. (1989). *Matrix computations* (2nd edn). Johns Hopkins University Press, Baltimore.

79. Green, W.L. and Kamen, E.W. (1985). Stabilizability of linear systems over a commutative normed algebra with applications to spatially distributed and parameter dependent systems. *SIAM J. Control and Optim.*, **23**, 1–18.

80. Green, M., Glover, K., Limebeer, D., and Doyle, J. (1990). A J-spectral factorization approach to H_∞ control. *SIAM J. Control and Optim.*, **28**, 1350–1371.

81. Gurvits, L.N. (1990). The criterion of convergence of the iterations of a discrete Riccati equation to a unique nonnegatively defined solution. *Cybernetics and Computation*, **85**, 79–87 (translation from Russian).

82. Hager, W.W. and Horowitz, L.L. (1976). Convergence and stability properties of the discrete Riccati operator equation and the associated optimal control and filtering problems. *SIAM J. Control and Optim.*, **14**, 295–312.

83. Hammarling, S.J. and Singer, M.A. (1984). A canonical form for the algebraic Riccati equation. *Mathematical Theory of Networks and Systems* (ed. P.A. Fuhrmann), pp. 389–405. Springer Verlag, Berlin.

84. Hautus, M.L.J. (1969). Controllability and observability conditions of linear autonomous systems. Koninklijke Nederlandse Akademie van Wetenschappen. *Indagationes Mathematicae*, **12**, 443–448.

85. Helton, J.W. (1976). A spectral factorization approach to the distributed stable regular problem: The algebraic Riccati equation. *SIAM J. Control and Optim.*, **14**, 639–661.

86. Hewer, G.A. (1971). An iterative technique for the computation of steady-state gains for the discrete optimal regulator. *IEEE Trans. Autom. Control*, **16**, 382–383.

87. Hewer, G. (1993). Existence theorems for positive semidefinite and sign indefinite stabilizing solutions of the Riccati equation. *SIAM J. Control and Optim.*, **31**, 16–29.

88. Hijab, O. (1987). *Stabilization of control systems*. Springer Verlag, New York.

89. Hille, E. and Phillips, R.S. (1957) *Functional analysis and semi-groups*. Amer. Math. Society, Providence.

90. Hinrichsen, D. and Pritchard, A.J. (1986a). Stability radii of linear systems. *Systems & Control Letters*, **7**, 1–10.

91. Hinrichsen, D. and Pritchard, A.J. (1986b). Stability radius for structured perturbations and the algebraic Riccati equation. *Systems & Control Letters*, **8**, 105–113.

92. Hinrichsen, D. and Pritchard, A.J. (1990). Rubustness measures for linear state systems under complex and real parameter perturbations. In *Perspectives in Control Theory (Sielpia), Progress in Systems and Control Theory*, **2**, Birkhauser, Boston, 56–74.

93. Hinrichsen, D., Kelb, B., and Linnemann, A. (1989). An algorithm for the computation of the structured complex stability radius. *Automatica-J. IFAC*, **25**, 771–775.

94. Hinrichsen, D., Pritchard, A.J., and Townley, S.B. (1990). Riccati equation approach to maximizing the complex stability radius by state feedback. *Intern. J. Control*, **52**, 769–794.

95. Horn, R.A. and Johnson, C.R. (1985). *Matrix analysis*. Cambridge University Press.

96. Horn, R.A. and Johnson C.R. (1991). *Topics in matrix analysis*, Cambridge University Press.

97. Ionescu, V. and Weiss, M. (1992). On computing the stabilizing solution of the discrete-time Riccati equation. *Linear Algebra and its Applications*, **174**, 229–238.

98. Kaashoek, M.A. and Ran, A.C.M. (1991). *State space methods for H_∞-control problems*, notes of a course for the Dutch Graduate Network on Systems and Control.

99. Kailath, T. (1980). *Linear systems*, Prentice Hall, Englewood Cliffs, N.J.

100. Kalman, R.E. (1960a). A new approach to linear filtering and prediction problems. *Journal of Basic Engineering (ASME)*, **82D**, 35–45.

101. Kalman, R.E. (1960b). Contributions to the theory of optimal control. *Boletin Sociedad Matematica Mexicana*, **5**, 102–119.

102. Kalman, R.E. and Bucy, R. (1961). New results in linear filtering and prediction. *Journal of Basic Engineering (ASME)*, **83D**, 366–368.

103. Kalman, R.E., Falb, P.L. and Arbib, M.A. (1969). *Topics in mathematical systems theory*. McGraw-Hill, New York.

104. Kamen, E.W. and Khargonekar, P.P. (1984). On the control of linear systems whose coefficients are functions of parameters. *IEEE Trans. Autom. Control*, **29**, 25–33.

105. Kenney, C. and Hewer, G. (1990). The sensitivity of the algebraic and differential Riccati equations. *SIAM J. Control and Optim.*, **28**, 50–69.

106. Kenney, C. and Laub, A.J. (1991). Rational iterative methods for the matrix sign function. *SIAM J. Matrix Anal. and Appl.*, **12**, 273–291.

107. Kenney, C. and Laub, A.J. (1992). On scaling Netwon's method for polar decomposition and the matrix sign function. *SIAM J. Matrix Anal. and Appl.*, **13**, 688–706.

108. Kenney, C., Laub, A.J., and Jonckheere, E.A. (1989). Positive and negative solutions of dual Riccati equations by matrix sign function iteration. *Systems and Control Letters*, **13**, 109–116.

109. Khargonekar, P.P., Petersen, I.R., and Rotea, M.A. (1988). H_∞ optimal control with state feedback. *IEEE Trans. Autom. Control*, **33**, 786–788.

110. Kleinman, D.L. (1968). On an iterative technique for Riccati equation computation. *IEEE Trans. Autom. Control*, **13**, 114–115.

111. Konstantinov, M.M., Petkov, P.Hr., and Christov, N.D. (1993). Perturbation analysis of the discrete Riccati equation. *Kybernetika*, **29**, 18–29.

112. Kučera, V. (1972a). A contribution to matrix quadratic equations. *IEEE Trans. Autom. Control*, **17**, 344–347.

113. Kučera, V. (1972b). The discrete Riccati equation of optimal control. *Kybernetika*, **8**, 430–447.

114. Kučera, V. (1973). A review of the matrix Riccati equation. *Kybernetika*, **9**, 42–61.

115. Kučera, V. (1979). *Discrete linear control*. J. Wiley & Sons, Chichester.

116. Kučera, V. (1991). Algebraic Riccati equations: Hermitian and definite solutions. Chapter 3 of *The Riccati equation* (ed. S. Bittanti, A.J. Laub, and J.C. Willems), Springer Verlag, Berlin.

117. Kwakernaak, H. and Sivan, R. (1972). *Linear optimal control systems*. Wiley Interscience, New York.

118. Lancaster, P. and Rodman, L. (1980). Existence and uniqueness theorems for algebraic Riccati equations. *International Journal of Control*, **32**, 285–309.

119. Lancaster, P. and Rodman, L. (1991). Solutions of the continuous and discrete time algebraic Riccati equations: A review. Chapter 2 of *The Riccati equation* (ed. S. Bittanti, A.J. Lamb and J.C. Willems), Springer Verlag, Berlin.

120. Lancaster P. and Rodman, L. (1994). Invariant neutral subspaces for symmetric and skew real matrix pairs. *Canadian Jour. Math.*, **46**, 602–618.

121. Lancaster, P. and Tismenetsky, M. (1985). *The theory of matrices (second edition) with applications*. Academic Press, Orlando.

122. Lancaster, P., Ran, A.C.M., and Rodman, L. (1986). Hermitian solutions of the discrete algebraic Riccati equation. *International Journal of Control*, **44**, 777–802.

123. Lancaster, P., Ran, A.C.M., and Rodman, L. (1987). An existence and monotonicity theorem for the discrete algebraic Riccati equation. *Linear and Multilinear Algebra*, **20**, 353–361.

124. Lancaster, P., Markus, A.S., and Ye, Q. (1994). Low rank perturbations of strongly definitizable transformations and matrix polynomials. *Linear Algebra & its Applications*, **197/198**, 3–29.

125. Laub A.J. (1991). Invariant subspace methods for the numerical solution of Riccati equations. In *The Riccati equation* (ed. S. Bittanti, A.J. Laub and J.C. Willems), Springer Verlag, Berlin, pp. 163–196.

126. Lindquist, A. and Picci, G. (1991). A geometric approach to modelling and estimation of linear stochastic systems. *J. of Math. Systems, Estimation and Control*, **1**, 241–333.

127. Lindquist, A., Michaletzky, G., and Picci, G. (1994). Zeros of spectral factors, the geometry of splitting subspaces, and the algebraic Riccati inequality (to appear).

128. Malcev, A.I. (1963). *Foundations of linear algebra.* W.H. Freeman, San Francisco and London.

129. Markushevich, A.I. (1965). *Theory of functions of a complex variable,* Vol. III. Prentice Hall, Englewood Cliffs, N.J.

130. Mårtensson, K. (1971). On the matrix Riccati equation. *Information Sciences,* **3**, 17–49.

131. Mehrmann, V.L. (1988a). *The linear-quadratic control problem.* Habilitationsschrift, University of Bielefeld.

132. Mehrmann, V.L. (1988b). A symplectic orthogonal method for single input or single output discrete time optimal quadratic control problems. *SIAM J. Matrix Anal. and Appl.,* **9**, 221–247.

133. Mehrmann V.L. (1991). *The autonomous linear quadratic control problem.* Lecture Notes in Control and Information Sciences, Vol. 163. Springer Verlag, Berlin.

134. Molinari, B.P. (1973a). The stabilizing solution of the algebraic Riccati equation. *SIAM J. Control and Optim.,* **11**, 262–271.

135. Molinari, B.P. (1973b). Equivalence relations for the algebraic Riccati equation. *SIAM J. Control and Optim.,* **11**, 272–285.

136. Molinari, B.P. (1975). The stabilizing solution of the discrete algebraic Riccati equation. *IEEE Trans. Autom. Control,* **20**, 396–399.

137. Noble, B. (1969). *Applied linear algebra.* Prentice Hall, Englewood Cliffs, N.J.

138. Ostrowski, A.M. (1973). *Solution of equations in Euclidean and Banach spaces.* Academic Press, New York.

139. Ortega, J.M. and Rheinboldt, W.C. (1970). *Iterative solution of nonlinear equations in several variables.* Academic Press, New York.

140. Pappas, T., Laub, A.J., and Sandell, N.R. (1980). On the numerical solution of the discrete-time algebraic Riccati equations. *IEEE Trans. Autom. Control,* **25**, 631–641.

141. Payne, H.J. and Silverman, L.M. (1973). On the discrete-time algebraic Riccati equation. *IEEE Trans. Autom. Control,* **18**, 226–234.

142. Petersen, I.R. (1987). Disturbance attenuation and H^∞-optimization: A design method based on the algebraic Riccati equations. *IEEE Trans. Autom. Control,* **32**, 427–429.

143. Polderman, J.W. (1986). A note on the structure of two subsets of the parameter space in adaptive control problems. *Systems and Control Letters,* **7**, 25–34.

144. Popov, V.M. (1973). *Hyperstability of control systems.* Springer Verlag, Berlin.

145. Potter, J.E. (1966). Matrix quadratic solutions. *SIAM J. Appl. Math.*, **14**, 496–501.

146. Ran, A.C.M. (1982). Minimal factorization of self-adjoint rational matrix functions. *Integral Equations and Operator Theory*, **5**, 850–869.

147. Ran, A.C.M. and Rodman, L. (1984a). The algebraic matrix Riccati equation. In *Operator Theory: Advances and Applications*, **12**, (ed. H. Dym and I. Gohberg), Birkhauser, Basel, pp. 351–381.

148. Ran, A.C.M. and Rodman, L. (1984b). Stability of invariant maximal semidefinite subspaces II, Applications: Selfadjoint rational matrix functions, algebraic Riccati equation. *Linear Algebra and its Applications*, **63**, 133–173.

149. Ran, A.C.M. and Rodman, L. (1984c). Stability of invariant maximal semidefinite subspaces I. *Linear Algebra and its Applications*, **62**, 51–86.

150. Ran, A.C.M. and Rodman, L. (1988a). On parameter dependence of solutions of algebraic Riccati equations. *Mathematics of Control, Signals and Systems*, **1**, 269–284.

151. Ran, A.C.M. and Rodman, L. (1988b). Stability of invariant lagrangian subspaces I. *Operator Theory: Advances and Applications*, **32**, (ed. I. Gohberg), 181–218.

152. Ran, A.C.M. and Rodman, L. (1989). Stability of invariant lagrangian subspaces II, *Operator Theory: Advances and Applications*, **40**, (ed. H. Dym, S. Goldberg, M.A. Kaashoek and P. Lancaster), 391–425.

153. Ran, A.C.M. and Rodman, L. (1990). Stable invariant lagrangian subspaces: Factorization of symmetric rational matrix functions and other applications. *Linear Algebra and its Applications*, **137/138**, 575–620.

154. Ran, A.C.M. and Rodman, L. (1992a). Stable solutions of real algebraic matrix Riccati equations. *SIAM J. Control and Optim.*, **30**, 63–81.

155. Ran, A.C.M. and Rodman, L. (1992b). Stable hermitian solutions of discrete algebraic Riccati equations. *Mathematics of Control, Signals and Systems*, **5**, 165–193.

156. Ran, A.C.M. and Trentelman, H.L. (1993). Linear quadratic problems with indefinite cost for discrete time systems. *SIAM J. Matrix Anal. and Appl.*, **14**, 776–797.

157. Ran, A.C.M. and Vreugdenhil, R. (1988). Existence and comparison theorems for algebraic Riccati equations for continuous-and discrete-time systems. *Linear Algebra and its Applications*, **99**, 63–83.

158. Reid, W.T. (1972). *Riccati differential equations.* Academic Press, New York.

159. Roberts, J.D. (1980). Linear model reduction and solution of the algebraic Riccati equation by use of the sign function. *Intern. J. of Control*, **32**, 677–687.

160. Rodman, L. (1980). On extremal solutions of the algebraic Riccati equation. In *Lectures in Applied Mathematics*, Vol 18 (ed. C.I. Byrnes and C.F. Martin), pp. 311–327. AMS, Providence, Rhode Island.

161. Rodman, L. (1981). On non-negative invariant subspaces in indefinite scalar product spaces. *Linear and Multilinear Algebra*, **10**, 1–14.

162. Rodman, L. (1983). Maximal invariant neutral subspaces and an application to the algebraic Riccati equation. *Manuscripta Math.*, **43**, 1–12.

163. Rodriguez-Canabal, J. (1973). The geometry of the Riccati equation. *Stochastics*, **1**, 129–149.

164. Rosenbrock, H.H. (1970). *State-space and multivariable theory.* Nelson, London.

165. Russell, D.L. (1979). *Mathematics of finite-dimensional control systems.* Marcel Dekker, New York.

166. Ruymgaart, P.A. and Soong, T.A. (1988). *Mathematics of Kalman–Bucy filtering*, (2nd edn). Springer Verlag, Berlin.

167. Salgado, M., Middleton, R., and Goodwin, G.C. (1988). Connection between continuous and discrete Riccati equations with applications to Kalman filtering. *IEE Proceedings*, **135**, 28–34.

168. Scherer, C. (1991). The solution set of the algebraic Riccati equation and the algebraic Riccati inequality. *Linear Algebra and Appl.*, **153**, 99–122.

169. Shayman, M. (1983). Geometry of the algebraic Riccati equation, Parts I and II. *SIAM J. Control and Optim.*, **21**, 375–394 and 395–409.

170. Shayman, M.A. (1985). On the phase portrait of the matrix Riccati equation arising from the periodic control problem. *SIAM J. Control and Optim.*, **23**, 717–751.

171. Shilov, G.E. (1961). *Linear spaces.* Prentice Hall, Englewood Cliffs, N.J.

172. Silverman, L.M. (1976). Discrete Riccati equations: Alternative algorithms, asymptotic properties and system theory interpretations. *Control and Dynamic Systems*, **12**, 313–386.

173. Singer, M.A. and Hammarling, S.J. (1983). The algebraic Riccati equation: A summary review of some available results. National Physical Laboratory, Teddington, Report DITC 23/83.

174. Smith, L. (1984). *Linear algebra*, (2nd edn). Springer Verlag, New York.

175. Sontag, E.D. (1990). *Mathematical control theory*. Springer Verlag, New York.

176. Stewart, G.W. (1972). On the sensitivity of the eigenvalue problem $Ax = \lambda Bx$. *SIAM J. Numer. Anal.*, **4**, 669–686.

177. Stoorvogel, A.A. and Trentelman, H.L. (1990). The quadratic matrix inequality in singular H_∞ control with state feedback. *SIAM J. Control and Optim.*, **28**, 1190–1208.

178. Stummel, F. (1971). Diskreter Konvergenz linearer operatoren, II. *Mathematische Zeitschrift*, **120**, 231–264.

179. Tadmor, G. (1990). Worst-case design in the time domain: The maximum principle and the standard H_∞ problem. *Mathematics of Control, Signals, and Systems*, **3**, 301–324.

180. Thompson, R.C. (1990). Pencils of complex and real symmetric and skew matrices. *Linear Algebra and its Applications*, **147**, 323–371.

181. Trentelman, H.L. and Willems, J.C. (1991). The dissipation inequality and the algebraic Riccati equation. In *The Riccati equation* (ed. S. Bittanti, A.J. Laub, and J.C. Willems), Springer Verlag, Berlin, pp. 197–242.

182. Tugnait, J.K. (1985). Continuous-time system identification on compact parameter sets. *IEEE Trans. Inform. Theory*, **31**, 652–659.

183. Van Dooren, P. (1981). A generalized eigenvalue approach for solving Riccati equations. *SIAM J. Scient. Stat. Comput.*, **2**, 121–135.

184. Van Huffel, S. and Vandewalle, J. (1988). Analysis and solution of the nongeneric total least squares problem. *SIAM J. Matrix Anal. and Appl.*, **9**, 360–372.

185. Van Huffel, S. and Vandewalle, J. (1991). *The total least square problem: Computational aspects and analysis*. Society for Industrial and Applied Math., Philadelphia, PA.

186. Vandewalle, J.P. and Dewilde, P. (1977). On the irreducible cascade synthesis of a system with a real rational transfer matrix. *IEEE Trans. Circuits and Systems*, **CAS–24**, 481–494.

187. Vaughan, D.R. (1970). A nonrecursive algebraic solution for the discrete Riccati equation. *IEEE Trans. Autom. Control*, **15**, 597–599.

188. Willems, J.C. (1971). Least squares stationary optimal control and the algebraic Riccati equation. *IEEE Trans. Autom. Control*, **AC–16**, 621–634.

189. Wimmer, H.K. (1974). Inertia theorems for matrices, controllability, and linear vibrations. *Linear Algebra and Appl.*, **8**, 337–343.

190. Wimmer, H.K. (1976). On the algebraic Riccati equation. *Bull. Australian Math. Soc.*, **14**, 457–461.

191. Wimmer, H.K. (1982). The algebraic Riccati equation without complete controllability. *SIAM J. Algebraic and Discrete Methods*, **3**, 1–12.

192. Wimmer, H.K. (1984). The algebraic Riccati equation: Conditions for the existence and uniqueness of solutions. *Linear Algebra and Appl.*, **58**, 441–452.

193. Wimmer, H.K. (1985). Monotonicity of maximal solutions of algebraic Riccati equations. *Systems and Control Letters*, **5**, 317–319.

194. Wimmer H.K. (1989). Strong solutions of the discrete-time algebraic Riccati equation. *Systems Control Letters*, **13**, 455-457.

195. Wimmer, H.K. (1991). Normal forms of symplectic pencils and the discrete-time algebraic Riccati equation. *Linear Algebra and Appl.*, **147**, 411–440.

196. Wimmer, H.K. (1992a). Geometry of the discrete-time algebraic Riccati equation. *J. of Math. Systems, Estimation and Control*, **2**, 123–132.

197. Wimmer, H.K. (1992b). Monotonicity and maximality of solutions of discrete-time algebraic Riccati equations. *J. of Math. Systems, Estimation and Control*, **2**, 219–235.

198. Wimmer, H.K. (1994a). Unmixed solutions of the discrete-time algebraic Riccati equations. *SIAM J. Control and Optim.* (to appear).

199. Wimmer, H.K. (1994b). Existence of positive-definite and semidefinite solutions of discrete-time algebraic Riccati equations. *International J. Control*, **59**, 463–471.

200. Wonham, W.M. (1968). On a matrix Riccati equation of stochastic control. *SIAM J. Control*, **6**, 681–697 (erratum: *SIAM J. Control*, **7**, (1969), 365).

201. Wonham, W.M. (1970 and 1979). *Linear multivariable control*. Springer Verlag, Berlin.

202. Yanushevsky, R.T. (1993). Robust control of delay feedback systems with bounded uncertainty. *SIAM J. Matrix Anal. and Appl.*, **14**, 978–990.

203. Zames, G. (1981). Feedback and optimal sensitivity: Model reference transformations, multiplicative seminorms, and approximate inverses. *IEEE Trans. Autom. Control*, **26**, 301–320.

204. Zhou, K. and Khargonekar, P.P. (1988). An algebraic Riccati equation approach to H^∞ optimization. *Systems and Control Letters*, **11**, 85–91.

List of Notation and Conventions

1. GENERAL

CARE	continuous algebraic Riccati equation
DARE	discrete algebraic Riccati equation
LQR	linear-quadratic regulator
$A \subseteq B$	containment of a set A in a set B ($A = B$ not excluded)
$A \subset B$	containment of a set A in a set B ($A = B$ excluded)
$\delta_{ij} = 1$ if $i = j$, $\delta_{ij} = 0$ if $i \neq j$	Kronecker symbol

2. NUMBERS AND VECTOR SPACES

$\mathbb{R}$	the set of real numbers
$\mathbb{C}$	the set of complex numbers
$\mathbb{T} = \{z \in \mathbb{C} \mid |z| = 1\}$	the unit circle
$\operatorname{Re} z$	the real part of $z \in \mathbb{C}$
$\mathbb{C}_+ = \{z \in \mathbb{C} \mid \operatorname{Re} z > 0\}$	the open right half-plane
$\mathbb{R}^{m \times n}$	the real vector space of $m \times n$ real matrices
$\mathbb{C}^{m \times n}$	the complex vector space of $m \times n$ complex matrices
$\mathbb{R}^{n \times 1}$, $\mathbb{C}^{n \times 1}$	are abbreviated $\mathbb{R}^n$, $\mathbb{C}^n$, respectively
$\langle x, y \rangle$	the standard scalar product in $\mathbb{R}^n$ or $\mathbb{C}^n$
$\|x\| = \langle x, x \rangle^{1/2}$	the euclidean norm (if not specified otherwise)
$[x, y]$	an indefinite scalar product on $\mathbb{C}^n$ or $\mathbb{R}^n$
$e_1, \ldots, e_n$	the standard basis in $\mathbb{R}^n$ or $\mathbb{C}^n$; the n columns of the $n \times n$ identity matrix
$[x]_X$	the coordinate vector of $x \in \mathbb{R}^n$ or $\mathbb{C}^n$ relative to a basis X.
$\operatorname{span}\{x_1, \ldots, x_r\}$	the subspace spanned by $x_1, \ldots, x_r$
$\mathcal{M} \dotplus \mathcal{N}$	direct sum of subspaces
$\mathcal{M} \oplus \mathcal{N}$	orthogonal direct sum of subspaces
$\mathcal{M}^\perp$	the orthogonal complement to a subspace $\mathcal{M}$
$\mathcal{M}^{[\perp]} = \{x \in \mathbb{R}^n \text{ or } \mathbb{C}^n \mid [x, y] = 0$ for all $y \in \mathcal{M}\}$	the *orthogonal companion* of a subspace $\mathcal{M}$ with respect to the indefinite scalar product $[., .]$ in $\mathbb{R}^n$ or $\mathbb{C}^n$
$\mathcal{C}_{A,B}$	the controllable subspace of the pair of matrices (A, B)
$\mathcal{U}_{C,A}$	the unobservable subspace of the pair of matrices (C, A)

3. MATRICES

All matrices are assumed to have real or complex entries. When convenient, an $m \times n$ matrix A is also understood as the linear transformation $\mathbb{R}^n \to \mathbb{R}^m$ (or $\mathbb{C}^n \to \mathbb{C}^m$, as the case may be) represented by A in the standard bases in $\mathbb{R}^m$ (or $\mathbb{C}^m$) and $\mathbb{R}^n$ (or $\mathbb{C}^n$).

I_n, I	$n \times n$ identity matrix
$0_n, 0$	$n \times n$ zero matrix
λI ($\lambda \in \mathbb{C}$)	is often abbreviated to λ
$\sigma(A)$	the spectrum of A; the set of distinct eigenvalues of matrix A
$J_r(\lambda)$	the $r \times r$ upper triangular Jordan block with eigenvalue λ
$J_{2k}(\lambda \neq i\mu)$	the $2k \times 2k$ almost upper triangular real Jordan block with eigenvalues $\lambda \pm i\mu$ ($\lambda \in \mathbb{R}$, $\mu \in \mathbb{R}$, $\mu \neq 0$)
$\bar{X}$	the complex conjugate (entrywise) of a matrix X
X^T	the transpose of X
X^*	the conjugate transpose of X
$(X^*)^{-1} = (X^{-1})^*$	is sometimes abbreviated X^{-*}
$A^{[*]}$	the adjoint of matrix A in the scalar product $[.,.]$
$\mathrm{diag}\,[Z_1, \ldots, Z_p]$ or $z_1 \oplus \cdots \oplus Z_p$	the block diagonal matrix with the diagonal blocks $Z_1, \ldots, Z_p$
$\mathrm{Ker}\, X = \{x \in \mathbb{R}^n \text{ or } \mathbb{C}^n \mid Xx = 0\}$	the kernel, or nullspace, of an $n \times n$ matrix X
$\mathrm{Im}\, X = \{Xx \mid x \in \mathbb{R}^n \text{ or } \mathbb{C}^n\}$	the image, or range, of an $m \times n$ matrix X

For $\lambda_0 \in \sigma(A)$, where A is $n \times n$ the *algebraic multiplicity* of λ_0 is $\dim \mathrm{Ker}\,(A - \lambda_0 I)^n$; the *geometric multiplicity* of λ_0 is $\dim \mathrm{Ker}\,(A - \lambda_0 I)$; the *partial multiplicities* of λ_0 are the sizes of the Jordan blocks with the eigenvalue λ_0 in the Jordan form of A (these definitions apply also for real matrices A and their nonreal eigenvalues λ_0; then $A - \lambda_0 I$ is understood as a linear transformation from $\mathbb{C}^n$ into $\mathbb{C}^n$).

Her$(\,,\,)$	the set of hermitian solutions of a CARE
Ric$(\,)$	a function from subspaces to matrices

LIST OF NOTATION AND CONVENTIONS

$\mathcal{R}_{\lambda_0}(A) = \text{Ker}\,(A - \lambda_0 I)^n$	the *root subspace* of an $n \times n$ matrix A corresponding to $\lambda_0 \in \sigma(A)$
$\mathcal{R}_{\lambda \pm i\mu}(A) = \text{Ker}\,(A^2 - 2\lambda A + (\lambda^2 + \mu^2)I)^n$	the *real root subspace* of a real $n \times n$ matrix A corresponding to $\lambda \pm i\mu \in \sigma(A)$, ($\lambda \in \mathbb{R}, \mu \in \mathbb{R}, \mu \neq 0$)
$\mathcal{R}(A, \Omega)$	the sum of root subspaces of A corresponding to its eigenvalues in the set Ω
$A|\mathcal{N}$	restriction of A to its invariant subspace $\mathcal{N}$
$\text{Inv}\,(A)$	the set of all invariant subspaces of A
$A > B$	means that $A - B$ is positive definite (for hermitian matrices A and B)
$A \geq B$	means that $A - B$ is positive semidefinite (for hermitian matrices A and B)
$A \otimes B$	the right Kronecker product of matrices A and B
For a hermitian matrix H (or a matrix which is similar to a hermitian matrix), we define $\nu(H) = \min$ (the number of positive eigenvalues of H, the number of negative eigenvalues of H), where the eigenvalues are counted with their algebraic multiplicities.	
The inertia of a hermitian matrix H is defined by $\text{In}(H) = $(# of positive eigenvalues of H, # of eigenvalues of H, # of zero eigenvalues of H).	
$R^{1/2}$	the positive semidefinite hermitian square root of a positive semidefinite hermitian matrix R
$\delta(W)$	the McMillan degree of a rational matrix function $W(\lambda)$

Index

adjoint 36, 56
analytic matrix function 189
analyticity, of solutions
    of CARE 189–192, 198, 263–266
    of DARE 283
angle 13

bounded real lemma 410–412

c-set 46, 47, 178
c-stabilizable pair 90, 93–95, 103, 155
canonical form 26, 41–46, 56, 60
CARE 151, 357, 440
Cayley transform 39, 64, 281, 444
characteristic polynomial 3
column space 9
companion matrix 69
comparison theorems 231–236, 241–243
continuity
    of solutions of CARE 184, 224, 260–263
    of solutions of DARE 282
contraction 421
    strict 421, 425–435
control function 85, 349
control normal form 92, 96
controllability Gramian 87
controllable pair 83–85, 87–89,
    conjugate 428
cost functional 350, 363
    nondegenerate 350, 355–357, 365
covariance 373

d-set 141
d-sign controllable 291, 330
d-stabilizable pair 90, 94, 292, 311, 325,
    335, 365–368
DARE 271, 366, 382, 444
deflating subspace 22–25, 64, 67
    spectral 25, 67, 340
detectable pair 91, 158, 222, 234

diagonable matrix 4
direct sum 11

eigenspace, generalized 10, 17, 18
eigenvalue 3, 22
    infinite 22
    undetectable 327
eigenvector 3, 22
    generalized 4
expected value 372
extremal solutions 168–176, 222, 223, 260–266, 330–332

factorization
    associated 138
    canonical 397–407
    minimal 122–129, 251–253, 255
    real canonical 403
    spectral 404–407
feedback 91, 299, 350, 409, 412
field of values 305
filter 377
    H^∞ 417–418
    time invariant 380–385
functions of matrices 26–28

gain matrix 379
gap
    between solutions 171
    between subspaces 184, 206
graph, of a matrix 149, 154, 274
graph subspace 149, 155, 215
    extended 202, 337

H^∞ control 397, 409–419
H-orthogonal matrix 56
H-orthogonal pencil 80, 342
H-self-adjoint matrix 37–43, 46, 52
H-self-adjoint pencil 62–63, 444
H-skew-symmetric matrix 60–62, 76, 215
H-symmetric matrix 55–59

H-symplectic pencil 334, 337, 444
H-unitary matrix 38–46
H-unitary pencil 64–70, 334
Her(..) 182

image 9
index (of an eigenvalue) 8
inertia 46, 426–427
invariant subspace 9, 13–20, 29
   neutral 52–55
   nonnegative 46–51
   spectral 18
isolated solutions 187–188, 305

Jordan block 4, 16, 29
Jordan chain 4, 150
Jordan form 7, 16, 29
Jordan matrix 6, 17
Jordan subspace 5

Kalman filter 371–386
Kalman normal form 92
kernel 8
Kronecker product 97–100

Lagrangian subspace 274
linearization 69
LQR form, for the CARE 200–205, 241–243
LQR problem 298, 337, 349–369
   discrete 362–368
Lyapunov equation 101–105

McMillan degree 122, 430
Markov process 376, 379
matrix
   c-stable 90, 91
   d-stable 90
   H-self-adjoint 37–43, 46, 52
   H-skew-symmetric 60–62, 76
   adjoint 36
   diagonable 4
   sip 41
   skew-adjoint 74
   skew-symmetric 76–78, 207
   stable 90
   state characteristic 422
matrix pencil, see pencil

matrix polynomial 69, 70
matrix sign function 437–446
mean 373
metric
   on subspaces 205
Moore–Penrose inverse 319
multiplicity
   algebraic 7
   geometric 7, 27
   partial 7, 28, 52, 113, 130, 159–163, 211, 285

negative definite 20
negative semidefinite 20
Newton–Kantorovich method 236–238, 244, 327
nonderogatory 29
norm 13
   of a matrix 205
   of a vector 206
numerical range 305

observable pair 86, 87, 89, 214, 355–357, 365
optimal control 350, 357–359, 369
optimal cost 350, 363
orthogonal companion 32
orthogonal complement 13
orthogonal equivalence 81

partial order
   on solutions of the CARE 168, 177–183, 223
   on solutions of the DARE 308
pdf 372
   conditional 373
   joint 372
   marginal 372
pencil 22, 227, 333–345
   dilated 201, 336, 343
   H-orthogonal 80–82
   H-self-adjoint 62–63, 69, 70
   H-skew-symmetric 80–82
   H-symplectic 334
   H-unitary 64–70
   regular 22, 201, 334
polarization identity 35
positive definite 20, 21

positive semidefinite 20, 21
probability distribution 372
    conditional 373
    joint 372
    marginal 372
projector 13
    complementary 14
    orthogonal 14
    Riesz 15

random process 375
random vector 372
range 9
rate of convergence 236–238, 313–315
rational matrix function 107
    contractive 421–435
    hermitian 131, 132, 137–141, 284
    nonnegative 130–137, 143, 247–256
    pole of 115
    real 141–143
    regular 115
    unitary completion 430–434
    zero of 115
realization 107–146
    dilation of 111, 457
    existence of 110
    locally minimal 116–121
    minimal 113, 115, 122–145
    reduction of 111
    similarity of 110
Rellich theorem 130, 426
Ric (.), Ric (.; ..)
    for the CARE 178, 184–187, 189–192, 224
    for the DARE 282
Riccati equation
    almost stabilizing solutions 194, 238–241
    antistabilizing solutions 244, 441
    differential 169, 360
    maximal solution of 171–176, 195, 196, 260, 307–312
    minimal solution of 171–176, 199, 263
    nonsymmetric 149–151
    number of solutions 179, 210, 212, 226
    perturbed 257–267, 329–332
    real 215–229, 244, 266, 301–303
    regular 197, 199, 293
    robustly solvable 267
    special solution of 167, 222
    stabilizing solutions 194, 226, 238–241, 324–326, 339–342, 440
Riccati inequality 231–236, 307
root subspace 10, 17, 18, 36

scalar product 12, 20
    indefinite 31, 55
    skew-symmetric 71–82
sign-characteristic 42–44, 57, 66, 78–80, 163
sign condition 48, 167
sign controllable pair 155–158, 166, 193, 205, 210, 257, 429
    real 219–221
sign function 437–446
similar matrices 3
similar realizations 110
sip matrix 41
skew-adjoint 74
Smith form, local 113, 211
spectral mapping theorem 28
spectral subspace 10, 15, 18, 25
stability radius
    structured 447–457
stabilizable pair 90, 93, 103, 158, 194, 220, 226, 232, 354
statistical independence 373
Stein equation 100–106
    symmetric 101–106
strict equivalence 25, 26
structured stability radius 447–458
subspace
    antistable 439
    complementary 13
    controllable 84
    deflating, see deflating subspace
    degenerate 33, 35
    H-negative 35
    H-neutral 35, 48, 52–55, 58, 74
    H-nonnegative 33, 46–52, 58
    H-nonpositive 35, 46–52
    H-positive 33
    isotropic 35
    neutral 35, 52–55, 58, 74, 152, 155
    nondegenerate 33
    nonnegative 33–35, 46–52, 58, 152
    nonpositive 35, 46–52
    orthogonal 13

positive 33–35
reachable 85
unobservable 86
supporting decomposition 124
supporting projector 124, 132, 138
Sylvester equation 100–106
system 85, 89, 349, 363, 371, 375, 412, 417
   observer 384–385
   state estimator 417

tensor product 99

total least squares 387–395

undetectable eigenvalue 327
unitary completion 430
   minimal 430
unitary equivalence 65, 66

variance 373
vec-function 99

white noise 375
   normal process 375